Teacher's Annotated Edition

Technology
Today and Tomorrow

Fourth Edition

Teacher's Manual

Eric Thompson
Technology Instructor
Onalaska High School
Onalaska, WI

Glencoe
McGraw-Hill

New York, New York   Columbus, Ohio   Woodland Hills, California   Peoria, Illinois

Glencoe/McGraw-Hill

A Division of The McGraw·Hill Companies

Copyright © 1999, 1997, 1993, and 1988 by Glencoe/McGraw-Hill. All rights reserved. Except as permitted under the United States Copyright Act, no part of this publication may be reproduced or distributed in any form or by any means, or stored in a database or retrieval system, without prior written permission from the publisher.

Send all inquiries to:
Glencoe/McGraw-Hill
3008 W. Willow Knolls Drive
Peoria, IL 61614

ISBN 0-02-658570-7 (Teacher's Annotated Edition)

  2 3 4 5 6 7 8 9 10   087   02 01 00

Printed in the United States of America.

Table of Contents

Overview **TM-4**

Course Outlines **TM-7**

Resources **TM-9**

Scope and Sequence Chart **TM-15**

General Preparation Suggestions **TM-20**

Lesson Plans **TM-22**
    Chapter 1 . . . . TM-22
    Chapter 2 . . . . TM-24
    Chapter 3 . . . . TM-27
    Chapter 4 . . . . TM-29
    Chapter 5 . . . . TM-31
    Chapter 6 . . . . TM-33
    Chapter 7 . . . . TM-35
    Chapter 8 . . . . TM-38
    Chapter 9 . . . . TM-41
    Chapter 10 . . . . TM-43
    Chapter 11 . . . . TM-45
    Chapter 12 . . . . TM-47
    Chapter 13 . . . . TM-49
    Chapter 14 . . . . TM-52
    Chapter 15 . . . . TM-55
    Chapter 16 . . . . TM-57
    Chapter 17 . . . . TM-59
    Chapter 18 . . . . TM-62
    Chapter 19 . . . . TM-65
    Chapter 20 . . . . TM-68
    Chapter 21 . . . . TM-71
    Chapter 22 . . . . TM-73
    Chapter 23 . . . . TM-76
    Chapter 24 . . . . TM-79

Overview

Technology: Today and Tomorrow is a four-component program for teaching and learning. It includes:
- **Student Text**—a textbook for learning about technology in today's world.
- **Teacher's Annotated Edition**—the Student Text and hundreds of teaching tips included on the text pages, plus a bound-in Teacher's Manual.
- **Teacher's Resource Binder**—a large notebook containing teaching strategies, section tests, chapter tests, transparency masters, handout masters, lesson plans, and other helpful information.
- **Student Workbook**—a soft-cover book containing worksheets keyed to the textbook chapters, cross-curricular activities, and hands-on technology activities.

Each of the components listed above is described in detail on the following pages. Learning about the components will enable you to realize their full value for your students.

The Student Text

The *Technology: Today and Tomorrow* student text is the principal learning tool in the program. It is an information resource that will aid learning and hold students' interest. Combining solid data, strong illustrations, and special features, the student text is the core unit in the learning program.

Sections

The *Technology: Today and Tomorrow* text is divided into six sections.

Section I: Introduction to Technology introduces technology, defines the problem-solving process, and explains the parts of a technological system (input, process, output, and feedback). Each of the other five sections deals with a major division in technology.

Section II: Communication focuses on what communication technology is all about, the communication system, telecommunication, graphic communication, and computers.

Section III: Manufacturing examines different types of manufacturing systems, product development, production planning, production, marketing, and computer-integrated manufacturing.

Section IV: Construction discusses different types of construction systems, planning construction, managing construction, constructing houses and other buildings, and other types of construction projects.

Section V: Transportation highlights different types of transportation systems, the types and modes of transportation, intermodal transportation, and power in transportation.

Section VI: Bio-Related Technology introduces bio-related technology systems and gives an overview of the many applications used today so that students gain an appreciation of their growing importance.

Each section ends with technology activities that relate to topics covered in the chapters. Many of these activities are open-ended, so they allow students to be creative and use their critical thinking skills.

Chapters

Technology: Today and Tomorrow has twenty-four chapters. Each chapter presents the topic information in a logical, easily understood manner. Each chapter includes the following:

Objectives. The chapter objectives present concise statements of the principal themes discussed in the chapter.

Terms. A list of vocabulary terms is also located at the front of each chapter. Terms are printed in boldface type and defined where first used in the chapter.

Technology Focus. At the beginning of each chapter is a feature titled "Technology Focus." These articles describe a current or future development in technology, with emphasis on the people and companies that create technology.

Text. Each chapter's text is carefully organized and information clearly presented. The distinctive heads and subheads indicate the major divisions of information in the text. The vocabulary of technology is defined on first use.

Review. Included in each chapter is a review. This consists of a brief chapter summary, entitled "Reviewing Main Ideas;" five objective questions, entitled "Understanding Concepts;" and five questions for "Thinking Critically." Also included in each chapter's review are cross-curricular activities entitled "Applying Concepts and Solving Problems." In these, students apply skills in language arts, social studies, science, and mathematics to technology-related problems.

Special Highlights

Full-Color Illustrations. The many full-color photographs will attract students' interest, motivating them to learn about the new technologies explored in the text. The photographs and drawings provide a strong visual reinforcement for the themes and concepts discussed.

Fascinating Facts. These short bits of interesting information are sprinkled throughout the text. They pertain to the particular type of technology being discussed at the time, and they are mostly historical items about that technology.

Science Connections. These articles focus on various principles of science as they pertain to the technology under discussion. Their purpose is to show students that the pure sciences have practical value. Each ends with a brief activity suggestion.

Mathematics Connections. Like the Science Connections, these articles bring home the everyday benefits of learning mathematics principles. Each ends with a brief activity suggestion.

Career Files. Set up in a question-and-answer format, these files give basic information about many of the diverse occupations related to technology.

Safety Comes First. This article introduces the text to help emphasize the importance of following all safety rules while working in the technology lab. Students should read this article *before* working in the lab.

Glossary. The glossary presents a complete, alphabetized inventory of the important terms in the text. Each of the terms is defined.

Index. In a technical book, the index is invaluable. Comprehensive, and carefully constructed, this index is a guide to the main themes and topics of the text.

The Teacher's Annotated Edition

Technology: Today and Tomorrow is also published in a *Teacher's Annotated Edition*. The TAE differs from the student text in the following two ways:

1. The textbook pages include teaching suggestions in the bottom page margins. These annotations are keyed by number to passages in the text and are printed in blue.

2. A comprehensive Teacher's Manual is bound in with the annotated text pages.

Annotations

Annotations take various forms. Each annotation has a title to indicate its purpose, such as **Review, Enrichment, Critical Thinking**, etc. The annotations provide:
- background information about the chapter
- suggested questions and activities to reteach, enrich, or extend the chapter
- points to stress and additional relevant information
- cross-references to activities or text passages that discuss related information

The Teacher's Manual

The Teacher's Manual helps the teacher plan and conduct the course. It includes a program overview, scope and sequence chart, course outlines for 9, 18, and 36 weeks, a resource list, and chapter-by-chapter lesson plans.

Lesson Plans

For each chapter of the textbook, there is a corresponding lesson plan in the Teacher's Manual. Each lesson plan includes the following items:

Chapter Outline and Main Goal. An outline of the main topics covered in the chapter and a brief statement of the main learning goal of the chapter.

Objectives. A list of objectives from the text.

Preparation. Tips for the teacher on how to prepare for the lesson.

Teaching. General advice on teaching the chapter.

Evaluation. Suggestions for evaluating student performance.

Answers to Chapter Questions. Answers for the questions that appear at the end of each chapter.

The Teacher's Resource Binder

The *Teacher's Resource Binder* provides a valuable supplement to the text. It includes a program overview, standards correlations, expanded lesson plans with teaching suggestions for the reproducibles (transparency and handout masters) that accompany each chapter, and sections on student diversity, interpersonal skills, careers, portfolios, and using the program in a modular setting. In addition, the *Teacher's Resource Binder* provides the following:

Spanish Handbook. This handbook provides Spanish definitions of all of the textbook's glossary terms plus Spanish versions of each textbook chapter's summary.

Section Tests. Based on material discussed in the six sections of the text, these objective tests check student mastery.

Answer Keys. This section of the binder includes answers for the chapter reviews and section tests.

Transparency and Handout Masters. Located toward the end of this binder, these reproducible masters are keyed to important elements in the text and are designed to be used with an overhead projector or photocopied and handed out to students.

Chapter Reviews. These reproducible quizzes review students' mastery of factual material.

Color Transparencies. These transparencies are designed to help students understand the concepts presented in the text. Teaching tips accompany them.

Media Correlations. This section correlates selected video and laser discs to the student text. Teachers who want to enrich their courses with multimedia will find the correlations useful.

Safety Handbook. This handbook includes an introductory essay, handouts of safety rules and safety tests.

Science and Mathematics Handbook. This handbook consists of reproducible handouts that review basic science and mathematics principles. The handbook may be reproduced as a whole and given to students to keep, or individual handouts may be given to students when a related subject is being studied.

Teacher's Manual for the Student Workbook. This contains tips for using the cross-curricular activities in the workbook plus answer keys for the workbook.

The Student Workbook

The *Student Workbook* includes worksheets keyed to each chapter and numerous hands-on technology activities that encourage students to apply what they have learned from the text. Also included are cross-curricular activities for science, mathematics, language arts, and social studies.

Testmaker

A computer program and test bank allow teachers to create customized tests. The program provides 1500 questions that can be used as is, modified, or added to. It is Windows and Mac compatible.

Course Outlines

Many school districts are trying various modes of course scheduling, including year-round school, block scheduling, and trimesters. Because of this variety of schedules, it is impossible to provide course outlines that will be a close match in every case.

The course outlines presented below are based upon traditional class lengths, meeting times, and school years. The model is a 45-50 minute class that meets every day. This model provides approximately four class hours (instructional time) per week. Furthermore, a school year is assumed to consist of 175-180 instructional days. At the beginning of each outline, the total possible number of instructional hours on which that outline is based is provided.

Teachers can calculate the total instructional hours in their specific class schedule and, with modifications for number of weeks, use the schedule that most closely resembles theirs.

Chapters shown in parentheses are optional and may be assigned if time permits. If time does not permit students to read the optional chapters, be sure to include summarized information from those chapters during lectures and discussions. Another possibility is to place students in teams, assign one optional chapter to each team, and have that team present a summary of the chapter to the class.

9 Weeks (36 Class Hours)

Weeks 1-2 (8 Class Hours)

Section I: Introduction to Technology
    Chapter 1: Exploring Technology
    Chapter 2: Problem Solving
    Chapter 3: Technology Systems

Week 3 (4 Class Hours)

Section II: Communication Technology
    Chapter 4: Communication Systems
    (Chapter 5: Computers)
    (Chapter 6: Telecommunication)
    (Chapter 7: Graphic Communication)

Weeks 4-5 (8 Class Hours)

Section III: Manufacturing Technology
    Chapter 8: Manufacturing Systems
    (Chapter 9: Product Development)
    (Chapter 10: Production Planning)
    (Chapter 11: Production)
    (Chapter 12: Marketing)
    (Chapter 13: Computer-Integrated Manufacturing)

Weeks 6-7 (8 Class Hours)

Section IV: Construction Technology
    Chapter 14: Construction Systems
    (Chapter 15: Planning Construction)
    (Chapter 16: Managing Construction)
    (Chapter 17: Constructing Homes and Other Buildings)
    (Chapter 18: Other Construction Projects)

Week 8 (4 Class Hours)

Section V: Transportation Technology
    Chapter 19: Transportation Systems
    (Chapter 20: Types and Modes of Transportation)
    (Chapter 21: Intermodal Transportation)
    (Chapter 22: Power in Transportation)

Week 9 (4 Class Hours)

Section VI: Bio-Related Technology
    Chapter 23: Bio-Related Systems
    (Chapter 24: Applications of Bio-Related Technology)

18 Weeks (72 Class Hours)

Weeks 1-2 (8 Class Hours)

Section I: Introduction to Technology
    Chapter 1: Exploring Technology
    Chapter 2: Problem Solving
    Chapter 3: Technology Systems

Weeks 3-5 (12 Class Hours)

Section II: Communication Technology
   Chapter 4: Communication Systems
   Chapter 5: Computers
   Chapter 6: Telecommunication
   Chapter 7: Graphic Communication

Weeks 6-10 (20 Class Hours)

Section III: Manufacturing Technology
   Chapter 8: Manufacturing Systems
   Chapter 9: Product Development
   Chapter 10: Production Planning
   Chapter 11: Production
   Chapter 12: Marketing
   Chapter 13: Computer-Integrated Manufacturing

Weeks 11-14 (16 Class Hours)

Section IV: Construction Technology
   Chapter 14: Construction Systems
   Chapter 15: Planning Construction
   Chapter 16: Managing Construction
   Chapter 17: Constructing Homes and Other Buildings
   Chapter 18: Other Construction Projects

Weeks 15-17 (12 Class Hours)

Section V: Transportation Technology
   Chapter 19: Transportation Systems
   Chapter 20: Types and Modes of Transportation
   Chapter 21: Intermodal Transportation
   Chapter 22: Power in Transportation

Week 18 (4 Class Hours)

Section VI: Bio-Related Technology
   Chapter 23: Bio-Related Systems
   Chapter 24: Applications of Bio-Related Technology

36 Weeks (144 Class Hours)

Weeks 1-4 (16 Class Hours)

Section I: Introduction to Technology
   Chapter 1: Exploring Technology
   Chapter 2: Problem Solving
   Chapter 3: Technology Systems

Weeks 5-10 (24 Class Hours)

Section II: Communication Technology
   Chapter 4: Communication Systems
   Chapter 5: Computers
   Chapter 6: Telecommunication
   Chapter 7: Graphic Communication

Weeks 11-18 (32 Class Hours)

Section III: Manufacturing Technology
   Chapter 8: Manufacturing Systems
   Chapter 9: Product Development
   Chapter 10: Production Planning
   Chapter 11: Production
   Chapter 12: Marketing
   Chapter 13: Computer-Integrated Manufacturing

Weeks 19-25 (28 Class Hours)

Section IV: Construction Technology
   Chapter 14: Construction Systems
   Chapter 15: Planning Construction
   Chapter 16: Managing Construction
   Chapter 17: Constructing Homes and Other Buildings
   Chapter 18: Other Construction Projects

Weeks 26-31 (24 Class Hours)

Section V: Transportation Technology
   Chapter 19: Transportation Systems
   Chapter 20: Types and Modes of Transportation
   Chapter 21: Intermodal Transportation
   Chapter 22: Power in Transportation

Weeks 32-34 (12 Class Hours)

Section VI: Bio-Related Technology
   Chapter 23: Bio-Related Systems
   Chapter 24: Applications of Bio-Related Technology

Weeks 35-36 (8 Class Hours)

Review all chapters and sections

Resources

Books

Aaker, David A., and Day, George S. *Marketing Research*. 4th ed. New York, NY: John Wiley and Sons, 1990.

Adams, Jack A. *Human Factors Engineering*. New York, NY: Macmillan, 1989.

Adams, Russ, and Lane, Joyce. *The Black and White Solution: Bar Code and the IBM PC*. St. Peterborough, NH: Helmers Publishing, Inc., 1987.

Ahlgren, A. and Rutherford, J. *Science for All Americans*. New York, NY: Oxford University Press, 1991.

Anderson, J. M., and Mikhail, E. M. *Introduction to Surveying*. New York, NY: McGraw-Hill, 1985.

Asfahl, C. Ray. *Industrial Safety and Health Management*, 2nd ed. Englewood Cliffs, NJ: Prentice-Hall, 1990.

Back to Basics. Pleasantville, NY: Reader's Digest Association, Inc., 1981.

Beaver, Patrick. *A History of Tunnels*. Secaucus, NJ: The Citadel Press, 1973.

Bentil, Kwekuk. *Fundamentals of the Construction Process*. Kingston, MA: R. S. Means, Co., 1989.

Berkowitz; Hartely; Kerin; and Rudelius. *Marketing*. 5th ed. Homewood, IL: Irwin, 1996.

Besterfield, Dale. *Quality Control*. 3rd ed. Englewood Cliffs, NJ: Prentice-Hall, 1990.

Betts, R. M.; Hauenstein, A. Dean; and Fannin, John W. *Exploring the Construction Industry*. Bloomington, IL: McKnight, 1976.

Bohn, Ralph C., and MacDonald, Angus. *Energy Technology: Power and Transportation*. Columbus, OH: Glencoe/McGraw-Hill, 1992.

Bohn, Ralph C.; MacDonald, Angus J.; Fales, James F.; and Kuetemeyer, Vincent F. *Energy, Power, and Transportation Technology*. Columbus, OH: Glencoe/McGraw-Hill, 1986.

Brady, George S., and Clauser, Henry R. *Materials Handbook*, 13th ed. New York, NY: McGraw-Hill, 1991.

Brennan, R. P. *Levitating Trains and Kamikaze Genes: Technological Literacy for the Future*. New York, NY: John Wiley & Sons, 1994.

Broekhuizen, Richard J. *Graphic Communications*. Columbus, OH: Glencoe/McGraw-Hill, 1995.

Collins, David J., and Whipple, Nancy N. *Using Bar Code: Why It's Taking Over*. Duxbury, MA: Data Capture Institute, 1990.

Commercial Biotechnology: An International Analysis. (Office of Technology Assessment) New York, NY: Pergamon Press, 1984.

Daiber, Robert A., and Erekson, Thomas L. *Manufacturing Technology: Today and Tomorrow*. Columbus, OH: Glencoe/McGraw-Hill, 1991.

DeGarmo, E. P. *Materials and Processes in Manufacturing*. New York, NY: Macmillan Publishing, 1979.

Dieter, George E. *Engineering Design: A Materials and Processing Approach*. 2nd ed. New York, NY: McGraw-Hill, 1991.

Fales, James F. *Construction Technology: Today and Tomorrow*. Columbus, OH: Glencoe/McGraw-Hill, 1991.

Feirer, John L. *Wood Technology and Processes*. Columbus, OH: Glencoe/McGraw Hill, 1994.

Feirer, John L., and Hutchings, Gilbert R. *Residential Masonry*. Columbus, OH: Glencoe/McGraw-Hill, 1989.

Fitzpatrick, Mike. *Computer Numerical Control Simplified*. Columbus, OH: Glencoe/McGraw-Hill, 1989.

Freeman, Harry M., ed. *Standard Handbook of Hazardous Waste Treatment and Disposal*. New York, NY: Glencoe/McGraw-Hill, 1998.

Fuchs, Jerome H. *The Prentice-Hall Illustrated Handbook of Advanced Manufacturing Methods*. Englewood Cliffs, NJ: Prentice-Hall, 1988.

Groover, Mikell P. *Industrial Robotics: Technology, Programming, and Applications*. New York, NY: McGraw-Hill, 1986.

Hammer, Willie. *Product Safety Management and Engineering*. Englewood Cliffs, NJ: Prentice-Hall, 1980.

Harmon, Craig K., and Adams, Russ. *Reading Between the Lines: An Introduction to Bar Code Technology*. St. Peterborough, NH: Helmers Publishing, Inc., 1984.

Harms, Henry R., and Kroon, Dennis K. *Production Systems Technology*. Columbus, OH: Glencoe/McGraw-Hill, 1992.

Havrella, Raymond A. *Heating, Ventilating, and Air Conditioning Fundamentals*. Columbus, OH: Glencoe/McGraw-Hill, 1981.

Hepler, Donald; Wallach, Paul R.; and Hepler, Dana J. *Architecture: Drafting and Design.*, Columbus, OH: Glencoe/McGraw-Hill, 1998.

Hernandez, Arnaldo. *Just-In-Time Manufacturing: A Practical Approach*. Englewood Cliffs, NJ: Prentice-Hall, 1989.

Hutchinson, John, and Karsnitz, John. *Design and Problem Solving in Technology*. Columbus, OH: Glencoe/McGraw-Hill, 1994.

Huth, Mark W. *Construction Technology*. Columbus, OH: Glencoe/McGraw-Hill, 1996.

Jensen, Cecil H., and Helsel, Jay D. *Engineering Drawing and Design*. Columbus, OH: Glencoe/McGraw-Hill, 1996.

Johnson, D. W.; Roger, T.; and Holubec, F. *Circles of Learning*. Alexandria, VA: ASCD, 1993.

Johnson, D. W.; Roger, T.; and Holubec, F. *Nuts & Bolts of Cooperative Learning*. Alexandria, VA: ASCD, 1994.

Johnson, D. W., and Roger, T. *Learning Together and Alone*. Englewood Cliffs, NJ: Prentice-Hall, 1994.

Kagan, S. *Cooperative Learning Resources for Teachers*. San Juan Capistrano, CA: Resources for Teachers, 1991.

Kimbrell, Grady, and Vineyard, Ben S. *Entering the World of Work*. Columbus, OH: Glencoe/McGraw-Hill, 1989.

Kimbrell, Grady, and Vineyard, Ben S. *Succeeding in the World of Work*. Columbus, OH: Glencoe/McGraw-Hill, 1998.

Kissam, Jerry A., and Kissam, P. *Surveying Practice*. 4th ed. New York, NY: McGraw-Hill, 1988.

Konz, Stephen. *Work Design: Industrial Ergonomics*. 3rd ed. Worthington, OH: Publishing Horizons, 1990.

Krar, Stephen F., and Gill, Arthur. *CNC: Technology and Programming*. Columbus, OH: Glencoe/McGraw-Hill, 1990.

Krar, Stephen F., and Check, Albert F. *Technology of Machine Tools*. Columbus, OH: Glencoe/McGraw-Hill, 1997.

Leape, Martha P., and Vacca, Susan M. *Harvard Guide to Careers*. Cambridge, MA: Harvard University Press, 1991.

Lindbeck, John R.; Williams, Molly W.; and Wygant, Robert M. *Manufacturing Technology*. Englewood Cliffs, NJ: Prentice-Hall, 1990.

Lovejoy's College Guide. New York, NY: Prentice-Hall, 1991.

Maley, D. *Answers to Questions on Teaching Technology Education*. Reston, VA: ITEA, 1988.

Macaulay, D. *The Way Things Work*. Boston, MA: Houghton Mifflin, 1988.

Manners, David X. *The Great Tool Emporium*. New York, NY: Book Division, Times Mirror Magazine, Inc., 1979.

Meyer, Earl C., and Allen, Kathleen R. *Entrepreneurship and Small Business Management*. Columbus, OH: Glencoe/McGraw-Hill, 1994.

Mundel, Marven E. *Motion and Time Study: Improving Productivity*. 6th ed. Englewood Cliffs, NJ: Prentice-Hall, 1985.

Oberg; Jones; Horton; Ryffel; Green; and McCauley. *Machinery's Handbook*. 25th ed. New York, NY: Industrial Press, Inc., 1996.

Outerbridge, David. *Bridges*. New York, NY: Harry N. Abrams, Inc., 1989.

Page, John S. *Estimator's General Construction Man-Hour Manual*. 2nd ed. Houston, TX: Gulf Publishing, 1977.

Parmley, Robert O. *Standard Handbook of Joining and Fastening*. New York, NY: McGraw-Hill, 1977.

Pelsma, Kimberlie H., ed. *Ergonomics Sourcebook: A Guide to Human Factors Information*. Lawrence, KS: The Report Store, A Division of Ergosyst Associates, Inc., 1987.

Peterson's Guide to 2 Year Colleges. Princeton, NJ: Peterson's Guides, 1991.

Peterson's Job Opportunity Series. Princeton, NJ: Peterson's Guides, 1991.

Repp, Victor E. *Metalwork: Technology and Practice*. Columbus, OH: Glencoe/McGraw-Hill, 1994.

Richardson, Terry L. *Industrial Plastics: Theory and Application*. Albany, NY: Delmar, Inc., 1997.

Rosenberg, Paul. *Installation Requirements of the 1990 National Electrical Code*. New York, NY: Macmillan, 1991.

Salvadori, M. *Why Buildings Stand Up*. New York, NY: W.W. Norton & Co., 1994.

Savage, E. and Sterry, L. *A Conceptual Framework for Technology Education*. Reston, VA: ITEA, 1991.

Slavin, R. E. *Cooperative Learning: Theory, Research, and Practice*. Englewood Cliffs, NJ: Prentice-Hall, 1995.

Smith, R. C., and Andres, C. K. *Materials of Construction*. Columbus, OH: Glencoe/McGraw-Hill, 1988.

Source Book 2: Learning By Design. New York, NY: American Institute of Architects Press, 1988.

Stewart, J. *Construction Glossary*. New York, NY: John Wiley and Sons, 1980.

Sule, D. R. *Manufacturing Facilities*. Boston, MA: PWS-Kent Publishing, 1988.

Todd, Ronald D.; Todd, Karen Rhone; and McCrory, David L. *Introduction to Design and Technology*. Columbus, OH: Glencoe/McGraw-Hill, 1996.

Troitsky, M. S. *Cable-Stayed Bridges: Theory and Design*. London: Granada Publishing Ltd., 1977.

Wallach, Paul, and Hepler, Donald. *Reading Construction Drawings*. Columbus, OH: Glencoe/McGraw-Hill, 1990.

Watson, Donald A. Construction Materials and Processes. Columbus, OH: Glencoe/McGraw-Hill, 1986.

Periodicals

Consumer Reports Magazine, 101 Truman Ave., Yonkers, NY 15238

Fine Homebuilding Magazine, The Taunton Press, Inc., 63 S. Main St., P.O. Box 5506, Newtown, CT 06470-5506

Futurist, The, World Future Society, 4916 St. Elmo Ave., Bethesda, MD 20814-5089

High Technology, P.O. Box 2808, Boulder, CO 80321

Popular Science, Times Mirror Magazines, Inc., Subscription Department, Boulder, CO 80322

Popular Mechanics, P.O. Box 7170, Red Oak, IA 51591

Technology and Learning, Computer Center, 650 Clark St., Chicago, IL 60605-9626

Videos, CDs, and/or Filmstrips

Agency for Instructional Technology, Box A, Bloomington, IN 47402, (800) 457-4509

Bergwall Productions, Inc., P.O. Box 2400, Chad Falls, PA 19031-2400, (800) 645-3565 (videos)

Career Aids Division, Opportunities for Learning, Inc., 20417 Nordhoff St., Chatsworth, CA 91311, (818) 341-2535

Career Development Software, Vancouver, WA

Catskill Mountain Lumber Co., Swan Lake, NY

Center for Implementing Technology Education, Ball State University, Muncie, IN

Encyclopaedia Britannica Educational Corp., Chicago, IL

Finley-Holiday Film Corporation, Whittier, CA

Glencoe/McGraw-Hill, Columbus, OH, (800) 334-7344

Hobar Publications, 3943 Meadowbrook Rd., Minneapolis, MN 55426, (800) 846-7027

Hometime, Eden Prairie, MN

IBM Film Library, 4705-F Babers Ferry Rd. SW, Atlanta, GA 30336

INC Magazine Video Series, Bala Cynwyd, PA

Mid-America Vocational Curriculum Consortium, 1500 W. 7th Ave., Stillwater, OK 74074, (800) 654-3988

National Audiovisual Center, National Archives and Records, 8700 Edgeworth Dr., Capital Heights, MD 20734-3701, (301) 763-1896

PBS Video, Public Broadcasting Service, 1320 Braddock Place, Alexandria, VA 22314-1698, (800) 344-3337

Portland Cement Association, 5420 Old Orchard Rd., Skokie, IL 60077

School Company, Vancouver, WA

Schoolworks, 712 Bancroft Rd., Ste 264, Walnut Creek, CA 94598, (800) 3YOURKIDS, www.choicemall.com/schoolworks

Society for Visual Education, Chicago, IL

Taunton Press, Newtown, CT

Software

Bloc Corporation, Miami, FL

Broderbund Software, Inc., P. O. Box 12947, San Rafael, CA 94913, (800) 521-6263

Career Aids Division, Opportunities for Learning, Inc., 20417 Nordhoff St., Chatsworth, CA 91311, (818) 341-2535

Career Development Software, Inc., Vancouver, WA

Claris Corporation, 5201 Patrick Henry Dr., Box 58168, Santa Clara, CA 95052, (800) 747-7483, www.claris.com

CompTech Systems Design, P. O. Box 516, Hastings, MN 55033, (612) 437-1350

Glencoe/McGraw-Hill, Columbus, OH (800) 334-7344

Hearlihy & Co., 714 W. Columbia St., P. O. Box 8690, Springfield, OH 45504, (800) 622-1000

Hobar Publications, 3943 Meadowbrook Rd., Minneapolis, MN 55426, (800) 846-7027

Intuit, Menlo Park, CA

Kitchen Sink Software, Inc., 1169 Stroud Ct., Westerville, OH 43081, (800) 235-5502, www. kitchen-sink.com

Modern School Supplies, Inc., P. O. Box 958, Hartford, CT 06143, (800) 243-2329

Pre-Engineering Software, 5800 One Perkins Place Dr., Ste 10-D, Baton Rouge, LA 70808, (504) 769-3728, www.pre-engineering.com

Radio Shack, 500 One Tandy Center, Fort Worth, TX 76102

School Company, Vancouver, WA

Shopware Educational Systems, 101 Hill Rd., Aberdeen, WA 98520-9984, (206) 532-3392

Sunburst Communications, 101 Castleton St., Pleasantville, NY 10570-3498, (800) 321-7511, www.sunburst.com

Tharo Systems, Inc., Brunswick, OH

Laboratory Equipment and Materials

Badge A Minit, 348 North 30th Road, Box 800, La Salle, IL 61301, (800) 223-4103

CompTech Systems Design, P.O. Box 516, Hastings, MN 55033, (612) 437-1350 (simple robots)

Creative Learning Systems, Inc., 9899 Hibert St. Suite C, San Diego, CA 92131, (800) 458-2880

D & M Computing Inc., 1010 32nd Ave, So., Moorhead, MN 56560, (218) 233-5172 (robots)

Edmund Scientific Co., 101 E. Gloucester Pike, Barrington, NJ 08001-1380, (609) 547-8880

Energy concepts, Inc., 595 Bond St., Lincolnshire, IL 60069, (800) 621-1247

EKI Inc., 16634 Noyes Ave., Irving, CA 92714, (800) 453-1708 (electronic kits and robotics)

fischer america inc., 1904 Oostburg Ct., Modesto, CA, 95356 (fischertechnik computer interface and systems)

Graymark International, Inc., Box 51020, Santa Ana, CA 62704, (800) 854-7393 (electronic kits)

Hearlihy & Co., 714 W. Columbia St., P.O. Box 8690, Springfield, OH 45504, (800) 622-1000 (software, drafting supplies and robotics)

Heathkit Co., P.O. Box 1288, Benton Harbor, MI 49022, (800) 253-0570 (electronic kits)

IASCO, 5724 W. 36th St., Minneapolis, MN 55416-2594, (800) 328-4824 ext. 2662

Kelvin, 10 Hub Dr., Melville, NY 11747, (800) 535-8469 (electronics, kits, and general supplies)

Lab-Volt Systems, Inc., P.O. Box 686, Farmingdale, NJ 07727, (800) 522-8658 (modular systems)

LEGO Dacta, 555 Taylor Road, Enfield, CT 06082, (800) 527-8389

Mid-America Co., Inc., 2309 S. Archer Ave., Chicago, IL 60616, (800) 621-1530 (electronic supplies)

Midwest Technology Products and Services, Inc., 400 S. Indiana St., P.O. Box 564, Hobart, IN 46342, (219) 942-1134 (model airplane kits)

Midwest Shop Supplies, P.O. Box 3717, Sioux City, IA 51102, (800) 831-5904

Modern School Supplies, Inc., P.O. Box 958, Hartford, CT 06143, (800) 243-2329 (software, robotics, and computer drafting supplies)

PITSCO, P.O. Box 1328, Pittsburg, KS 66762, (800) 835-0686

Radio Shack, 500 One Tandy Center, Fort Worth, TX 76102 (electronics, software, and computers)

Scholastic Futures Publishing, Division of Theta Industrial Products, Inc., P.O. Box 70, Mound MN 55364, (612) 471-8569

Theta Industrial Products, Inc. P.O. Box 70, Mound, MN 55364

The Teacher's Laboratory, Inc., 104 Canal St., P.O. Box 6480, Brattleboro, VT 05301-6480, (802) 254-3457 (robotics and applied science)

Television Programs

NOVA (PBS)
Science and Technology and *Future Watch* (CNN)
Beyond 2000 (Discovery Channel)

Other Sources of Information

Alternative Energy Resources Organization
324 Fuller C4
Helena, Montana 59601

American Congress on Surveying and Mapping
210 Little Falls St.
Falls Church, VA 22046

American Institute of Building Design
1412 19th St.
Sacramento, CA 95814

American National Standards Institute (ANSI)
1430 Broadway
New York, NY 10018

American Plywood Association
P. O. Box 11700
Tacoma, WA 98411-0700

American Society of Testing Materials (ASTM)
655 15th St., NW
Washington, DC 20005

American Solar Energy Society
2030 17th Street
Boulder, CO 80302

American Telephone and Telegraph
Public Relations Office
10 S. Canal St.
Chicago IL 60606

American-Gas Association
1515 Wilson Blvd.
Arlington, VA 22209

American Wood Council
1619 Massachusetts Ave, NW
Washington, DC 20036

Automatic Identification Manufacturers
1326 Freeport Road
Pittsburgh, PA 15238

Building Officials and Code Administration International (BOCA)
4051 W. Flossmoor Rd.
Country Club Hills, IL 60477

Cooperative Learning Center
60 Peik Hall
University of Minnesota
Minneapolis, MN 55455 (612) 624-7031
www.clcrc.com (cooperative learning materials)

Environmental Protection Agency
401 M St. SW
Washington, DC 20460

Federation of Societies for Coatings Technology
1315 Walnut
Philadelphia, PA 19107

General Robotics Corp.
14618 E. 6th Ave. Ste. 15
Golden, CO 80401

Gypsum Association
1603 Orrington Ave.
Evanston, IL 60201

Industrial Designers Society of America
1360 Beverly Rd.
McLean, VA 22101

International Technology Education Association (ITEA)
1914 Association Drive
Reston, VA 22091-1502, (703) 860-2100
(Professional organization—books, other publications, and video discounts for members)

Kelvin Electronics
7 Fairchild Ave.
Plainview, NY 11803

Landscape Architecture Foundation
   1733 Connecticut Ave., NW
   Washington, DC 20009

LEGO Dacta
   555 Taylor Rd.
   Enfield, CT 06082

Lumbermate Company
   10443 Baur Blvd.
   St. Louis, MO 63132

Methods-Time-Measurement Association
   for Standards and Research
   1601 Broadway
   Fair Lawn, NJ 07410

Mid-America Vocational Curriculum
   Consortium
   1500 West Seventh Ave.
   Stillwater, OK 74074-4364

Midwest Products Co., Inc.
   400 S. Indiana St.
   P. O. Box 564
   Hobart, IN 46342

NASA Core
   Lorain County Joint Vocational School
   15181 Route 58 South
   Oberlin, OH 44074

NASA Jet Propulsion Laboratory
   Attn: Teacher Resource Center
   JPL Educational Outreach
   4800 Oak Drive
   Mail Code CS-530
   Pasadena, CA 91109

NASA Spacelink
   NASA Headquarters
   Educational Publications Services, XEP
   Washington, DC 20546

National Aeronautics and Space Administration
   Kennedy Space Center Film Library
   Kennedy Space Center, FL 32899

National Aeronautics and Space Administration
   Lewis Research Center
   Office of Education Programs
   21000 Brook Park Rd.,
   Cleveland, OH 44135

National Air and Space Museum
   Smithsonian Institution
   Education Resource Center, P-700
   Washington, DC 20560

National Association of Home Builders
   Remodeler's Council
   15th and M Streets, NW
   Washington, DC 20005

National Fire Protection Association
   1 Battery March Park
   P. O. Box 9146
   Quincy, MA 02269-9959

National Forest Products Association
   1250 Connecticut Ave., NW
   Suite 200
   Washington, DC 20036

National Junior Achievement, Inc.
   45 East Clubhouse Dr.
   Colorado Springs, CO 80906

National Science Foundation
   4201 Wilson Blvd.
   Arlington, VA 22230, (703) 306-1130

Occupational Safety and Hazards
   Administration
   200 Constitution Ave.
   Washington, D.C. 20210

Solar Energy Institute of North America
   3404 Connecticut Ave., NW
   Washington, DC 20008

Southern Forest Products Association
   P. O. Box 52468
   New Orleans, LA 70152

The Construction Specifications Institute
   601 Madison St.
   Alexandria, VA 22314-1791

Trus-Joist Corp.
   Div. of T. J. International
   P. O. Box 60
   Boise, Idaho 83707

Water Environment Federation
   601 Wythe St.
   Alexandria, VA 22314-1994

Western Wood Products Association
   Yeon Building
   552 SW Fifth Ave.
   Portland, OR 97204-2122

Weyerhauser Building Products
   700 Emerald St.
   St. Paul, MN 55114

Scope and Sequence Chart

Themes and Concepts	Chapter 1	Chapter 2	Chapter 3	Section 1 Activities	Chapter 4	Chapter 5
Technological Principles and Systems	pp. 17, 18, 22-27		Entire chapter		Entire chapter	Entire chapter
Identifying Evolving Technologies	pp. 20-21		p. 51		pp. 73, 75-76	pp. 99, 109-117
Social/Cultural Impacts of Technology	pp. 18-21, 29				pp. 74, 76, 97	pp. 112-114, 116, 119
Design and Problem Solving		Entire chapter		Activities 2 and 4		
Finding, Processing, and Using Information	pp. 21, 31	pp. 34-35, 37		Activities 1 and 2		pp. 99, 112-114, 119
Tools, Materials, and Processes			pp. 51-54, 56, 57	Activity 4	pp. 84-86, 90-91	
Using Computers		pp. 33, 45			p. 97	Entire chapter
Using Resources	p. 31	pp. 34-35	pp. 48-49, 51-56	Activities 2, 3, and 4	pp. 81-89	pp. 99, 112-114, 119
Safety		p. 44		pp. 64, 68		
Skills						
Critical Thinking	p. 31	p. 45	p. 59	Activities 2 and 4	pp. 73, 97	p. 119
Science	p. 21	p. 41	pp. 50, 59	pp. 66, 69	p. 78	pp. 106, 119
Mathematics		p. 40	p. 59	pp. 61, 63; Activities 3 and 4	p. 97	
Language Arts	p. 31		pp. 50, 59	p. 61		
Social Studies	pp. 25, 31		p. 59	p. 61		
Technology Skills		pp. 33, 45		Activity 4	p. 97	pp. 99, 119
Investigating Careers	p. 30	p. 44	p. 58	Activity 1	p. 96	p. 118
Interpersonal Skills		p. 45	p. 59	Activities 1, 3, and 4		

Scope and Sequence Chart *(cont'd)*

Themes and Concepts	Chapter 6	Chapter 7	Section 2 Activities	Chapter 8	Chapter 9	Chapter 10
Technological Principles and Systems	Entire chapter			Entire chapter		
Identifying Evolving Technologies	pp. 121, 126-127, 136-137	pp. 143, 155, 166		p. 214	pp. 209, 216, 227	
Social/Cultural Impacts of Technology	pp. 121, 132, 134, 138, 141	p. 149		pp. 186-187		p. 231
Design and Problem Solving		pp. 144-148, 169	Activities 2 and 3		Entire chapter	p. 231
Finding, Processing, and Using Information			Activities 1, 2, and 3		pp. 225-227	
Tools, Materials, and Processes	Entire chapter	pp. 149-167		pp. 196-199, 201, 203	pp. 215-227	pp. 233-234, 240-243
Using Computers		pp. 143, 147-148, 166-167	Activities 1, 3, and 4	p. 207		
Using Resources			Activities 1, 2, and 3	pp. 195-201, 207		pp. 233-234, 240-243
Safety			p. 173			
Skills						
Critical Thinking	p. 141	p. 169	Activities 2 and 3	p. 207	pp. 209, 211, 229	pp. 231, 245
Science	pp. 123, 141	p. 169			pp. 211, 229	p. 245
Mathematics	p. 141	pp. 150, 169	pp. 172, 178, 180	p. 194	p. 229	pp. 242, 245
Language Arts	p. 121	p. 169	Activity 3	pp. 185, 207		
Social Studies				pp. 186-187, 191-193, 207		
Technology Skills		pp. 143, 150, 158-166, 169	All activities	p. 207	p. 229	
Investigating Careers	p. 140	p. 168		p. 206	p. 228	p. 244
Interpersonal Skills			Activity 3			

Scope and Sequence Chart (cont'd)

Themes and Concepts	Chapter 11	Chapter 12	Chapter 13	Section 3 Activities	Chapter 14	Chapter 15
Technological Principles and Systems					Entire chapter	pp. 337-345
Identifying Evolving Technologies	p. 254-255		Entire chapter		pp. 317, 319, 320	pp. 334, 341
Social/Cultural Impacts of Technology					pp. 308, 323	p. 331
Design and Problem Solving				Activities 1, 2, 3, and 4	p. 307	pp. 329, 337-341
Finding, Processing, and Using Information	pp. 263-264			All activities		pp. 333-336, 337-340
Tools, Materials, and Processes	pp. 253-257		pp. 277-289	Activities 1, 2, and 4	pp. 313, 314-322, 323	pp. 334-335
Using Computers			Entire chapter	Activities 3 and 4		pp. 341, 347
Using Resources	pp. 250-251	pp. 263-264		All activities	pp. 307, 311-323, 327	pp. 337-345
Safety				pp. 292, 295		
Skills						
Critical Thinking	pp. 252, 259	pp. 261, 269	p. 291	Activities 1, 2, 3, and 4	p. 237	pp. 329, 347
Science		pp. 267, 269	pp. 273, 291	pp. 294, 298, 300, 303 Activity 2	pp. 317, 325, 327	pp. 336, 347
Mathematics	pp. 252, 259	p. 269		pp. 294, 300 Activities 3 and 5		
Language Arts	p. 259	p. 269		Activity 4		
Social Studies		p. 269			p. 309, 327	pp. 332, 337
Technology Skills	pp. 247, 259		p. 273, 291	All activities	pp. 307, 327	pp. 329, 347
Investigating Careers	p. 258	p. 268	p. 290		p. 326	p. 346
Interpersonal Skills				Activities 1, 2, 3, and 4		

Scope and Sequence Chart *(cont'd)*

Themes and Concepts	Chapter 16	Chapter 17	Chapter 18	Section 4 Activities	Chapter 19	Chapter 20
Technological Principles and Systems					Entire chapter	Entire chapter
Identifying Evolving Technologies	pp. 334, 335	pp. 361, 374-375	pp. 397-399			pp. 441, 446, 454-455, 460-461
Social/Cultural Impacts of Technology		p. 361	p. 381	Activity 1	pp. 421, 435-436	pp. 441, 461
Design and Problem Solving		p. 379		Activities 3, 4, and 5		p. 441
Finding, Processing, and Using Information				Activities 2, 4, and 5		
Tools, Materials, and Processes	p. 359	Entire chapter	Entire chapter	Activities 3, 4, and 5	pp. 426, 431, 432-434	
Using Computers	p. 349			Activities 4 and 5 (optional)		p. 463
Using Resources	pp. 351-353, 354	p. 379		All activities	pp. 421, 427-432	p. 463
Safety	p. 357			pp. 410, 412, 415		
Skills						
Critical Thinking	p. 359	pp. 361, 379	p. 401		p. 439	pp. 441, 463
Science		pp. 371, 397	p. 401	pp. 409, 411, 417	p. 439	pp. 443, 463
Mathematics	pp. 354, 359	p. 379	pp. 394, 401	pp. 403, 414 Activity 2	pp. 430, 439	p. 453
Language Arts				pp. 403, 414		
Social Studies	pp. 352, 356		pp. 381, 397-399	Activity 1	pp. 430, 439	p. 463
Technology Skills	pp. 349, 358	379		All activities		p. 463
Investigating Careers	pp. 349, 358	p. 378	p. 400		p. 438	p. 462
Interpersonal Skills				Activity 3		

Scope and Sequence Chart (cont'd)

Themes and Concepts	Chapter 21	Chapter 22	Section 5 Activities	Chapter 23	Chapter 24	Section 6 Activities
Technological Principles and Systems	Entire chapter	Entire chapter	Activities 1 and 2	Entire chapter	Entire chapter	Activity 2
Identifying Evolving Technologies	pp. 465, 477	pp. 496-497, 499-501, 505-507	Activity 1	p. 527	pp. 547-550, 553, 554 (illus.), 556-557, 559-561, 562, 564	
Social/Cultural Impacts of Technology	p. 477		Activity 1	pp. 530-535, 539-541	pp. 546, 548, 563, 567	
Design and Problem Solving	p. 465	p. 509	Activities 1 and 2			Activity 2
Finding, Processing, and Using Information		p. 481	Activities 1 and 3	p. 543	p. 545	Activity 2
Tools, Materials, and Processes	pp. 467, 470-471, 472-474, 475-477	pp. 491-494, 499-500, 501-502, 505-507	All activities		pp. 554-557, 560-562, 565	Both activities
Using Computers			Activity 1	p. 543		
Using Resources			Activity 1	p. 543		
Safety			pp. 510, 515, 519			pp. 570, 571
Skills						
Critical Thinking	pp. 465, 475, 479	p. 509	Activity 1	p. 543	p. 569	Activity 2
Science	p. 479	pp. 483-487, 490, 493-495, 497, 501-502, 504, 506 (illus.), 509	p. 513 Activities 2 and 3	Entire chapter	Entire chapter	p. 575
Mathematics	p. 475	pp. 481, 509	pp. 513, 518, 523	p. 543	pp. 558, 569	pp. 572, 575
Language Arts	p. 479		Activity 1	p. 529		p. 572
Social Studies	p. 477		Activity 1	pp. 530-535		
Technology Skills	p. 465	p. 509	All activities	p. 543		Both activities
Investigating Careers	p. 478	p. 508		p. 542	p. 568	
Interpersonal Skills			Activities 1 and 2			Activities 1 and 2

General Preparation Suggestions

General Teaching Methods and Strategies

1. Make use of cooperative learning techniques whenever possible and practical. Technology educators (and industrial arts teachers before them) have frequently directed students to work together in teams or groups. A major problem of this approach has always been determining how to make *all* the students in the team or group responsible for the work performed. Briefly, cooperative learning methods help enable the instructor to make sure that each student is responsible and participates in the learning process. Cooperative learning requires students to depend upon each other and to cooperate to complete the assignment. A more detailed description of cooperative learning is included in the Teacher's Resource Binder. Several suggested methods are included in the specific chapter teaching suggestions. There are several sources for more information about cooperative learning listed in the RESOURCES section.

2. Many technology educators are using a workstation approach to introductory technology education classes. This involves setting up several workstations where student teams perform specific activities/experiments relating to a specific concept or system of technology. An advantage of this approach is that equipment does not need to be purchased in quantities large enough for an entire class to do a specific activity at once. The number of workstations is determined by class size and how many concepts/activities the teacher wants the students to participate in. Disadvantages include: significant time investment in teacher created workbooks and/or activity books for each station, difficulty in supervision and lost parts. Also, since students are involved in many different concepts at the same time another disadvantage is increased preparation time and student questions. Most teachers who are successful with the workstation approach use cooperative learning methods of one form or another. Many technology instructors include a requirement that only one member of a team may ask questions of the instructor and then only after the entire team has attempted to answer the question first. Another helpful requirement is to make the team responsible for all the materials at the workstation they are currently using.

    Combining the workstation and cooperative learning methods may have advantages. Individual instructors experiment with both methods until they find a balance with which they are most comfortable and their students are the most successful.

3. The change from industrial arts to technology education has been the most difficult for those students who learn more slowly. In the past, industrial arts classes were frequently seen by guidance counselors as an appropriate placement for those students having difficulty in "academic" curriculums. Work with your building's special education teachers and counselors to help them understand the new requirements of technology education classes. Special education teachers can usually help you adapt your lessons, assignments and activities to enable special needs students to succeed. The cooperative methods mentioned above can considerably help special needs students.

General Suggestions for Preparing for All Chapters

1. Read the chapter and the lesson plans for the chapter. Decide which activities you plan to use with the chapter. (There are cross-curricular and other activities at the end of each textbook chapter. More activities can be found in the *Teacher's Resource Binder* and in the *Student Workbook*.)

2. Obtain the necessary materials and supplies to conduct the activities. Duplicate any transparency masters or handouts needed for the chapter. Obtain/reserve any videotapes that you will use with the chapter.

3. Check with your school building and/or district media specialist for a list of computer programs available to you. Ask him/her to make recommendations of any software that may apply to the chapter. Frequently, software recommended for science, math, and social studies can have applications to chapter concepts. Preview the software and make any special instruction sheets that you want the students to use with the software.

4. Use the ideas supplied with each chapter to make at least one bulletin board for the chapter. Frequently, the chapter suggestions will include ideas for student-made bulletin boards.

5. On a large piece of poster board, draw a diagram of the system model shown in Figure 3-1. Save this poster to put up as a review of the system model at the beginning of each major section of the text.

6. Contact any speakers that you plan to use in conjunction with the chapter. Arrange any field trips that you plan to take with your class. Local industries are often an excellent source of speakers and will often provide interesting field trip locales. Many of those industries are run by the parents of your past, present, or future students and they often welcome a chance to help technology educators.

7. Contact the local media for any especially newsworthy projects that your students will be completing during this chapter. Never underestimate the importance of public relations as part of promoting and planning technology education.

8. Obtain a supply of colorful, interesting posters that can be changed regularly. Changing posters can often be a good activity for students, and it helps them feel some ownership of their classroom environment.

9. Share the cross-curricular activities with teachers of other curricular areas in your school. Discuss with them the possibility of assigning interrelated projects. An example might be a report about the history of a specific technological object. Such a report would be a cross-curricular activity that could receive credit from the technology, social studies, and language arts teachers. Many technology teachers are team teaching or working closely with teachers in other disciplines to help students understand the ways in which science, technology, and society are closely tied together.

Lesson Plans

Chapter 1
Lesson Plan:
Exploring Technology

Chapter Goal and Outline

The main goal of this chapter is to make students aware of technology's importance in our lives and to introduce them to the five families of technology.

I. What Is Technology?
   A. A History of Technology
   B. The Pace of Change
II. Using Technology
   A. Communication Technology
   B. Manufacturing Technology
   C. Construction Technology
   D. Transportation Technology
   E. Bio-Related Technology
   F. Combined Technology
III. Technology and Human Knowledge
   A. Technology and Science
   B. Technology and Math
   C. Technology and Language Arts
   D. Technology and Social Studies
IV. Technology Assessment

Objectives

After studying this chapter, students should be able to:
- define technology.
- explain why the pace of technological change has increased.
- define five areas in which technology is used to solve problems.
- describe the relationship between technology and other areas of knowledge.

Preparation

1. Review "General Suggestions for Preparing for All Chapters" on page 21 of this Teacher's Manual.

2. Make a bulletin board appropriate to the chapter content. On a blank bulletin board, put only the title "What Is Technology?". Have students bring in pictures that show technology in use. Divide the board into separate sections for each type of technology—communication, manufacturing, construction, transportation, and bio-related technology—and put appropriate photos or drawings in each section.

Teaching

1. **Open Meeting.** Group the students in a tight circle. Explain that you are going to brainstorm together and that there are no "right" or "wrong" answers, that all answers are valid. Establish ground rules that no one may make fun of anyone for any answers so that everyone will feel at ease to reply. Everyone must reply, but "pass" can be a valid reply.

   Ask the question: "What is technology?" and go around the circle until all students have answered or passed. Using the same method but starting with a different student first, ask: "What effect does technology have on your life?" After these two questions have been answered, indicate that the students should remember their answers so that in future weeks, they can see if they get new or different ideas about these questions.

2. **Overview.** Discuss your overall outline and goals for the quarter (or semester or year) with the students.

3. **Reading and Writing.** Assign reading of the chapter and completion of the Understanding Concepts and Thinking Critically questions at the end of the chapter. Answers to these end-of-chapter questions can be found at the end of this Lesson Plan.

4. **Reinforcement.** Discuss the five main systems into which we divide technology (communication, manufacturing, construction, transportation and bio-related). Help the students understand the meaning of each. Explain that we divide technology into these separate categories in order to make technology more easily understood; reinforce the text's statements about the inter-relatedness of all the categories.

5. **Reteaching.** In a discussion, have the students imagine that suddenly one of the main systems (e.g., transportation) completely breaks down. Lead the discussion so that the students see that technological systems cannot function without the support of all the other systems. (In the example of the breakdown of the transportation system, materials to make products could not be delivered to the factory and manufactured goods could not be shipped from the factory.)

6. **Cross-Curricular Connections.** Have students complete the Applying Concepts and Solving Problems activities at the end of the chapters.

Evaluation

1. Check students' answers to the Understanding Concepts and Thinking Critically questions. Review any topics that students do not understand.

2. Make copies of the Chapter Review in the *Teacher's Resource Binder* to administer as an open-book test or as a scored chapter test. Answers to the questions in this Chapter Review can also be found in the *Teacher's Resource Binder*.

Answers to Textbook Questions

Understanding Concepts

1. Technology is the generation of knowledge and processes to develop systems that solve problems and extend human capabilities.

2. The pace has increased because of faster communication and the build-up of technology over centuries.

3. The categories are (1) communication—sending and receiving messages; (2) manufacturing—changing raw materials into finished products; (3) construction—building structures; (4) transportation—carrying things and/or people from one place to another; (5) bio-related technology—technology connected with plant and animal life.

4. (Examples for each area will vary.) Science—telescope to learn about stars and planets; microscope for examining atoms. Math—calculation of weight structure can withstand; manufacturing components to exact specifications. Language Arts—instruction manuals; advertising and selling. Social Studies—determining the impact of a technology.

5. Examining the cost, safety, and economic and environmental impacts of a technology.

Thinking Critically

1. It represents a means of solving problems and extending human capabilities. For example, stone tools were used to kill animals for food, to remove skin from animals for clothing, to cut larger animals into smaller sections, to make utensils, etc.

2. It affects the ways people live and work. For example, technology may bring about job specialization, as did the waterwheel.

3. Computers, television, cellular telephones, satellites, optical fibers, fax machines, the Internet.

4. Communication technology is necessary to convey instructions during manufacturing; equipment used for communication (e.g., telephones, computers, even pens and paper) must be manufactured.

5. Transportation technology: improved transportation vs. smog and other pollution problems. Communication technology: faster, expanded information at home and at work via computers vs. social isolation, invasion of privacy, and violation of copyright. Bio-related technology: improved treatment of diseases vs. genetic control of humans.

Chapter 2 Lesson Plan: Problem Solving

Chapter Goal and Outline

The main goal of this chapter is to acquaint students with the problem-solving process and the scientific method.
I. The Problem-Solving Process
   A. State the Problem Clearly
   B. Collect Information
   C. Develop Possible Solutions
      1. Brainstorming
      2. Thinking Skills
   D. Select the Best Solution
   E. Implement the Solution
   F. Evaluate the Solution
II. The Scientific Method
III. Entrepreneurs
IV. The Technology Student Association

Objectives

After studying this chapter, students should be able to:
- describe six steps of the problem-solving process.
- define three types of thinking skills.
- describe the six steps of the scientific method.

Preparation

1. Review the "General Suggestions for Preparing for All Chapters" on page 21 of this Teacher's Manual. (***Note:*** *Henceforth, these will be referred to simply as "General Suggestions."*)

2. Make a bulletin board with the six steps of the problem-solving process (state the problem clearly, collect information, develop possible solutions, select the best solution, implement the solution, and evaluate the solution) as separate columns or areas.

3. As an alternative to #2 above, make a bulletin board that lists problems facing our society (pollution, overpopulation, crime, etc.). Then, after you have discussed the problem-solving process in class, ask students to bring in newspaper clippings that illustrate attempts to solve these problems using technological means.

4. Ask a science teacher in your building what students at your class's grade level should have already learned about the scientific methodology. Use this information to help you with #5 in the Teaching Suggestions section below.

Teaching

1. **Reading and Writing.** Assign reading of the chapter and completion of the Understanding Concepts, Thinking Critically, Applying Concepts and Solving Problems questions at the end of the chapter. Answers to these end-of-chapter questions can be found at the end of this Lesson Plan.

2. **Cooperative Learning Strategy/Task Groups.** A task group is a team established to perform a specific task. Cooperative learning teams work best when the teams are heterogeneous groups—a mixture of students with varying abilities and skills. It is *very* important that students feel that you are using some "fair" technique to assign teams and roles. Organized/random selection of teams usually works best. This can be accomplished in a variety of ways such as counting off, or dividing according to the compass directions of seats.

   Divide the class into task group teams of four. (Some teams may need to have five members, depending upon class sizes.) Assign these roles: *Leader, Recorder, Facilitator,* and *Reporter.* (In five member teams, you may want to assign two Facilitators.) Again, make the selections by some organized/random technique, such as suggested above.

   Note: It is important that all students get the opportunity to perform different roles and that the students are in different task groups throughout the term. The same students should *not always* be Leaders (or any other role).

Discuss the duties and responsibilities of students as they perform their roles: *Leader*—responsible for making sure that the team stays on track and accomplishes its task; *Recorder*—writes down all instructions and writes any necessary written team output; *Facilitator*—helps everyone work cooperatively and helps prevent conflict or impasses; and *Reporter*—reports the outcome of the team's problem solving to the teacher or other groups (depending on the assignment).

Note: Only one team member (Reporter or the Leader) is allowed to ask questions of the teacher after the teams are put onto the task. This duty is usually assigned by the teacher to either the Reporter or the Leader. The team *must have tried* to find the answer to the question before asking the teacher. (Some teachers allow each team only a specific number of questions.)

Ask teams: "Are we using too much fossil fuel?" The teams must come up with a clear statement of the problem within ten minutes. At the end of this time limit, each Reporter will read his/her team's problem statement to the class.

Lead a discussion to help the students understand why each team might have come to a different conclusion about the problem (because of the complexities and value judgments involved in answering the question).

3. **Reteaching.** Lead discussions to make sure that all students understand thinking skills.

4. **Review.** Review the scientific method with students, based on this text and what they have already studied in science classes.

5. **Cross-Curricular Connections.** Have students complete the Applying Concepts and Solving Problems activities at the end of the chapter. Answers can be found at the end of this Lesson Plan.

Evaluation

1. Check students' answers to the Understanding Concepts and Thinking Critically questions. Review any topics that students do not understand.

2. Make copies of the Chapter Review in the *Teacher's Resource Binder* to administer as an open-book test or as a scored chapter test. Answers to the questions in this Chapter Review can also be found in the *Teacher's Resource Binder.*

Answers to Textbook Questions

Understanding Concepts

1. State the problem clearly in order to understand exactly what the problem is. Gather information in order to know as much about the problem as possible to use in developing the solution. Develop possible solutions through brainstorming, etc. Since most problems have more than one solution—the more ideas, the more options from which to choose. Select the best solution; advantages and disadvantages of each possible solution must be analyzed to determine which best solves the problem. Implement the solution; models are made and ideas are tested to make sure the solution is workable and to work out any "bugs." Evaluate the solution to make sure there are no problems with product (or other output) and that it doesn't become outdated.

2. Critical thinking: analyze problems; make judgments. Creative thinking: develop or improve original ideas. Decision making: ability to select, make choices.

3. Steps of scientific method: (1) make an observation; (2) collect information; (3) form a hypothesis; (4) perform experiment to test hypothesis; (5) analyze results; and (6) repeat to verify results.

4. Entrepreneurs need to be good problem solvers, able to think critically, creative, willing to take risks, ambitious, and hard-working.

5. Any three: aerospace, architecture, construction, electronics, manufacturing, computer applications.

Thinking Critically

1. Possible answers—*Alike:* Each would choose a house that would provide the most comfort and most protection from weather conditions in his/her particular area. *Different:* Person in north would want a site that offered protection from cold winds, and the person in the tropics would want a site in an open area that allowed cooling breezes to pass through the home.

2. Answers will vary.

3. Only steps 1 and 2 are essentially the same. Step 3 of each is similar, but they begin to show the difference in approach to a question. Step 4 differs most—selection of an option vs. setting up an experiment to test a hypothesis. Steps 5 and 6 differ in terms of action. The problem-solving process applies to an action to be performed; the scientific method applies to research and confirmation of a theory.

4. Answers will vary.

5. Answers will vary.

Chapter 3 Lesson Plan: Technology Systems

Chapter Goal and Outline

The main goal of this chapter is to provide students with a general introduction to the parts of a system and the systems model.
  I. What Is a System?
 II. Input
     A. People
     B. Information
     C. Materials
        1. Natural and Synthetic
        2. Raw and Industrial
     D. Tools and Machines
     E. Energy
        1. Sources of Energy
        2. Forms of Energy
     F. Capital
     G. Time
III. Process
 IV. Output
  V. Feedback

Objectives

After studying this chapter, students should be able to:
- define system.
- describe the four parts of a system.
- name seven types of resources that provide input for a system.
- compare three different sources of energy.

Preparation

1. Review the "General Suggestions" listed on page 21 of this Teacher's Manual.
2. Make a bulletin board with the seven resources—people, information, materials, tools and machines, energy, capital, and time. Cut out magazine pictures that illustrate each resource and put them in the appropriate columns.
3. If you have not already done so, make the poster described in #5 of the "General Suggestions." (This is a diagram of the system model, as shown in Fig. 3-1 in the textbook.)

Teaching

1. **Reading and Writing.** Assign reading of the chapter and completion of the Understanding Concepts and Thinking Critically questions at the end of the chapter. Answers to these end-of-chapter questions can be found at the end of this Lesson Plan

2. **Reteaching.** Spend as much time as needed in lecture/discussions to assure that students are familiar and comfortable with the concept of systems in technology. Use as many examples as possible to enable students to view systems as central to all technological understanding.

3. **Reinforcement.** Lead discussions to make sure that all students understand the system model since it will be used in each of the five major sections of the text (Communication, Manufacturing, Construction, Transportation, and Bio-Related Technology).

4. **Critical Thinking.** Ask students if it would be possible to build any technological object without any one of the seven resources. Use this as the basis of a class discussion to help students understand why a society cannot afford to run out of any of the seven resources.

5. **Cross-Curricular Connections.** Have students complete the Applying Concepts and Solving Problems activities at the end of the chapter.

Evaluation

1. Check students' answers to the Understanding Concepts and Thinking Critically questions. Review any topics that students do not understand.

2. Make copies of the Chapter Review in the *Teacher's Resource Binder* to administer as an open-book test or as a scored chapter test. Answers to the questions in this Chapter Review can also be found in the *Teacher's Resource Binder*.

3. Administer the Section I Test, which you will find in the *Teacher's Resource Binder*.

Answers to Textbook Questions

Understanding Concepts

1. It is a group of parts that work together to achieve a goal. Input: what goes in to the system. Process: activities that need to take place for the system to produce the expected result. Output: everything that results when input and process go into effect; includes product or creation plus impacts. Feedback: information about the output that helps evaluate the system.

2. Seven resources are: people, information, materials, tools and machines, energy, capital, and time. (Examples will vary.)

3. Inexhaustible energy: not used up by human action; possible example—sun, wind, or tides. Renewable energy—can be replenished; possible example—human energy, animal energy, trees, or plants. Nonrenewable energy: cannot be resupplied; possible example—any of the fossil fuels (coal, petroleum, or natural gas) or nuclear energy sources.

4. Input includes anything that is put into the system. Feedback becomes input when it is sent back into the system to evaluate or improve it.

5. Management plans, organizes, and controls the process.

Thinking Critically

1. *People* design, make the parts for, assemble, inspect, and finish the cars; they also manage and supervise operations and market the cars. *Information* about past designs, people's wants and needs, fuel, etc. needed to design cars; workers need blueprints and specifications to make and assemble the parts. *Materials* such as metal, plastic, glass, wiring, and paint needed. *Tools and machines* are needed to process materials and make and assemble cars. *Energy* is needed to power tools and machines; human energy also needed. *Capital* is needed to buy materials and tools and machines, to pay workers, and to pay energy. *Time* is needed to design and make the car and to transport it to dealer.

2. Vehicles have positive impacts (provide transportation; help the economy) and negative impacts (cause air pollution, traffic deaths, and injury).

3. Thermostat senses temperature. When it is too low, thermostat sends message to furnace, which sends heat through house. When thermostat senses temperature is correct, it sends this message to furnace, which quits running.

4. Answers will vary.

5. Possible answers might refer to the manufacture of hardware in a TV that could control access to particular stations, could search programs according to an owner's preferences, computerize programming, etc.

Chapter 4 Lesson Plan: Communication Systems

Chapter Goal and Outline

The main goal of this chapter is for students to understand the four parts of communication systems, the purposes of communication, and the impact of communication technology on our daily lives.

  I. What Is Communication Technology?
  II. Communication Innovations and Trends
  III. Why We Need Communication Technology
  IV. Parts of the Communication System
  V. Inputs
    A. People
    B. Information
    C. Materials
    D. Tools and Machines
    E. Energy
    F. Capital
    G. Time
  VI. Processes
  VII. Outputs
    A. Positive Impacts
      1. Long Distance Communication
      2. Communication in Schools
    B. Negative Impacts
      1. Results of Viewing TV
      2. Concerns about Privacy
      3. Effects on Health and Safety
      4. Pollution
  VIII. Feedback

Objectives

After studying this chapter, students should be able to:

- define communication technology.
- identify trends in communication systems.
- explain why we need communication systems.
- describe how inputs, processes, outputs, and feedback apply to a communication system.
- explain some impacts of communication systems.

Preparation

1. Review the "General Suggestions" listed on page 21 of this Teacher's Manual.

2. As described in #5 of these "General Suggestions," put up the poster of the system model. Use this to help students fit communication systems into the systems model.

3. On a bulletin board, put up the headings of the basic reasons for communicating (inform, educate, persuade, entertain, and control). Cut out pictures that illustrate each of these.

4. As an alternate to #3 above, on a bulletin board, put the large title "Communication" and have each student bring in a picture of a communication device or an example of communication. Have each student tack his/her picture onto the bulletin board.

Teaching

1. **Pretest.** Use the Section II Test in the *Teacher's Resource Binder* as a pretest for this Communication section. Do NOT count this as a grade for students. Rather, use this to accurately gauge students' learning by comparing these test scores with the scores obtained after students have studied this section.

2. **Reinforcement.** Review the basic system model (discussed in Chapter 3) with the students. Refer to the model frequently throughout this section to help students understand the communication system.

3. **Overview.** Since this chapter introduces the Communication section, discuss with students any projects that they will be doing pertaining to communication.

4. **Reading and Writing.** Assign reading of the chapter and completion of the Understanding Concepts and Thinking Critically questions at the end of the chapter. Answers to these end-of-chapter questions can be found at the end of this Lesson Plan.

5. Extension. The communication system presents teachers with probably the widest variety of possible student activities of any of the five systems (communication, construction, manufacturing, transportation, and bio-related technology). Anything from desktop publishing to CAD, photography to video production, student newsletters to student radio stations, platen press operation to electronics can be incorporated into the study of communication. Whatever resources and expertise you have (or have access to) can be built into excellent student learning experiences.

6. Reteaching. Communication devices such as telephones, televisions, radios, CD players, etc. are probably some of your students' favorite technological devices. Consequently, this chapter provides many opportunities to enable students to see direct impacts of technology on their lives. Students have trouble even imagining a world without communication systems. (Imagine any of your teenage students without access to a telephone!) Use discussions about negative impacts to help students understand that even the telephone (a seemingly benign device) can have harmful effects. All students have probably been part of family arguments over using the telephone. Most have probably received annoying phone calls.

7. Cross-Curricular Connections. Here students complete the Applying Concepts and Solving Problems activities at the end of the chapter.

Evaluation

1. Check students' answers to the Understanding Concepts and Thinking Critically questions. Review any topics that students do not understand.

2. Make copies of the Chapter Review in the *Teacher's Resource Binder* to administer as an open-book test or as a scored chapter test. Answers to the questions in this Chapter Review can also be found in the *Teacher's Resource Binder*.

Answers to Textbook Questions

Understanding Concepts

1. Communication technology is all the things people make and do to send and receive messages.

2. Innovation: introduction of something new; e.g., portable TV phone. Trend: general movement or pattern; e.g., CD-ROMs for computers.

3. Inform (possible examples: news broadcast, report); educate (possible examples: textbook, video); persuade (possible examples: advertising poster, newspaper ad, or radio or TV commercial); entertain (possible examples: video games, movies); control (possible examples: traffic light sensors, timed VCR machines).

4. a) feedback, b) output, c) input, d) process.

5. Answers will vary, depending on communication system chosen.

Thinking Critically

1. Answers will vary. Possible disadvantages of TV: too little physical exercise, too isolating, possible influences anti-social behavior, detracts from reading time. Some possible advantages: educational benefits; entertainment; fast, worldwide information; expands horizons.

2. Answers will vary. Some possible advantages of computers in classrooms: Internet for immediate access to worldwide information, new skills learned, more individualized instruction possible. Some possible disadvantages: expensive, requires updating and maintenance of equipment, possible problems of theft, possibly less interaction/discussions in classroom.

3. Answers will vary.

4. Answers will vary.

5. Answers will vary.

Chapter 5 Lesson Plan: Computers

Chapter Goal and Outline

The main goal of this chapter is for students to understand how each part of the computer works, how we communicate with computers, and how computers are used to communicate in various communication systems as well as via the Internet and on-line services.

I. Computers in Communication
   A. Parts of a Computer
      1. Input Units
      2. Central Processing Unit
      3. Memory or Storage Units
      4. Output Units
   B. The Language of Computers
      1. Programming Language
      2. Machine Language
      3. Electrical Signals
II. The Amazing Integrated Circuit
III. How the CPU Works
IV. Sending and Altering Information
   A. Altering Information
   B. Sending Voices
   C. Sending Images
   D. Retrieving Information
   E. Using the Bar Code
V. The Internet and On-Line Services
VI. Artificial Intelligence
   A. Expert Systems
   B. Speech Recognition
VII. Digital Libraries
VIII. Virtual Reality

Objectives

After studying this chapter, students should be able to:
- describe how computers are used to communicate.
- identify the basic components of computer systems.
- tell what integrated circuits are and identify their functions.
- discuss the Internet, artificial intelligence, digital libraries, and virtual reality.

Preparation

1. Review the "General Suggestions" listed on page 21 of this Teacher's Manual.
2. Put up the poster of the system model. (See #5 of the "General Suggestions.") Use this to help students see how computer systems fit into the system model.
3. Try to obtain a couple of obsolete and/or broken computers. One can be taken apart and used to point out various components of the computer, especially the main processor chip. Another can be used for discussions about obsolesce in computer technology.
4. Obtain a handful of obsolete or damaged IC chips (check local electronic repair shops or computer stores) that can be passed around for students to hold and examine closely during discussions.
5. If your school has a computer lab (or other setting) with Internet access, arrange to take your students into it for several class periods. If your school has no computer lab, see if there are any computers in the building (or in another building in your district) that can access the Internet. Arrange to use the Internet at least once.

Teaching

1. **Reading and Writing.** Assign reading of the chapter and completion of the Understanding Concepts and Thinking Critically questions at the end of the chapter. Answers to these end-of-chapter questions can be found at the end of this Lesson Plan.

2. **Enrichment.** Many of your students have grown up viewing computers as just another technological object. Some will have used a computer since they were small children. Help students understand the extremely rapid development that computer technology has experienced and that, historically speaking, personal computers are a very recent invention. If available, use a computer that is 10 or more years old as an example of how quickly computers become "obsolete."

3. **Extension.** If your school has a computer lab with Internet access, show students how to use a browser program and demonstrate how to effectively use search engines to find information. Create an Internet Treasure Hunt in which students work in teams to find specified information and record the URL of the site where they found it.

4. **Enrichment.** If only an individual computer with Internet access is available, demonstrate a browser program and various search engines to your students.

5. **Cross-Curricular Connections.** Have students complete the Applying Concepts and Solving Problems activities at the end of the chapter.

Evaluation

1. Check students' answers to the Understanding Concepts and Thinking Critically questions. Review any topics that students do not understand.

2. Make copies of the Chapter Review in the *Teacher's Resource Binder* to administer as an open-book test or as a scored chapter test. Answers to the questions in this Chapter Review can also be found in the *Teacher's Resource Binder*.

Answers to Textbook Questions

Understanding Concepts

1. Input units feed information into the computer; output units decode information from the computer; CPU fetches, decodes, and executes instructions; and memory or storage units help the CPU remember instructions and store information.

2. Often called a microchip, an integrated circuit is a tiny piece of silicon that contains thousands of interconnected electrical circuits (pathways for electric current or pulses) that work together.

3. They enable people to give instructions to computers.

4. Each product has its own pattern of this striped code. A laser scanner reads light/no light from the stripes; this information is sent to a computer, which locates information about the item in its memory and sends that data back. The computer also writes the product's code on an inventory file.

5. It provides prompt access to large quantities of information and connects people from all over the world quickly and easily.

Thinking Critically

1. Answers will vary. Some possible advantages: will save money; may be more accurate than a human. Some possible disadvantages: school nurses do more than diagnose health problems; will need to hire someone to run the program and deal with sick children.

2. Answers will vary.

3. Answers will vary.

4. Answers will vary.

5. Answers will vary.

Chapter 6 Lesson Plan: Telecommunication

Chapter Goal and Outline

The main goal of this chapter is to introduce students to the media that use long-distance signal transmission and reception.

I. What Is Telecommunication?
II. The Telegraph
   A. How the Telegraph Works
III. The Telephone
   A. How the Telephone Works
   B. Electronic Telephones
   C. Cellular Telephones
   D. Transmission Channels
      1. Copper Wire
      2. Coaxial Cable
      3. Optical Fibers
      4. Microwaves
      5. Radio Waves
   E. Getting a Message to the Right Place
      1. Early Switching Systems
      2. Automated Telephone Exchanges
IV. Radio
   A. How Radio Works
   B. Digital Radio
V. Television
   A. How Television Developed
   B. How Television Works
      1. The Television Camera
      2. Transmission
      3. The Television Receiver
   C. High-Definition Television
   D. Digital Television
VI. Satellite Communication Systems
   A. How the System Works

Objectives

After studying this chapter, students should be able to:
- explain the importance of telecommunication.
- show how signals are transmitted.
- describe how telegraphs, telephones, radios, televisions, and satellites operate.
- identify the effects of digital technology.

Preparation

1. Review the "General Suggestions" listed on page 21 of this Teacher's Manual.

2. At the top of a bulletin board, put up the title "Telecommunication." At the bottom, make a time line (divided into decades) from 1830 until the present. As your class progresses through the chapter, tack pictures of telecommunication devices above the date of their invention. Offer extra credit to any students willing to research how many homes had telephones, radios, or televisions at the beginning of each decade after their invention.

3. Display diagrams that show and explain the workings of popular telecommunication devices like CD players or VCRs. The book *The Way Things Work*, by David Macaulay, is one source for entertaining diagrams. Ask students to make enlarged copies of the drawings.

Teaching

1. **Reading and Writing.** Assign reading of the chapter and completion of the Understanding Concepts and Thinking Critically questions at the end of the chapter. Answers to these end-of-chapter questions can be found at the end of this Lesson Plan.

2. **Enrichment.** Using telecommunication devices is one of your students' favorite pastimes. It is unlikely, however, that more than a tiny percentage of them know how any of those devices function. Telecommunication devices are increasingly becoming "black boxes" whose operating principles, even when taken apart and examined, remain mysterious. Twenty years ago, a curious person could dismantle an electronic gadget and see mostly separate, distinct components—resistors, capacitors, diodes, transistor, etc.; the function of those individual parts could be readily learned. Modern systems, on the other hand, have more and more components integrated into increasingly complex (and tiny) microchips. Students need help in understanding the *basics* of electricity and electronic theory so that they do not assume that electronics is beyond their comprehension.

Review basic electric circuits and magnetism with your class. Dismantle an old radio or similar device with your students. Have your students build simple kits or etch their own circuit boards. Many students are fascinated with electronics once they see that it is not beyond their ability.

> **Safety Note:**
> Make sure your students understand that it is NOT safe for them to dismantle many electronic devices—especially televisions—on their own. Any warning panels or labels on electric devices MUST be followed.

3. **Enrichment.** Discuss the huge variety of careers available in telecommunication and electronics.

4. **Cross-Curricular Connections.** Have students complete the Applying Concepts and Solving Problems activities at the end of the chapter. Answers can be found at the end of this Lesson Plan.

Evaluation

1. Check students' answers to the Understanding Concepts and Thinking Critically questions. Review any topics that students do not understand.

2. Make copies of the Chapter Review in the *Teacher's Resource Binder* to administer as an open-book test or as a scored chapter test. Answers to the questions in this Chapter Review can also be found in the *Teacher's Resource Binder*.

Answers to Textbook Questions

Understanding Concepts

1. Telegraph key acts as a switch that opens and closes electrical circuit. When key is down, current flows to electromagnet at other end, which causes metal arm to make a click. The clicks can be varied to make a code.

2. (Any three examples.) Carries many more voice circuits; light and small so can be used where space is limited; light signals carried on optical fibers don't fade as fast as electrical signals sent over copper wire; eliminates many noises, such as clicking and buzzing.

3. Amplitude is the strength of an electromagnetic wave. Frequency is the number of waves that pass a given point in one second.

4. Setting radio dial at a number tunes in a particular radio station's frequency. Radio antenna picks up the waves of that frequency, and circuitry in the radio demodulates the waves and sends a boosted electrical signal to the speaker, which changes the signal back to sounds.

5. Signals are sent by means of earth stations.

Thinking Critically

1. Answers will vary.

2. Considered necessary to prevent obscenity and propaganda from pervading airways. Possible disadvantages: interferes with right to free speech; beliefs of those regulating may affect what is allowed and what is not.

3. Answers will vary.

4. Answers will vary.

5. Possible advantage: inspect military installations and activity at home and abroad. Possible disadvantages: can't see everything and may draw false conclusions; activity may be perceived as spying.

Applying Concepts and Solving Problems

4. **Mathematics.** The weight decreased by 98%; the price decreased by 94%.

Chapter 7 Lesson Plan: Graphic Communication

Chapter Goal and Outline

The main goal of this chapter is to introduce students to graphic message transmission and reception, including printing, photography, and drafting.

I. What Is Graphic Communication?
II. Visual Design
  A. What Makes a Design?
  B. Principles of Design
    1. Balance
    2. Proportion
    3. Contrast
    4. Variety
    5. Harmony
    6. Unity
III. The Creative Design Process
  A. Designing with a Computer
IV. Printed Communication
  A. The Importance of Printing
  B. Types of Printing Processes
    1. Porous Printing Processes
    2. Planographic Printing Processes
    3. Gravure Printing Processes
    4. Electrostatic Printing Processes
    5. Photographic Printing
    6. Ink-Jet Printing
    7. Laser Printing
  C. Digital Technology and Printing
V. Communicating through Photography
  A. The Photographic Process
  B. Using a Camera
  C. Photographic Film
  D. Digital Photography
  E. Photographic Composition
    1. Balance
    2. Framing
    3. Simplicity
    4. Leading Lines
  F. Holography
VI. Drafting: The Universal Language
  A. The Process of Drafting
  B. Multiview Drawings
  C. Pictorial Drawings
  D. Computer-Aided Drafting

Objectives

After studying this chapter, students should be able to:

- discuss the importance of graphic communication in our society.
- identify what makes a good graphic design.
- explain how printing, photography, and drafting work and are used in communication.

Preparation

1. Review the "General Suggestions" listed on page 21 of this Teacher's Manual.

2. Put the titles "Printing," "Photography," and "Drafting" on a bulletin board. Put a variety of examples under each title. Try to include as many different types as possible and find examples that demonstrate good visual design.

3. As an alternate to #2 above, have students bring in examples of computer-generated or computer-modified graphic image and put these examples on a bulletin board. If you have access to the Internet, many computer-generated graphic images are available for downloading.

Teaching

1. **Reading and Writing.** Assign reading of the chapter and completion of the Understanding Concepts and Thinking Critically questions at the end of the chapter. Answers to these end-of-chapter questions can be found at the end of this Lesson Plan.

Note: This chapter presents a large amount of new material to students. Students will need more time to study and absorb the material of this chapter than for most other chapters. You might want to divide the chapter into major sections and assign students to cover only one section at a time.

2. **Cooperative Learning.** Divide your class into teams and assign student roles using the methods described in #2 of the Teaching Suggestion of the Lesson Plan for Chapter 2. After reviewing with the class the elements that go into good graphic design (as discussed in Chapter 7), assign one specific power tool from your lab to each team. Each team's goal is to produce a safety poster for that tool. The poster will be evaluated on the strength of its graphic design. Each team may determine the size, shape, and all other variables including whether or not to use pictures, whether to use color or black and white graphics, etc. Display completed posters near the corresponding tools.

3. **Reteaching/Enrichment.** It is not necessary that students do projects using each of the printing techniques; simple demonstrations can help students quickly grasp the workings of the various printing methods. Take your class to look at your school's copy machine when discussing electrostatic printing. Make a simple screen printing device for them to see or even try. If you have a set of letter punches, demonstrate relief printing techniques by coating the letters with ink and pressing them onto paper. Ask a local printing shop for an old set of metal printing plates to show students as you explain their use. Have students use a magnifying glass to compare the type outputs of dot matrix, ink jet, and laser computer printers.

4. **Enrichment.** Students often have trouble grasping the concept that video tapes, cassette tapes, and computer disks *magnetically* store information. Bring in some old or discarded video tapes, some exposed film, expendable computer disks, and cassette tapes. Let students compare the film, tapes, and dissembled computer disks. Ask them to decide which store information magnetically. Record your voice on a cassette tape and play it for the students. Pass a powerful magnet over the cassette several times and then replay it. The loss of sound will prove to students that sounds are magnetically recorded. If you have access to a video camera and VCR, do the same experiment with the old video tape. Passing a magnet over the computer disk can also cause it to lose data stored on it.

5. **Reteaching.** Display and discuss posters or any other graphics that demonstrate good graphic design. (*Note*: Art teachers can often be used as a resource for graphic designs to borrow and display.)

6. **Cross-Curricular Connections.** Have students complete the Applying Concepts and Solving Problems activities at the end of the chapter.

Evaluation

1. Check students' answers to the Understanding Concepts and Thinking Critically questions. Review any topics that students do not understand.

2. Make copies of the Chapter Review in the *Teacher's Resource Binder* to administer as an open-book test or as a scored chapter test. Answers to the questions in this Chapter Review can also be found in the *Teacher's Resource Binder*.

3. Administer the Section II Test for this Communication section, which you will find in the *Teacher's Resource Binder,* as a scored test. Compare these test scores with the scores for the section pretest given during Chapter 4, the first chapter of this section. Carefully examine the tests for any questions consistently missed by students. This will help you pinpoint any specific areas that should be reviewed and help you determine any adjustments that you may need to make for teaching future classes.

Answers To Textbook Questions

Understanding Concepts

1. A symbol by which a company or organization is quickly and easily identified. Captures viewers' attention and leaves a lasting impression.

2. Thumbnail sketches are made to come up with many ideas. A rough layout is made of selected idea, showing more detail. Then a comprehensive layout showing details and specifications is made. Type and art are put into place and made camera-ready.

3. In relief printing, printing is done from a *raised* surface. In gravure printing, images are transferred from plates that have depressed areas.

4. *Light-tight space* to store film. *Lens* to focus image on the film. *Aperture with aperture diaphragm* to control the amount of light reaching the film. *Shutter* to control the amount of time light is allowed to reach film. *Filmholder* to hold film in position. *Viewfinder* to locate and position the subject.

5. To describe an object or structure so accurately that someone else can use the drawing to create an exact model of the idea.

Thinking Critically

1. Answers will vary.

2. Answers will vary.

3. Answers will vary.

4. Possible answers: holograms of plants and animals for science, landforms for geography, and sculptures for art.

5. For builders, working drawings might be most useful, especially if dimensions are included. For buyers, perspective drawings are most "lifelike" and give the best idea of a product's appearance.

Answers to Chapter 7 Mathematics Connection

14 point type; line length 16 picas

10 point type; line length 14 picas

Chapter 8 Lesson Plan: Manufacturing Systems

Chapter Goal and Outline

The main goal of this chapter is to help students understand the manufacturing system and its impact on their lives.

I. What Is Manufacturing?
II. The Development of Manufacturing
   A. Cottage Industries
   B. The Factory System
   C. Modern Manufacturing
      1. Custom Production
      2. Intermittent Production
      3. Continuous Production
III. Who Does Manufacturing?
   A. Companies and Corporations
   B. The Global Market
   C. International Competition
IV. A Managed Production System
   A. Inputs
      1. People
      2. Materials
      3. Tools and Machines
      4. Energy
      5. Capital
      6. Information
      7. Time
   B. Processes
      1. Management Processes
      2. Production Processes
   C. Outputs
   D. Feedback

Objectives

After studying this chapter, students should be able to:
- define manufacturing.
- describe how manufacturing developed.
- name the three basic types of modern production systems.
- explain how manufacturing and the economy influence each other.
- describe the four parts of a managed production system.

Preparation

1. Review the "General Suggestions" listed on page 21 of this Teacher's Manual.

2. Read the Lesson Plans for all of the chapters in this Manufacturing section. If you decide to have your students develop and build a product to solve a problem (as outlined in #5 of the Teaching section of the Lesson Plan for Chapter 10), make sure that you will have adequate space and materials by the time the students need them.

3. Put up the poster of the system model. (See #5 of the "General Suggestions" on pg. 21.) Use this as you help students fit manufacturing systems into the system model.

4. On a bulletin board. make a pictorial time line that shows the history of production systems. Begin with a picture of a very primitive person making a simple product like a stone axe. Next, show a representation of a cottage industry, then, an early factory, and, finally, an ultramodern factory with robots or CAM systems.

5. As an alternate to #4 above, on a bulletin board, put up the titles of the three modern production systems (*Custom, Intermittent and Continuous*). Under each title, put pictures of products produced by that system. Put the fewest pictures under Custom, more under Intermittent, and many under Continuous to represent the ratio of products produced by each type of system.

Teaching

1. **Overview.** This chapter begins the Manufacturing section of the text. Go over the general outline of your plans for this section with the students. Briefly discuss the planned activities and mention the main concepts that they will be studying.
   Note: If you plan for students to produce any products to sell, check with your school district's administration about policy implications *before* planning to allow students to sell their products. Many school districts have specific policies concerning students' sale of anything and how the profits must be handled.

2. **Pretest.** Use the Section III Test in the Teacher's *Resource Binder* as a pretest for this Manufacturing section. Do NOT count this as a grade for students. Rather, use this to accurately gauge students' learning by comparing these test scores with the scores obtained after students have studied this section.

3. **Review/Reinforcement.** Review the basic system model (discussed in Chapter 3) with the students. Refer to the model frequently throughout this section to help students understand how manufacturing operates as a system.

4. **Reading and Writing.** Assign reading of the chapter and completion of the Understanding Concepts and Thinking Critically questions at the end of the chapter. Answers to these end-of-chapter questions can be found at the end of this Lesson Plan.

5. **Enrichment.** This chapter provides an ideal opportunity to teach safe power tool operation to students. Assign a simple, quick, and inexpensive project that can be made individually by each student. In the process, each student will get the opportunity to acquire safe work habits and become familiar with your lab's power tools. Additionally, the individual projects can represent custom production methods; later, while mass producing another project, students quickly learn the vast time difference in custom vs. mass production manufacturing systems. This individual project can extend into Chapter 9.

6. **Cross-Curricular Connections.** Have students complete the Applying Concepts and Solving Problems activities at the end of the chapter. Answers can be found at the end of this Lesson Plan.

Evaluation

1. Check students' answers to the Understanding Concepts and Thinking Critically questions. Review any topics that students do not understand.

2. Make copies of the Chapter Review in the *Teacher's Resource Binder* to administer as an open-book test or as a scored chapter test. Answers to the questions in this Chapter Review can also be found in the *Teacher's Resource Binder*.

Answers to Textbook Questions

Understanding Concepts

1. Answers will vary.

2. Cottage industry: manufacturing at home by hand. Industrial Revolution: the factory system in which products are made in factories by machines. Modern manufacturing: more technical, specialized, efficient, and safer means of production.

3. Custom production: examples might include a personally designed necklace; furniture, cabinets, and drapery designed for a specific room; stained-glass window for an existing window shape. Intermittent production: examples might include 747 jets, lawnmowers, summer clothing, holiday items. Continuous production: examples might include packaging for food products, textiles, paper.

4. The need and demand for products determines what will be manufactured—e.g., if more people want audio CDs instead of records, more CDs and fewer records will be manufactured.

5. Answers will vary. Some examples are: Input—milling machines, saws, wood, metal. Process—A manager organizes employee tasks. Pieces of metal are welded together before being formed into a shape. Output—a mechanical device for opening gates; a wooden fence. Feedback—Concerned citizens object to pollution from nearby factory. A test of a product shows it to weaken under certain weather conditions.

Thinking Critically

1. This helps manufacturers design and produce what best suits the consumers' needs and price range, which helps ensure sales of the product.

2. Efficiency has several results: can save time, resources, and costs; can enable greater productivity and profit.

3. Answers will vary but should include both pros and cons.

4. Answers will vary.

5. Answers will vary, but answers must include an awareness of manufacturing byproducts and waste. Students may consider solving these problems by: changing materials, finding ways to recycle, creating laws that prohibit the manufacture of products that cause severe pollution, advertising campaigns for products manufactured without polluting the environment.

Answers to Chapter 8 Mathematics Connection

1. $500 - $101.17 - $21.11 = $377.72

2. 0.062 x $400 = $24.80

Chapter 9 Lesson Plan: Product Development

Chapter Goal and Outline

The main goal of this chapter is to help students understand how ideas are turned into plans and then into products.

I. New Products, New Ideas
   A. The Product Life Cycle
   B. Research and Development
      1. New Product Ideas
      2. Types of Research
II. Product Design
   A. Drawings
   B. Mock-Ups
   C. Management Approval
III. Product Engineering
   A. Functional Design
      1. Materials
      2. Structural and Mechanical Elements
      3. Power Sources
   B. Design for Manufacturability
      1. Interchangeability of Parts
      2. Standardization
      3. Simplification
      4. Modular Design
      5. Assembly
   C. Working Drawings
   D. Analyzing the Design
      1. Prototypes
      2. Types of Analysis

Objectives

After studying this chapter, students should be able to:
- explain what product engineers do to develop a product.
- explain design for manufacturability.
- name three types of drawings used in manufacturing.
- explain the importance of product testing and analysis.

Preparation

1. Review the "General Suggestions" listed on page 21 of this Teacher's Manual.

2. Put pictures of new inventions on a bulletin board. Ask students to look for other pictures of new inventions. As students bring in pictures, replace the pictures you put up with the students' examples.

Teaching

1. **Reading and Writing.** Assign reading of the chapter and completion of the Understanding Concepts and Thinking Critically questions at the end of the chapter. Answers to these end-of-chapter questions can be found at the end of this Lesson Plan.

2. **Review/Reteaching.** Review the problem solving process introduced in Chapter 2. Help students understand that product development steps relate to the problem solving process. The students discovered in Chapter 2 that identifying a problem correctly is essential to solving that problem. Product development begins when someone or some group gets an idea for a product that may solve a specific problem. If the idea develops into a product that does not solve the problem, it probably will not (or should not) be produced. A product that does not solve the specific problem that a consumer buys it to solve will not usually be successful—unless an advertising campaign convinces consumers otherwise!

   Discuss with students that even extensive planning and testing does not guarantee that consumers will purchase the product. This will help them understand that manufacturing any new product is risky and has a fair chance of failure. Mention that they will learn more about selling a product in Chapter 12.

3. **Cross-Curricular Connections.** Have students complete the Applying Concepts and Solving Problems activities at the end of the chapter. Answers can be found at the end of this Lesson Plan.

Evaluation

1. Check students' answers to the Understanding Concepts and Thinking Critically questions. Review any topics that students do not understand.
2. Make copies of the Chapter Review in the *Teacher's Resource Binder* to administer as an open-book test or as a scored chapter test. Answers to the questions in this Chapter Review can also be found in the *Teacher's Resource Binder*.

Answers to Textbook Questions

Understanding Concepts

1. Introduction of a new product, growth period, maturity stage, saturation, decline, development of a new product.
2. An idea comes from an individual or research and development department of a company. Drawings and mock-ups are made, management approves, product engineering begins, then manufacturing.
3. Product engineers plan and design to ensure that a product works properly, has durability, and will be manufactured with a minimum of problems.
4. 1) Mass production/interchangeability of parts; 2) standardization; 3) simplification; 4) modular design; and 5) design for assembly.
5. a) Detail drawings: dimensioned drawings of particular parts showing shape, special features, and pertinent information. b) Assembly: how parts fit together and proper placement; c) Schematic: position of parts drawn with symbols, such as electrical.

Thinking Critically

1. Products are readily available, more affordable. Parts or tools are easier to find and repair because of standardization.
2. Benefits of product testing: allows manufacturers to make sure product performs as expected before time and money are spent on mass production. Benefits to consumers: safer and more reliable products.
3. Answers will vary. Possible answers include fads: e.g., clothing, sport items, toys, etc.
4. Thought should be given to appearance in relation to a product's functional design.
5. Answers will vary, but require the students to think through the process of visualization of a product and communication of ideas; e.g., name types of drawings, describe function of product, etc.

Applying Concepts and Solving Problems

3. **Mathematics.**
   a. maximum length—13.18 in.
      minimum length—13.06 in.
   b. $17\frac{7}{8}$ in.
      $18\frac{1}{16}$ in.

Answer to Chapter 9 Science Connection

2. Answers will vary but might include tests for safety, durability, and toxicity.

Chapter 10 Lesson Plan: Production Planning

Chapter Goal and Outline

The main goal of this chapter is to help students understand the kinds of production planning required for production to be efficient.

I. Process Planning
   A. Production Approval
   B. Product Requirements
   C. Make-or-Buy Decisions
   D. Selecting the Process
   E. Group Technology
II. Facilities Planning
   A. Plant Location
   B. Plant Layout
   C. Materials-Handling Systems
III. Putting the Plan into Action
   A. Organizing People
   B. Tooling Up
   C. The Pilot Run

Objectives

After studying this chapter, students should be able to:

- describe planning procedures used to achieve production efficiency.
- describe how the layout of a manufacturing facility is developed.
- explain how machines and tools are changed for a new process.
- explain why a pilot run is important.

Preparation

1. Review the "General Suggestions" listed on page 21 of this Teacher's Manual.
2. Contact a local manufacturing company and ask if they will supply you with plan drawings that show how they planned for production flow. Put these up on a bulletin board and use them to help students understand the importance of careful planning.
3. As an alternative to #2 above, make a large drawing of your classroom's floor plan and put it on a bulletin board. As study of the chapter progresses, develop a large flow chart or plan for an upcoming production project with the students.
4. Review cooperative learning in the *Teacher's Resource Binder*.

Teaching

1. **Reading and Writing.** Assign reading of the chapter and completion of the Understanding Concepts and Thinking Critically questions at the end of the chapter. Answers to these questions can be found at the end of the Lesson Plan.

2. **Review/Enrichment.** This chapter presents the ideal opportunity to plan a small mass production project with your students. Refer students back to the problem-solving process discussed in Chapter 2. Explain that the problem-solving process can be used to invent a simple product that might solve a common student problem. (See the cooperative learning activity in #3, below.)

3. **Cooperative Learning.** Assign students into teams and assign roles. (See Step #2 in the Teaching section of the Chapter 2 Lesson Plan for recommendations on assigning teams and roles.) Pose questions to the teams to direct them to the problem area for which they must design a product that will remedy the situation. Examples might be: "What can be done about all the messy lockers in our school?"(locker organizer) or "Do you have cassette tapes lying around in your room at home?"(cassette racks)

   Have teams follow the problem solving process until they have designed a product that they feel solves the problem they identify. With careful, *nonintrusive* guidance, you can help each team come up with a design that is within their abilities to produce. (*Note*: You may want to impose cost, size, or other restrictions on the products.) Then have the teams design a production plan for their product and produce enough products for all team members. (*Note*. If available workspace is limited, have the entire class vote on which product best solves the problem. The whole class can then work on one assembly line.) If you have access to any cam equipment, try to incorporate it into the assembly line. (Students will be learning about CAM in Chapter 13.)

Continue this project into Chapter 11. If you have time and facilities, students can produce enough products to sell, thus continuing the project/text interrelationship into Chapter 12.

4. **Cross-Curricular Connections.** Have students complete the Applying Concepts and Solving Problems activities at the end of the chapter. Answers can be found at the end of this Lesson Plan.

Evaluation

1. Check students' answers to the Understanding Concepts and Thinking Critically questions. Review any topics that students do not understand.

2. Make copies of the Chapter Review in the *Teacher's Resource Binder* to administer as an open-book test or as a scored chapter test. Answers to the questions in this Chapter Review can also be found in the *Teacher's Resource Binder*.

Answers to Textbook Questions

Understanding Concepts

1. Manufacturing engineers study materials needed, drawings to analyze parts, and costs as to whether to make or buy parts.

2. Planning procedures include process charts (sequence of manufacturing) and group technology (existing records of parts that might be reused).

3. Engineers need to design efficient plans for the flow of production, as well as for shipping and receiving parts and materials.

4. Jigs and fixtures are used to adapt general purpose machines to do certain operations. Punch and die machines can be altered to punch or cut different shapes from materials.

5. A pilot run tests the system in order to find and correct problems, give workers a chance to learn new tasks, plus check and adjust timing along the production and assembly lines.

Thinking Critically

1. Production would be slowed and possibly even come to a halt. Workers would be laid off. Producer might fail to meet scheduled delivery times and lose customers.

2. Answers will vary.

3. Use of machines with molds and dyes generally reduces costs. The manufacturer can change a dye or mold, but does not have to replace machinery for each different operation.

4. Answers will vary but might include parts outside tolerance levels; defective or inferior parts, operation or equipment failures that could result in costly shutdown of production line, and wrong quantity causing problems at assembly line.

5. Answers will vary, but should consider the machinery, assembly, and the production process plans.

Applying Concepts and Solving Problems

5. **Mathematics.** a. $120.83; b. $128.68

Answer to Chapter 10 Mathematics Connection

11½ in.

Chapter 11 Lesson Plan: Production

Chapter Goal and Outline

The main goal of this chapter is to help students understand how components and assemblies are made into products and how production, inventory, and quality are controlled.

I. Producing Products
   A. Components
   B. Assemblies
   C. Packaging
II. Production and Inventory Control
   A. Production Control
   B. Inventory Control
III. Quality Assurance
   A. Standardization
   B. Process Improvement
   C. Inspection
      1. Inspection Tools
      2. Acceptance Sampling
   D. Burn In

Objectives

After studying this chapter, students should be able to:

- explain the difference between components and assemblies.
- explain the purposes of packaging.
- describe how production and product quality are controlled.
- explain how inventory is controlled.

Preparation

1. Review the "General Suggestions" listed on page 21 of this Teacher's Manual.

2. If, as suggested in Chapter 10's Lesson Plan, you made a flow chart of a production plan students are or will be working on, you may wish to retain the bulletin board with the developed flow chart.

Teaching

1. **Reading and Writing.** Assign reading of the chapter and completion of the Understanding Concepts and Thinking Critically questions at the end of the chapter. Answers to these questions can be found at the end of this Lesson Plan.

2. **Enrichment/Cooperative Learning.** If your students are producing products, help them design a quality control system to assure that they produce the best possible products. Discuss prevention and detection and as a class, decide if the students feel that they will need one or both of these methods. Define roles and goals of the quality assurance team.

    Discuss whether it will be possible to use tolerance gauges to inspect products. If so, using the methods described in Teaching Suggestion #2 in the Lesson Plan for Chapter 2, form teams to develop possible prototype inspection tools. As a class, decide which of the prototype inspection tools would be best to use for inspection of products.

    Have students try producing products with and without a prevention system in place. Students should keep careful records of the number of unsatisfactory products produced both with and without the prevention system in place. These records can be used later with Chapter 13.

3. **Cross-Curricular Connections.** Have students complete the Applying Concepts and Solving Problems activities at the end of the chapter.

Evaluation

1. Check students' answers to the Understanding Concepts and Thinking Critically questions. Review any topics that students do not understand. Answers can be found at the end of this Lesson Plan.

2. Make copies of the Chapter Review in the *Teacher's Resource Binder* to administer as an open-book test or as a scored chapter test. Answers to the questions in this Chapter Review can also be found in the *Teacher's Resource Binder*.

Answers to Textbook Questions

Understanding Concepts

1. A component is a single part; a subassembly is several parts that form a unit in a larger item; an assembly is several components put together.

2. Packaging materials can help prevent breakage or damage. Preprinted information can help sell the product and make handling more efficient.

3. A production schedule is a time chart for the production of parts; it predicts what will happen during manufacturing. Shop floor control involves keeping records during the process; it gives feedback about the production schedule for any necessary adjustments.

4. ANSI: coordinates groups to agree on standards. ISO: establishes and maintains quality control practices.

5. Methods include inspections, keeping records of progress (shop floor control) during WIP, statistical process control of machines, control charts; acceptance sampling with inspection tools; and "burn in" for electronic products.

Thinking Critically

1. Answers will vary. Students should consider the function and appearance of packaging: shipping/handling materials, advertising, important product information.

2. Answers will vary but should show an analysis of the basic parts of any product.

3. Answers will vary but might include paying higher costs for speedier delivery, checking customers' needs, communicating current status for possible schedule adjustments, considering alternatives for manufacturer, etc.

4. Possible consequences include producing inferior or defective parts or products; possible production slowdowns affecting production or assembly line, delays in shipping, loss of sales, etc.

5. Answers may vary, but should consider the following: quality control, inventory control, production control, and standardization.

Chapter 12 Lesson Plan: Marketing

Chapter Goal and Outline

The main goal of this chapter is to help students understand how companies promote and sell their products.

I. Markets
   A. The Industrial Market
   B. The Consumer Market
II. The Marketing Plan
   A. Market Research
   B. Advertising
III. Sales
   A. Direct Sales
   B. Wholesale Sales
   C. Retail Sales
IV. Distribution
   A. Warehousing
   B. Transportation

Objectives

After studying this chapter, students should be able to:

- define market and marketing.
- describe how market research is done.
- explain why advertising is important.
- name three major ways of selling products.
- explain how products are distributed.

Preparation

1. Review the "General Suggestions" listed on page 21 of this Teacher's Manual.
2. Construct a bulletin board entirely of advertisements cut from magazines and newspapers. Include as many large familiar trademarks and logos as possible.

Teaching

1. **Critical and Creative Thinking.** If your students have been manufacturing products as described in previous chapters, have each team devise a marketing plan based on what they learn in this chapter. Student teams should conduct marketing surveys, design an advertising campaign, and sell their product with direct sales techniques. (*Note*: Refer to the note that accompanies #1 in the Teaching section of the Chapter 8 Lesson Plan about students' sales of products.)

2. **Reading and Writing.** Assign reading of the chapter and completion of the Understanding Concepts and Thinking Critically questions at the end of the chapter. Answers to these questions can be found at the end of this Lesson Plan.

3. **Critical Thinking.** Review the ground rules and setup for open meetings described in #1 of the Teaching section of the Chapter 1 Lesson Plan. Open with the question to each student: "Have you ever bought a product because of an advertisement?" Then read the following statement to them: "A famous American historian was once asked what he would change about American society if he could change anything he wanted. He said, 'I would abolish advertising because it has convinced people that their wants are needs.' " Ask students to explain what they think this historian meant. Then have students discuss whether or not they agree with his statement.

4. **Cross-Curricular Connections.** Have students complete the Applying Concepts and Solving Problems activities at the end of the chapter.

Evaluation

1. Check students' answers to the Understanding Concepts and Thinking Critically questions. Review any topics that students do not understand. Answers can be found at the end of this Lesson Plan.

2. Make copies of the Chapter Review in the *Teacher's Resource Binder* to administer as an open-book test or as a scored chapter test. Answers to the questions in this Chapter Review can also be found in the *Teacher's Resource Binder*.

Answers to Textbook Questions

Understanding Concepts

1. A market is a specific group of people who might buy a product. Marketing refers to all the activities involved in trying to sell to that group of people, such as offering a free trial.

2. Market research includes interviews and test marketing of a product.

3. Advertising helps convince consumers that they need the product and should buy it.

4. Direct sales are made directly to the consumer or user. Retail sales involve a reseller who first buys the product from the manufacturer or a wholesaler.

5. The chain of distribution is the "path" that goods take in moving from the manufacturer to the consumer. It may include manufacturers, wholesalers, transportation, warehouses, retailers, consumers.

Thinking Critically

1. Specific answers may vary. For new products, the earlier the better, before manufacturers invest large amounts of money in production.

2. Advertising provides information about the product itself, its price, the store that sells it, and so on.

3. Possible answers include advantages of earning a greater income from commission, more independence. Disadvantages include possible lack of sales, earning no income, accumulating expenses.

4. Answers will vary. For example, an ad for athletic shoes on TV during a televised basketball game is aimed at people interested in sports. The style of advertising might reflect the atmosphere of a sporting event. (Emphasis is on students' awareness of how ads are placed for maximum marketing advantages.)

5. The more links in the chain, the higher the final price. Warehouse stores and outlet stores (direct sales from manufacturer to consumer) can save distribution costs that increase prices, for example.

Applying Concepts and Solving Problems

5. **Mathematics.** $320 commission.

Chapter 13 Lesson Plan: Computer-Integrated Manufacturing

Chapter Goal and Outline

The main goal of this chapter is to help students understand how computers are being integrated into the planning, production, and control of manufacturing.

I. What Is Computer-Integrated Manufacturing?
II. Computerized Planning
   A. Product Design
      1. Product Analysis
      2. Design for Assembly and Manufacturability
      3. Virtual Reality
   B. Product Engineering
   C. Production Planning
III. Computerized Production
   A. Computer Numerical-Control
   B. Cutting Technology
   C. Flexible Machining Centers
   D. Automated Assembly
   E. Robotics
IV. Computerized Control
   A. Inspection Devices and Methods
   B. JIT Manufacturing
   C. Automatic Identification
   D. Speech Recognition
   E. Programmable Controllers
   F. Computer Networks
   G. Automated Materials Handling
      1. Automatic Guided Vehicle System
      2. Automated Storage and Retrieval System
V. Computer-Integrated Manufacturing in Action
   A. The Automated Factory
   B. The Automatic Factory
VI. Factories in Space

Objectives

After studying this chapter, students should be able to:
- describe how different computer programs help product designers and production planners.
- describe computerized production methods.
- explain the differences between an automated factory and an automatic factory.

Preparation

1. Review the "General Suggestions" listed on page 21 of this Teacher's Manual.

2. The new terms for Chapter 13 include many acronyms. Place pictures showing only computer-integrated manufacturing around the boarder of a bulletin board. Make large copies of the acronyms AS/RS, AGVS, CAM, CAPP, CIM, CMM, FMC, JIT, MRP II, and SQC on separate sheets of paper. On the same size paper, copy the full name of each acronym. Place the acronym on top of the full name and tack them together onto the bulletin board so that the acronym can be lifted to see the full name underneath. As students pass the bulletin board they can lift the top sheet to see the full name as a constant simple review process.

Teaching

1. **Reading and Writing.** Assign reading of the chapter and completion of the Understanding Concepts and Thinking Critically questions at the end of the chapter. Answers to these questions can be found at the end of this Lesson Plan.

2. **Extension.** If you have access to Lego TC Logo, fischertechnik or similar systems use them with Chapter 13 to enable students to construct their own computer controlled machines and production systems. Computer interfaces are available for Apple II, Macintosh and IBM computers. These systems provide an excellent introduction to robotics and computer controlled systems. Plans for conveyor belts, robotic arms, optical sorters and other manufacturing systems are available for both Lego and Fischer systems. These materials allow students to construct their own working models of the latest manufacturing technologies. Capsella systems can also provide a simplified introduction to robotics.

3. **Enrichment.** Most application software can help show how the computer is used as a tool in manufacturing and increase student understanding of CIM systems. If your students are continuing a mass produced product from previous chapters, use a computer spreadsheet or data base to keep track of parts, production schedules, and sales. If (as suggested in #2 of the Teaching section of the Chapter 11 Lesson Plan) teams experimented with producing products both with and without prevention quality control methods, use the resulting data to produce graphs that show the effects of the quality control systems.

4. **Enrichment.** Chapter 13 provides many excellent opportunities to discuss the role of people in modern manufacturing operations. Discussions concerning robots replacing unskilled (and sometimes skilled) workers can help students understand the importance of continuing their education.

5. This chapter can also provide an opportunity for cross-curricular projects. An example might be a research paper written for credit in technology education, language arts, and social studies about the possible effects of CIM on society.

6. **Cross-Curricular Connections.** Have students complete the Applying Concepts and Solving Problems activities at the end of the chapter. Answers can be found at the end of this Lesson Plan.

Evaluation

1. Check students answers to the Understanding Concepts and Thinking Critically questions. Review any topics that students do not understand.

2. Make copies of the Chapter Review in the *Teacher's Resource Binder* to administer as an open-book test or as a scored chapter test. Answers to the questions in this Chapter Review can also be found in the *Teacher's Resource Binder*.

3. Administer the Section III Test for this Manufacturing section, which you will find in the *Teacher's Resource Binder*, as a scored test. Compare these test scores with the scores from the section pretest, given during Chapter 8, the first chapter in this section. Carefully examine the tests for any questions consistently missed by students. This will help you pinpoint any specific areas that should be reviewed and help you determine any adjustments that you may need to make for teaching future classes.

Answers to Textbook Questions

Understanding Concepts

1. CIM allows manufacturing simulations to test a system before beginning production, enables greater efficiency, helps reduce costs, helps engineers and planners make the best use of their resources. Disadvantages might include less employment for workers as more work is done by machines, and high cost of equipment.

2. They are used for drawings, calculations for assembly, rapid prototypes, and simulations.

3. CAPP is useful for determining best processes for scheduling production and for obtaining materials. MRP is used for determining materials, time, costs, and personnel requirements.

4. A robot.

5. **a.** CMM: coordinate measuring machine. It measures rounded or spherical parts and can be programmed to compare measurements with specifications stored in its memory.

   b. SQC: statistical quality control. This involves gathering information from a small number of parts and using statistics to predict the percentages of parts not meeting specifications.

   c. JIT: just-in-time delivery. Materials are delivered as they are needed to eliminate the need for inventory.

   d. LAN and WAN: local area network and wide area network. LANs can transmit instructions and information inside a plant; WANs can transmit communications around the world.

Thinking Critically

1. Answers will vary. Choices should be based on appearance and other surface qualities.
2. Answers will vary.
3. All of them would benefit. A LAN is local and would be useful within each plant.
4. Maintenance and repair, programming, janitorial work.
5. Answers will vary.

Chapter 14 Lesson Plan: Construction Systems

Chapter Goal and Outline

The main goal of this chapter is to help students understand the scope of construction—the people, materials, and technology needed to build projects.

I. What Is Construction?
   A. Our Constructed World
   B. Types of Construction
II. Part of the Construction System
III. Inputs
   A. People
   B. Information
   C. Materials
      1. Wood
      2. Metals
      3. Concrete
      4. Masonry
      5. Plastics
      6. Adhesives
      7. Other Materials
   D. Tools and Machines
   E. Energy
   F. Capital
   G. Time
IV. Processes
V. Output
VI. Feedback

Objectives

After studying this chapter, students should be able to:
- explain the importance of construction.
- define and give examples of the four basic types of construction.
- describe how the seven resources are used in construction systems.
- detail some advantages and disadvantages of several types of structural materials.

Preparation

1. Review the "General Suggestions" listed on page 21 of this Teacher's Manual.

2. Put up the system model poster described in #5 of the "General Suggestions" on page 21 of this Teacher's Manual. Use this as you help students fit construction systems into the system model.

3. This chapter introduces construction systems to students. Make a bulletin board that is a montage of construction pictures. To prepare simply put the title "Construction" across the top of the bulletin board. Once you begin discussing Chapter 14, have each student bring in at least one small picture that represents something about the construction system to him/her. Let the students tack their pictures on the board themselves.

4. As an alternative to #3 above, divide the bulletin board into four sections, one for each type of construction—residential, industrial, public works, and commercial. Put up titles for each section. Underneath each title, place pictures that illustrate that type of construction.

Teaching

1. **Pretest.** Use the Section IV test in the *Teacher's Resource Binder* as a pretest for this Construction section. Do NOT count this as a grade for students. Rather, use it to accurately gauge student learning by comparing these test scores with the scores obtained after students have studied this section.

2. **Reinforcement.** Review the basic system model (discussed in Chapter 3) with the students. Ask them to try to store a picture of the model in their mind to help them understand the construction system. Refer to the model frequently in discussions throughout this section.

3. **Overview.** Since this chapter introduces the Construction section, discuss with students any projects that they will be doing pertaining to construction.

4. **Reading and Writing.** Assign reading of the chapter and completion of the Understanding Concepts and Thinking Critically questions at the end of the chapter. Answers to these questions can be found at the end of this Lesson Plan.

5. **Real World Connections.** Of the five basic systems of technology (communication, manufacturing, construction transportation and bio-related), construction is the most essential to basic human survival. Your students have probably already learned in science class that all animals (including humans) need food, water, and shelter to survive. Help them understand that construction technology represents the ways that technological societies meet the need for shelter. Stress to your students (especially if your live in an area that experiences temperatures below freezing in the winter) that we could not live on much of the earth's surface without construction systems. Far too often students tend to take construction technology for granted and fail to fully realize how essential it is to their own survival.

6. **Cross-Curricular Connections.** Have students complete the Applying Concepts and Solving Problems activities at the end of the chapter.

Evaluation

1. Check student's answers to the Understanding Concepts and Thinking Critically questions. Review any topics that students do not understand.

2. Make copies of the Chapter Review in the *Teacher's Resource Binder* to administer as an open-book test or as a scored chapter test. Answers to the questions in this Chapter Review can also be found in the *Teacher's Resource Binder*.

Answers to Textbook Questions

Understanding Concepts

1. Residential construction is building structures in which people live. Examples may include: houses, apartments, town houses, condominiums, duplexes. Industrial construction is building or remodeling factories and industrial structures. Examples may include: factories, mills, warehouses, refineries. Commercial structures are those for businesses. Examples may include: supermarkets, malls, office buildings, restaurants. Public works construction is building structures for public use or benefit. Examples may include: hospitals, schools, highways, bridges, dams, airports, sewer systems, parks.

2. Any three of the following: Architects design structures and develop plans for building them; engineers make sure the design is structurally sound, how the structure will be built, and what materials should be used; heavy equipment operators operate cranes and earth moving equipment; electricians install wiring; plumbers install fresh and waste water systems; masons lay bricks and blocks or pour concrete; carpenters measure, saw, and fasten boards.

3. Structural materials are those used to support heavy loads or to hold the structure rigid. They must be strong and rigid.

4. Wood (solid-sawn lumber and engineered wood): advantages—can be a renewable resource, fairly durable in varying weather conditions, strong for its weight, and can be readily cut, shaped, and fastened; disadvantages—can be damaged by termites and other insects and may burn, warp, split, or rot.
   Metals: advantages—steel is very strong and rigid and can be formed into various shapes and aluminum is lightweight and corrosion resistant; disadvantages—steel may rust or corrode and loses its strength when exposed to high temperatures.

Concrete: advantages—very strong and can be molded into almost any shape; disadvantages—heavy and may crack or crumble if not prepared, poured, finished, and cured correctly.

Masonry: advantages—fireproof and sturdy; disadvantages—more costly than wood and requires a lot of skilled labor, and mortar holding materials tends to crumble over time.

Plastics: advantages—strong, lightweight, waterproof, economical, and resistant to rust and corrosion; possible disadvantages—will melt when exposed to too much heat and may increase intensity of a fire, may crack, especially when exposed to freeze/thaw conditions.

Adhesives: advantages—can be used to bond any material and can help save time.

5. Materials must be delivered on time so they are available when needed and work will not be delayed. Jobs are interdependent and must be scheduled so each is done on time so it does not delay next step in construction process.

Thinking Critically

1. Answers will vary but might include making neighborhood noisy, causing traffic to become heavy or congested, appearance unsightly, and causing pollution.

2. Answers will vary.

3. Answers will vary.

4. They help ensure that building materials are not of inferior quality and that the structure will be safer to use or inhabit.

5. Answers will vary.

Chapter 15
Lesson Plan:
Planning Construction

Chapter Goal and Outline

The main goal of this chapter is to help students understand that every detail of construction is planned so that the project will be usable and meet needs.

I. Initiating Construction
   A. Community Planning
II. Controlling Construction
III. Site Selection and Acquisition
   A. Selecting a Site
      1. The Survey
   B. Acquiring the Site
IV. The Design Process
   A. Identifying Specific Needs
   B. Developing Preliminary Ideas
   C. Refining Ideas and Analyzing the Plan
   D. Preparing Final Plans and Specifications
      1. Working Drawings
      2. Specifications

Objectives

After studying this chapter, students should be able to:

- explain how construction projects are started.
- discuss regulations that apply to construction.
- identify factors to consider when selecting a site.
- list the steps in the design process.
- describe architectural working drawings.

Preparation

1. Review the "General Suggestions" listed on page 21 of this Teacher's Manual.

2. Contact a local architect or construction contractor. Ask for any sets of blueprints that could be donated or loaned to you. If possible, obtain a site plan, floor plan, elevation drawing, and detail drawing. Display these on your bulletin board and refer to them as you discuss planning construction.

3. Contact a local survey company. Ask if a surveyor would be willing to visit your class and demonstrate how the survey process is accomplished. (Surveying methods dramatically demonstrate how technology and math are often connected.)

4. Review the discussion of cooperative learning in the *Teacher's Resource Binder*.

Teaching

1. **Reading and Writing.** Assign reading of the chapter and completion of the Understanding Concepts and Thinking Critically questions at the end of the chapter. Answers to these questions can be found at the end of this Lesson Plan.

2. **Cooperative Learning Method/Critical Thinking.** Using the methods described in #2 of the Teaching section of the Chapter 2 Lesson Plan, assign teams and roles to students. Each team will represent a different special interest group with an opinion about an imaginary, newly planned shopping mall, which is to be located in what is now a wooded area on the edge of town. Suggestions for teams include: mall investors/operators, planning commission, contractors, environmentalists, *downtown* business owners, property owners, hunters, or any other groups that might be appropriate to your area. Each team is to argue its position as persuasively as possible. Give the teams one class period to prepare; then use another class period to have a "community meeting" to discuss the mall plans.

3. **Enrichment.** If you have access to computers and CAD programs, Chapter 15 presents the ideal opportunity to introduce your students to computer-aided architectural design/drawing. Simple programs such as the *Design Your Own Home* series enable students to see why CAD systems are now used by most architectural firms. Many instructors assign simple paper drafting assignments to enable students to grasp the basic concepts. They then move on to simple CAD assignments that graphically demonstrate the advantages that CAD has over drawing on paper.

4. **Cross Curricular Connections.** Have students complete the Applying Concepts and Solving Problems activities at the end of the chapters.

Evaluation

1. Check students' answers to the Understanding Concepts and Thinking Critically questions. Review any topics that students do not understand.

2. Make copies of the Chapter Review in the *Teacher's Resource Binder* to administer as an open-book test or as a scored chapter test. Answers to the questions in this Chapter Review can also be found in the *Teacher's Resource Binder*.

Answers to Textbook Questions

Understanding Concepts

1. Community planners study the size and character of the population, the community's economy, the nature and quantity of natural resources, transportation facilities, and education facilities. Planning commissions consider impacts on the lives and property of those living in the community; the local economy; employment level; property values; taxes; and the health, safety, and general welfare of those living in the area.

2. Residential, commercial, and industrial zones. They might specify maximum property size, maximum building height, number of families that can occupy a house, number of parking spaces a commercial building must provide, and setbacks from property lines.

3. Any five: location, size, shape, topography, availability of utilities, zoning, cost.

4. A site plan shows where the structure is located on the site; it includes boundaries, roads, and utilities. The floor plan shows the locations of rooms, walls, windows, doors, stairs, and other features.

5. Specifications are written details about what materials are to be used for a project, as well as the standards and government regulations that must be followed. They spell out exactly which materials will be used for every aspect of construction, all regulations that must be followed, and standards that must be met so that they serve as a full set of detailed instructions for the building contractor. They also allow owners to know just what materials and services they are buying; they can use specifications as a checklist to make sure they are getting what they are paying for.

Thinking Critically

1. Answers will vary.
2. Answers will vary.
3. Answers will vary.
4. Possible answers include improved accuracy, less chance for human error, and quick measurement of much longer distances.
5. Answers will vary.

Chapter 16 Lesson Plan: Managing Construction

Chapter Goal and Outline

The main goal of this chapter is to help students understand that construction is a complicated process that requires careful management, scheduling, monitoring, and quality control.

I. Who Manages Construction Projects?
II. Scheduling Construction
   A. Work Schedule
   B. Materials Delivery Schedule
   C. Financial Schedule
III. Monitoring
   A. Monitoring Materials
   B. Monitoring Job Progress
   C. Monitoring Quality
IV. Inspecting Construction
   A. What Do Inspectors Look For?
   B. Who Inspects?

Objectives

After studying this chapter, students should be able to:
- describe how a contractor is selected.
- explain the responsibilities of a contractor.
- describe the three basic types of schedules.
- describe how construction is monitored.
- explain the importance of quality control.
- identify what inspectors look for and discuss the importance of inspections.

Preparation

1. Review the "General Suggestions" listed on page 21 of this Teacher's Manual.
2. On a bulletin board, list the job titles of as many construction-related jobs as you can fit. If space permits, attach a picture of someone performing the job next to the titles.
3. Contact a local building inspector and/or general contractor and ask him/her to visit your class and describe the inspection and management processes of construction to your students.

Teaching

1. **Reading and Writing.** Assign reading of the chapter and completion of the Understanding Concepts and Thinking Critically questions at the end of the chapter. Answers to these questions can be found at the end of this Lesson Plan.

2. **Real World Connections.** The suggestion in the teacher's annotations to invite a contractor to speak to your class is especially important. Local contractors have built many of your students' homes and other buildings with which students are familiar. If your school is fairly new, invite the contractor who built it. (He/she may even be able to bring in the plans.) Ask the contractor to address career topics, such as what he/she looks for in potential employees or whether most new employees come from local technical/trade schools.

3. **Cross-Curricular Connections.** Have students complete the Applying Concepts and Solving Problems activities at the end of the chapter. Answers can be found at the end of this Lesson Plan.

Evaluation

1. Check students' answers to the Understanding Concepts and Thinking Critically questions. Review any topics that students do not understand.
2. Make copies of the Chapter Review in the *Teacher's Resource Binder* to administer as an open-book test or as a scored chapter test. Answers to the questions in this Chapter Review can also be found in the *Teacher's Resource Binder*.

Answers to Textbook Questions

Understanding Concepts

1. Makes sure construction is done by certain date, controls costs, hires subcontractors, schedules workers, and materials delivery, makes sure high-quality work is done; makes sure everything is constructed according to the plans, meets all specifications, and encourages continuous quality control by workers.
2. The contractor gathers information and estimates how much materials and labor will cost and adds to this the amount of desired profit. The contractor submits this amount as a bid to project developers. *Usually*, the lowest bid wins the contract.
3. The contractor must know what must be done, and when, where, and how it must be done, so he or she can determine the number of workers needed, skills or crafts needed, equipment needed, time required for each process, and sequence of jobs.
4. The purchasing agent checks to see that materials are correct and in the right quantity, and delivered to the correct site, and are delivered on time and in good condition, of acceptable quality.
5. They check structural, electrical, and mechanical elements. They are concerned that the materials, methods, and quality meet or exceed set standards. Periodic inspections help spot problems early so they can be corrected without losing too much time and are less costly.

Thinking Critically

1. Possible qualities of a contractor include being a good leader, well-organized, a good planner, good with math, and adaptable.
2. Possible qualities of an inspector include thoroughness, an eye for detail, well-organized, good with math and measurement, conscientious, knowledgeable.
3. If estimates are too high, the contractor's bid will be too high. If estimates are too low, the contractor may not make a profit or may even lose money.
4. Answers will vary.
5. Answers will vary but might include that workers feel more involved and take more pride in their work so their work is high quality; possible problems are caught more quickly so there is less rework to be done, which saves time and money.

Applying Concepts and Solving Problems

2. **Mathematics.** $550

Answers to Chapter 16 Mathematics Connection

1. 8
2. $80
3. 25
4. $140

Chapter 17 Lesson Plan: Constructing Homes and Other Buildings

Chapter Goal and Outline

The main goal of this chapter is to help students understand the major processes used in constructing homes and buildings, from clearing the site to transferring the finished building to the owner.

I. Preparing the Construction Site
   A. Clearing the Site
      1. Demolition
      2. Earthmoving
   B. Laying Out the Site
II. The Foundation
   A. Excavating for the Foundation
   B. Parts of the Foundation
III. Building the Superstructure
   A. Types of Superstructures
      1. Framed Structures
      2. Load-Bearing Wall Structures
   B. Enclosing the Superstructure
   C. Installing Floors
   D. Installing Utilities
   E. Installing Insulation
   F. Finishing the Interior
      1. Ceilings and Walls
      2. Floor Coverings
      3. Trim and Other Finish Work
   G. Prefabricated Structures
      1. Panelized Construction
      2. Modular Construction
      3. Manufactured Housing
IV. Post-Construction
   A. Finishing the Site
      1. Paving
      2. Landscaping
   B. Transferring the Project

Objectives

After studying this chapter, students should be able to:
- describe preparation of a construction site.
- explain how foundations and superstructures are constructed.
- decide how interiors are finished.
- discuss finish work and landscaping.

Preparation

1. Review the General Suggestions listed on page 21 of this Teacher's Manual.

2. On a bulletin board, put pictures of a local construction project as it progresses through various stages. A local contractor or architect may be able to furnish you with pictures of a recent project. Ask your students if any of their families have recent pictures showing their home being built. (Have copies made so that the originals are not damaged. Making copies also allows you to have enlargements made so that they will make a more visible bulletin board.) If there is a construction site close to your school, take pictures of the process and put them on the bulletin board. (See #2 in the Teaching section below.)

Teaching

1. **Reading and Writing.** Assign reading of the chapter and completion of the Understanding Concepts and Thinking Critically questions at the end of the chapter. Answers to these questions can be found at the end of this Lesson Plan.

2. **Extension.** If possible, take your class on a field trip to a construction site. If you can find a nearby site, your class could visit it several times during the construction process to observe and photograph the progress. Different classes could visit the site on different days or weeks. Small groups of students from the visiting class could report on the progress to the other classes.

3. **Enrichment.** Invite a woman construction worker to speak to your class. Students of both sexes tend to stereotype the construction trade as male only. The construction worker might be willing to discuss both the positive aspects and difficulties she has experienced.

4. **Cross-Curricular Connections.** Have students complete the Applying Concepts and Solving Problems activities at the end of the chapter. Answers can be found at the end of this Lesson Plan.

Evaluation

1. Check students' answers to the Understanding Concepts and Thinking Critically questions. Review any topics that students do not understand.

2. Make copies of the Chapter Review in the *Teacher's Resource Binder* to administer as an open-book test or as a scored chapter test. Answers to the questions in this Chapter Review can also be found in the *Teacher's Resource Binder*.

Answers to Textbook Questions

Understanding Concepts

1. Workers take measurements from surveyors' stakes at each corner of the property to the corners and edges of the proposed building and mark new locations with stakes; the building's boundaries are marked with batter boards placed 4-5 feet from boundaries; string used to connect batter boards at opposite sides crosses directly over the boundary stakes, which then can be removed.

2. Footing: the part of the foundation on which the foundation wall sits; it distributes the structure's weight. Foundation wall: a wall made of concrete blocks or poured concrete; it transmits the weight of the superstructure to the foundation.

3. A frame structure has a main skeleton, or framework, that supports the weight of the building; the framework gives the building its shape. In a load-bearing wall structure, the heavy walls (usually concrete blocks, poured concrete, or precast concrete panels) support the weight of the structure; there is no frame. Studs: evenly spaced vertical boards that frame the walls; attached to horizontal plates. Joists: evenly spaced horizontal boards that form frames that support floors and ceilings. Trusses: preassembled triangular frames that form frame for roof. Rafters: sloping roof frame members cut from individual pieces of dimension lumber; extend from ridge downward over side walls.

4. Electrical; plumbing; HVAC (heating, ventilating, and air conditioning). In *roughing in*, parts such as wires, pipes, and ductwork are placed in walls and floors before interior surfaces are closed. *Finishing* is work done after walls are enclosed, such as putting in light switches and plumbing fixtures.

5. The earth is shaped, leveled, and smoothed; plantings are added. All of this is done according to a landscape plan.

Thinking Critically

1. Answers will vary. Possible answers might include that the structure could be positioned wrong and, in some cases, could violate zoning laws; if fences are erected or parking lots made that cross actual boundary lines, neighboring property owners could sue and have these removed.

2. Answers will vary; possible answers include being subject to insect and water damage; that damage is more extensive if fire occurs; and that it is more susceptible to damage from high winds.

3. Answers will vary.

4. Answers will vary.

5. Answers will vary. Possible answers include climate; size of plantings when mature; style of home; amount and type of sunlight (morning, late afternoon, etc.) available; owner's likes and dislikes; compatibility with neighboring yards.

Applying Concepts and Solving Problems

3. **Mathematics.** Current R-value is 12. Needs approximately 4 more inches.

Answer to Chapter 17 Science Connection

2. Other sources include oceans, lakes, rivers.

Chapter 18 Lesson Plan: Other Construction Projects

Chapter Goal and Outline

The main goal of this chapter is to help students understand the major processes used in constructing highways, bridges, tunnels, dams, pipelines, and air-supported structures and to understand the challenges of designing and building space structures.

   I. Other Types of Construction Projects
  II. Highway Construction
 III. Bridges
     A. Substructure of Bridges
     B. Types of Bridge Superstructures
  IV. Dams
   V. Canals
  VI. Tunnels
 VII. Pipelines
VIII. Air-Supported Structures
  IX. Construction in Space
     A. Space Stations
        1. *Skylab*
        2. *Mir*
        3. The International Space Station

Objectives

After studying this chapter, students should be able to:
- describe how highways, dams, canals, tunnels, pipelines, bridges, and air-supported structures are built.
- identify and describe the seven types of bridges.
- discuss construction in space.

Preparation

1. Review the "General Suggestions" listed on page 21 of this Teacher's Manual.
2. Read the Cooperative Learning Strategy in #2 of the Teaching section of this Lesson Plan. Acquire a supply of magazines (especially *Popular Science* and *Popular Mechanics*) for students to refer to and prepare a blank bulletin board for students' drawings.
3. Review the discussion of cooperative learning in the *Teacher's Resource Binder*.

Teaching

1. **Reading and Writing.** Assign reading of the chapter and completion of the Understanding Concepts and Thinking Critically questions at the end of this chapter. Answers to these questions can be found at the end of this Lesson Plan.

2. **Cooperative Learning.** Using the methods suggested in #2 of the Teaching section of the Chapter 2 Lesson Plan, divide the class into teams and assign roles. Assign one type of structure (e.g., bridges, tunnels, canals, pipelines, stadiums, or dams) to each team. The team goal is to draw a picture of a structure that is a good representative of the type they were assigned. They may either copy an actual structure (highest grade attainable: B) or design one of their own (highest grade attainable: A). (Each team member will receive the same grade for his/her drawing.) Give the teams a maximum and minimum size for their drawings so that all drawings will be able to fit on the bulletin board space or other wall space you have.

3. **Enrichment.** The Chapter 18 activity, Designing, Building, and Testing a Model Bridge, is an exciting activity for students. Software programs such as Bridge Builder allow students to first experiment with a computer simulation of the construction of a bridge. This closely resembles the actual methods frequently used today in construction planning and designing. Computer modeling is becoming more and more prevalent as computer software becomes more powerful and accurate at predicting the behavior of real structures. Discussions about bridge or other structural failures can be tied into discussions about the importance of computer modeling in preventing such disasters.

4. **Cross-Curricular Connections.** This chapter provides an ideal opportunity for cross-curricular assignments. Building a model of a space station or lunar dwelling could be assigned as a technology education and science project. During technology class, students could apply construction methods to airless or weightless environments. Students can use Lego/Logo and fischertechnik systems to design and build remote controlled robotic arms or other devices that might be used in space construction.

    In science, the same students might research how plants could provide the oxygen required for human survival. Social studies classes might try to predict what changes would occur in a society unable to survive outside of structures without spacesuits.

5. **Cross-Curricular Connections.** Have students complete the Applying Concepts and Solving Problems activities at the end of the chapter. Answers can be found at the end of this Lesson Plan.

Evaluation

1. Check students' answers to the Understanding Concepts and Thinking Critically questions. Review any topics that students do not understand.

2. Make copies of the Chapter Review in the *Teacher's Resource Binder* to administer as an open-book test or as a scored chapter test. Answers to the questions in this Chapter Review can also be found in the *Teacher's Resource Binder.*

3. Administer the Construction Section IV Test for this Construction section, which you will find in the *Teacher's Resource Binder* as a scored test. Compare these test scores with the scores from the section pretest, given in conjunction with Chapter 14, the first chapter in this section. Carefully examine the post-tests for any questions consistently missed by students. This will help you pinpoint any specific areas that should be reviewed and help you determine any adjustments that you may need to make for teaching future classes.

Answers to Textbook Questions

Understanding Concepts

1. Flexible: a thick bed of gravel or similar material is placed on the subgrade as a subbase, then asphalt pavement is laid over this; the gravel subbase transfers and spreads the loads on the highway into the sail below. PCC (or concrete): may be reinforced or nonreinforced; reinforced has steel bars on metal supports so that steel is in the middle when the concrete poured; nonreinforced has steel bars across joints to keep the joints from moving; loads are spread through the slab to the area below.

2. Beam: piers support beams that support spans of reinforced concrete slabs.

   Arch: the load is transferred along the arch to the abutments or piers at the end of the arch.

   Truss: triangular frameworks are used above or below the roadway to support bridge.

   Cantilever: beams called cantilevers extend from each end of the bridge and are connected by a suspension span.

   Suspension bridge: two tall towers support main cables that run the entire length of the bridge and are secured by anchorages at each end; suspender cables dropped from main cables are attached to the roadway.

   Cable-stayed: similar to suspension bridge except the cables are connected directly to the roadway.

   Movable: designed so a portion of the roadway can be moved to allow large water vessels to pass underneath.

3. A lake in which water is stored for use. Used to provide a dependable water supply for communities and also used for recreation.

4. A temporary watertight wall built to keep the construction site dry and water out of workers' way when building a dam.

5. Consists of three parallel 31-mile-long tunnels that run under the English Channel to connect England and France. Giant boring machines were used to cut through the rock below the channel; conveyors and small boxcars on rails carried the cut-away rock to the ends of the tunnels.

Thinking Critically

1. Answers will vary. Possible answers include that some businesses would relocate out of town, unemployment would rise and/or citizens would move to find jobs, other businesses would be hurt by reduced citizen income, tax revenues would fall, new businesses wouldn't be attracted to the area, less pollution, areas located near new highway would have business growth.

2. Answers will vary but might include the distance that must be spanned, the amount and type of traffic that would be using the bridge, and subsoil conditions.

3. Answers will vary. Possible answers include the need for the dam, effects on people and businesses nearby, effects on wildlife, geology and soil conditions, and history of flooding.

4. Possible positive impacts include the irrigation canals and reservoirs that supply water, electricity from hydroelectric power plants, and highways and bridges to transport crops and livestock to market. Possible negative impacts: construction of dams and resulting reservoir flood farmland, farmland taken away to build highways, highways can increase noise and pollution and cause areas to have heavy traffic congestion.

5. Answers will vary.

Applying Concepts and Solving Problems

2. **Mathematics.** 4 cubic yards.

Answers to Chapter 18 Mathematics Connection

Type of illness	Percentage of U.S. workers affected annually	Annual U.S. cost in lost work-days and productivity	Estimated number of BCC workers affected	Estimated cost to BCC in lost work-days and productivity
Skin diseases	Less than 1%	$1 billion	34	$726.88
Asthma	6%	$400 million	204	$4361.28
Accidental injuries	4%	$121 billion	136	$2907.52

Chapter 19 Lesson Plan: Transportation Systems

Chapter Goal and Outline

The main goal of this chapter is to help students develop a general understanding of transportation. This encompasses defining transportation, describing how transportation is used, and identifying the parts of a transportation system.

I. What Is Transportation?
II. Using Transportation
   A. Ways and Routes
   B. Support Facilities
   C. Passengers and Cargo
   D. Economic Value of Using Transportation
III. Parts of the Transportation System
IV. Inputs
   A. People
   B. Information
   C. Materials
   D. Tools and Machines
   E. Energy
   F. Capital
   G. Time
V. Processes
   A. Management Processes
   B. Production Processes
VI. Outputs
   A. Impacts of Transportation
      1. Positive Impacts—The Rewards
      2. Negative Impacts—The Challenges
VII. Feedback

Objectives

After studying this chapter, students should be able to:
- define transportation.
- describe how transportation is used to provide services.
- explain how transportation affects the economic value of products.
- the parts of transportation systems.
- identify some of the positive and negative impacts of transportation.

Preparation

1. Review the "General Suggestions" listed on page 21 of this Teacher's Manual.

2. Put up the system model poster described in the "General Suggestions" on page 21 of this Teacher's Manual. Use this as you help students fit transportation systems into the systems model.

3. This chapter introduces transportation systems to students. Make a bulletin board that is a montage of transportation pictures. To prepare, simply put the title "Transportation" across the top of the bulletin board. Once you begin discussing Chapter 19, have each student bring in at least one small picture that represents transportation to him/her. Let the students tack their pictures on the board themselves.

4. Make a bulletin board to illustrate the vast changes that have taken place in the speed of human transportation systems.

Teaching

1. **Pretest.** Use the Section V Test in the *Teacher's Resource Binder* as a pretest for this Transportation section. Do NOT count this as a grade for students. Rather, use it to accurately gauge student learning by comparing these test scores with the scores obtained after students have studied this section.

2. **Reinforcement.** Review the basic system model (discussed in Chapter 3) with the students. Ask them to try to store a picture of the model in their mind to help them understand the transportation system. Refer to the model frequently in discussions throughout this section.

3. **Reading and Writing.** Assign reading of the chapter and completion of the Understanding Concepts and Thinking Critically questions at the end of the chapter. Answers to these end-of-chapter questions can be found at the end of this Lesson Plan.

4. **Reinforcement.** In discussions help the students understand the degree to which all other systems of technology depend upon transportation systems. Lead the students in a discussion that helps them see the worldwide scope of our transportation systems.

5. **Critical Thinking.** This chapter provides an excellent vehicle for discussions about the tradeoffs involved in any technological system. People's desire for the convenience of automobiles as individual transportation vehicles has caused air pollution and gridlock problems currently facing large cities like Los Angeles. The easy access to and cheap production costs of fossil fuel energy sources have lead to the present and future necessity of spending billions of dollars worldwide to combat the greenhouse effect. Ask students to compare the use of the bicycle as a personal transportation vehicle as compared to the use of an automobile; discuss the advantages and disadvantages of each.

6. **Cross-Curricular Connections.** Have students complete the Applying Concepts and Solving Problems activities at the end of the chapter. Answers can be found at the end of this Lesson Plan.

Evaluation

1. Check students' answers to the Understanding Concepts and Thinking Critically questions. Review any topics that students do not understand.

2. Make copies of the Chapter Review in the *Teacher's Resource Binder* to administer as an open-book test or as a scored chapter test. Answers to the questions in this Chapter Review can also be found in the *Teacher's Resource Binder*.

Answers to Textbook Questions

Understanding Concepts

1. The movement of people, animals or things from one place to another using vehicles.

2. Ways are the actual spaces set aside especially for use by the transportation system. Routes are particular courses traveled.

3. Bulk cargo is loose cargo; break bulk consists of single units or cartons of freight. Examples of bulk may include: sand, gravel, cement, oil, coal, grain. An example of break bulk could be any packaged good.

4. An item is transported from where it is produced or commonly grown to another area where it is in demand. For example, oranges grown in California can be sold at a much higher price in Michigan than they can in California.

5. Positive impacts (any three): development of the country; exploration; helps make new discoveries; expanded markets; increased sales; business growth; competitive pricing allows world trade; people not as separated by distances; emergency vehicles as health benefit. Disadvantages (any three): traffic jams; air pollution; higher taxes to build and maintain ways, routes, and facilities; lives lost in accidents; land used up for roads.

Thinking Critically

1. Answers will vary.
2. Answers will vary.
3. Answers will vary.
4. Answers will vary.
5. Answers will vary.

Applying Concepts and Solving Problems

3. **Mathematics.** 32 miles/gallon

Answers to Chapter 19 Mathematics Connection

1. 44 miles
2. 654 miles

Chapter 20 Lesson Plan: Types and Modes of Transportation

Chapter Goal and Outline

The main goal of this chapter is to help students identify and describe the five basic modes of transportation. The students should understand the similarities and differences of each transportation mode, plus identify the advantages and disadvantages of each mode.

I. Air Transportation
  A. Types of Air Transportation
  B. Aircraft
  C. Airways
  D. Facilities and Support Services
II. Rail Transportation
  A. Passenger Service
  B. Freight Service
  C. Rolling Stock—the Vehicles
  D. Facilities
III. Water Transportation
  A. Types of Water Transportation
  B. Vehicles
  C. Waterways and Sea-lanes
  D. Facilities
IV. Highway Transportation
  A. Personal Transportation
  B. Commercial Transportation
  C. Types of Freight Operations
  D. Freight-Hauling Vehicles
    1. Computers in Trucks
    2. Improved Communication
  E. Facilities
V. Pipeline Transportation
  A. Using Pipeline Transportation
  B. Pipelines
  C. Facilities
VI. Space Transportation
VII. Computerized Navigation
  A. Global Positioning System
  B. Smart Highways/Smart Cars

Objectives

After studying this chapter, students should be able to:
- name the five modes of transportation.
- explain the reasons for selecting one mode over another.
- name different types of vehicles used to move people and cargo.
- tell which facilities are needed for each mode.

Preparation

1. Review the "General Suggestions" listed on page 21 of this Teacher's Manual.

2. Divide a bulletin board into five columns or areas. Label each area with the title of one of the modes of transportation (air, rail, highway, water and pipeline). For each mode, make one label for "Advantages" and "Disadvantages"; save these until you are ready to do the cooperative learning activity described in the Teaching section below.

3. Review the discussion of cooperative learning in the *Teacher's Resource Binder*.

Teaching

1. **Reading and Writing.** Assign reading of the chapter and completion of the Understanding Concepts and Thinking Critically questions at the end of the chapter. Answers to these questions can be found at the end of this Lesson Plan.

2. **Cooperative Learning.** Divide the class into five teams. Assign each team a different mode of transportation. Assign a role to each team member: (See #2 in the Teaching section of the Chapter 2 Lesson Plan for suggestions about dividing the class into teams and about role assignments.)

The goal for each team is to search the chapter and make a complete list of all the advantages and disadvantages of their assigned mode of transportation, print these advantages and disadvantages neatly on paper, and staple them to the bulletin board. Each team has only 15 minutes to complete their task. As soon as students begin work, staple the "Advantages" and "Disadvantages" labels you already made to the bulletin board. At the end of the designated time, each team must explain and justify their work to the other teams.

3. **Real World Connections.** This chapter helps students understand why a specific mode of transportation is selected to transport a specific cargo. As appropriate for your locale (since the mode(s) used to transport a given product to a given area depend on that area's geographic location and accessibility to transportation routes) tie-in a discussion of local transportation systems with the modes of transportation discussed in the text. In the process, help students to realize that many other systems are also affected and sometimes even determined by the mode, costs, and/or availability of transportation. For example, transportation costs and local resources usually determine how electricity is generated. Is your electricity generated by plants burning coal or oil, or is it supplied by nuclear or hydroelectric plants? Is fuel for home heating supplied by a local pipeline or by propane that is trucked in? If your area has a significant manufacturing sector, is it partly due to access to major rail line, a large navigable waterway, and/or major highways?

4. **Enrichment/Science Connections.** This chapter provides many excellent opportunities for classroom discussions. It will be a favorite of those students interested in space travel and the future. However, as most students have seen futuristic science fiction movie portrayals of space travel, many of your students will have serious misconceptions about space travel, including the belief that faster-than-light travel is possible in the near future. Current scientific theories and experiments all seem to indicate that no object with any mass traveling in our "ordinary" universe can travel at (or even very near) the speed of light. The only remotely possible means of ever "traveling" faster than light discussed by physicist theoreticians involves entering time/space anomalies called *wormholes* and *black holes*—not a particularly inviting means of travel. Discussions about the actual difficulties of space transportation and current systems will help students separate fantasy from reality. Help students understand that, for the foreseeable future, human space travel will probably be limited to our own solar system.

5. **Cross-Curricular Connections.** Have students complete the Applying Concepts and Solving Problems activities at the end of the chapter.

Evaluation

1. Check students' answers to the Understanding Concepts and Thinking Critically questions. Review any topics that students do not understand.

2. Make copies of the Chapter Review in the *Teacher's Resource Binder* to administer as an open-book test or as a scored chapter test. Answers to the questions in this Chapter Review can also be found in the *Teacher's Resource Binder*.

Answers to Textbook Questions

Understanding Concepts

1. a) water transportation, b) highway, c) pipeline.

2. Air: fast, saves time, but expensive means of travel.

    Rail: protects cargo, transports heavy, large loads, but limited to routes of tracks.

    Water: less expensive per ton mile, but slow.

    Highway: greater independence of operator, but overcrowded, traffic jams, air pollution.

    Pipeline: safer means of transporting certain materials, but vulnerable to leaks and possible damage to surrounding environment.

3. Answers may vary. Location, weight, time, cost, and bulk would affect the decision. For example, the railroad uses less fuel than several trucks, travels inland where ships could not, transports heavy loads that can be enclosed in boxcars to protect cargo, avoids highway traffic jams.

4. Air: (commercial, general, military) airplanes, helicopters, airships, rockets, hot-air balloons, blimps, dirigibles.

    Rail: high-speed passenger trains, freight trains (unit trains, regular freight trains), engines (diesel-electric locomotive), maintenance vehicles.

    Water: general cargo ships, supertankers, bulk carriers, cruise ships, paddlewheel boats, ferries, barges, towboats, tugboats.

    Highway: trucks (single unit, tractor-semi-trailer rigs, trailer units) cars, vans, buses.

    Pipeline: cargo ships, trucks, trains to pipeline stations.

5. Terminals, places for passengers to meet and purchase tickets, docks or warehouses, places to load and unload cargo, nearby fuel and storage areas, offices or stations for exchange and management of cargo and for services to passengers.

Thinking Critically

1. Answers will vary. As a plane moves down the runway, air flows over and under the wings. Shape of the wing increases speed of air flow on its upper surface, reducing pressure. At same time, pressure on lower surface increases, causing upward movement called lift.

2. Answers will vary.

3. a. Coal: water (barge) transportation because it is cheaper.

    b. Oil: water or pipeline; economical, efficient.

    c. Medicines: air, since it is fastest.

4. Answers will vary.

5. Answers will vary but should include evidence of the student's awareness of protecting the environment. Example: electric trains have replaced coal-operated trains that polluted the air.

Answer to Chapter 20 Mathematics Connection

9 hp

Chapter 21 Lesson Plan: Intermodal Transportation

Chapter Goal and Outline

The main goal of this chapter is to help students understand that an efficient transportation system is based upon cooperation and exchange among the different modes.

I. Advantages of Intermodal Transportation
II. Intermodal Cargo Transportation
   A. Trailer on Flatcar
   B. Containerized Shipping
   C. Moving Liquids
   D. Moving Coal and Gravel
   E. Other Intermodal Cargo Transportation
III. Intermodal Passenger Transportation
   A. Improved Airport

Objectives

After studying this chapter, students should be able to:
- define intermodal transportation.
- describe how intermodal semi-trailer systems combine rail and highway transportation modes.
- define containerization and explain its advantages.
- describe how liquid and solid bulk cargo are transported intermodally.
- describe how people can be transported on intermodal systems.

Preparation

1. Review the "General Suggestions" listed on page 21 of this Teacher's Manual.
2. Pick a specific product such as a CD player. Find pictures in magazines or other sources that show the various modes of transportation on which the product would travel in order to reach a consumer. Use string to connect the pictures together to represent intermodal transportation paths.
3. Review the discussion of cooperative learning in the *Teacher's Resource Binder*.

Teaching

1. **Real World Connections.** Intermodal transportation can, at first, sound like a confusing concept to students. Students first need to understand that the term *intermodal* simply means that more than one transportation mode is used to transport a product. Most students have seen trailers-on-flatcars; the use of this example helps students quickly understand that intermodal transportation is not a difficult concept to grasp.

2. **Reading and Writing.** Assign reading of the chapter and completion of the Understanding Concepts and Thinking Critically questions at the end of the chapter. Answers to these questions can be found at the end of this Lesson Plan.

3. **Cooperative Learning.** Divide the class into task groups and assign roles. (See #2 in the Teaching section of the Chapter 2 Lesson Plan for suggestions on assigning teams and roles and for descriptions of role responsibilities.) Instruct each team to vote to pick a "favorite object". Each team should trace the various modes of transportation that the team thinks that their object traveled upon. Discuss the bulletin board described in Preparation step #2 above. The teams may use drawings or pictures to illustrate the journey, but they may not spend any money on the project. Poster board, cardboard, pictures, and other materials that any team member finds at home may be used, but nothing may be purchased. Display the posters when completed.

4. **Cross-Curricular Connections.** Have students complete the Applying Concepts and Solving Problems activities at the end of the chapter.

Evaluation

1. Check students' answers to the Understanding Concepts and Thinking Critically questions. Review any topics that students do not understand.

2. Make copies of the Chapter Review in the *Teacher's Resource Binder* to administer as an open-book test or as a scored chapter test. Answers to the questions in this Chapter Review can also be found in the *Teacher's Resource Binder*.

Answers to Textbook Questions

Understanding Concepts

1. Intermodal transportation combines two or more modes of transportation to increase efficiency. Advantages: saves time and costs, reduces possible damages from handling cargo.

2. Trailer on flatcar (TOFC) is an intermodal method of cargo transportation whereby trucking companies' semi-trailers travel on railroad flatcars. The trailers are then unloaded from the train and attached to tractors to be driven to another destination. Its common name is piggyback service.

3. Containerization (containerized shipping) is a method of shipping cargo in a standardized box. Without containerization, there might be damage from too much loading and unloading of contents en route; cargo could be exposed to the elements, goods might be left in storage for longer periods of time and delivery of products could be slower.

4. A rolling pipeline refers to a tank truck or railroad tank car that transports liquids.

5. Answers will vary.

Thinking Critically

1. It would help reduce their expenses; they would only be paying the actual expense of transporting goods, not the added-on costs that would make up the transporting company's profits. They could schedule a more efficient and desirable timetable for delivery of needed materials and of their finished products. They could keep better track of just where materials, supplies, and products are when enroute.

2. Answers will vary.

3. Airplanes this large would be very expensive and would use lots of fuel, adding to transportation expense. Such containers would be too heavy and awkward for lighter-than-air vehicles.

4. They used wagons and horses to travel across the land. The wagons and horses were loaded onto ferries to cross large lakes and rivers. Sometimes, on smaller rivers, the wagons themselves became types of water vehicles and floated across or down the waterway. Railroads gradually spread across the West. Settlers might take these to a certain point, then travel farther by stagecoach or horse. Horses were also transported by train and later ridden to desired destination(s).

5. The more changes, the more chances for passenger or cargo to be lost, misdirected, or delayed. (Students may provide other ideas.)

Chapter 22 Lesson Plan: Power in Transportation

Chapter Goal and Outline

The main goal of this chapter is for students to expand their knowledge of power and engines used for transportation and to understand the potential for development in modern transportation systems.

I. Prime Movers
II. Internal-Combustion Engines
   A. Types of Motion
      1. Reciprocating Motion
      2. Rotary Motion
      3. Linear Motion
   B. Engine Systems
   C. Engine Operation
      1. Four-Stroke Engines
      2. Two-Stroke Engines
      3. Diesel Engines
III. External-Combustion Engines
   A. Steam-Driven Engines
   B. Modern Uses for Steam Power
IV. Motors
   A. Electric Motors
      1. Electric Vehicles
      2. Pipeline Motors
   B. Fluid-Powered Motors
   C. Motor and Diesel-Engine Combinations
V. Alternative Power
   A. Wind Power
   B. Solar Power
   C. Alternative Fuels
      1. Hydrogen Gas
      2. Methanol
      3. Methane Gas
      4. Recycled Fuels
   D. Maglev Propulsion
   E. Fuel Cells
   F. Hybrid Power
VI. The Effect of Vehicle Design
   A. Aerodynamics
   B. Size and Materials
      1. Plastics
      2. Ceramics
      3. Composites
   C. Electronics
      1. Fiber Optics

Objectives

After studying this chapter, students should be able to:
- describe the two main types of engines and tell how they work.
- describe electric and fluid-powered motors.
- describe several alternative sources of power.
- explain how design affects vehicle performance.

Preparation

1. Review the "General Suggestions" listed on page 21 of this Teacher's Manual.

2. Locate wall or bulletin board space to be ready to display the posters produced by student teams in the cooperative learning activity described in the Teaching section below.

3. Construct a bulletin board with pictures of the latest technological advances in transportation. Pictures of maglev trains, experimental automobiles and aircraft, and high-tech racing bicycles would all be appropriate. Ask students to bring in pictures that they find and replace the pictures you put up with the students' examples. Award extra credit points to anyone who can bring in an example of a transportation or power technology that no one else in the class has yet heard of.

4. Review cooperative learning in the *Teacher's Resource Binder*.

Teaching

1. **Reading and Writing.** Assign reading of the chapter (in sections) and completion of the Understanding Concepts and Thinking Critically questions at the end of the chapter. Answers to these questions can be found at the end of this Lesson Plan.

   Note: This chapter contains more than the usual amount of new information for students to grasp, more time will be needed to present and discuss this chapter. Assigning portions of the chapter for reading rather than the whole chapter will help students assimilate new concepts more easily.

2. **Reinforcement/Reteaching.** Make full use of the transparencies provided in the *Teacher's Resource Binder* to help students understand the many new concepts presented in this chapter. One of the most important concepts of this chapter that students need to understand is the fact that many different power sources are used in transportation systems. Stressing the reasons why a particular type of power source is used will help students understand all the engines and motors. The various engines and engine cycles can be demonstrated with "visible" model engines (available at most hobby shops). If possible, show students disassembled small two- and four-stroke cycle engines so that they can see actual examples of the parts.

3. **Cooperative Learning.** Assign students to task group teams and assign roles. (See #2 in the Teaching section of the Chapter 2 lesson plan for help on team assignment and roles.) The task for each team will be to produce a poster that demonstrates one of the specific engines discussed in this chapter. Assign an engine type to each team and give a due date for the posters. Each team member is responsible for producing and understanding the team's finished poster. Explain to the class that the poster will be graded on two criteria: (1) the quality of the poster and (2) the explanation—presented by *one team member selected at random*—of how the engine functions. Make sure that the students understand that no one will know which team member will be called upon to explain the engine. On the due date use a random selection method for selecting which team member will explain the engine function. (See #2 in the Teaching section of the Chapter 2 Lesson Plan for suggestions on making random selections.)

4. **Cross-Curricular Connections.** Have students complete the Applying Concepts and Solving Problems activities at the end of the chapter. Answers can be found at the end of this Lesson Plan.

Evaluation

1. Check students' answers to the Understanding Concepts and Thinking Critically questions. Review any topics that students do not understand.

2. Make copies of the Chapter Review in the *Teacher's Resource Binder* to use as an open-book test or as a scored chapter test. Answers to the questions in this Chapter Review can also be found in the *Teacher's Resource Binder*.

3. Administer the Transportation Section V Test for this Transportation section, which you will find in the *Teacher's Resource Binder,* as a scored test. Compare these test scores with the scores from the section pretest, given in conjunction with Chapter 19, the first chapter in this section. Carefully examine the post-tests for any questions consistently missed by students. This will help you pinpoint any specific areas that should be reviewed and help you determine any adjustments that you may need to make for teaching future classes.

Answers to Textbook Questions

Understanding Concepts

1. Internal-combustion engines: fuel is burned within the engine. External-combustion engines: fuel is burned outside the engine.

2. Reciprocating engine: An up-and-down or back-and-forth motion is changed into a rotating motion of a crankshaft with a piston and connecting rods.

   Rotary engine: A turbine or rotor mounted on a shaft rotates in a combustion chamber.

   Linear engine: Expanding gas of internal combustion is aimed behind the vehicle, making the vehicle move in the opposite forward direction.

3. Answers will vary.

4. Heavy materials require more energy to push them; quantity of materials used also affects weight.

5. That like poles of a magnet repel one another.

Thinking Critically

1. It's probably not practical because an electric-powered truck would require a huge battery as a power source. Most trucks make long hauls and batteries would need frequent recharging.

2. In engines, friction is created by parts rubbing together, creating resistance to motion between the parts as well as heat. In pipelines, friction between contents and pipeline walls reduces pressure and slows down movement of contents, so pumping stations must be set up along the pipeline to add pressure at regular intervals. Vehicles need friction to help them slow down and stop.

3. Answers will vary but should consider a variety of topics such as fuel source, engine design, motor, size, and materials.

4. Answers will vary but should reflect an understanding of power sources and engine types.

5. Students may express a variety of concerns in their answers, such as safety, the environment, costs, feasibility, etc.

Applying Concepts and Solving Problems

4. **Mathematics.** The amperage is 0.001.

Chapter 23 Lesson Plan: Bio-Related Systems

Chapter Goal and Outline

The main goal of this chapter is to provide students with a general introduction to bio-related technology systems.

I. What Is Bio-Related Technology?
   A. The Development of Bio-Related Technologies
      1. Louis Pasteur
      2. Joseph Lister
      3. Gregor Mendel
      4. Watson, Crick, and Wilkins
   B. How Does Bio-Related Technology Affect Your Life?
II. Inputs
III. Processes
   A. Propagation
   B. Growth
   C. Maintenance
   D. Harvesting
   E. Adaptation
   F. Treatment
   G. Conversion
IV. Outputs
V. Feedback
   A. Bioethics
      1. Medical Knowledge
      2. Commercial Use of Living Beings
      3. Environmental Interactions

Objectives

After studying this chapter, students should be able to:
- define bio-related technology.
- explain how bio-related technology developed.
- discuss the inputs, processes, outputs, and feedback of several bio-related technologies.
- discuss the impacts of the newer bio-related technologies.
- define bioethics.

Preparation

1. Review the "General Suggestions" listed on page 21 of this Teacher's Manual.

2. Bio-related technology is the most recently introduced area of study in technology education. Many (if not most) technology education instructors have little background/training in this field. Careful prereading of the chapter by the instructor will undoubtedly be necessary in most cases. (Biological science teachers in your building may be valuable resources for Chapters 23 and 24.)

3. Put up the poster of the system model. (See #5 of the "General Suggestions" on page 21.) Use this to help students fit bio-related systems into the system model.

4. Prepare a bulletin board for the cooperative learning activity suggested in #6 of the Teaching section below by dividing the board into four equal sections. In the first section, put up the name of Louis Pasteur; in the second, Joseph Lister; in the third, Gregor Mendel, and in the fourth, the team of James Watson, Francis Crick, and Maurice Wilkins.

5. Bio-related technology presents technology education teachers with an excellent opportunity to work with science teachers on cross-curricular assignments. Discuss with science teachers in your building possible adaptations of #5 and #6 in the Teaching section below (or other activities) and/or joint assignments.

Teaching

1. **Pretest.** Use the Section VI Test in the *Teacher's Resource Binder* as a pretest for this Bio-Related Technology section. Do NOT count this as a grade for students. Rather, use this to accurately gauge student learning by comparing these test scores with the scores obtained after students have studied this section.

2. **Reinforcement.** Review the basic system model (discussed in Chapter 3) with the students. Refer to the model frequently throughout the section to help students understand how bio-related technology operates as a system.

3. **Overview.** Since this chapter introduces the bio-related technology section, discuss with students any projects that they will be doing pertaining to bio-related technology.

4. **Reading and Writing.** Assign reading of the chapter and completion of the Understanding Concepts and Thinking Critically questions at the end of the chapter. Answers to these questions can be found at the end of this Lesson Plan.

5. **Cooperative Learning/Extension.** Divide the class into four teams. (See #2 in the Teaching section of the Chapter 2 Lesson Plan for recommendations for assigning teams.) Assign each team one of the following: Louis Pasteur; Joseph Lister; Gregor Mendel; or the team of James Watson, Francis Crick, and Maurice Wilkins. To facilitate random selection of team roles, have students make slips of paper equal to the number of students in their team; four of these slips should have a role—Leader, Recorder, Facilitator, or Reporter—written on them, and the remainder should be blank. Each team member draws a folded strip; those who pick a slip indicating a role should assume that role. Teams then decide how to use their ¼ bulletin board to illustrate what their assigned scientist or team of scientists contributed to bio-related technology. Each team should be graded as a group rather than individually.

6. **Enrichment/Critical Thinking.** Bio-related technology is frequently in the news. Ethical implications of recent cloning experiments provide excellent opportunities to enable students to see important ethical impacts of technology on societies. Use discussions about potential impacts to help reinforce students' understanding that all technologies pose societies with problems as well as solutions. This can help students further understand why all citizens in a democratic society need to be technologically literate. (*Note*: While this chapter presents many excellent opportunities for spirited class discussions, extreme care needs to be exercised to prevent any student from feeling that his/her religious beliefs or lack thereof are assaulted.)

7. **Extension.** Bring information about the Chernobyl disaster to class. Have students track the path taken by the radioactive dust. Discuss the effects the incident had on the surrounding area and people. Locate the nuclear plant closest to your town. Have students research prevailing winds and other conditions and then track the path the dust would take if an accident occurred.

8. **Cross-Curricular Connections.** Have students complete the Applying Concepts and Solving Problems activities at the end of the chapter. Answers can be found at the end of this Lesson Plan.

Evaluation

1. Check students' answers to the Understanding Concepts and Thinking Critically questions. Review any topics that students do not understand.

2. Make copies of the Chapter Review in the *Teacher's Resource Binder* to administer as an open-book test or as a scored chapter test. Answers to the questions in this Chapter Review can also be found in the *Teacher's Resource Binder*.

Answers to Textbook Questions

Understanding Concepts

1. Bio-related technology is any technology having a strong relationship to living organisms.

2. Pasteur discovered that bacteria were responsible for many diseases; he developed vaccines to keep people well.

3. All the DNA in an organism.

4. Any three: propagation, growth, maintenance, harvesting, adaptation, treatment, conversion.

5. The technique for reproducing a living organism.

Thinking Critically

1. Earlier scientists were trying to identify organisms; today's scientists have the ability to change organisms.

2. Answers will vary.

3. Under propagation, answers might include replenishing the stock of fish; under harvesting, they might include netting the fish and preparing it for market.

4. Answers will vary.

5. Answers will vary.

Applying Concepts and Solving Problems

3. **Mathematics.** Of the 30 million estimated viruses, only 0.0005% are known to be harmful to humans.

Chapter 24 Lesson Plan: Applications of Bio-Related Technology

Chapter Goal and Outline

The main goal of this chapter is to make students aware of some of the many applications of bio-related technology.

I. Agriculture
   A. Genetic Engineering
      1. Benefits and Risks
   B. Cloning
   C. Controlled-Environment Agriculture
      1. Hydroponics
      2. Aquaculture
   D. Integrated Pest Management
II. Health Care
   A. Prevention of Disease
      1. DNA Vaccines
   B. Diagnosis of Disease
      1. Laboratory Tests
      2. Instrumental Screening
      3. Medical Imaging
      4. Surgical Examination
      5. Genetic Testing
   C. Treatment of Disease
      1. Recombinant DNA
      2. Monoclonal Antibodies
      3. Gene Therapy
      4. Tissue Engineering
      5. Surgical Treatments
   D. Prosthetics
      1. Biomaterials
III. Human Factors Engineering
   A. Machine Design
   B. Work Methods
   C. The Built Environment
IV. Waste Management
   A. Solid Waste
   B. Sewage
   C. Hazardous Waste

Objectives

After studying this chapter, students should be able to:
- describe technologies used to control plant and animal production.
- discuss healthy care technologies.
- describe human factors engineering.
- describe methods for disposing of solid, sewage, and hazardous wastes.

Preparation

1. Review the "General Suggestions" listed on page 21 of this Teacher's Manual.

2. Bio-related technology applications are becoming more common in many areas of the country. Check your phone book to see if there are any firms in your area who would be willing to send a guest speaker or be willing to host a field trip.

3. Contact your state's agricultural extension agency or local university, to see if an expert is available to be a guest speaker to discuss what bio-related technology applications are in use in your state.

4. Arrange a field trip to your local wastewater treatment facility and/or solid waste disposal facility.

5. Monsanto Company in St. Louis has made a definite commitment to biotechnology. Their life sciences research center houses a series of growth chambers whose "microlimes" mimic conditions ranging from deserts to rain forests. Contact them for information regarding human health, animal nutrition, and plant agricultural products.

Teaching

1. **Reading and Writing.** Assign reading of the chapter and completion of the Understanding Concepts and Thinking Critically questions at the end of the chapter. Answers to these questions can be found at the end of this Lesson Plan.

2. **Reinforcement.** Spend extra time with the new vocabulary in Chapter 24. Students who have not yet studied biology will be unfamiliar with most of these terms and may not understand underlying concepts.

3. **Enrichment.** Contact the Sierra Club, the American Medical Association, the American Water Resource Association, and the American Farm Bureau to learn their stand on various aspects of biotechnology. Use this information as a springboard for class discussions.

4. **Extension.** Have certain students research one of the following "good ideas gone bad":

   a) wild hogs in South Carolina

   b) walking catfish in Florida

   c) Kudzu, the vine that has captured the South

   Have students report their findings to the class. Then ask the class to suggest steps that might have been taken to prevent the problems that resulted.

5. **Cross-Curricular Connections.** Have students complete the Applying Concepts and Solving Problems activities at the end of the chapter. Answers can be found at the end of this Lesson Plan.

Evaluation

1. Check students' answers to the Understanding Concepts and Thinking Critically questions. Review any topics that students do not understand.

2. Make copies of the Chapter Review in the *Teacher's Resource Binder* to administer as an open-book test or as a scored chapter test. Answers to the questions in this Chapter Review can also be found in the *Teacher's Resource Binder*.

3. Administer the Section VI Test for this Bio-Related Technology section, which you will find in the *Teacher's Resource Binder*, as a scored test. Compare these text scores with the scores for the section project given in Chapter 23, the first chapter of this section. Carefully examine the tests of any questions consistently missed by students. This will help you pinpoint any specific areas that need to be reviewed and help you determine any adjustments that you may need to make for teaching future classes.

Answers to Textbook Questions

Understanding Concepts

1. Gene splicing is the use of enzymes to move genes from one organism to another; it is done to give one organism the traits of the other.

2. In aeroponics, a plant's roots hang in the air and are sprayed with a nutrient solution.

3. A vaccine stimulates the body's immune system to produce antibodies for a particular disease; if the person is later exposed, the antibodies recognize and attack the disease germs.

4. Sound waves.

5. The landfill has a liner at the bottom to prevent the waste from polluting surface or ground water.

Thinking Critically

1. It could monitor temperature, humidity, light levels, and nutrient quality.

2. It has to do with the methods of agriculture used by a particular group of people that are often passed down from generation to generation.

3. Eliminating even one species of pest will have an effect on the ecosystem as a whole and may do more harm than good.

4. In addition to the occasional man-eating tiger, crocodile, and mosquito, humans are routinely attacked by microbes.

5. Answers will vary.

Applying Concepts and Solving Problems

2. **Mathematics.** 0.705 liters

Answers to Chapter 24 Mathematics Connection

a) 3.82×10^2 meters

b) 2.75×10^{-4} meters

Technology
Today and Tomorrow

Fourth Edition

Sharon A. Brusic, Ed.D.
Asst. Professor of Technology Education
Department of Teaching and Learning
Virginia Tech
Blacksburg, Virginia

James F. Fales, Ed.D., CMfgE
Loehr Professor and Chairman
Department of Industrial Technology
Russ College of Engineering and Technology
Ohio University
Athens, Ohio

Vincent F. Kuetemeyer, Ed.D.
School of Vocational Education
Louisiana State University
Baton Rouge, Louisiana

Glencoe
McGraw-Hill

New York, New York   Columbus, Ohio   Woodland Hills, California   Peoria, Illinois

Glencoe/McGraw-Hill
*A Division of The **McGraw·Hill** Companies*

Copyright © 1999, 1997, 1993, and 1988 by Glencoe/McGraw-Hill. All rights reserved. Except as permitted under the United States Copyright Act, no part of this publication may be reproduced or distributed in any form or by any means, or stored in a database or retrieval system, without prior written permission from the publisher.

Send all inquiries to:
Glencoe/McGraw-Hill
3008 W. Willow Knolls Drive
Peoria, IL 61614

ISBN 0-02-658569-3 (Student text)

    4 5 6 7 8 9 10   071/071   02 01 00

Printed in the United States of America.

ISBN 0-02-658570-7 (Teacher's Annotated Edition)

    2 3 4 5 6 7 8 9 10   071/071   02 01 00

Contents in Brief

Safety Comes First, 11

Section 1: Introduction to Technology 14

Chapter 1 Exploring Technology, 16; **Chapter 2** Problem Solving, 32; **Chapter 3** Technology Systems, 46

Section 2: Communication Technology 70

Chapter 4 Communication Systems, 72; **Chapter 5** Computers, 98; **Chapter 6** Telecommunication, 120; **Chapter 7** Graphic Communication, 142

Section 3: Manufacturing Technology 182

Chapter 8 Manufacturing Systems, 184; **Chapter 9** Product Development, 208; **Chapter 10** Production Planning, 230; **Chapter 11** Production, 246; **Chapter 12** Marketing, 260; **Chapter 13** Computer-Integrated Manufacturing, 270

Section 4: Construction Technology 304

Chapter 14 Construction Systems, 306; **Chapter 15** Planning Construction, 328; **Chapter 16** Managing Construction, 348; **Chapter 17** Constructing Homes and Other Buildings, 360; **Chapter 18** Other Construction Projects, 380

Section 5: Transportation Technology 418

Chapter 19 Transportation Systems, 420; **Chapter 20** Types and Modes of Transportation, 440; **Chapter 21** Intermodal Transportation, 464; **Chapter 22** Power in Transportation, 480

Section 6: Bio-Related Technology 524

Chapter 23 Bio-Related Systems, 526
Chapter 24 Applications of Bio-Related Technology, 544

Glossary, 576

Acknowledgments

The publisher gratefully acknowledges the cooperation and assistance received from the following teachers, writers, reviewers and businesses during the development of *Technology: Today and Tomorrow*. Numerous teachers have contributed activities. Their names are listed within this book, following their activities. In addition, many individuals and corporations provided drawings and photographs. Their names are listed in the credits at the back of this book.

Reviewers

James A. Bailey,
North Parkway Middle School, Jackson, TN

Joel D. Berger Subject Area Specialist/Teacher
Technology Education, Grades 6-12
Souderton Area School District
Souderton, PA

Mark Brehmer, Technology Teacher
Edgar High School, Edgar, WI

Larry Davis, Technology Teacher
Northwest Middle School
Winston Salem, NC

Michael R. Gargiulo,
Technology Education Teacher
North High School, Valley Stream CHSD
Valley Stream, NY

John E. Gray, Technology Education Teacher
Klondike Middle School
West Lafayette, IN

Barry E. Haley, Technology Teacher
Licking Valley High School, Newark, OH

Tina E. Hayden, Technology Teacher
Owen Valley High School, Spencer, IN

John Klock
Critical Thinking & Multimedia
   Development
T. A. Dugger Jr. High School
Elizabethton, TN

Ronald F. Logan
Drafting/Design Technology Instructor
Vocational Dept. Co. Chair
Austin-East Magnet High School
Knoxville, TN

Brad Moore
Winamac Community High School
Winamac, IN

L. D. Skarzinski,
Technology Education Teacher
East Fairmont High School, Fairmont, WV

Steven L. Wash
Batesburg-Leesville High School
Batesburg, SC

Henry L. Webb
Oak Grove High School, North Little Rock, AR

Contributors

Grace Bachmann, Joliet, IL

Center for Implementing Technology
   Education
Ball State University, Muncie, IN

Jeff Colvin, Blacksburg, VA

Ellen Credille, Chicago, IL

Doug Cropper, Philmont, NY

Robert A. Daiber, Ed. D, St. Jacob, IL

Terrie Greenberg, Saugus, CA

Kramer & Kramer, Niles, IL

Jacquelyn W. Rozman, Tampa, FL

Karl Stull, Peoria, IL

Cathy Ann Tell, Crystal Lake, IL

Eric Thompson, Onalaska, WI

Peter A. Tucker, St. Jacob, IL

Brigitte Valesey, Bethesda, MD

Marlene Weigel, Joliet, IL

Table of Contents

Safety Comes First 11
Preventing and Putting Out Fires, 11; Hazardous Substances and Devices, 11; Eight Guidelines for Safety on the Job, 12

Section 1 — Introduction to Technology 14

Chapter 1   Exploring Technology 16
Technology Focus: Spinning into the Future, 17
What Is Technology?, 18; Using Technology, 22; Technology and Human Knowledge, 27; Technology Assessment, 29.
Science Connection: Making Copies of Living Things, 21
Career File: Chemical Technician, 30

Chapter 2   Problem Solving 32
Technology Focus: Odyssey of the Mind, 33
The Problem-Solving Process, 34; The Scientific Method, 41.Entrepreneurs: Turning Problem Solving into Profit, 42; The Technology Student Association, 43.
Mathematics Connection: How To Read Graphs, 40
Career File: Consumer Safety Inspector, 44

Chapter 3   Technology Systems 46
Technology Focus: Stonehenge—A Prehistoric Computer?, 47
What Is a System?, 48; Input, 48; Process, 56; Output, 56; Feedback, 57.
Science Connection: Physical Fitness and Job Performance, 50
Career File: Electronics Technician, 58

Section 1 Activities 60
Making a Technology Occupations Survey, 60; Investigating Your Environment, 62; Developing a School Recycling Program, 64; Using Problem Solving to Make a Chair, 68.

Section 2: Communication Technology 70

Chapter 4 Communication Systems 72
Technology Focus: From the Comic Pages to Real Life, 73
What Is Communication Technology?, 74; Communication Innovations and Trends, 75; Why We Need Communication Technology, 77; Parts of the Communication System, 79; Inputs, 80; Processes, 90; Outputs, 91; Feedback, 95.
Science Connection: Human Sound Reception, 78
Career File: Computer Systems Analyst, 96

Chapter 5 Computers 98
Technology Focus: The Information Superhighway, 99
Computers in Communication, 100; The Amazing Integrated Circuit, 107; How the Microprocessor Works, 108; Sending and Altering Information, 109; The Internet and On-Line Services, 112; Artificial Intelligence, 114; Digital Libraries, 116; Virtual Reality.117
Science Connection: Semiconductors, 106
Career File: Computer Programmer, 118

Chapter 6 Telecommunication 120
Technology Focus: Telecommuting, 121
What Is Telecommunication?, 122; The Telegraph, 123; The Telephone, 125; Radio, 132; Television, 133; Satellite Communication Systems, 137.
Science Connection: Magnetism and Magnetic Media, 123
Career File: Computer Repairer, 140

Chapter 7 Graphic Communication 142
Technology Focus: Seeing Is Believing—or Is It?, 143
What Is Graphic Communication?, 144; Visual Design, 144; The Creative Design Process, 146; Printed Communication, 149; Communicating Through Photography, 155; Drafting: The Universal Language, 161.
Mathematics Connection: Measuring Type, 150
Career File: Technical Writer, 168

Section 2 Activities 170
Finding Information on the Internet, 170; Designing a Cellular Telephone Network, 173; Using a Digital Editor, 176; Making an Internet Home Page, 179.

Section 3: Manufacturing Technology 182

Chapter 8   Manufacturing Systems 184
Technology Focus: JIT—A New Strategy, 185
   What Is Manufacturing?, 186; The Development of Manufacturing, 186; Who Does Manufacturing?, 190; A Managed Production System, 195.
Mathematics Connection: How Earnings Become a Paycheck, 194
Career File: Industrial Engineer, 206

Chapter 9   Product Development 208
Technology Focus: CAD and CAM, Together at Last, 209
   New Products, New Ideas, 210; Product Design, 215; Product Engineering, 218.
Science Connection: Designing a Scientific Experiment, 211
Career File: CAD Drafter/Technician, 228

Chapter 10   Production Planning 230
Technology Focus: Awards for Innovation, 231
   Process Planning, 232; Facilities Planning, 235; Putting the Plan into Action, 240.
Mathematics Connection: Working with Fractions on Product Drawings, 242
Career File: Employee Training Specialist, 244

Chapter 11   Production 246
Technology Focus: Quality in Electric Guitars, 247
   Producing Products, 248; Production and Inventory Control, 250; Quality Assurance, 253.
Mathematics Connection: Using a Balance Sheet, 252
Career File: Painting and Coating Machine Operator, 258

Chapter 12   Marketing 260
Technology Focus: Snowboarding: A Case of Uphill Marketing, 261
   Markets, 262; The Marketing Plan, 263; Sales, 264; Distribution, 266.
Science Connection: Effects of Storage on Product Quality, 267
Career File: Manufactuerer's Sales Representative, 268

Chapter 13   Computer-Integrated Manufacturing 270
Technology Focus: Stereo Lithography—Turning Ideas into Plastic Reality, 271
   What Is Computer-Integrated Manufacturing?, 272; Computerized Planning, 272; Computerized Production, 277; Computerized Control, 281; Computer-Integrated Manufacturing in Action, 286; Factories in Space, 288.
Science Connection: Can Computers Think?, 273
Career File: Industrial Machinery Repairer, 290

Section 3 Activities 292
Researching and Developing a Composite Material and Product, 292; Testing Composite Materials and Prototypes, 295; Creating an Inventory Management System for the Manufacture of a Product, 297; Creating a Multimedia Advertisement, 299; Selecting and Buying Standard Stock, 301.

Section 4 — Construction Technology 304

Chapter 14 Construction Systems 306
Technology Focus: Shopping Malls Invest in New Business, 307
What Is Construction?, 308; Inputs, 311; Processes, 323; Outputs, 323; Feedback, 324.
Science Connection: The Chemistry of Plastics, 313
Career File: Architect, 326

Chapter 15 Planning Construction 328
Technology Focus: The New Old Ball Parks, 329
Initiating Construction, 330; Controlling Construction, 332; Site Selection and Acquisition, 332; The Design Process, 337.
Science Connection: How a Laser Works, 336
Career File: Cost Estimator, 346

Chapter 16 Managing Construction 348
Technology Focus: Breaking Tradition: Women in Construction, 349
Who Manages Construction Projects?, 350; Scheduling Construction, 351; Monitoring Construction, 353; Inspecting Construction, 356.
Mathematics Connection: Using Inventory Records, 354
Career File: Construction Supervisor, 358

Chapter 17 Constructing Homes and Other Buildings 360
Technology Focus: Smart Houses—The Future Is Now, 361
Preparing the Construction Site, 363; The Foundation, 365; Building the Superstructure, 366; Post-Construction, 376.
Science Connection: Water Down Under, 371
Career File: Carpenter, 378

Chapter 18 Other Construction Projects 380
Technology Focus: A Giant among Dams, 381
Highways, 382; Bridges, 384; Dams, 387; Canals, 389; Tunnels, 389; Pipelines, 392; Air-Supported Structures, 395; Construction in Space, 397.

Mathematics Connection: Costs of Workplace Illness, 394
Career File: Civil Engineering Technician, 400

Section 4 Activities 402

Learning about Construction in Your Community, 402; Estimating and Bidding a Construction Project, 404; Designing and Building Modular Hotel Rooms, 410; Investigating and Designing Highway Construction, 412; Designing and Building Model Space Structures, 415.

Section 5 Transportation Technology 418

Chapter 19 Transportation Systems 420

Technology Focus: Transportation Lifelines, 421
   What Is Transportation?, 422; Using Transportation, 423; Inputs, 427; Processes, 432; Outputs, 434; Feedback, 437.
Mathematics Connection: Calculating Distances on a Map, 430
Career File: Truck Driver, 438

Chapter 20 Types and Modes of Transportation 440

Technology Focus: A High-Tech Nightmare, 441
   Air Transportation, 442; Rail Transportation, 445; Water Transportation, 449; Highway Transportation, 451; Pipeline Transportation, 455; Space Transportation, 458; Computerized Navigation, 460.
Mathematics Connection: Horsepower, 453
Career File: Billing Clerk, 462

Chapter 21 Intermodal Transportation 464

Technology Focus: RR Stands for RoadRailer, 465
   What Is Intermodal Transportation?, 466; Intermodal Cargo Transportation, 467; Intermodal Passenger Transportation, 475.
Mathematics Connection: How Far, How Fast, How Expensive?, 475
Career File: Dispatcher, 478

Chapter 22 Power in Transportation 480

Technology Focus: World's Biggest Trucks, 481
   Prime Movers, 482; Internal-Combustion Engines, 482; External-Combustion Engines, 493; Motors, 495; Alternative Power, 499; The Effect of Vehicle Design on Power, 504.
Science Connection: Friction, 490
Career File: Aircraft Mechanic, 508

Section 5 Activities 510
Designing a Transportation System for the 21st Century, 510; Designing and Building a Solar-Powered Vehicle, 514; Investigating the Aerodynamics of Gliders and Airplanes, 519.

Section 6 — Bio-Related Technology 524

Chapter 23 Bio-Related Systems 526
Technology Focus: Fighting Pests with Technology, 527
What Is Bio-Related Technology?, 528; Inputs, 535; Processes, 536; Outputs, 538; Feedback, 538.
Science Connection: Emerging Viruses—A Deadly Mystery, 529
Career File: Clinical Laboratory Technician, 542

Chapter 24 Applications of Bio-Related Technology 544
Technology Focus: Medical Orphans—Help at Last, 545
Agriculture, 546; Health Care, 553; Human Factors Engineering, 563; Waste Management, 564.
Mathematics Connection: Using Scientific Notation To Write Large Numbers, 558
Career File: Forester, 568

Section 6 Activities 570
Creating and Marketing Composted Soil, 570; Designing and Making New Body Joints, 573.

Glossary 576
Credits 590
Index 594

Safety Comes First

More than two million people are injured in work accidents each year. Over 11,500 die from their injuries. About half of work accidents are caused by carelessness. The people did not use the equipment properly, or they failed to wear safety gear. Safety starts with a safe attitude. It means having the maturity to take responsibility for your own safety and for the safety of those around you. See pages 12-13 for Eight Guidelines for Safety on the Job or in the Lab.

Six colors are used on labels and signs to indicate different kinds of hazards. The colors are shown in Table A.

Preventing and Putting Out Fires

During your lab experiences, you may work with some potential fire hazards. Always know where the fire alarm and fire extinguisher are located. If a fire starts, turn on the alarm immediately. Small fires can be put out with the fire extinguisher. If the fire is large or spreading, leave the building right away. If your clothing catches on fire, drop to the ground and roll around to put out the flames. Do not run.

Fire extinguishers are designed for different types of fires. "A" types are used for ordinary burning materials, such as paper or cloth. "B" types are for gas or oil fires. "C" types are for electrical fires. "ABC" extinguishers can be used on all fires.

The best kind of fire safety is prevention. Learn the rules for handling flammable materials. Keep oily or paint-covered rags in ventilated containers. Store flammable materials away from heat in ventilated areas. Keep electrical cords in good condition, and replace those that need repair.

Hazardous Substances and Devices

Hazardous substances come in many forms. They may be toxic (poisonous), cause burns, produce fumes, or cause disease. Hazardous devices may have cutting edges or produce shocks. During this course, you will be asked to use certain tools and materials. The tools may be those used in industry, with sharp edges and points. Materials may include chemicals or the resins and catalysts used to create plastics. In every case, they require your respect.

Never use any materials or equipment without prior instruction and your teacher's permission. Failure to abide by safety rules can have devastating results.

Table A. Colors for Safety

Color	Meaning
Red	Danger or emergency
Orange	Be on guard
Yellow	Watch out
White	Storage
Green	First aid
Blue	Information or caution

Eight Guidelines for

 Develop a safe attitude.
How many times have you said, "I just wasn't thinking," or "I knew it was dumb before I did it"? Of course, by then the harm was already done. A safe worker thinks about safety and then acts on it.

 Stop, look, listen!
Be aware of your surroundings. What do you see? What do you hear? Where are you located in relation to tools, machines, and other workers?

 Keep your work area clean and neat.
Clutter, spills, and dirt create hazards. A tool left carelessly on the floor can trip someone. A knife hidden by a paint rag can cut a hand. Put tools and equipment away when you're finished with them. Clean up any mess you've made.

 Anticipate problems before they occur.
When you see the potential for trouble, do something about it. If you see old paint rags stuffed in a wastebasket, don't wait for the rags to burst into flames. Take responsibility for disposing of them properly.

Safety on the Job

 Take the time to do it right.
Workers who take pride in their jobs are usually safer than those who rush. Workers who select the right tool or equipment for the job and learn the right way to use it are less likely to be injured.

 Keep tools and equipment in good working order.
Be sure that equipment is set up and adjusted properly. A tool or machine that isn't working right is a dangerous tool or machine. If you use it anyway, not only do you risk injury, but it will also take you longer to do the job and may cause damage to your project.

Dress for the job.
This means wearing the right safety equipment, such as glasses, gloves, and hard hats. It also means wearing clothing appropriate to the work involved. Keep pants and overalls at the right length, keep sleeves and jackets buttoned, and keep long hair tied back.

Follow the rules.
Most accidents can be avoided if everyone follows the rules. Thoughtlessness can put not only you, but also other people, in danger. It's not cool or funny to be a rule breaker in situations where people can get hurt.

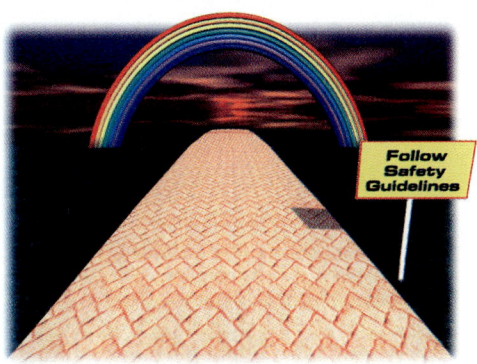

SECTION 1

CHAPTERS

1. Exploring Technology
2. Problem Solving
3. Technology Systems

TECHNOLOGY TIME LINE

8000 B.C.	3114 B.C.	2700 B.C.	700	800	1232	1454	1800
Agriculture begins.	The Maya develop a precise calendar.	Egyptians begin building pyramids.	Waterwheels are common in Europe.	Arabs introduce the Hindu concept of zero to their mathematics.	Chinese develop the first rockets.	The Bible is printed with movable type.	Industrial Revolution spreads to North America.

Introduction to Technology

You use technology every day, but do you know what it is? Technology is much more than computers or space stations.

In this section, you'll learn what technology is and how it relates to other areas of human knowledge, such as science. You'll find out how to solve problems, one step at a time. You'll discover that any technology, no matter how complex, can be described as a four-part system.

1939	1945	1962	1969	1971	1997
NBC begins commercial television broadcasts.	The first nuclear bomb is exploded.	James D. Watson and two British colleagues discover the structure of DNA.	First humans walk on the moon.	Intel Corporation of the United States introduces the first microprocessor.	The first adult mammal, a sheep, is cloned.

SECTION 1: Introduction to Technology

CHAPTER 1

Exploring Technology

Objectives

After studying this chapter, you should be able to:
- define technology.
- explain why the pace of technological change has increased.
- define five areas in which technology is used to solve problems.
- describe the relationship between technology and other areas of knowledge.

Terms

bio-related technology
communication technology
construction technology
emerging technologies
manufacturing technology
technology
technology assessment
transportation technology

Technology Focus
Spinning into the Future

What do you think of when you hear the word *technology*? Computers? Robots? Space shuttles? They are all products of technology but are not themselves technology. Technology is a process—the way human beings use their resources to solve problems and meet their needs and wants.

Music and Technology

Let's look at one familiar product of technology—a CD, or compact disc. You may have a CD of your favorite band's music, but how does that music get from the band to you? Let's look at just a few of the many technological steps needed to get that new music to you.

First, the musicians use technology—their artistry with words and musical instruments—to compose the songs. Their instruments had to first be designed, then materials had to be harvested and processed to make them—all processes of technology.

At a recording studio with specially constructed rooms, the band plays their new music into very high-quality microphones. Recording engineers use special tape recorders and other equipment to store the new songs onto recording tape, which is made from plastic coated with fine metal particles. It takes many technological processes to create the plastic and process the metal and then manufacture these materials into tape form.

After the recorded tapes, called *master tapes*, are carefully checked for possible defects, a laser beam is used to cut the music into a special CD master. Individual CDs made from special lightweight plastic are produced from the CD master in an ultra-clean, dust-free manufacturing plant. A design created by graphic designers and artists is then painted onto the top of the CD, using a process called *screen printing*.

Wrapping It Up

The CD is then packed with its paper liner notes into a plastic case. Once encased, several CDs are packed together into cardboard boxes to protect them during shipping. The boxes of CDs are then shipped by trucks, trains, boats, and airplanes—all products of technology—to warehouses. From there, they are shipped to stores like the one where you shop.

As you can see from this one example, objects that we tend to think of as technology are actually the result of many different technological processes. By studying technology, you will come to understand that those processes are parts of larger technological systems. By understanding these systems, you will become aware of technology's roles in our society.

Take Action!

Do you know anyone who still uses a turntable and vinyl records? If so, ask that person about the advantages and disadvantages of records versus compact discs.

Resources
- Chapter 1 Lesson Plan in the *Teacher's Annotated Edition* and in the *Teacher's Resource Binder.*
- Chapter 1 Review in the *Teacher's Resource Binder.*
1. **Extension.** Ask students what technology they think might replace today's music CDs. One possibility: downloading music from the Internet.

What Is Technology?

Technology is all around you. It has always been a part of your daily life. Yet, could you define technology? The *Technology for All Americans* project defines **technology** as "the generation of knowledge and processes to develop systems that solve problems and extend human capabilities." In other words, people create technology to solve problems and to make it possible to do new things.

People needed a way to keep food cold during hot weather. They invented refrigerators. People wanted to explore the ocean. They invented diving gear. What other examples of technology can you name? What problems has the technology solved? How has it extended human capabilities? Fig. 1-1.

A History of Technology

Technology has been around as long as the human race. When we think of modern technology, we often think of computers, complex machines, and space shuttles. However, people had to find ways of solving problems and meeting needs—in other words, develop technology—way back in history. Back in the Stone Age (beginning around two million years ago), people were using technology when they made tools out of natural materials like stone, wood, and bone. In the Bronze Age (beginning around 3000 B.C.), people learned how to make bronze out of copper and tin. Using this technology, they could make better tools and weapons. In the Iron Age (beginning around 1200 B.C.), people learned how to mine and use iron, which is harder than bronze, and improved their tools even more. Fig. 1-2.

FIG. 1-1 How does the technology of diving equipment extend human capability?

2. **Resource.** For more information about Technology for All Americans, check the ITEA Web site.

3. **Reinforcement.** Have students identify several examples of technology found in your classroom. Ask them how the technology extends human capabilities.

FIG. 1-2 Stones such as flint can be chipped and flaked to make tools. This scraper was made between 70,000 and 50,000 B. C. It was used to dig up roots and cut meat.

As people began to farm land to provide their food, they developed technologies like the plow, carts and carriages, and the waterwheel to help them to plant, harvest, and prepare their food.

The waterwheel uses the power of water to help people do their work. A flour mill, for example, would be built on the bank of a stream or river. A chute directed the water from the stream into the buckets on the wheel, and the weight of the water in the buckets turned the wheel. The wheel was connected by shafts and gears to huge millstones. As the wheel turned, it caused these shafts and gears to turn, which caused one of the millstones to turn. This movement crushed the grain into flour.

Because of the waterwheel, people didn't have to grind their grain by hand—a slow, hard process. They used technology to help them. Later, the waterwheel was adapted (changed to make usable) for many mechanical jobs, like sawing wood or running machines.

The technologies people create change their lives. Take the example of the waterwheel. Instead of people building their own waterwheels, it made sense for one person in a community to own and operate a mill. The other people could then pay the miller money or goods in exchange for grinding their grain. The social life of the whole community changed because of the waterwheel. The waterwheel was only one of many technologies, each of which caused its own changes in the way people lived.

The Pace of Change

If technology has been around for so long, you may wonder why people are so concerned about technology in our century. The answer has to do with the amount of technology around us, how quickly it changes and grows, and how much it affects our lives.

Even though people have used technology throughout history, the rate of technological change in early times was slow. New technologies came into use gradually over the course of many years. This gave people a chance to adjust to the new technologies they created. It took a long time for a technology to spread from one place to another. During the Iron Age, for example, many people were still making tools from stone because they did not know about ironmaking technology. Before the printing press was invented, information about a new technology could only be spread by the handwritten word, such as letters, or by word-of-mouth. This meant it took a long time for other people to learn about any new technology.

After the invention of the printing press, people were able to share their ideas about technology more easily and quickly. For example, someone in one place could print several copies of papers or books telling about the design of a certain machine. (The more copies made, the quicker the information could be spread.) A person in another place who knew a lot about metals and other materials could read that information and put his or her

4

4. **Extension.** Set up a bulletin board of early technologies. Have students bring in pictures or drawings of examples.

CHAPTER 1: Exploring Technology   **19**

own knowledge together with the first person's to come up with a new idea. In this way, people's knowledge of technology began to spread and build on itself.

Different technologies are like different foods in a kitchen. Each time you add another food to your supply, you have a greater ability to create new and different foods. If you just have apples, for example, all you can do is eat raw or cooked apples. If you add flour, sugar, and butter to your kitchen supply, you can combine these to make something new, like an apple pie. The more technologies humans invent, the more they are able to combine these technologies to make new technologies.

Nowadays, it has become easier and easier for human beings to share their knowledge with one another. People can send information from one side of the globe to the other in a matter of seconds. Because of this ease of communication and the buildup of technology over centuries, we live in a time that is full of many kinds of technology that are changing rapidly.

Emerging technologies are new technologies that are just coming into use. Cellular phones were once an emerging technology. Today they are common.

FIG. 1-3 "Dolly" was the first mammal to be cloned from an adult animal. A clone is a genetic duplicate of an existing animal or plant. Most mammals have genes from two parents. Dolly's genes supposedly came from the cells of only one sheep.

Their use has brought about changes in the way we live and work. The cloning of animals is an emerging technology. If this technology becomes common, it could have far-reaching effects on our lives. Fig. 1-3.

Technology, as well as its resulting impacts on our lives, has a snowball effect. Like a snowball rolling downhill, it grows rapidly and at times seems out of control. We do not have as much time to adjust to the new technologies as people in earlier times had. Because of this, we need to be even more careful to manage our technologies well.

Fascinating Facts

Some new technologies catch on quickly. Alexander Graham Bell patented his telephone in 1876. By 1895, there were 339,000 telephones in use in the United States. By 1915, there were 10,475,000.

5. **Reteaching.** Point out that most products are the result of combined technologies. Example: The cellular telephone combined radio and telephone technologies.

6. **Enrichment.** Have students trace the development of the personal computer.

7. **Critical Thinking.** Ask students what effects cloning technology might have on our lives.

SCIENCE CONNECTION
Making Copies of Living Things

In 1997, Dr. Ian Wilmut, a scientific researcher in Scotland, did what many people had believed was impossible. He cloned an adult mammal to produce a sheep he named Dolly.

A clone is an exact copy of a creature made from one of its own cells. To create Dolly, a cell was taken from the udder of a female sheep—we'll call her sheep A—and placed in a culture having few nutrients. During a normal life cycle, cells divide. Because this cell was starving from lack of food, it stopped dividing. It also switched off its active genes, the bits of hereditary material that determine a cell's characteristics.

The scientists then took an unfertilized egg cell from another female sheep (let's call her B) and removed all the genes it contained. Although the egg cell had no genes, it had the other biological equipment necessary to produce an embryo, or baby.

Then the two cells were placed next to each other and an electric pulse was sent through them. The pulse caused them to fuse together as one. Then another burst of electricity was used to wake up the genetic material originally in the cell from sheep A and start normal cell division.

Six days later, a tiny embryo had formed and was placed in the uterus of a third sheep (C). After a normal gestation period, the lamb, Dolly, was born. Now an adult, Dolly is identical to sheep A, the original donor, because sheep A's genes were used.

Although the process is simple, cloning Dolly was not easy. Dr. Wilmut tried the process 277 times. Out of those tries, only 29 embryos survived longer than six days. The other 28 died. Dolly herself may have problems no one yet suspects.

The excitement Dolly caused will earn her a place in history because cloning technology offers many advantages. Animals with special characteristics can be cloned for medical purposes. Animals valuable for food purposes can also be cloned. For example, a champion milk cow can be cloned to create an entire herd that produces more milk.

Try It!

> Select a food animal or one that is popular as a pet. Do some research to determine how breeding has changed that animal over time.

---

8. **Extension.** Twins are natural clones. Have students research articles about identical twins who were separated at birth. Do differences in environment appear to create differences in personality?

CHAPTER 1: Exploring Technology

Using Technology

In this book, you will learn about five areas of life in which technology is used to solve problems and extend capabilities: communication, manufacturing, construction, transportation, and bio-related technology. Fig. 1-4.

Communication Technology

Communication is the process of sending and receiving messages. We send messages for many reasons. Sometimes, we want to inform other people about something. For example, a newspaper informs people about important events in the world. We communicate to educate, or teach, each other about things. We may communicate to persuade others, such as when political candidates give speeches to try to persuade others to vote for them. Sometimes, communication is used to entertain, such as when people perform a play or make a movie. People also use communication to control machines and tools. We want to tell machines how to do the work we want them to do.

Communication technology includes all the ways people have developed to send and receive messages. Telephones, radios, television, and computers are all examples of technologies that help us communicate with one

FIG. 1-4 The technologies you will learn about are listed here. Note that some of these technologies can be grouped into a larger category called physical technology.

PHYSICAL

MANUFACTURING TECHNOLOGY--MAKING PRODUCTS

CONSTRUCTION TECHNOLOGY--BUILDING STRUCTURES

TRANSPORTATION TECHNOLOGY--MOVING PEOPLE AND GOODS

POWER & ENERGY TECHNOLOGY--CONVERTING AND TRANSMITTING ENERGY TO DO WORK

INFORMATION

COMMUNICATION TECHNOLOGY--COMMUNICATING WITH PEOPLE AND MACHINES

BIO-RELATED

BIO-RELATED TECHNOLOGY--IMPROVING HEALTH, FOOD PRODUCTION, AND THE ENVIRONMENT

9. **Reinforcement.** Have students give examples of communication technologies that correspond to the five purposes of communication.

FIG. 1-5 A kitchen contains many examples of manufacturing technology.

another. In addition to communicating with other people, communication technology can be used to communicate with machines and to help machines communicate with each other, such as when a computer directs a factory cutting machine to cut a piece of metal in a certain way.

Manufacturing Technology

Another way humans use technology is to help them make the material things they need and want in their lives, such as clothing, furniture, cars, and even toothbrushes. This process of making products is called manufacturing. Just by looking around you, you can see the many things people manufacture. Think of your kitchen, for example. Fig. 1-5. All of the objects there, from the range to the dishes to the plumbing, are manufactured goods.

Manufacturing technology is all the technologies people use to make the things they want and need. It includes the equipment and machines used to change the *raw materials* (materials as they occur in nature), such as changing trees into finished products like furniture. It also includes the technologies needed to design the product, ensure its quality, and sell it.

Manufacturing technology doesn't have to be complex. A custom weaver, for example, may make woven cloth and blankets on a loom powered by his or her own feet. A clothing factory, by contrast, may use much electrical power to run complicated machinery to make shirts and dresses and slacks. However, both the weaver and the clothing company are using manufacturing technology.

Construction Technology

People need buildings and structures for shelter and other purposes. Construction is the process of building structures. We build houses and apartment buildings to live in and skyscrapers for business offices. Fig. 1-6. We build bridges to help us cross waterways and roads for travel.

Construction technology is all the technology used in designing and building structures. It can range from something as simple as a hammer used to drive nails to something as complex as a bulldozer for leveling earth or a crane for lifting heavy objects to build a tall structure.

10. **Extension.** Explain how manufactured goods vary among cultures by discussing clothing. How is the clothing worn by northern Canadians, for example, different from clothing worn by people in India. Why?

11. **Critical Thinking.** Have students compare their school building with a residence. How are the buildings the same and how are they different? Why?

FIG. 1-6 Construction technology is used to build the structures in which people live and work.

Transportation Technology

We also need ways to move ourselves and other things. Transportation is the business of carrying goods or passengers from one place to another. Whether you fly from New York to Los Angeles, drive from home to school, or send a package to a friend in another state, you are using transportation. Manufacturers need transportation to get raw materials to their factories and their finished products to stores so people can buy them.

Transportation technology includes all the means we use to help us move through the air, in water, or over land. Planes, boats, trains, and cars as well as the engines for those vehicles are all examples of transportation technology. Other examples include escalators, elevators, the locks that help ships move along a river, and pipeline systems that move oil from one part of the country to another.

Bio-Related Technology

People also need to take care of their own bodies and the living environment (surroundings) around them. If they become ill, they need to be cured. They need to know how best to grow and harvest food. They want to understand how life works.

Bio-related technology is all the technology connected with plant and ani-

Fascinating Facts

The Panama Canal runs through the narrowest part of Central America and connects the Atlantic Ocean with the Pacific Ocean. About 12,000 ships pass through the canal every year. Tolls to pass through the canal are decided by the weight of the ship. The lowest toll ever paid was 36 cents. It was paid by a swimmer who swam through the canal in 1928.

12. **Reinforcement.** Identify the kinds of transportation facilities and companies that exist in your community.

SECTION 1: Introduction to Technology

FASCINATING FACTS

The Alaska Highway, which is a winding 1,523 miles long, was completed on October 20, 1942. Its purpose was to link Alaska with the northwestern United States, so that Alaska would not be so vulnerable to attack from Japan during World War II. Amazingly, the highway was completed in about eight months.

Combined Technology

We have looked briefly at five areas of life where people use technology to meet their needs. Even though we talk about these areas as if they were separate, in reality they blend together. Each area depends on the others.

Think, for example, of an airplane. In one sense an airplane is an example of transportation technology, because it is used to move people and things from one place to another. However, an airplane depends on communication technology. It has radio equipment for communicating with air traffic controllers. It has several instrument panels designed to communicate the status of the plane's systems to the pilot. A communication system allows the pilot and flight attendants to speak to the passengers. Some airplanes have phones so passengers can call people on the ground. On some trips, the airlines show movies for the passengers' entertainment.

mal life. It includes medical technologies like X-ray machines and MRI machines that help us see into our bodies. Fig. 1-7. It includes the machinery we use for plowing, sowing seed, and harvesting food. It includes the processes like canning, drying, freezing, and curing that we use to preserve our food. It also includes using living organisms to produce things we need. For example,

13 genetic engineering uses special techniques to create improved plant varieties in order to provide more food.

FIG. 1-7 The technology of magnetic resonance imaging (MRI) allows us to examine the human body. MRI is an example of bio-related technology.

13. **Reinforcement.** Explain to students that genetic engineering involves moving genes from one organism into another. The organism that receives the genes will thus have characteristics that it would not have by nature.

CHAPTER 1: Exploring Technology **25**

FIG. 1-8 Most of the time, the areas of technology overlap and combine with one another. An airplane is an example of transportation technology, but it also depends on other areas of technology.

An airplane is also an example of manufacturing technology. It takes more than 700,000 rivets just to hold all the parts of a medium-sized airplane together. These rivets, along with the engines, flight instruments, seats, and thousands of other parts, all need to be manufactured before the plane can be assembled. It takes the combined efforts of thousands of different companies to make all the parts needed for a single plane, as well as one company to put the parts together to make the finished aircraft.

You might think that an airplane has nothing to do with construction technology. Imagine, for a moment, however, that there were no airports, air traffic control towers, hangars, or runways. What good would it do to make airplanes without constructing the necessary buildings and structures to use those planes? An airplane would be a useless and expensive mass of parts without the structures that support its use.

How does bio-related technology affect an airplane? The meals you eat on plane trips have been processed and preserved using the knowledge provided by such technology. Support systems such as oxygen masks depend upon technology related to the human body and its functions.

14 As you can see, even though you may put an example of technology into one of the five areas of technology, it most likely includes elements of some of the other areas. Fig. 1-8.

Technology and Human Knowledge

15 You have seen how the areas of technology overlap with one another and are dependent on one another. In addition, technology is not separate from other forms of human knowledge. All of the subjects you study in school are related to technology because they affect technology and technology affects them. Fig. 1-9.

Technology and Science

Technology and science are interrelated. Scientists spend their time doing research and making discoveries. A *discovery* is an observation of something in the world around us that no one has seen in quite the same way before. For example, Isaac Newton said he discovered gravity by watching an apple fall in his garden while he was drinking tea. Newton noticed certain patterns about the way objects fall. People had seen things fall before, but Newton figured out that the force of gravity makes them fall in certain ways.

Technologists use the knowledge discovered by scientists to develop better technologies. Knowing about gravity means technologists can design airplane engines that are able to overcome the force of gravity holding a plane to the ground.

The relationship between science and technology is greatly interwoven. Technologists use science to do their work

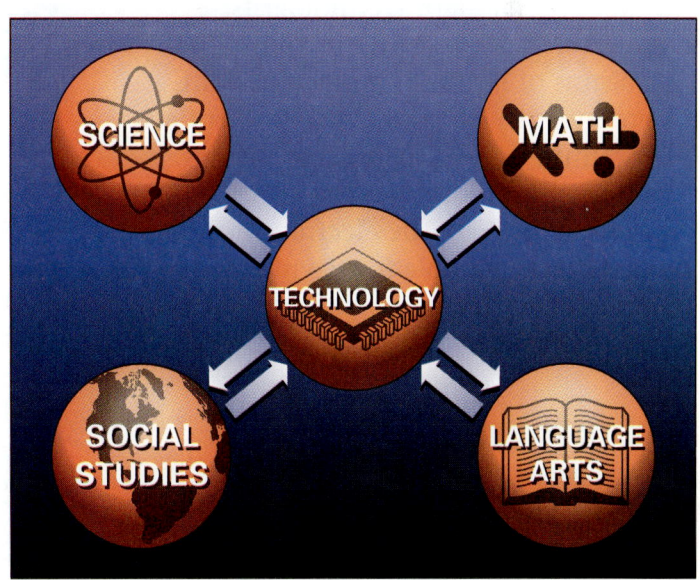

FIG. 1-9 Technology affects—and is affected by—other areas of human knowledge.

14. **Reteaching.** Have students describe other examples of combined technology. For example, what technologies are involved in the clothing we wear?
15. **Reinforcement.** Have students name examples of how technology is used or taught in their own science, mathematics, language arts, or social studies classes.

CHAPTER 1: Exploring Technology

FIG. 1-10 Science helps technology, and technology helps science. The technology of the electron microscope helps scientists make important discoveries.

of invention, but scientists also use technology to do their work of discovery. In 1608, for example, Hans Lippershey invented the telescope. A year later, Galileo heard about the telescope and made one of his own. Telescope technology made Galileo able to discover the moons of Jupiter and the rings of Saturn.

In 1931, modern technologists invented the electron microscope, which uses beams of electrons instead of light rays to magnify objects. With it, scientists can examine objects as small as viruses and bacteria and discover more about how they work. In 1970, the electron microscope made it possible for scientists to examine and learn more about atoms. Fig. 1-10.

Technology and Mathematics

Mathematics is an important knowledge resource for technology. Engineers use mathematical formulas when calculating the amount of weight a structure can withstand—a very important piece of information when you are designing bridges or buildings. In manufacturing, measurements need to be very precise. If the part being manufactured is not extremely close to the measurement needed, it will not fit where it is supposed to fit. Imagine that you bought a do-it-yourself bicycle kit in which the factory had mismeasured and now the teeth in the gears are much too large to fit into the chain. You would not be able to assemble or use the bike.

Technology and Language Arts

Without language resources available to them, technologists would have more difficulty doing their job. Suppose a company that manufactured computers had no way of telling customers how to use them. The company would quickly go out of business if other computer companies were making instruction booklets and sending salespeople out to explain their computers to customers. Without language, companies would have no way even to let customers know their product exists.

16. **Extension.** As an additional example of discovery, tell students how Sir Alexander Fleming discovered penicillin when he noticed a mold had accidentally grown on some bacteria cultures and killed the bacteria.

17. **Reinforcement.** Emphasize how scientific discoveries can lead to technological innovations. For example, by understanding the penicillin mold, technologists can develop ways to process it into a medication in pill, liquid, or injection form.

Technology and Social Studies

Social studies helps us understand the impacts technology has on human culture. Earlier in this chapter, we talked about how the technology of the waterwheel changed the way people in communities related to one another. For a more modern example, think of how the telephone has changed human communication. People used to have to write letters or go to a telegraph office to send messages to each other. Now, they only need to pick up the phone and punch in or dial a series of numbers. They can speak directly to the other party instead of going through the long process of sending messages back and forth.

Technology Assessment

Technology helps human beings solve their problems and meet their needs. We've already looked at the many benefits technology can provide—shelter; ways of producing, harvesting, and preparing food; manufactured goods; transportation; and many other useful things.

However, sometimes the technologies we develop and use create problems. For example, compounds called chlorofluorocarbons (CFCs for short) were used in refrigeration to help keep food cold. The technology of CFCs thus helped human beings preserve food. The CFCs were also used in air conditioners and in aerosol spray cans of deodorants, hair sprays, and cleaning products. Then it was learned that CFCs can contribute to the destruction of the earth's ozone layer. The ozone layer helps prevent too many of the sun's harmful ultraviolet rays from penetrating the earth's atmosphere. People began to realize that the benefits of CFCs were not worth the harm they caused to the environment. People found other, less harmful ways of meeting their needs.

As you can see, technology is a mixed blessing. Technology, in solving one problem, can help people in many ways. However, technology sometimes creates other problems.

Studying the effects of a technology is called **technology assessment**. To assess something means to determine its importance, size, or value. Technology assessment involves looking at cost, safety, and economic and environmental impacts. As a consumer and as a member of your community, you may need to make some difficult decisions about technology. By understanding how technology works and studying the effects it has, you can make good decisions about how to use technology to make our world a better place in which to live.

Fascinating Facts
Low pressure in Earth's upper atmosphere lets the ozone layer spread out. It's about 40 kilometers thick. If all that ozone were under sea-level pressure, the layer world be only a few millimeters thick.

18. **Extension.** Survey students. How many of them use e-mail? What are the benefits and drawbacks of e-mail versus phone conversation?
19. **Reteaching.** Explain that exposure to harmful ultraviolet rays can increase the risk of skin cancer.
20. **Critical Thinking.** Ask students whether technology's overall role in human life is good, bad, or neutral. Have them explain their answer.

CHAPTER 1: Exploring Technology

Career File—Chemical Technician

EDUCATION AND TRAINING
At least two years of specialized training is required, although many chemical technicians hold a bachelor's degree in chemistry, a related science, or mathematics.

AVERAGE STARTING SALARY
Starting salaries range from $15,000 to $19,000 per year, depending on education, experience, and industry.

OUTLOOK
Job opportunities are expected to increase at an average rate through 2005, according to the Bureau of Labor Statistics. Although job openings in chemical manufacturing may decline, more technicians will be needed to help regulate waste products, collect polluted air and water samples, and clean up contaminated sites.

What does a chemical technician do?
Chemical technicians do testing and laboratory experiments using chemicals. Most do research and development in chemical manufacturing firms, where they improve and create new plastics, drugs, soaps, paints, cosmetics, and other chemical products. Other chemical technicians test packaging for environmental regulations or develop processes that save energy and reduce pollution.

Does most of the work involve data, people, or things?
Most of a technician's work is done with chemicals, laboratory equipment, computers, and possibly robots. Often, a technician is part of a team.

What is a typical day like?
A technician's day might include setting up laboratory instruments, recording observations during experiments, interpreting data, or routinely testing chemical products for quality, strength, or durability.

What are the working conditions?
Most chemical technicians work indoors, usually in laboratories. Technicians sometimes work with toxic chemicals or radioactive isotopes. However, there is little risk if proper safety procedures are followed.

Which skills and aptitudes are needed for this career?
Perseverance, curiosity, the ability to concentrate on detail, and the ability to work independently are essential qualities.

What careers are related to this one?
Chemists and chemical engineers do work that is similar to that of chemical technicians. However, they have more responsibility. Other types of technicians who work in laboratories include biological technicians and engineering technicians.

One chemical technician talks about her job:
What I'm doing in the lab helps, in a small way, to provide better medicines for people who are sick. I'm proud to be part of the medical team.

Chapter 1 REVIEW

 ## Reviewing Main Ideas

- People create technology to solve problems and to make it possible to do new things.
- In earlier times, technology changed slowly and people had time to adjust to its effects. Today technology is changing and growing at an ever-increasing pace.
- Five areas of life that use technology are communication, manufacturing, construction, transportation, and bio-related technology.
- Technology depends upon human knowledge. Science, mathematics, social studies, and language skills are an important foundation for technology.
- Technology has both positive and negative effects.

 ## Understanding Concepts

1. Define technology.
2. Why has the pace of technological change increased?
3. Briefly define the five categories of technology.
4. What are four areas of human knowledge that support technology? Give an example of how each one relates to technology.
5. Explain what technology assessment means.

 ## Thinking Critically

1. Why is a stone tool an example of technology?

2. How can technology change people's lives?
3. The invention of the printing press made it possible for technology to spread faster. Name some modern inventions that make it possible for technology to spread quickly.
4. Describe some ways in which two technologies depend on each other. For example, how is communication technology needed in manufacturing, and how does communication depend on manufacturing?
5. What are some important issues (benefits and problems) caused by technology?

 ## Applying Concepts and Solving Problems

1. **Technology.** Make a column for each of the five areas of technology: communication, manufacturing, construction, transportation, bio-related. List the technologies you use in a normal day in the appropriate columns. Select ten items from the list you made. Describe the positive and negative effects of each.

2. **Language Arts.** Write a report about how computers are used in your school. What are the advantages and disadvantages? Survey your classmates. What changes would they like to see?

3. **Social Studies.** Find out about the first manufacturing plant built in the United States. Where was it built and who built it? What did it manufacture?

21. **Resource.** Answers to "Understanding Concepts" and "Thinking Critically" can be found in the Teacher's Manual at the front of this *Teacher's Annotated Edition* and in the *Teacher's Resource Binder*.

CHAPTER 2

Problem Solving

Objectives

After studying this chapter, you should be able to:
- describe six steps of the problem-solving process.
- define three types of thinking skills.
- describe the six steps of the scientific method.

Terms

brainstorming
entrepreneur
hypothesis
problem-solving process
scientific method
simulation

Technology Focus
Odyssey of the Mind

Suppose that you were asked to build a car that pops balloons. What would your ideas be for a mobile balloon-popper?

Creative solutions to such problems as a balloon-popping car are the goal in Odyssey of the Mind (OM), an annual competition involving teams of five to seven students from schools around the world. Contestants get their choice of several problems. Some problems are oriented toward technology, others toward the humanities.

Unusual Challenges

Consider the challenge of building a car that pops balloons. Team members were asked to build not one, but a total of three, cars. Some of the cars towed balloons that were used as moving targets.

The propulsion system for each car had to be based on elasticity—that is, involving a material that rebounds to its original form. Rubber bands were an obvious choice. A mousetrap was legal under a loose definition of elasticity. The judges said no to using compressed natural gas, but other creative solutions were encouraged.

There is a limit as to how much money can be spent for an OM project. Performance of a skit that demonstrates the solution is also required. In one memorable skit, a team presenting an android that had to be capable of one human-like action staged a night-club scene. In the scene, their android played the harmonica through its nose.

No Instructors, Just Coaches

The adult coach primarily helps students function as a team. The skits and solutions are solely the students' work. They brainstorm, plan, critique, model-build, implement, and evaluate their solutions. If a solution doesn't work, the team may have to start over from the beginning. Students learn to bring out the best in one another.

How OM Began

OM got started in the classroom of technology teacher Sam Micklus. Word spread about the fun his students were having as they grappled with unusual problems in industrial design at Glassboro State College (now Rowan University). In 1978 the first OM competition brought together 28 teams from New Jersey schools.

The OM approach relates closely to problem solving in the business world and has attracted major corporate supporters.

Take Action!
Visit Odyssey of the Mind on the World Wide Web:
http://www.odyssey.org/

Resources
- Chapter 2 Lesson Plan in the *Teacher's Annotated Edition* and in the *Teacher's Resource Binder.*
- Chapter 2 Review in the *Teacher's Resource Binder.*

The Problem-Solving Process

The **problem-solving process** is a multi-step procedure (series of actions) used to develop workable solutions to problems. It includes the following steps:
- State the problem clearly.
- Collect information.
- Develop possible solutions.
- Select the best solution.
- Implement the solution.
- Evaluate the solution.

State the Problem Clearly

Solving any problem starts with knowing what that problem is. Stating the problem clearly in a sentence or two often helps to identify just what the problem is. In fact, sometimes a clear statement of the problem actually suggests a possible solution. Some people say that figuring out the problem is half the job of solving it.

For a long time, for example, people whose legs were paralyzed couldn't drive cars. They couldn't work the brake and gas pedals. The problem was to design a car so that people would not have to use their legs to operate it. Once the problem is clearly stated, it is easy to see that the solution must involve developing controls that disabled people can operate with other parts of their body other than legs and feet. This was the first step in developing cars with hand-operated gas and brake pedals. Fig. 2-1.

Other problems are not so easy to identify. What about a situation like the problem of too much garbage? Is the problem (1) to design ways of removing the garbage, or (2) to design ways of reusing the garbage, or (3) to design ways of making products that create less garbage? A fourth possibility might be a combination of these three possibilities. Defining exactly which problem to work on is important in order to avoid wasting time and money.

FIG. 2-1 Stating the problem clearly was the first step toward developing a car that could be operated by people whose legs were paralyzed. Here, gas pedal and brakes are hand operated.

Collect Information

Once the problem is thoroughly understood, information that can be used to develop a good solution must be gathered. Sources of information depend, of course, on the nature of the problem.

1. **Enrichment.** Have students choose any technology and identify the problem that encouraged the development of that technology.

2. **Brainstorming.** Divide students into three groups and have them brainstorm possible solutions to each of these three problems. Discuss how stating the problem differently results in different solutions.

34 SECTION 1: Introduction to Technology

They could include libraries, museums, interviews with people who have worked in that particular area, and one's own lab or shop research. In solving a problem with toxic waste disposal, for example, a company might use a library or computer files to research information about toxic wastes. The company might also interview experts on safety in the workplace and scientists knowledgeable about toxic wastes. It could even set up a laboratory of its own to experiment with effective ways of dealing with waste.

Develop Possible Solutions

Most problems have more than one possible solution. At first, the more possibilities people can come up with, the better. That way, there are more options from which to choose.

Brainstorming

One way of coming up with solutions is brainstorming. In **brainstorming**, people try to think of as many possible solutions as they can. They don't stop to evaluate the possible solutions at this point. The idea is just to come up with as many ideas as possible. Then all the solutions are discussed to select the ones that show the greatest promise.

Another way of developing alternative solutions is through trial and error. For example, when Thomas Edison was inventing the lightbulb, he tried many types of materials for the lightbulb filament, including strands of red hair. He failed many times. Finally, he found that carbonized thread worked.

Sometimes a person will get really lucky and a solution to a problem will present itself accidentally. In 1974, this happened to 3M chemist Arthur Fry. As church choir director, Mr. Fry had been using slips of paper to mark songs in his hymnal, but found that they rarely stayed in place. Fry remembered an earlier discovery by a fellow 3M chemist, Spencer Silver, who had been experimenting with adhesives. Silver had discovered a "not-too-sticky" glue, but it just wasn't sticky enough for the purpose for which it was intended. Nobody could think of a good use for Silver's "invention." Fry decided to try a little of Silver's adhesive. Fry's solution to the problem of the "ever-disappearing bookmarks" paved the way for the development of 3M's Post-it® note pads. Today this product is widely used in homes and offices. Fig. 2-2.

FIG. 2-2 Post-it® notes are a good example of how an accidental discovery can lead to success. Chemist Arthur Fry put two ideas together for a winning combination.

3. **Extension.** Point out that when conducting trial and error research, it is important to keep a record of what has been tried and the results.
4. **Extension.** Discuss the role of failure in the development of technology. Point out that when something does not work, the failure provides information that can help people develop a better solution.

CHAPTER 2: Problem Solving 35

Thinking Skills

All of the methods for developing solutions have one thing in common. They all require thinking skills. Did you know there are different kinds of thinking skills?

- *Critical thinking skills* are used to analyze problems and make judgments.
- *Creative thinking skills* are used to develop original ideas or improve other people's ideas.
- *Decision-making skills* are used to make a choice among several possibilities.

When you use a problem-solving method, you apply all these thinking skills. Critical thinking, for example, will help you identify the problem. Creative thinking is especially useful when brainstorming solutions. When choosing a solution, you will again need critical thinking, as well as decision-making skills.

How do you develop thinking skills? One way is to take a technology course. 5 The technology activities you do, alone and in groups, will give you a chance to learn and practice your thinking skills. As you apply the problem-solving process, you'll learn to share ideas, analyze them, try them out, and improve them.

FIG. 2-3 There can be many possible solutions to a problem. All of these examples of homes are solutions to the problem of finding a place to live.

5. **Reinforcement.** Point out that technology is changing rapidly. Many job skills will become obsolete, but the ability to think clearly and solve problems will help people learn new skills.

Employers are looking for people with good thinking skills. Such people are better problem solvers and decision makers. Use your school years to develop your thinking skills. Doing so will not only help you get good grades; it will also help you succeed on the job.

Select the Best Solution

In order to choose the best solution, all the possible solutions must be evaluated. Evaluating involves looking at all the advantages and disadvantages of each possible solution to determine which one best solves the problem. Many factors must be considered, and a good decision must be based on your goals and your particular situation.

As an example, suppose the problem is that a couple wants to build a house. They must look at the advantages and disadvantages of building materials and house styles and weigh these along with their goal before deciding what type of house to build. A small Cape Cod house is cozy and economical, but may not be suitable for a couple starting a family. Wood is a cheaper exterior finish than brick, but it requires painting every few years. A single-level, ranch-style house would be ideal for people who have trouble climbing stairs. Ranch-style homes take up a lot of yard space, however, so people who enjoy the outdoors or like to garden may wish to build a two-story house so they will have more yard space. Fig. 2-3.

Part of good problem solving is being able to recognize which factors are most important. There is rarely a "perfect" solution to any problem.

Sometimes, more than one solution to a problem will be chosen because there is seldom one best solution that fits all circumstances. Seat belts, for example, are a good solution to the problem of passenger safety in cars.

Air bags provide additional protection. However, neither seat belts nor air bags are appropriate for babies and young children. In a collision, their fragile bodies can actually be harmed by seat belts and air bags. For them, the best solution is a special seat fastened to the car's rear seat. Fig. 2-4.

Fascinating Facts

Skyscrapers were made possible only after safe elevators were invented. Elisha Otis made the first safe elevator, which he demonstrated in New York in 1853. He was so certain of its safety that he rode his elevator to a great height, then had his assistant cut the elevator cable. Instead of crashing, the elevator simply stopped. Otis had proved his elevator was safe.

Implement the Solution

After the best solution has been selected, the next step is to implement it (put it into effect). During the implementation process, models are made and ideas are tested to make sure the solution is workable.

6. **Extension.** Have students write to several car manufacturers to ask about safety features. Discuss the benefits and drawbacks of each type of safety feature.

7. **Extension.** Discuss with students the controversy created when airbags actually killed several women and children.

FIG. 2-4 Here are three solutions to the problem of protecting automobile drivers and passengers during a collision. What are the pros and cons of each solution?

 Often, simulations are a good way to test solutions. In a **simulation**, equipment is set up in a lab or testing area in a way that imitates as closely as possible the real-life circumstances for which the solution is designed. For example, when testing seat belt and air bag designs for cars, auto manufacturers actually simulate crashes, using dummies in the cars. This way, they can see what would happen to a human being in an actual accident. Fig. 2-5.

Often a model of the proposed solution is built to aid in testing. You will learn about such models in Chapter 9.

The information people get from testing a solution helps them refine the solution. In a crash test for a seat belt, for example, perhaps the fastener designed for the belt broke loose at a certain speed. The designers would want to improve the fastener before manufacturing it.

Once the "bugs" of a new technology have been worked out, the technology can be put into effect. Improvement of a solution is part of the implementation process.

Evaluate the Solution

Problem solving doesn't end once the solution is put into effect. Consumers may

8. **Extension.** Computer simulations will be discussed in Chapter 9, but you may want to introduce the topic here.

SECTION 1: Introduction to Technology

report that they are dissatisfied with the way a product works or that it wears out easily, or they may have had trouble assembling it in the first place.

9   Things that happen in the world may make a technological solution *obsolete* (out-of-date) or make a better solution possible. For example, typewriters used to be a standard piece of equipment in business offices. Now computers do everything a typewriter does and more, and at a reasonable cost. Does the fact that the typewriter is being replaced by computers mean that the typewriter was a bad solution? Not at all. It's just that new information and events have made a better solution available. Problem solving is a never-ending process. Fig. 2-6.

FIG. 2-5 Testing a solution can revel its strengths and weaknesses. If a seat belt design doesn't work well enough in a crash simulation, it can still be improved.

FIG. 2-6 Sometimes one solution replaces another. In most offices, computers have replaced typewriters. What might be some advantages of using a computer instead of a typewriter?

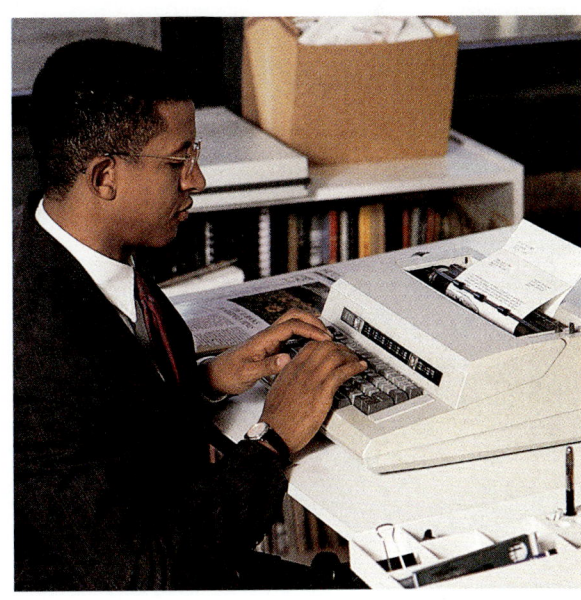

9. **Enrichment.** Have each student make a poster displaying a technology that has replaced another by solving the problem better. Display the posters around the room.

CHAPTER 2: Problem Solving

MATHEMATICS CONNECTION
How To Read Graphs

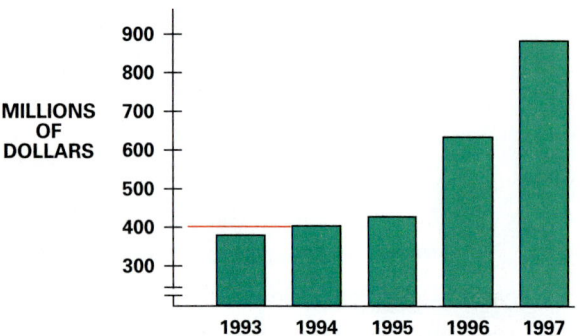

BAR GRAPH
ANNUAL SALES OF THE RED DOT COMPANY (1993-97)

A bar graph is used to compare different quantities. The height of each bar is proportional to the size of the quantity represented. Look at the bar graph shown here. Find the bar showing sales for 1994. A dotted line has been drawn from the top of the bar to the dollar amounts at the left. In this example, the amount even with the top of the bar is $400 million, which is sales for that year.

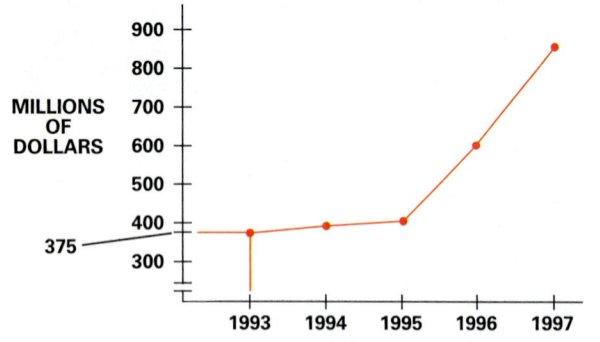

LINE GRAPH
ANNUAL SALES OF THE RED DOT COMPANY (1993-97)

This line graph shows the same data used to make the bar graph in a different form. Years are shown along the bottom. Dollar amounts are shown at the left. Find 1993 along the bottom. Look along the dotted line to the dot that indicates sales for that year. Then look to the left. The dot is in line with the mark indicating $375 million.

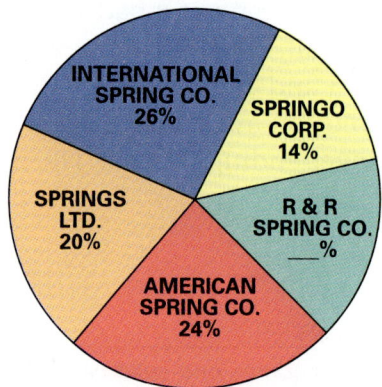

CIRCLE GRAPH
1997 LEADING SPRING MANUFACTURERS

A circle graph is divided into sectors, or pieces. The size of each sector is directly proportional to the size of the quantity represented. Each piece of the circle has a percentage associated with it, and the sum of all the pieces is 100%. In this example, the leading spring manufacturer in 1997 was International Spring Company at 26%.

Try It!

Create a bar graph using the information shown on the circle graph.

Answers to these questions can be found in the Lesson Plans in the Teacher's Manual at the beginning of this book.

SECTION 1: Introduction to Technology

The Scientific Method

You read in Chapter 1 that technology and science are interrelated. Scientists are trying to find out how the universe works. In their research, they use a process called the **scientific method**. As you read about the scientific method, think about how this method is similar to the problem-solving process.

- Make an observation. You might notice, for example, that certain green plants in your yard grow better in sunlight than in shade.
- Collect information. Does a particular green plant (Plant A) only grow well in direct sunlight? What kinds of plants grow in shade? How much sunlight do plants need to thrive?
- Form a hypothesis. A **hypothesis** is an explanation that can be tested. Your hypothesis might be "Green plant A needs direct sunlight to grow well."
- Perform an experiment to test the hypothesis. You might, for example, try growing several small samples of this plant. You would put some of the samples in sunlight. You would keep others in the shade. All of the other factors, such as the amount of water and fertilizer you use, should be the same for all the plants.
- Analyze the results after two weeks. Which plants did better?
- Repeat the process with other samples of the same plant to make sure the results are consistent.

Can you see the similarities between the scientific method and the problem-solving process? Both involve observing, thinking about solutions, trying them out, and analyzing the results.

Technologists apply science knowledge to develop new technologies. For example, technologists used the knowledge that plants need sunlight. They developed a lamp that provides the same wavelengths of light as does sunlight. Such a lamp is used for growing plants indoors, where sunlight coming through a window may not be enough for the plant to thrive. Fig. 2-7.

FIG. 2-7 Lamps have been developed that provide simulated sunlight for indoor plants

10. **Critical Thinking.** Ask students how the scientific method might be applied to solving a technological problem. One example: forming a hypothesis about a design (such as "A round building will be less likely to be damaged by high winds") and then testing the design.

Fascinating Facts

Human DNA, the carrier of heredity, is helping law enforcement solve some of their problems. It can do so because DNA is like a fingerprint. Everyone, except identical twins, has his or her own DNA. Because DNA is unique, it can help convict criminals. DNA from blood found at a crime scene can be compared to a suspect's DNA and used as evidence in the courtroom.

Entrepreneurs: Turning Problem Solving into Profit

Most of us can identify lots of everyday problems. We often say, "Why don't they make. . ." or "There should be somebody who can do . . ." For most of us, that's where it ends. Some people, however, don't just identify such problems. They think of solutions and implement them by starting their own business.

A person who starts a business is an **entrepreneur** (AHN´-truh-pruh-NUR). The entrepreneur might have an idea for a new product or a new service. For example, a common problem is that people have trouble programming their VCRs. One solution would be to invent a device that makes programming easier. Another solution would be to program people's VCRs for them. Still another would be to create books or videos that teach people how to program VCRs. Any of these solutions might be the beginnings of a business.

Entrepreneurs must be creative and able to think critically to solve problems. The solution they develop must be a good one, and it must fill a real need.

They must also be willing to take risks. Most entrepreneurs have to borrow money to start their company. They may also sell *shares of stock* in their company. The worth of the company is divided into a number of portions, or shares. Each person who buys a share thus owns a portion of the company. The more shares a person buys, the more of the company he or she owns. If the company is successful, all the shareholders will benefit from the profits. If the company fails, all of them will lose money.

Making a company successful is not easy. Entrepreneurs must be willing to work hard. Often, they have to work harder at their own business than they would if they were employed by someone else. Still, many entrepreneurs prefer to be their own boss. Fig. 2-8.

People who are creative, ambitious, and hard-working often make good entrepreneurs. If you have these qualities, you may want to study business and marketing courses in school to help you prepare for owning and running your own business.

11 **Extension.** Invite an entrepreneur to speak to the class about the challenges and rewards of starting a business.

FIG. 2-8 Entrepreneur Jan Benson Wright is owner, editor, and publisher of three magazines—*Inter-Business Issues*, *Peoria Woman*, and *Arts Alive*.

The Technology Student Association

12 The Technology Student Association (TSA) is dedicated to the needs of technology education students of all ages, throughout the country. Founded in 1994, TSA's mission is to prepare its members for the challenges they will face. This mission is carried out by promoting leadership, problem solving, and technological literacy.

The TSA high school programs consist of activities and competitions that cover a broad spectrum. These topics include the latest technology in aerospace, architecture, construction, electronics, manufacturing, and computer applications. In addition to competitions, TSA provides students with opportunities to participate in conferences, the TSA National Service Project, and recognition programs. High school students also have the opportunity to compete for scholarships and awards.

TSA is currently 2,000 chapters strong, with a membership of about 150,000 elementary, middle, and high school students. With a sense of purpose and a desire to make the world a better place, TSA is an organization that is helping students prepare to step into the twenty-first century.

12. **Resource.** Use a search engine to locate TSA's Web site.

CHAPTER 2: Problem Solving

Career File—Consumer Safety Inspector

EDUCATION AND TRAINING
Although some employers require inspectors to have college training, most inspectors are trained on the job to learn applicable laws or inspection procedures.

AVERAGE STARTING SALARY
The beginning salary for a consumer safety inspector is about $19,000 per year.

OUTLOOK
According to the Bureau of Labor Statistics, jobs for consumer safety inspectors are expected to grow at an average rate through the year 2005. Most jobs will come from federal, state, and local governments as the public continues to demand a safe environment and quality products.

What does a consumer safety inspector do?
Consumer safety inspectors ensure that health and safety standards are maintained in the manufacture and distribution of consumer products. They inspect such things as foods, pesticides, cosmetics, drugs, and medical equipment. They see that laws and regulations are obeyed so that retailers and consumers know that products are healthy and safe. They also check weights and measures to be sure the amounts are correct.

Does most of the work involve data, people, or things?
Consumer safety inspectors work mostly with people and a variety of measuring instruments. When responding to consumer complaints, they question employees, vendors, and others to obtain evidence. On an inspection project, they may work individually or as part of a team. They use portable scales, cameras, ultraviolet lights, thermometers, chemical testing kits, radiation monitors, or other equipment.

What is a typical day like?
Making routine inspections of manufacturing or distribution sites, consumer safety inspectors look for inaccurate product labeling and contamination. They discuss needed corrections with the firm's management or other officials, write reports of their findings, and compile evidence that could be used in a court of law, if necessary.

What are the working conditions?
Much of a consumer safety inspector's work is in the field, visiting processing plants, distribution warehouses, and sales outlets. Some of these places, such as a slaughterhouse, may be unpleasant and dangerous.

Which skills and aptitudes are needed for this career?
Consumer safety inspectors should like detailed work. They should be able to speak and write effectively and like to work with people.

What careers are related to this one?
Other inspection occupations include fire marshals, fish and game wardens, aviation safety inspectors, and highway safety inspectors.

One consumer safety inspector talks about his job:
This job has a lot of responsibility. I feel like I'm helping keep our community healthier and safer.

Chapter 2 REVIEW

 ## Reviewing Main Ideas

- People can solve problems by following six steps of a problem-solving process: state the problem clearly; collect information; develop possible solutions; select the best solution; implement the solution; evaluate the solution.
- Critical thinking skills, creative thinking skills, and decision-making skills are necessary to finding solutions.
- To understand how our universe works, scientists use the scientific method, a multi-step process that is similar to the problem-solving process.
- Entrepreneurs are generally innovative problem solvers who turn solutions into a profitable business.
- The Technology Student Association promotes leadership, problem solving, and technological literacy for technology education students.

 ## Thinking Critically

1. Suppose you lived in a tropical climate and your best friend lived in a cold, northern climate. How might your solutions to the problem of a place to live be alike? How would they be different?
2. Describe an example from your own life where you tried one solution to a problem and failed, then tried something different and succeeded.
3. Compare each of the steps in the problem-solving process with the steps in the scientific method. Which ones are different? Which steps are most similar?
4. Give four examples of problems: two that would best be solved by using the problem-solving process and two that would most appropriately be answered by using the scientific method.
5. Describe two people you know who are good entrepreneurs. Explain why.

 ## Understanding Concepts

1. Briefly describe the six steps of the problem-solving process and explain the importance of each step.
2. Describe three types of thinking skills used to develop solutions.
3. Describe the six steps of the scientific method.
4. What is an entrepreneur? Identify several characteristics of an entrepreneur.
5. Name at least three technology areas in which the TSA offers activities and competitions.

 ## Applying Concepts and Solving Problems

1. **Language Arts.** Divide into groups and brainstorm ways of improving the use of your school's food service, main office, library, and your classroom. Present your best ideas to the class.

2. **Internet.** Access the following web sites to learn what they have available and how you might use them during this course:
   http://ipl.sils.umich.edu/
   http://www.pueblo.gsa.gov/

13. **Resource.** Answers to "Understanding Concepts," "Thinking Critically," and "Applying Concepts and Solving Problems" can be found in the Teacher's Manual at the front of this *Teacher's Annotated Edition* and in the *Teacher's Resource Binder*.

CHAPTER 2: Problem Solving 45

CHAPTER 3

Technology Systems

Objectives

After studying this chapter, you should be able to:
- define system.
- describe the four parts of a system.
- name seven types of resources that provide input for a system.
- compare three different sources of energy.

Terms

capital
energy
feedback
input
output
process
productivity
resource
system

Technology Focus

Stonehenge: A Prehistoric Computer?

Mystery veils the origins of Stonehenge, a ring of giant stones arranged in a perfect circle 100 feet across. Radiocarbon dating puts the earliest construction at 2800 B.C. Theories of who built Stonehenge and why tend to vary with the background and yearnings of the onlooker. Some say the wizard Merlin had a hand in it. Others say it was a temple of the Druids, keepers of Nature's secret laws and lore. Still others say Stonehenge was designed by visitors from outer space.

An Ancient Technological System

Whatever else it may have been, Stonehenge was and is a work of technology. Its giant sandstones came from a quarry 20 miles away from its site in Wiltshire, England. These stones were probably dragged on sleds, perhaps with the help of rollers. Tools were used to shape the stones and set them upright. The upright columns lock into holes in the crosspieces overhead, called lintels.

The lintels may have been raised to their position in small steps, with one end being levered up a bit and a timber placed beneath. Then the other end was levered up, and so on, until the timbers stacked up thirteen-and-a-half feet above the ground. Within a tolerance of one inch, the top of Stonehenge is level throughout its 300-foot circumference.

The inputs of a technological system are all visible in the remains of Stonehenge—people, information, materials, tools, energy, capital, and time. The capital almost certainly was not money but rather an exchange of food, goods, protection—things that we now obtain with money.

We can only guess about the process and output of Stonehenge, but feedback most definitely is there. About 500 years after the initial layout, workers made slight adjustments in the position of the stones.

What Was Its Purpose?

Why should the exact placement of the stones be so important? In 1960 Gerald Hawkins, an astronomer, got the idea that the designers of Stonehenge must have been astronomers themselves. Hawkins used a computer to correlate precise positions of Stonehenge landmarks with the moon's path across the sky. He found that Stonehenge could be used to predict lunar eclipses.

His discovery remains controversial. Most archeologists doubt his findings. All we can say for certain is that a technological system without information about its people or its purpose must remain a mystery.

Take Action!

Can you draw a perfect circle 100 feet across using only materials found in nature? Try your idea and report on the results.

Resources
- Chapter 3 Lesson Plan in the *Teacher's Annotated Edition* and in the *Teacher's Resource Binder*.
- Chapter 3 Review in the *Teacher's Resource Binder*.

What Is a System?

To solve problems, people create technology. Every technology that is invented as a solution to a problem can be thought of as a system. A **system** is a group of parts that work together to achieve a goal. These parts are: input, process, output, and feedback. Fig. 3-1.

Input

Input includes anything that is put into the system. Input comes from the system's resources. A **resource** is anything that provides support or supplies for the system. Fig. 3-2. There are seven types of resources that provide input for all technological systems: people, information, materials, tools and machines, energy, capital, and time.

Every system has these seven types of resources, but which specific resources are chosen depends on cost, availability, and appropriateness. For example, if CD players are being produced, then plastic would probably be one of the several material resources. It would be an appropriate choice because plastic is lightweight and durable. It is also easily available and not very expensive.

People

People are the most important resource. Without people, technologies wouldn't even begin to exist. People decide that problems exist and use the problem-solving process to develop workable solutions.

For example, people who wanted to listen to music that had a high-quality, realistic sound, spurred on the invention of CD players. If no one wanted to hear music, there would never have been a need for such a product in the first place. People created the demand, but it also

FIG. 3-1 A system's parts are input, process, output, and feedback.

1. **Reteaching.** On the chalkboard, write the four parts of the systems model. Select a system, such as a computer, and have students identify inputs, processes, outputs, and feedback.
2. **Brainstorming.** Ask students how the seven resources relate to a technological system they use, such as a computer or an automobile.

SECTION 1: Introduction to Technology

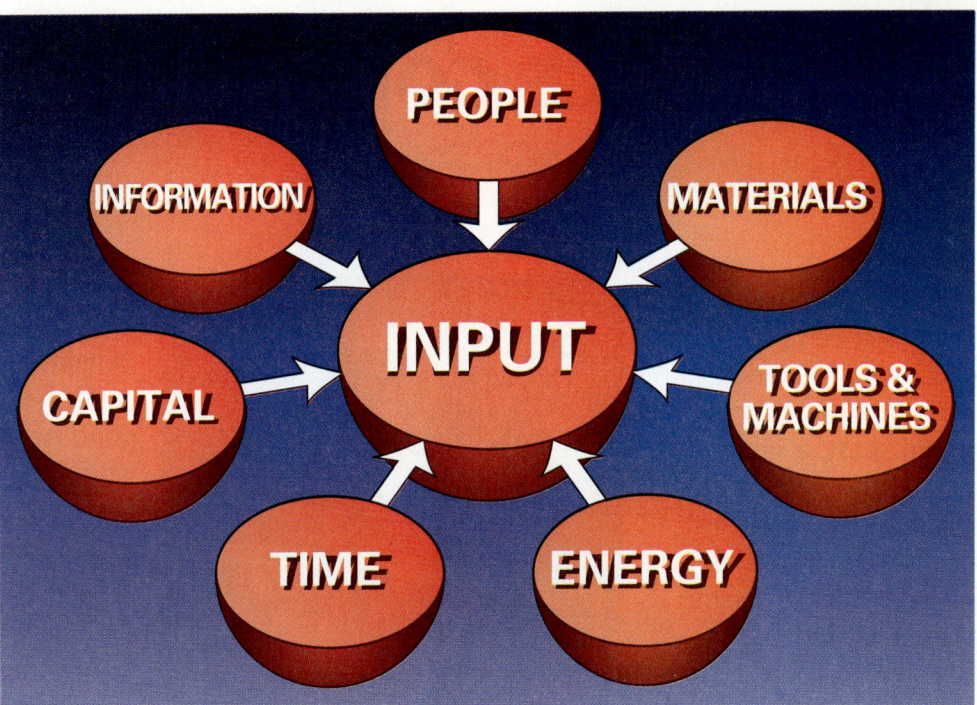

FIG. 3-2 The input to a system comes from its resources.

took people to figure out how to make something that would provide the desired sound quality. People are also needed to actually make the players and the discs. Fig. 3-3.

Fascinating Facts

People are the most valuable resource. One of the seven wonders of the modern world, the Panama Canal separates North America and South America and was completed in 1914. During building, 42,000 workers moved enough rubble and earth to open a 16-foot-wide tunnel to the center of the earth.

FIG. 3-3 Without people to make them, we would not be able to enjoy all the electronic devices that provide music for us.

CHAPTER 3: Technology Systems   **49**

SCIENCE CONNECTION
Physical Fitness and Job Performance

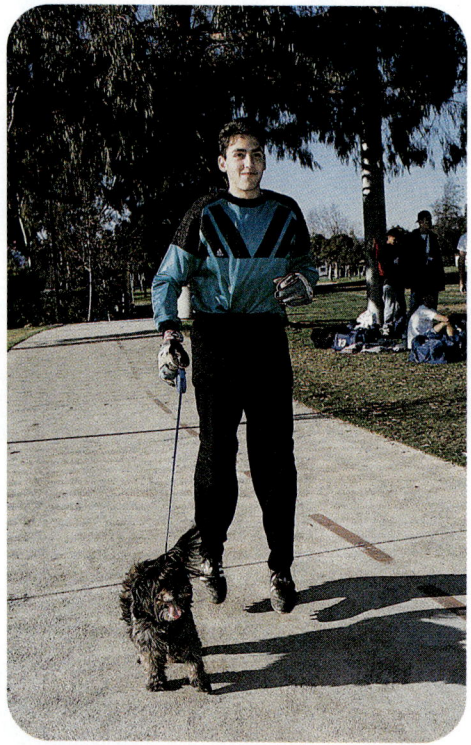

Did you know that being a couch potato can affect your job (and school) performance? In a recent study, a series of mental and physical tests were given to men and women who were fairly inactive. Then the same people were put on a four-month aerobic training program. At the end of the four months, their performance on mental tests improved as much as 10 percent. Researchers believe that the stress of exercise on the body leads to changes that also benefit the brain.

Exercise also improves the immune system, which is responsible for fighting disease. In another study, people who exercised moderately suffered only half as long from respiratory infections as those who were sedentary. In a third study, elderly women who were very fit had immune cells that were 85 percent more active than the cells in a group of college-age women who were couch potatoes.

Employers are beginning to pay attention to the effects of worker health on productivity and insurance costs. A recent study involving employees of Chrysler Corporation found that those with poor health habits generate 41 percent higher health-claim costs than workers who have good health habits. Those whose weight is 20 percent or more above or below the recommended range use hospitals 143 percent more than workers of normal weight. Companies are finding that people who exercise regularly feel better, think more clearly, and are sick less often. The quality of their work improves, they lose less time from work, and they are less likely to be injured in accidents.

Try It!

1. Assume that you work for a company employing a number of overweight, sedentary people. Write a persuasive letter to your company asking them to consider starting a health-promotion program.
2. Check out this site for health and fitness on the Internet: http://www.wellness.ucdavis.edu/index.html

Information

In order for people to design and make what is needed to solve a problem, they first need information. In Chapter 1, you read about how technologists need to use science, math, social studies, and language arts when they solve problems. To develop CD players, technologists used information about phonograph technology, laser technology, and digital technology. They came up with a new form of recording system where the "records" (CDs) don't wear out and can record sound more sensitively than traditional phonograph systems. Manufacturing workers need information, such as quantities, sizes, shapes, and types of parts, as well as how the parts are to be assembled in order to make CD players. Even machines need information. In today's factories, many machines are controlled by computers. The computers send information to the machines, telling them what to do.

Fascinating Facts

As a product is changed and improved, the amount of materials used may also change. An early computer, the Mark I, built in 1944, weighted five tons and was 51 feet long and eight feet tall. Compare that to one of today's laptop computers that weigh less than two pounds!

Materials

Materials are all the things that make up a product. A compact disc used in a CD player, for example, is made of plastic that has been coated with a thin layer of aluminum and covered with acrylic. The player itself has both plastic and metal parts.

Natural and Synthetic

Materials are either natural or synthetic. *Natural materials* are those that are found in nature. Metal, wood, and petroleum are some examples of natural materials. Some natural materials are *renewable*. This means that new amounts can be produced. Wood is a renewable material because new trees can grow to replace those that are cut down. *Nonrenewable* materials cannot be replaced. We have only a certain amount and once that amount is used up, there won't be any more. Petroleum is an example.

Synthetic materials are human-made. Plastics, ceramics, and composites are examples of synthetic materials. Fig. 3-4. You won't find these materials in a mine or growing on a tree. However, they are made from materials that came from nature. Plastics, for example, are made from petroleum, natural gas, or plant fibers.

Raw and Industrial

Raw materials are natural materials that have not been processed. Raw materials may be obtained by harvesting, mining, or drilling. To make them usable, raw materials are changed into *industrial materials*. The industrial materials are then made into finished products.

---

3. **Resource.** Computer-controlled machines are discussed in Chapters 11 and 13.
4. **Extension.** Point out that *renewable* does not mean *inexhaustible.* It's possible to run out of a renewable resource if we use it faster than it can be replaced.
5. **Extension.** Have students make a list of synthetic materials and indicate whether they are made from renewable or nonrenewable resources.
6. **Reteaching.** Make a list of raw and industrial materials and bring some samples to class.

CHAPTER 3: Technology Systems

MATERIAL	DESCRIPTION	TYPES	EXAMPLES
PLASTICS	SOURCES INCLUDE PETROLEUM, NATURAL GAS, AND PLANT FIBERS. MOLECULES OF CARBON COMPOUNDS FROM THESE SOURCES ARE COMBINED TO FORM LONG CHAINS CALLED POLYMERS.	THERMOPLASTICS-- CAN BE REPEATEDLY HEATED, SOFTENED, AND RESHAPED.	ACRYLIC PAINTS, POLYETHYLENE BOTTLES, POLYSTYRENE TOYS
		THERMOSETS-- CANNOT BE MELTED AND RESHAPED.	EPOXY ADHESIVES, URETHANE PLASTIC FOAMS AND INSULATION, SILICONE LUBRICANTS
CERAMICS	INGREDIENTS CAN INCLUDE SILICON, OXYGEN, ALUMINUM, AND MANY OTHER ELEMENTS. CLAY, SAND, AND LIMESTONE ARE SOME OF THE RAW MATERIALS FROM WHICH CERAMICS ARE MADE.	SOME CERAMICS ARE MOLDED INTO SHAPE AFTER BEING HEATED.	POTTERY, BRICKS
		OTHERS ARE HEATED FIRST AND THEN SHAPED.	GLASS, CEMENT
COMPOSITES	COMPOSITES ARE MADE BY COMBINING TWO OR MORE MATERIALS. EACH MATERIAL RETAINS ITS OWN PROPERTIES. THE COMPOSITE THUS HAS THE COMBINED PROPERTIES OF ITS INGREDIENTS, RATHER THAN COMPLETELY NEW PROPERTIES.	CONCRETE--A MIXTURE OF CEMENT, SAND, AND GRAVEL	ROADS, BRIDGES, BUILDINGS, AND OTHER STRUCTURES
		CEMENT--A MIXTURE OF METAL AND CERAMIC PARTICLES	JET TURBINE BLADES, ROCKET NOZZLES, HEAT-SHIELDING TILES FOR SPACE SHUTTLE
		FIBERGLASS-REINFORCED PLASTIC-- MADE OF SPUN GLASS REINFORCED WITH PLASTIC	BATHTUBS, BOAT HULLS, SOME CAR BODIES

FIG. 3-4 Synthetic materials are those that do not occur in nature.

The processes used to convert raw materials into industrial materials are called *primary processes*. Processes that convert industrial materials to finished products are *secondary processes*. Fig. 3-5.

PRIMARY PROCESSES | SECONDARY PROCESSES

TREES ARE CUT DOWN AND TAKEN TO SAWMILL.	BARK IS REMOVED FROM LOGS, AND THEY ARE CUT INTO BOARDS.	AFTER DRYING, BOARDS ARE READY FOR USE.	BOARDS ARE MADE INTO FURNITURE.

A HOLE IS DRILLED INTO ROCK FORMATIONS THAT HOLD PETROLEUM. A PIPE IS PUT IN PLACE, AND THE PETROLEUM IS PUMPED OUT.	THE PETROLEUM IS REFINED TO EXTRACT CHEMICALS FOR MAKING PLASTICS.	THE CHEMICALS ARE PROCESSED INTO PELLETS.	THE PELLETS ARE MELTED, COLORED, AND FORMED INTO PRODUCTS.

IRON ORE IS REMOVED FROM THE EARTH.	THE ORE IS MELTED AND PURIFIED TO MAKE PIG IRON. THE PIG IRON IS MADE INTO STEEL.	THE STEEL IS FORMED INTO STANDARD SHAPES SUCH AS SHEETS OR RODS.	THE STEEL IS USED TO MAKE A FINISHED PRODUCT.

FIG. 3-5 Shown here are some typical primary and secondary processes.

CHAPTER 3: Technology Systems **53**

Tools and Machines

Tools and machines are used to make materials take on the size and form needed to make the product. They are also used to transport materials, parts, and the final product. This category includes all the things people use to help them do their work, from the simple screwdriver and saw to a complex assembly-line robot.

There are thousands of tools and machines. They can be classified according to their purpose into the following groups. Fig. 3-6.

- Measuring tools are used to find out the size and shape of an item.
- Layout tools are used to make lines, angles, and circles on material to show how it is to be cut or bent. There are also layout tools for determining straight lines and angles for construction work.
- Separating tools and machines are used to cut materials to a certain size or shape.
- Forming tools and machines are used to change the shape of materials.
- Combining tools and machines are used to join parts.

Energy

Energy is needed to power the tools and machines. **Energy** can be defined as the ability to do work. Energy may come from inexhaustible, renewable, or nonrenewable sources.

Sources of Energy

- Inexhaustible sources are those that can never be used up by human action. The

FIG. 3-6 These are examples of tools for measuring, laying out, separating, forming, and combining.

7. **Resource.** See the Science and Mathematics Handbook in the *Teacher's Resource Binder* for information about the six simple machines. Point out that most of the tools and machines we use are some combination of these simple machines.
8. **Resource.** Additional information about energy can be found in the Science and Mathematics Handbook.

sun is an inexhaustible source of energy. We can use the sun's energy directly, for example, in solar heating. The sun's energy also creates wind and influences tides. Each of these can be an energy source for technology systems.

- Renewable energy sources are those that can be replenished (resupplied). Human and animal energy sources are renewable—so are trees and plants.
- Nonrenewable energy sources are those that cannot be replenished by human action. These include fossil fuels (coal, petroleum, and natural gas) and nuclear energy sources.

Forms of Energy

Energy comes in several forms. Heat, light, sound, chemical, nuclear, mechanical, and electrical energy are all used by technology systems. The energy provides movement, heat, or light. An electric drill, a truck, a furnace, a lightbulb—none of these would work without energy.

Energy cannot be destroyed, and it can be created only in a nuclear reaction. However, energy can be converted (changed) from one form to another with ordinary technology. It can also be transmitted (moved) from one place to another.

Many technological systems include the conversion or transmission of energy. Consider what happens when you turn on a lamp. The electrical energy moving through the bulb is converted into light and heat. That electrical energy was transmitted to your home from the power plant. The electrical energy sent out by the power plant might have started as heat energy produced by the burning of fossil fuel. The heat was used to produce steam, which turned the blades of a turbine. A

9. **Critical Thinking.** Ask students to suggest ideas for machines powered by the sun or wind.
10. **Reinforcement.** Remind students that one way to raise capital is to sell shares of stock in the company. This was discussed in Chapter 2.

FIG. 3-7 Steam turns the blades of these turbines. As the blades turn, so does the shaft. The shaft is attached to a generator. The generator converts the energy of motion (mechanical energy) into electrical energy.

generator attached to the turbine converted the mechanical energy from the turbine into electrical energy. Fig. 3-7.

Capital

Capital is another important resource for technology. **Capital** includes the money, land, and equipment needed to set up the technological system. Without money to build factories and hire workers, CD players would remain only an interesting idea instead of an actual technology.

CHAPTER 3: Technology Systems

Time

A final technological resource is time. When a company decides to make CD players, it invests time in that particular system. The company could be spending that same time making computers or furniture. Time is a valuable resource, so the ways people decide to use it are very important.

Time is one factor in determining productivity. **Productivity** is the amount of goods or services compared to the amount of resources that produced them. The more goods or services a worker can produce in one hour, the higher the productivity of that worker.

Process

The second part of a system is the process. The **process** includes all of the activities that need to take place for the system to give the expected result. For a CD player, the process might include designing the product itself and its individual parts, making the various parts out of appropriate materials, and putting all the parts together to make the player.

The process part of a system also includes management. In order for the system to produce the expected result, the process must be well planned, organized, and controlled. No process works perfectly. Both people and machines are needed to identify and solve problems and to make improvements. Fig. 3-8.

Output

The third part of a system is the output. The **output** includes everything that

PROCESS
- DESIGNING THE PRODUCT AND ITS PARTS
- MAKING THE PARTS
- ASSEMBLING THE PARTS
- MANAGEMENT

FIG. 3-8 The process part of a system includes all of the activities that are needed to solve the problem.

results when the input and process parts of the system go into effect. In all systems, there is, of course, the intended output, such as CD players. There are also, however, outputs that may not have been intended—such as waste that may have been created during the process, or changes in society caused by the product. Fig. 3-9.

CD players, for example, have had a great impact, or effect, on the recording industry. In many record stores, you no longer find vinyl phonograph records. These changes in the recording industry have an impact on other parts of the economy. Think of all the materials that were used to make vinyl records. Some of these are no longer needed. On the other hand, more of other types of materials may be needed to make compact discs.

FIG. 3-9 The outputs of a system can be expected and unexpected, desirable and undesirable.

11. **Enrichment.** Discuss the relationship between productivity and a nation's economic welfare.
12. **Reteaching.** Explaining the processes of separating, forming, conditioning, and combining may help give students a sense of what *process* is.
13. **Critical Thinking.** Choose a technology and have students identify outputs. Have them classify the outputs as expected or unexpected, desirable or undesirable.

Feedback

The outputs of a system must be closely watched to be sure that the system is solving the problem it was intended to solve and to be sure that the system is not creating new, greater problems. **Feedback** is information about output that is sent back to the system to help determine whether the system is doing what it is supposed to do. Fig. 3-10.

Feedback can take many forms. Some kinds of feedback are built into a system. For example, home CD units have error-checking features. Often, systems check themselves to make sure all the parts are functioning properly.

In addition, there is feedback from people who use the system. Suppose people liked the sound emitted from CDs, but did not like the size or bulk of the unit that played them. Manufacturers would listen to this feedback and redesign the unit so people would like it better.

Feedback becomes a form of input into the system. Fig. 3-11. It often leads to improvements in a system or to the creation of new systems.

FEEDBACK

- INFORMATION ABOUT OUTPUT SENT BACK TO SYSTEM

FIG. 3-10 Feedback occurs when information about the outputs of the system is sent back to the system.

FIG. 3-11 System for making a CD player.

INPUT
- PEOPLE TO DESIGN AND CONSTRUCT THE PLAYER
- INFORMATION GATHERED AND USED
- MATERIALS TO MAKE THE PRODUCT
- TOOLS AND MACHINES TO SHAPE AND FORM MATERIALS
- ENERGY TO POWER TOOLS AND MACHINES
- CAPITAL TO FINANCE THE COMPANY
- TIME SPENT TO DESIGN AND MAKE PLAYERS

PROCESS
- DESIGNING PLAYER AND ITS PARTS
- MAKING PARTS
- ASSEMBLING PARTS
- MANAGING THE PROCESS

OUTPUT
- CD PLAYER
- WASTE MATERIAL
- IMPACTS ON SOCIETY

FEEDBACK
- SELF-CHECKING OF THE PLAYER
- CONSUMER COMPLAINTS

14. **Reteaching.** Use sports coaching as an example of feedback. Many students can relate to this.
15. **Critical Thinking.** Discuss how certain technologies may remain in use if they have advantages over newer options. Example: Bicycles are still used for transportation. Why?

CHAPTER 3: Technology Systems

Career File—Electronics Technician

EDUCATION AND TRAINING

Most employers prefer to hire people with vocational or technical training. Persons with college courses in science, engineering, and mathematics may qualify for a beginning position, but some on-the-job training is usually required.

AVERAGE STARTING SALARY

Beginning wages for electronics technicians can vary from $15,000 to $18,500 per year, depending on education, experience, and area of the country.

OUTLOOK

According to the Bureau of Labor Statistics, job opportunities for electronics technicians are expected to grow more slowly than other occupations through 2005. Prospects will be best for technicians who keep up to date on the latest technologies.

What does an electronics technician do?
Electronics technicians help design and manufacture electronic devices such as radios, televisions, computers, industrial equipment, medical instruments, and navigation equipment. They also test, adjust, and repair electronic devices.

Does most of the work involve data, people, or things?
Electronics technicians work mostly with things—electronic equipment and the measuring and diagnostic devices used to test and repair it.

What is a typical day like?
Technicians who work in research and development might build or set up equipment, conduct experiments, record results, and make prototypes of newly designed products. Technicians who work in manufacturing plants might run tests to ensure product quality, study ways to improve manufacturing efficiency, or supervise production workers to make sure they follow the proper procedures.

What are the working conditions?
Most electronics technicians work regular hours in laboratories, offices, or electronics and industrial plants. Some may be exposed to hazards from equipment, chemicals, or toxic materials.

Which skills and aptitudes are needed for this career?
Creativity and an active imagination are desirable for technicians who are involved in research, development, and design work. Good communication skills and the ability to work well with others are important because technicians often work as part of a team.

What careers are related to this one?
Other occupations that require similar training and use similar scientific principles include drafters, surveyors, broadcast technicians, health technologists, and science technicians.

One computer systems analyst talks about his job:
Ever since I was a little kid, I've enjoyed working with electronics. My job is more fun than work!

Chapter 3 REVIEW

Reviewing Main Ideas

- A system is a group of parts that work together to achieve a goal. These parts are: input, process, output, feedback.
- Input includes resources that are put into the system: people, information, materials, tools and machines, energy, capital, and time.
- Energy from inexhaustible, renewable, or nonrenewable sources can be converted or moved from one form to another.
- Improvements in a system often depend upon feedback as input.

Understanding Concepts

1. What is a system?
2. Identify the seven resources of technology.
3. Identify and describe the three general categories of energy sources.
4. Tell how the first and fourth parts of a system are related.
5. Explain the role of management in the process part of the system.

Thinking Critically

1. Describe how the seven resources of technology are used in a car manufacturing system.
2. What are some of the outputs of a car manufacturing system? Be sure to include both negative and positive impacts.
3. Describe how a home heating system uses feedback to keep itself working properly. (Hint: think of how the thermostat works.)
4. What are some examples of technology systems that have replaced other technology systems? Explain how the changes affected people in their everyday activities.
5. How might the process of manufacturing a TV be changed if feedback from customers indicated that too many of the programs were unsuitable for viewing?

Applying Concepts and Solving Problems

1. **Mathematics.** Choose a booklet that describes how to make something. Then describe the mathematics skills you will need to make this product. (For example, making measurements for a cabinet.)
2. **Language Arts.** Select an object from home (e.g., a portable radio, a kitchen utensil or pan, an electronic item, a lightbulb, etc.) and describe how the four parts of a system apply to the manufacture of this product. Evaluate the product you chose and relate the feedback to a description of how you would go about improving the product.
3. **Science.** Describe the materials in your favorite chair. Tell which are natural or synthetic, which are raw (if any) or industrial. Explain what forms of energy were used to produce these materials.
4. **Social Studies.** Divide into groups and discuss current technology systems (e.g., computers, e-mail, cable television). Tell which systems you think are most important and why. Discuss what new technology systems you think will or should be developed in the future.

16. **Resource.** Answers to "Understanding Concepts" and "Thinking Critically" can be found in the Teacher's Manual at the front of this *Teacher's Annotated Edition* and in the *Teacher's Resource Binder*.

1 DIRECTED ACTIVITY

Making a Technology Occupations Survey

Context

What kinds of jobs do the people in your community and the surrounding area do? In which of the five major areas of technology do most of the jobs fall? Do most of the jobs involve performing a service or producing a product? Finding the answers to those questions can help you learn more about your community and the career opportunities within it.

Goal

For this activity, you will make a survey of as many students as possible in your school to learn about their parents' (or adult caregivers') occupations. As a result you will learn more about the work that takes place in your community and its relationship to technology.

Procedure

1. As a class, create a list of questions you would like to ask. Limit the number to no more than 10.

2. Using your questions, write a questionnaire similar to the one in Fig. A-1. Notice that the answers required in the sample questionnaire are short; the survey-taker should not have to do a lot of writing.

3. Decide when and where you will make your survey. For example, will survey-takers with clipboards walk around the cafeteria during lunch asking questions while students eat? Will they question students in the halls between classes?

4. Decide how much time to allow for the survey. Two days? A week?

5. How large do you want your survey to be? (The larger the survey, the more accurate the results in relation to the entire community.) Decide how many respondents each survey-taker will be responsible for.

6. Tabulate the responses. What percentage of the student body was questioned? In which of the five areas of technology do the majority of adults work? What information about your community can you gather from the survey?

7. Display the results of your survey.

Evaluation

1. If you were to repeat this survey, what would you do differently?
2. Which part of this activity did you enjoy most? Why?
3. Did you learn anything from this experience that you did not expect to learn? If so, what was it?

Useful Skills across the Curriculum

1. Language arts. Write an article describing the results of your survey for your school newspaper.

2. Mathematics. Find out the population of your state and your community. Your community makes up what percentage of the state population?

Sample Questionnaire

1. What is your father's job title? _____

2. In which of the following areas does your father work?

   _____ Communications   _____ Transportation

   _____ Manufacturing   _____ Bio-Related Technology

   _____ Construction

3. Does your father make a product (such as build houses)?

4. Or does he perform a service (such as repair computers)? _____

5. What is your mother's job title? _____

6. In which of the following areas does your mother work?

   _____ Communications

   _____ Manufacturing

   _____ Construction

   _____ Transportation

   _____ Bio-Related Technology

7. Does your mother make a product (such as build houses)? _____

8. Or does she perform a service (such as repair computers)? _____

Fig. A. These questions require short answers.

2 DESIGN AND PROBLEM-SOLVING ACTIVITY

Investigating Your Environment

Context

Although technology has made our lives easier in many ways, technological systems have often produced undesirable impacts on the environment. Radioactive waste, CFCs, acid rain, noise pollution, air pollution, and the deterioration of the ozone layer are just a few. It is normal to feel helpless and upset by all the environmental problems. However, there are ways you can make a difference, both in your community and in your school.

Goal

For this activity you will investigate environmental problems in your school or community and apply the problem-solving process in order to help solve one of them.

1. State the Problem

Start by deciding what environmental problem you wish to solve.
Eliminating oil spills would not be a problem you and your classmates could handle. However, what about paper waste in your school or noise pollution in the community? Perhaps if your community doesn't already have a recycling program, you could find out how to get one started. Looking in the newspaper or talking to your parents may give you some good ideas.

After you have identified the problem you wish to solve, state it clearly. Stating the problem clearly helps identify exactly what the problem is and may suggest possible solutions.

SECTION 1: Introduction to Technology

2. Collect Information

After you have identified the problem, you and your classmates need to gather all the information you can. You may wish to interview people, do research in the library, and find out what other schools or communities are doing to handle similar problems.

3. Develop Alternative Solutions

The next step is to brainstorm with your classmates about possible solutions. Use your imaginations and try not to criticize at this stage. Even if some of the solutions seem ridiculous, they may lead you to other more practical ones. Make a list of all the ideas.

4. Select the Best Solution

From the list of possible solutions, you and your classmates must choose the best solution. List and weigh the advantages and disadvantages of each. Some of your ideas may be impossible to carry out. For example, a solution must work within your school schedule. If it is an in-school project, will the rest of the students in the school be willing to cooperate and pitch in? Must equipment be purchased and, if so, can the school afford it? Questions such as these must be considered before choosing the best solution.

5. Implement the Solution

Try your solution for a specified period of time. Set a date for discussing progress. Is it working? Do you need to change anything? Make any needed changes.

6. Evaluate the Solution

Think of possible improvements. As a class, create a display of some kind that describes the problem and the steps you took.

Useful Skills across the Curriculum

1. Mathematics. The average American produces about 180 pounds of garbage each year. Find out the population of your school and your community. How many pounds of garbage are produced in each place each year?

2. Social Studies. What was the population of your community 100 years ago? Find out how waste management systems have changed during that time.

③ DIRECTED ACTIVITY

Developing a School Recycling Program

Context

One important reason for recycling is to preserve our natural resources, of which the earth contains a limited amount. Another reason for recycling is that it cuts down on the amount of waste deposited in landfills. In some large cities, the amount of waste is so great that it is loaded into railroad cars and shipped to landfills in other states. You and your classmates can help save natural resources, not only by recycling at home but also by working in school recycling programs.

Health and Safety Notes: Sharp metal edges can cause cuts. Be sure to wear work gloves when handling empty cans. Always wash your hands thoroughly when you are finished.

Goal

As a class, you will set up and operate an aluminum can recycling program. This program will not only save space in landfills, it will also earn your class money.

Procedure

Your instructor will divide the class into teams. Each team will be responsible for some of the tasks listed below. Be sure to maintain careful records.

Equipment and Materials
- recycling containers (trash cans, bins, etc.)
- trash bags
- scale
- holding bin

1. Determine the population in your school. This will allow you to estimate such things as numbers of containers. It will also help determine how successful your program is.

2. Obtain recycling containers. Contact your local soda distributors and ask if they will provide them. If they will, order one container for every 50 students. If they will not provide containers, you may have to purchase plastic trashcans having removable lids. For sanitary reasons, plan to line your containers with trash bags. During the collection process, the

---

1. **Extension.** Some states have made it a law that communities must have recycling programs. Other states have added a five- to seven-cent deposit on the sale of beverages in aluminum cans. People take the empty cans to can banks or recycling centers and get the deposit back.

64 SECTION 1: Introduction to Technology

bag can simply be lifted from the container and tied shut and a new bag installed.

3. Keep a record of all money spent on trash cans, trash bags, and so on; this money should be repaid later out of your profits.

4. Contact a recycling company in your area to inquire about selling the cans. If you collect a large number, the company may be willing to come to your school and pick up the cans. If they do not pick up the cans, you will have to plan for transportation.

5. Locate a storage bin in which to dump the cans as they are collected. You may wish to use large cardboard appliance boxes or contact a trash company and obtain an old dumpster. Another alternative is to fence in an area 10 ft. by 10 ft. outside the school.

6. Identify areas of the school where students gather to drink soda. Position the can collectors in those areas by doorways or near ordinary trash cans. Create a sign that identifies each collector. Assign each one a number, and give its location. This will help ensure that all collectors are emptied during each collection period. The sign could look something like this:

Aluminum Can Recycling Program
Collector No. 3
Lunchroom doorway

7. Develop a collection schedule for emptying the containers on a regular basis—at least once a week. Table A shows an example. This schedule should specify the collector number, location, students assigned to emptying, and the weight of the cans taken from each location.

8. Inform students, teachers, and staff members about the recycling program. You may want to make an announcement on the public address system, create posters, or develop an advertising campaign promoting the program.

9. Empty the collectors regularly and weigh the contents. A bathroom scale or large balance scale can be used. Enter the data on the collection record.

10. After each collection, wash the collector thoroughly so it does not develop an odor or attract flies and ants. Install a clean trash bag.

11. Empty the weekly collection into the storage bin.

12. After three or four weeks, analyze the collection record as to the quantities gathered at each location. You may wish to reposition some collectors.

13. When the storage bin is full, it is time for a shipment. If you do not have an agreement with a particular company, contact several local recycling companies and ask for a quote regarding the current price per pound. If possible, sell for the best price.

14. After the sale, record on a shipping schedule the date, students responsible for shipment, weight, price per pound, and total amount received. Table B shows an example.

15. Deposit the money from the sale in a bank account and keep track of the earnings.

16. Use the profits from recycling to pay any debts for supplies, increase the collection capacity, create a better storage area, or improve the facilities at your school. Some students have used money earned from recycling to build an outside picnic area and buy new drinking fountains.

Evaluation

1. Did you alter any routines for the sake of efficiency as you gained experience in the recycling program? Explain.

2. Calculate the square footage of your storage bin. Using that figure and the number of shipments made over a semester, estimate how many square feet of landfill space are being saved by your program.

Useful Skills across the Curriculum

1. Mathematics. Call a local bank and ask for their interest rates. Calculate the interest your program would earn in a year if the same amount of money was deposited each month.

2. Science. Do some research on the properties of aluminum. Why is it the material of choice for beverage cans?

Table A. Collection Schedule

Week	Collector No.	Location	Students Responsible	Weight

Table B. Shipping Schedule

Date	Students Responsible	Recycling Company	Total Shipment Weight	Price per Pound	Total Sale
				$	$
				$	$
				$	$
				$	$
				$	$
				$	$
				$	$

Credit: Robert A. Daiber

4 DESIGN AND PROBLEM-SOLVING ACTIVITY

Using Problem Solving to Make a Chair

Health and Safety Notes: Before doing this activity, make sure you understand how to use the tools and materials safely. Have your teacher demonstrate their proper use. Be sure knives and scissors are sharp. Follow all safety rules.

Context

You have learned in Chapter 2 about the six steps in problem solving. Reading about problem solving, however, does not teach you all there is to know about it. The best way to learn about its usefulness is to use it in solving a real problem.

Goal

For this activity, you will use the problem-solving process to design and build a chair out of cardboard. A person should be able to sit on this chair. You will be competing with your classmates to see who can find the best solution to the problem.

1. State the Problem

Design and build a full-size, usable chair out of cardboard.

Specifications and Limits

- The chair must be made completely of corrugated cardboard and masking tape. No glue, wood, metal, plastic, or other materials can be used.
- It must have a seat and back. The seat must be between 16 in. and 18 in. from the floor. The top of the back must be no less than 30 in. from the floor.
- It must support your instructor. He or she must be able to sit in the chair in a relaxed, comfortable position.
- The chair should weigh as little as possible. In the final competition, its weight will be measured and compared to that of other chairs made in class.
- The chair should be as attractive as possible.
- All chairs will be judged and awarded points. A total of 15 points is a perfect score. Points will be obtained as follows:

Equipment and Materials
- pencils and paper
- measuring instruments
- cutting tools, such as scissors and utility knife
- corrugated cardboard
- masking tape
- cutting boards or cutting surface

Following specifications—5 points
Weight—5 points
Design—5 points

Fig. A. These unusual chair designs have all been used at various times.

2. Collect Information

You will work in teams. With your teammates, read through this entire activity. You may want to copy the specifications outlined above. This will help you remember them later.

Study the designs of chairs you see around you. Look up chairs in an encyclopedia and learn what designs have been used in the past. Then study your materials. How can cardboard be made strong?

3. Develop Alternative Solutions

Talk with your teammates about possible chair designs. Each of you should make sketches of two or three that might work. Remember, attractiveness is also important. Fig. A.

4. Select the Best Solution

As a team, select the design you like best. Check it against the specifications. Is anything missing?

5. Implement the Solution

Create a set of working drawings for your chair. There should be a top, front, and side view. (See Fig. 7-22 on page 163 for an example of a working drawing.)

Build your chair using the cardboard and masking tape.

6. Evaluate the Solution

Evaluate your chair. Ask someone to sit on it. Did the chair hold the weight? Is the person comfortable? Look at its overall design. Is it attractive? Is the tape placed neatly?

If you were to build another chair, what would you do differently? What did you learn about materials from this activity?

Useful Skills across the Curriculum

1. Science. Determine which forces act upon chairs. Which mechanical properties does cardboard have?

2. Social Studies. Research how chair design has been influenced by social customs in the past.

Credit: Gene Stemmann

SECTION 2

CHAPTERS

4 Communication Systems

5 Computers

6 Telecommunication

7 Graphic Communication

TECHNOLOGY TIME LINE

3100 B.C. Cuneiform writing used in ancient Sumer.

780 B.C. Chinese perfect woodblock printing.

1045 Chinese printer Pi Sheng experiments with movable type.

1826 Joseph Niepce of France produces the world's first photographic image.

1866 First telegraph cable laid across the Atlantic.

1876 Alexander Graham Bell invents telephone.

1895 Guglielmo Marconi of Italy develops the wireless telegraph.

SECTION 2: Communication Technology

Communication Technology

You live in a society that depends on the communication of information. Communication technology is diverse; it includes everything from computers to space satellites to graphics. All communication technology deals with sending and receiving messages. In this section, you will learn how communication technology affects complex systems and everyday life.

1929	1947	1967	1989	1990	1997
Vladmir K. Zworykin demonstrates the first practical television system.	The transistor is invented.	A battery-operated cordless telephone is tested in the U.S.	U.S. leads the world in TV sets with 98% of households having at least one TV.	World Wide Web is created in Europe.	World chess champion defeated by IBM's chess computer.

SECTION 2: Communication Technology

CHAPTER 4

Communication Systems

Objectives

After studying this chapter, you should be able to:

- define communication technology.
- identify trends in communication systems.
- explain why we need communication systems.
- describe how inputs, processes, outputs, and feedback apply to a communication system.
- explain some impacts of communication systems.

Terms

CD-ROM
communication channel
electromagnetic compatibility (EMC)
electromagnetic interference (EMI)
electromagnetic radiation
laptop computers
modem
sensors
telemarketing
trend

Technology Focus

From the Comic Pages to Real Life

In 1931, Chester Gould created a comic strip centered around a detective hero named Dick Tracy. Dick had a wrist radio that worked almost like a telephone. In 1931, it was a dream, far from the ability of the technology of that time to produce. What about today?

Combining Radios and Telephones

Cellular telephones, often small enough to fit into a shirt pocket or purse, are just one example of the portable communication devices available today. A cellular phone is really a combination of a radio transmitter and a telephone. Radio signals from the phone are sent to networks of radio/telephone receivers. A user can make and receive phone calls from a car, from an airplane, or even while walking outdoors.

You may know someone who carries a *pager*, which is a tiny radio receiver that enables its user to receive messages anytime, anywhere. The pager usually has its own phone number. When someone dials that number, the pager may beep, display the phone number of the caller, or even play the voice of the person calling. The person carrying the pager then goes to a telephone to return the call.

Recently developed is a *wristwatch phone*, a tiny cell phone having a speech recognition system. No need to dial. You say the number, and you're connected. Just like Dick Tracy!

Computers Get into the Act

Many people now carry portable **laptop computers**, which are about the size of a small briefcase. Some laptops have a built-in **modem**. This device sends computer data over telephone lines to other computers. Wireless modems, like cellular phones, can be used in many locations. This allows laptop users to communicate with any other accessible computer or computer service.

What existed only in a comic strip in the 1930s has become real. In the future, the functions of phones, pagers, computers, and modems may also be combined into one small device.

Take Action!

Think of stories, movies, or TV shows that have presented technological devices of the future. Describe several of these devices that you think will one day be used in real life. Explain your choices.

1. **Extension.** Have students name some inventions that were first mentioned in comics, science fiction books, and/or television programs.

Resources
- Chapter 4 Lesson Plan in the *Teacher's Annotated Edition* and in the *Teacher's Resource Binder*.
- Chapter 4 Review in the *Teacher's Resource Binder*.

CHAPTER 4: Communication Systems

What Is Communication Technology?

Imagine a world without books, signs, computers, radios, and newspapers. What would life be like without a telephone or television? It may be hard to picture yourself making it through one day without using some of these tools, yet you probably take them for granted. We depend on the tools and equipment that help us to communicate—to send and receive messages. We depend on communication technology.

As you learned in Chapter 1, communication technology is all the things people make and do to send and receive messages. It's the knowledge, tools, machines, and skills that go into communicating.

Messages come in many different forms. They reach their destinations in various ways. For example, you often send and receive messages from your friends by using the telephone. Some companies send messages to you by advertising on television or radio. A buzzer or light in a car reminds passengers to fasten their seat belts. When we study communication technology, we are exploring the ways people use their knowledge and skills to send and receive messages.

Communication technology is always changing as techniques and devices are invented or improved. The invention of the telegraph in the mid-1800s gave people the opportunity to send and receive messages over long distances almost instantly. Today's methods of communication are of better quality and more convenient. Many kinds of communication technology are available, such as laptop computers, cellular telephones, and even portable televisions. Fig. 4-1.

> **FASCINATING FACTS**
> During World War II, Germany used a machine known as *Enigma* to send coded messages. The British cracked the code by building the most powerful computer of the time. Known as *Colossus*, it helped the Allies win the war.

FIG. 4-1 Would a battery-operated miniature television be more convenient for you? How?

2. **Brainstorming.** Have students respond to this question for a class discussion.

3. **Extension.** Ask students to suggest other forms of communication.

74 SECTION 2: Communication Technology

Communication Innovations and Trends

An *innovation* is something new. It can be a new device, process, or idea. Cellular telephones combined with computer technology may be considered innovative devices. Using satellites to help navigate automobiles is an innovative process. Making communication tools smaller and portable is an innovative idea. Fig. 4-2. Many innovative ideas lead to the development of trends.

A **trend** is a general movement or inclination toward something. It is not usually a specific device, product, or idea. Trends are often represented by several innovative devices, processes, or ideas.

There are a number of trends in communication technology today. However, most of them can be categorized as being aimed at improving one of the three following areas:

- quality of communications
- convenience and portability of devices
- speed and efficiency of devices and systems

Several trends in communication technology started as innovations. For example, the introduction of the superior sound *quality* of audio compact discs (CDs) began a new trend. These compact discs are so resistant to scratches and warpage that they have almost replaced vinyl records. Another type is a special compact disc for computers, called a **CD-ROM** (*c*ompact *d*isc-*r*ead *o*nly *m*emory). CD-ROMs have nearly replaced diskettes as the primary means for storing commercially-produced software. Fig. 4-3. High definition television (HDTV) demonstrates a marked improvement in television picture quality. The increased use of optical fibers in place of copper cables improves the transmission quality of sounds, images, and data sent over telephone lines.

FIG.4-2 This computer is so small that it can fit into the palm of your hand. It weighs less than one pound, but it can do many of the tasks of larger computers, such as word processing, data analysis, and telecommunications.

4. **Brainstorming.** Have students predict some innovations based on perceived needs.
5. **Extension.** Ask students to identify some other innovations and trends.

CHAPTER 4: Communication Systems

Fascinating Facts

Did you notice that compact discs seem to shine with rainbows of colors? The colors are reflections from the more than 3 billion tiny pits in the CD's surface. Each pit measures 1/100 of the width of a human hair.

FIG. 4-3 Because CDs can hold so much data, there is a trend toward using them to hold large computer programs. Their other qualities make them perfect for recording sound.

6. Convenience and portability are demonstrated by innovative devices such as miniature, battery-operated televisions, cordless telephones, and lightweight computers. People now have greater access to information, whether they are in their cars, on airplanes, or at the beach.

Advanced computer systems continue to increase the speed and efficiency of sending, receiving, and storing information. This trend in communication technology can be seen by the increased use of the Internet. The Internet is the name given to the connection of computers from around the world through telephone lines. Information can be sent back and forth through computers. The Internet makes it easier and faster to access information. Fig. 4-4. In this chapter and throughout the next three chapters, you will examine many other communication innovations and trends.

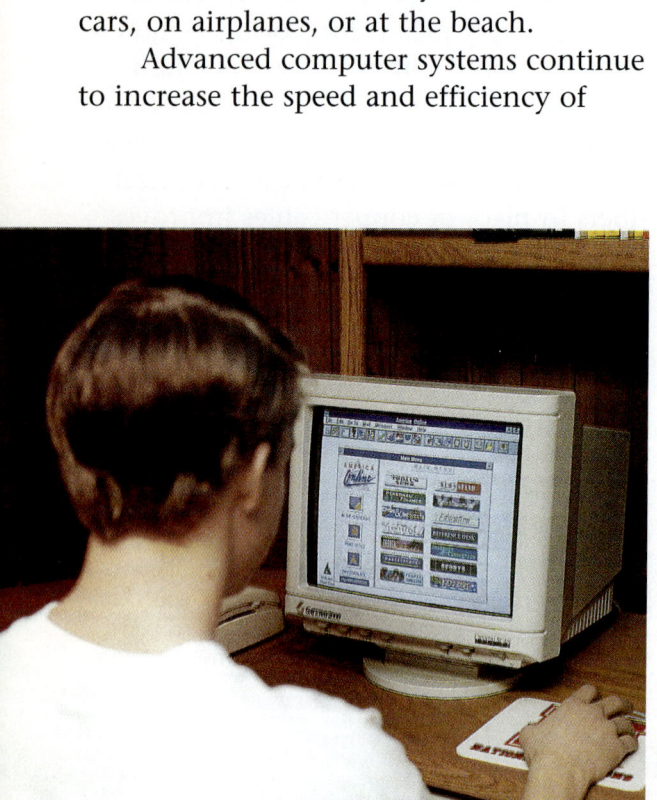

FIG. 4-4 With access to the Internet, people can use computers to quickly get information from around the world.

6. **Extension.** In the future, people may carry pocket phones and have a personal calling number (PCN), so they can be reached anywhere.

Why We Need Communication Technology

People send messages for many reasons. Communication technology extends our ability to send and receive these messages. We all know that it's impossible to talk to someone ten miles away without some special device. However, this task is quite easy with a telephone, television, or computer. Using these tools, we can send our voices, pictures, or other data over many, many miles. As you learned in Chapter 1, the reasons we communicate are to inform, educate, persuade, entertain, and control. Communication technology enables people to do these things faster and better:

- *Inform.* People read newspapers, watch television, and listen to the radio to stay informed about a wide variety of things, such as international politics, local sports, weather, and traffic. When salespeople are traveling, they often use cellular phones to quickly communicate with their offices and keep their customers informed.

- *Educate.* In addition to using textbooks, teachers use video and computer programs to help you learn about many subjects. Scientists rely on other communication devices to explore and learn about nature. For example, some marine scientists use a *hydrophone* to detect sound waves from whales and sea lions. This device helps them learn about sea creatures. Geologists use devices called *seismometers* that measure vibrations within the earth in order to study earthquakes and volcanic activity. As scientists become better educated about these natural events, they will know how to predict them more accurately and be able to warn people in the area. Fig. 4-5.

FIG. 4-5 Seismometers can record data about the earth that can be sent to scientists and others via radio or satellite.

7. **Reinforcement.** Have students list all the communication systems that are used today in schools. Encourage them to think of other systems that are not used but that might help them learn.

CHAPTER 4: Communication Systems

SCIENCE CONNECTION
Human Sound Reception

Did you know that you have your own built-in communication devices? You use them every moment you're awake (and sometimes even when you're asleep). They're your ears, and without them, communication would be much more difficult. Although they don't depend upon fiber optics or electricity, human ears have their own highly efficient mechanisms.

Sound is transferred through the air by means of sound waves. When a sound wave reaches your ear, it vibrates your ear drum and starts a chain reaction. The ear drum is a membrane stretched in front of three bones—the malleus, incus, and stapes. The vibration travels through the malleus and incus to the stapes, which then beats against the opening of the cochlea, a spiral-shaped part of the inner ear. Inside the cochlea are little hairs that move from the vibration. The cochlear hairs transfer the vibration to the cochlear nerve, which takes the sound's message to your brain. The brain is like your own private telephone operator in this communication process. When the message finally reaches your brain, the brain interprets it. Your brain may then send new messages to other parts of your body, telling them to respond.

Try It!

Look up hearing aids in an encyclopedia. Find out how they work and report your findings to the class.

- *Persuade*. Advertising is an example of using communication to persuade. Fig. 4-6. You can probably remember seeing a television commercial that made you think you wanted a certain product. Billboards along the highway may influence you to take a certain exit in order to refuel your car or satisfy your thirst.
- *Entertain*. Perhaps you play computer games for entertainment or you listen to the radio to hear music. If you're like most teens, you also watch television often to relax and be entertained. The telephone serves the dual communication purposes of entertaining and informing. It's fun to talk with friends on the phone. You can also use the phone to call your parents to tell them when you will be late getting home.
- *Control*. Communication technology plays an important role in controlling machines and tools. Traffic signals are a

8. **Reinforcement/Critical Thinking.** Ask students to describe ads they especially liked or disliked.

   Discuss what makes certain ads more persuasive than others.
9. **Critical Thinking.** Have students name some other communication devices or methods that can serve more than one of the purposes of communication.

SECTION 2: Communication Technology

FIG. 4-6 Advertisers use many different tactics to persuade people to do or buy something. What kind of strategy is being used in this ad?

Parts of the Communication System

All communication systems include a message, a sender, a communication channel, and a receiver. Fig. 4-7. A **communication channel** is the path over which a message must travel to get from the sender to the receiver. The channel might be a telephone line, the integrated circuit of a computer chip, or sound waves traveling through the air.

common example of using communication to control things. Computers and **sensors** (devices that sense things) send messages to traffic signals. These devices control when the lights change from red to green and back to red again. In turn, the traffic signals send a message to drivers, thus controlling the flow of traffic.

FIG. 4-7 Every communication system includes a message, sender, channel, and receiver. Sometimes the receiver may also be the sender, as in this picture.

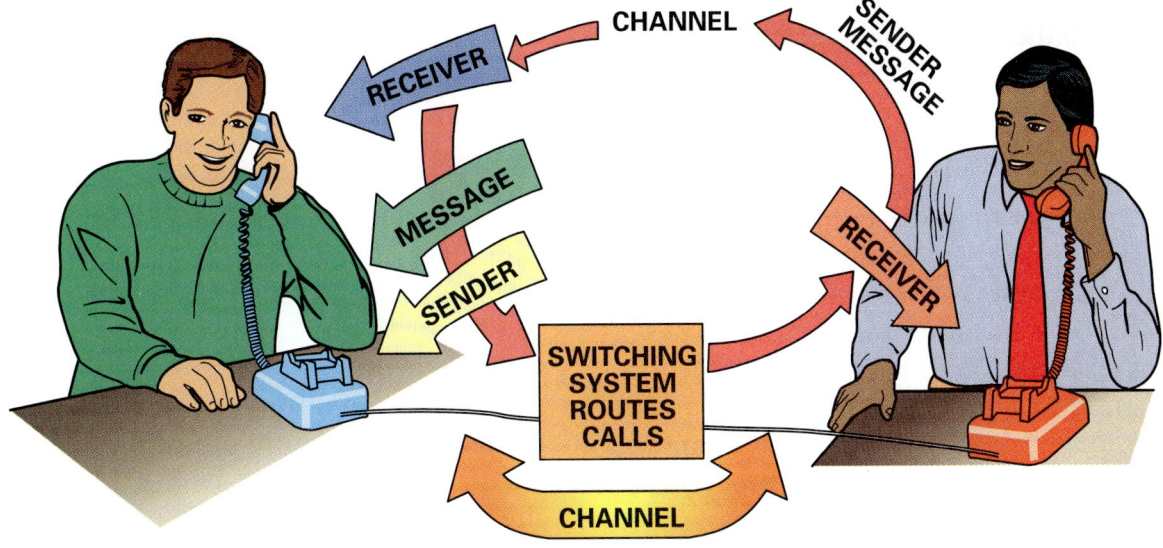

CHAPTER 4: Communication Systems 79

FIG. 4-8 This diagram illustrates the parts of the communication system.

In order to understand how the communication system works, it is useful to take a closer look at each of its basic parts. Like all systems, the communication system consists of four basic parts that work together to achieve a goal. In communication, the goal is to send or receive a message. The parts are the inputs, processes, outputs, and feedback. Fig. 4-8.

Inputs

As you learned in Chapter 3, the inputs are the seven resources that provide support or supplies for the system. In the communication system, these include all the things that are needed to start or to create a message and to provide a means for the communication processes to be carried out. The communication processes we use and the outputs we get depend on the resources available.

10. **Reinforcement.** Point out that systems are a part of every aspect of our lives. Ask students about the types of systems that are involved in sports, in school, and in other aspects of their lives.

SECTION 2: Communication Technology

The seven resources that are used in communication systems include: people, information, materials, tools and machines, energy, capital, and time. Let's look at how these resources are used in communication.

People

It is people who usually create the message that is to be sent. In most cases, it will be people who will receive and use the message. It is people who provide the knowledge, skills, and creativity needed to develop the technologies for sending and receiving messages. Many people operate communication devices as part of their jobs. Fig. 4-9.

People with a wide variety of skills and interests are needed to work in the communication industry. Many different types of careers are possible in recording, publishing, broadcasting, and manufacturing companies. Workers are needed in administrative, technical, specialty, support, and other occupations in the communications field. For example, people with strong technical skills are needed to design and develop new televisions and telephones. Trained technicians are needed to keep photocopiers, computers, and VCRs in good operating condition. Computer-aided drafting (CAD) operators and specialist engineers are hired to develop building plans, technical drawings, and *schematics* (drawings of electronic circuits). Technical expertise is also required to perform various production tasks in the communication industry. Printing press operators, color scanner operators, sound and video technicians, computer users, and cable installers are just some of the workers needed in these industries.

FIG. 4-9 Many types of communication systems are used every working day at the New York Stock Exchange. Here, people move information quickly to process orders and make trades.

11. **Enrichment.** Invite communications workers from several different areas of the field (technical, creative, management) to talk with your class about their jobs and the training they have.

CHAPTER 4: Communication Systems

Creative people are involved in all aspects of the communication field. Graphic designers prepare layouts for magazines, newspapers, and brochures. Copywriters and photographers collect and prepare material for television programs, commercials, and print advertisements. Computer graphic specialists develop multimedia presentations and Internet web sites (locations on the Internet). Fig. 4-10. (The term *media* refers to the means of communication; multimedia means more than one means were used.) Technology teachers and industrial trainers plan new and better ways of educating people about the fast-changing world of communications and electronics.

People are also needed to manage communication systems and sell tools and equipment. Network administrators manage computer systems installed in schools, corporations, and manufacturing plants. Store managers work in a wide variety of communication-related businesses such as printing shops, electronics distribution centers, and video outlets. Sales consultants are needed for all types of communication products including electronic pagers, computers, and printing presses.

Information

The people who design communication equipment and devices need information about a variety of technologies. People who operate these devices need information about how to use them to properly send or receive messages.

Designers must understand the advanced technologies that are involved in such things as microprocessors, digital components, photonics, electronic imaging, and mechanical systems. Fig. 4-11. In addition to good graphic skills, graphic designers need to know about page layout systems, digitization, laser scanning for color separations, and computer software and hardware. To install and repair communication devices, technicians rely on an understanding of electronic circuitry, mechanical controls, signal modulation, and signal transmission.

FIG. 4-10 This worker is creating a web site "home" page for her company. It will include advertising for products the company makes.

12. **Extension.** Have students name other jobs and careers in the communication field. You may want to have them research some of these careers in the *Dictionary of Occupational Titles*.

SECTION 2: Communication Technology

FIG. 4-11 In this research lab, new products are designed.

However, technical knowledge is not the only source of information. People often forget about the extensive pool of general information that they draw upon each day to do their jobs. People in the communication industry need knowledge and skills in reading, writing, mathematics, science, and human relations. This information will help them to perform their jobs well. They need to make decisions, solve problems, access more information and relate to coworkers.

Communication technology is constantly changing. New information is created each day and people must keep themselves updated in order to be wise consumers, responsible citizens, and efficient workers. Consider computers, for example. New developments change how computers work and what you can do with them. It is likely that you will buy one or more computers in your lifetime. As a consumer, you need to know how to make a good decision about which computer is best for your needs and budget. As a citizen, you need to consider how government regulations may affect your computer privacy and security. In the workplace, you may be expected to learn new computer systems and applications in order to do your job effectively. You need a constant supply of accurate information to do these things. Fig. 4-12.

FIG. 4-12 Salespeople in this store must stay up to date regarding all the new computer equipment and software that is bought by families each year.

13. **Reinforcement.** Have students suggest as many other ways as they can that indicate they are dependent on the communication of information.

CHAPTER 4: Communication Systems

Materials

Throughout history, communication devices have required abundant natural resources. Without these resources, communication systems would not be possible. Many natural resources such as trees, chemicals, oil, ore, and silicon are used to manufacture paper, inks, plastics, metal, and electronic parts. Enormous amounts of these manufactured products provide the means to send and receive messages.

Millions of gallons of water are used to make paper. We rely heavily on paper products for printing newspapers, pamphlets, books, and magazines. Fig. 4-13. Consider the millions of pages that run through presses and photocopiers every week! The ink used for printing consists mainly of linseed oil and chemically manufactured pigments.

Barrels of petroleum products are used to make plastics that are manufactured into camera films, cable insulation, photocopier toner, and circuit boards. Plastic materials are used in nearly all areas of communication, including video tapes, diskettes, photographic film, and compact discs. Think about all the plastic that goes into making telephones, computers, and radios, too.

Miles of metal wires and glass fiber optic cables are laid each year to connect communication devices together. Large metal towers are often constructed for transmitting radio signals. Fig. 4-14. Millions of electronic circuits are made from silicon, hydrogen, aluminum, and other raw materials. These manufactured circuits help us to control communication machines and systems.

FIG. 4-13 How many newspapers are printed on a daily basis in your area?

14. **Extension.** Have the students list the natural resources used to produce and receive a television program.

15. **Extension.** Many new materials have resulted from NASA research. An annual report, *Spinoff*, can be purchased from the U.S. Printing Office.

SECTION 2: Communication Technology

Recycling a wide variety of materials used in communication systems is important for our environment. Many of these materials are recycled in order to save valuable natural resources that could be depleted from overuse. For example, some companies can salvage copper from wires and cables. Recycling paper and plastics is a common practice in many communities. Fig. 4-15.

FIG. 4-14 Metal is used to construct antennas, such as this tower, which sends and receives radio waves.

FIG. 4-15 Advances in technology cause concern for and protection of our natural resources. Recycling services, such as this one, are a result.

FIG. 4-16 Drawing tools range from triangles, dividers, and compasses (above) to a plotter (right). The plotter prints a CAD drawing that has been drawn using a computer.

Tools and Machines

Consider the many kinds of tools and machines that are needed to communicate today. It takes both simple and complex tools and machines to send and receive messages. For example, simple drawing tools are often used to make very detailed, technical drawings that communicate ideas about how things should be made or built. These types of drawings can also be made using the more complex tools of a computer and a device called a plotter. Fig. 4-16. People rely on many electronic devices for home and business uses. Pagers and cellular phones are becoming more popular as personal communication tools. Innovative tools and machines are always being introduced into the marketplace. As the technologies are improved, more people want the convenience of these devices.

Other types of tools and machines include those used in printing such things as newspapers and magazines. Photocopiers, printing presses, scanners, and many types of cameras are used in graphic communications. Special tools and machines are needed to change voice signals into electric pulses capable of being sent over telephone lines. Other devices convert signals into waves that can be sent around the world. Radios, televisions, compact disc players, and VCRs (videocassette recorders) are other examples of tools and machines used to carry out electronic communication processes.

Energy

Energy is fundamental to any technological system, and communication is no exception. People involved in communica-

16. **Extension.** Ask students, "Does communication improve as we use more complex tools and machines?"

SECTION 2: Communication Technology

FIG. 4-17 High-voltage power lines (left) carry energy in the form of electricity. Dish antennas (above) are used to receive signals in the form of electromagnetic waves from orbiting satellites.

tion use their mental and physical energy to provide their individual input into the system. No form of electronic communication would be able to begin without some kind of energy.

Some forms of energy—such as mechanical energy and light energy—must be changed into electrical energy in order to power most communication devices. Communication satellites rely on solar energy to broadcast signals to every location on earth. Radiant energy, in the form of electromagnetic waves, is produced by radio waves, television waves, and microwaves. With electromagnetic waves, messages can be sent and received over long distances. Fig. 4-17.

Capital

As you know, capital includes the money, land, and equipment needed to set up and operate a technological system. Money is needed to buy equipment, pay workers, and also pay for the energy needed to operate communication systems.

FIG. 4-18 Most schools now have computers in the classroom.

Capital investments in communication technology are necessary in order to keep voice, video, and data systems operating properly. Such investments also help business and industry stay competitive in a global economy. Many companies invest money in communication products to meet customers' expectations. Today's customers and companies regularly communicate using facsimile machines, paging systems, voice mail, phones for teleconferencing, and Internet services. Since the mid-1990s, many companies and organizations have established Internet sites in order to maintain communications with customers.

Capital investments in communication technology have also been necessary for elementary schools, middle schools, high schools, and universities. More and more computers and Internet services are being used in education. Fig. 4-18. Costly instructional technology products have been purchased in order to improve teaching and learning environments. Many buildings needed to be rewired to accommodate the increasing number of computers and Internet connections. Teachers and other staff require training. Bringing more communication technologies into educational programs, businesses, and industries requires large amounts of capital.

Time

Time is needed to design and develop new communication technologies. Often, the time spent on development has resulted in technology that saves time, a very valuable resource. For example, computers can make millions of math calculations in seconds. Think how long it would take humans to make *millions* of calculations!

FIG. 4-19 Some customers who call a company may prefer talking to a receptionist such as this one. They find it friendlier and sometimes more helpful.

Time-saving communication technologies can improve efficiency and productivity. However, saving time may not be a priority for everyone. For example, some companies like using automatic telephone answering systems because it saves workers' time. Customers, however, may prefer talking to a human being rather than a machine. Fig. 4-19.

New products can also give people more options on how they use their time. For instance, with a portable, lightweight computer a manager can complete a report while traveling on a plane. A technician can call an office using a cellular phone without having to leave a job site. Fig. 4-20.

FIG. 4-20 Convenient and portable communication devices can help improve efficiency and save time.

CHAPTER 4: Communication Systems

Processes

Processes are all the things done to or with the inputs in a communication system in order to get the desired result, or output. Processes are all the activities involved in creating the messages, sending them, and receiving them.

This part of the communication system may include many types of activities. Using a computer to access information on the Internet, designing advertisements, and sketching ideas for new products are all communication processes. All the tasks involved in getting newspapers and magazines printed, from the photography to the writing to the printing, are communication processes. Turning sounds or pictures into electrical signals that can be sent through the air and space is a communication process.

Making a movie is an example of a complex communication process. When you go to the theater to see a film, do you think you see every move that the actors are making? The truth is, you don't! The process of making a movie is actually the creation of an illusion. When you watch a movie, you are actually seeing about 30 *still* pictures flashing in front of your eye every second. Each frame records a different position of the object or person in motion. When these frames are run through a projector at the same 30 frames per second, your eye and your brain interpret the series of pictures as one continuous picture. Fig. 4-21.

Communication processes—and the tools and machines used with them—constantly change as new technologies are developed. Consider printing, for example. Johann Gutenberg developed movable type in the 1400s. The first book he printed was the *Bible*. It took over two years to print 200 copies. Today, using new print-

FIG. 4-21 Cartoons, too, are a series of still photographs shown in rapid progression. The computer operator creates the beginning and ending positions of a character's motion, and the computer creates the different images in between. Note the movement of Bugs Bunny's left arm.

17. **Reteaching.** Use drawings on a small stack of index cards to illustrate what a movie actually is.

18. **Brainstorming.** Have students think of other great advances in communication resulting from the development of new technologies.

FIG. 4-22 The pilot and co-pilot rely on output received from an airplane's instrument panel. The many types of dials, indicators, and lights are communication devices that tell them such things as fuel supply, altitude, speed, engine problems, and whether there is an obstacle ahead.

ing processes, the same number of copies can be printed in less than a day. As another example, back in the 1800s, it often took several weeks to get a letter from one state to another. Today, letters and printed documents can be sent instantly from China to the United States, for example, using electronic mail and fax machines.

Several trends in communication processes continue to be developed and improved. These developments reflect the wants and needs of people. Every day millions of people access information through their computers linked to the Internet. New computer software programs and computer equipment continue to make information access faster and more efficient. Digital television advancements are expected to vastly improve the clarity and quality of video broadcasts in the future. Two-way pagers and cellular telephones that depend on radio waves represent still another trend in communication processes. They are fast becoming standard communication devices in many businesses and households.

Outputs

The message is created, coded if necessary, and transmitted (sent). Once the message is received, the desired output of the communication system has been achieved—information has been communicated. The received message is the output.

Outputs of a communication system come in many forms. The form depends on the inputs and processes used. The outputs can be images, words, sounds, or other forms of information. Fig. 4-22.

19. **Critical Thinking.** Suggest an input and ask students to name a process and an output. For example, input-paper, process-printing, output-books.

CHAPTER 4: Communication Systems **91**

Consider a telephone communication system. One person may initiate the phone call. That person is providing input to the system. The message is transmitted using communication processes. When you receive the message, you are receiving output of the system. In this case, the output is in the form of sounds. Your brain interprets these sounds. They may be voices, noises, static, or music. Everything you receive—or hear, in this case—is called output. Output may not always be what you want. For example, you may hear static during your phone call. Static is an output, but not one that you like.

Positive Impacts

The output of a system includes not only the desired result of the system, but also any other effects, or impacts, that the system has on people and the environment, or even the system itself. Communication systems have both positive and negative impacts and desired and undesired results.

Long Distance Communication

Today, television and satellite technology enable us to see and hear what is happening on the other side of the world, even as it happens. We take it for granted today that we can quickly place a phone call and talk to a friend or relative who is almost anyplace else in the world. Think how the telephone has changed and continues to change our world.

Communication in Schools

Computers enable administrators and teachers to manage incredible amounts of information, such as enrollment records, scheduling, grade reporting, and lesson planning. It's difficult to imagine how all this information was handled *before* computers were invented!

Computers that can use CD-ROMs enable students to search through large data bases for information related to specific topics of interest. This is especially useful when it comes time for writing reports. Fig. 4-23.

Computers connected with modems can send computer data over telephone lines to other computers. This method enables students to access a large amount of information outside of their school without actually leaving the building.

Special telephone systems used in some schools help to improve the communication between home and school. Computerized phone calls are made to homes of students who fail to show up at school. Parents can call a special school number during the evening to hear a tape-recorded message about school news.

Negative Impacts

When the output of a system creates unwanted results, it has negative impacts. Such impacts from a communication system can range from social behavior to concerns about safety.

Results of Viewing TV

When television was first invented, people never expected that it would become so popular. They also did not expect the problems that have come along with increased television use. Many studies have been conducted in recent years to determine the effects of television on children. Violent television shows have been highly criticized as encouraging violent or antisocial behavior.

---

20. **Reteaching.** Point out components of a circuit board (transistors, diodes, capacitors, relays). Explain how these components work together as a system.
21. **Extension.** Explore with the students how the telephone has changed and continues to change our world.

FIG. 4-23 New information systems and computers are gradually replacing stacks of files, records, and books. All volumes of this encyclopedia have been recorded on the single CD-ROM.

22 Many people also feel that we depend too much on television. They think it affects how much people read and could therefore affect their reading ability. However, is it television as a communication medium that is being questioned, or is it the way television is used?

Concerns about Privacy

23 There is a great deal of personal information regarding such things as finances and medical conditions stored in computers. This information can often be easily accessed by many businesses and agencies. This raises the question of the *right to privacy*.

The increasing use of telephones for sales and surveys has also raised a question. Using the telephone to sell goods and services is called **telemarketing**. Many people prefer to receive only personal calls and do not appreciate people calling them at home and asking them questions or trying to sell them something. In many cases, people have to pay extra to keep their phone numbers from being published in the phone book.

Effects on Health and Safety

Some people fear that the increased use of headphones could affect hearing. Others worry that prolonged viewing of computer screens could affect eyes.

22. **Extension.** Have students name some other criticisms that have been made about the effects of watching TV.
23. **Extension.** Explain in greater detail about the types of personal information that are easily accessible by almost anyone.

CHAPTER 4: Communication Systems

Growing numbers of computer users are reporting physical problems associated with computer use. Common complaints include blurred vision, eyestrain, and headaches.

All electric devices (computers, televisions, electric blankets) emit some amount of electromagnetic radiation. **Electromagnetic radiation** is an invisible source of energy given off by the movement of electrons. You cannot see electromagnetic radiation but it exists in many places around you. For example, radio and television signals are constantly bouncing around the atmosphere in the form of electromagnetic waves that give off a small amount of radiation. This amount of electromagnetic radiation is considered safe by most people. However, it is not without problems.

Some researchers feel that people are being exposed to greater amounts of electromagnetic radiation all the time. More people are using electronic devices, especially cellular phones, computers, and pagers. There have not been enough studies to find out how much electromagnetic radiation is "safe."

Electromagnetic radiation can also wreak havoc on other electronic devices. It can cause unwanted **electromagnetic interference (EMI)** that can disrupt how these devices work. For example, some medical researchers are concerned about the effects of EMI on patient care. There have been documented cases where signals

FIG. 4-24 This patient is connected to many machines that monitor body functions. Proper shielding and product testing minimize the chance that the devices will malfunction.

from nearby radio transmitters, police radios, or cellular phones have caused patient monitors to not work properly. Pacemakers (electrical devices for maintaining the heartbeat) and other medical appliances have been known to malfunction. Medical devices such as these are quite sensitive, because they must measure very low-level signals, such as heartbeat, from the patient, making the machines more susceptible to interference. In addition, the wires connecting the patient to these devices and the human body itself act as antennas.

Researchers and product developers can address these problems by shielding the devices. Shielding involves adding coverings that block unwanted signals. Fig. 4-24. There are also researchers studying **electromagnetic compatibility (EMC)**. This field of study focuses on understanding how to protect devices from EMI and how to develop components that are not affected by EMI.

Pollution

Communication systems also affect our natural environment. Consider the impact on the aesthetic beauty of your surroundings when new antennas and radio towers are erected.

Every year, thousands of acres of trees are cut down to make the paper for newspapers, magazines, and computer printouts. Paper and pulp manufacturing requires abundant water resources for processing. As a result, lakes and rivers are affected by mills built near them. Some communities must also cope with excessive odors that these facilities emit.

Feedback

What happens when you receive output in a communication system? Do you do or say anything? Do you respond? Most likely, you do. Your response would be the feedback. Feedback is the part of the system that checks the output. For example, feedback in a communication system may simply let the sender know that the message was received. On the telephone, your immediate feedback might be a statement such as, "Yes, I'll take care of that," or you might say, "What? I cannot hear you very well. There's too much static." Information about positive and negative impacts of the output is considered as part of the feedback. If the output is not what is desired, or there are negative impacts, then the inputs or the processes of the system will need to be changed.

Some forms of feedback occur right away. Other feedback results in actions taken later. Some feedback causes a delay. For example, static during phone calls is abnormal; you may decide to get your phone repaired. Repairs may take a while. You may choose to buy a new phone. You might even decide to cancel or change your phone service. Whether immediate or delayed, each of these responses is a form of feedback.

---

24. **Extension.** Point out other social and environmental impacts of communication.

25. **Reteaching.** Explain to the students why feedback is a necessary part of a communication system.

CHAPTER 4: Communication Systems

Career File—Computer Systems Analyst

EDUCATION AND TRAINING

Most employers prefer applicants who have at least a bachelors degree in computer science, information systems, or a related field. Continuous study after graduation is necessary to keep skills up to date.

AVERAGE STARTING SALARY

Earnings of systems analysts vary widely, depending on the background of the applicant and the size of the company. Starting salaries can range from $19,000 in the federal government to over $50,000 in a large company.

OUTLOOK

Systems analyst jobs are expected to grow faster than average through the year 2005, according to the Bureau of Labor Statistics.

What does a computer systems analyst do?

Computer systems analysts solve problems by designing solutions that use computers. The solution may be an entirely new computer system that replaces operations still done manually, or it could be an enhancement to, or replacement for, an old computer system.

Does most of the work involve data, people, or things?

Systems analysts work with all three. They are the link between the people who use the system and the technical staff who program and implement the system. They design how data can be manipulated to give the users what they want, and they use computer equipment extensively.

What is a typical day like?

The activities of any one day vary with the phase of a project's development. Systems analysts begin a project by gathering information from managers and users to determine the exact problem. They design a workable solution, prepare system specifications for the technical staff, and assist in testing and correcting problems. The analyst releases the system when it works properly and the user is provided with written instructions on its use.

What are the working conditions?

Systems analysts normally work in offices with comfortable surroundings. Evening or weekend work may be necessary to meet deadlines or solve problems.

Which skills and aptitudes are needed for this career?

Systems analysts must be able to concentrate, think logically, and pay close attention to detail. People with strong computer skills, as well as good interpersonal and business skills, make good systems analysts.

What careers are related to this one?

Other workers who use logic, research, and creativity to solve business problems include computer programmers, urban planners, engineers, operations research analysts, and financial analysts.

One computer systems analyst talks about his job:

I enjoy figuring out the requirements for a new process and then applying technologies to meet those requirements. It's exciting to see a new system develop.

Chapter 4 REVIEW

Reviewing Main Ideas

- Communication technology is all the things people make and do to send and receive messages.
- Communication systems inform, educate, persuade, entertain, and control.
- Positive impacts of output include worldwide communication and ability to handle large quantities of information.
- Negative impacts of output include misuse of TV time, threats to the right of privacy, human exposure to electromagnetic radiation, and environmental pollution.

Understanding Concepts

1. What is communication technology?
2. What is the difference between an innovation and a trend? Give an example of each.
3. Name the five reasons why we communicate. Describe a situation in your daily life that demonstrates each one.
4. Tell into which part of a system the following belong:
   a. "I can't hear you."
   b. a delivered message
   c. people, materials, energy
   d. printing a newspaper
5. Select one communication system. Describe two positive and two negative impacts of that system.

Thinking Critically

1. Do children and young adults watch too much television? Explain your position.

26. **Resource.** Answers to "Understanding Concepts" and "Thinking Critically" can be found in the Teacher's Manual at the front of this *Teacher's Annotated Edition* and in the *Teacher's Resource Binder*.

2. List the impacts of computers in the classroom.
3. How would business be conducted today without telephones? How many people do you think are employed by telephone companies?
4. In what ways do you think telephone systems between home and school have a positive impact or negative impact?
5. How have changes in communications in the past 20 years affected jobs?

Applying Concepts and Solving Problems

1. **Language Arts.** Keep a daily log of your television viewing for a week. In a brief report, summarize your viewing habits. Be sure to include the number of hours spent and types of programs watched. Were all of the programs for the purpose of entertainment, or were any other purposes of communication served?

2. **Internet.** Many web sites about foreign languages exist. Access the following site and report to the class on what you find:
   http://www.kli.org/KLIhome.html

3. **Mathematics.** Obtain a copy of your family's telephone bill.
   a. Check its accuracy by totaling all usage charges, fees, and taxes. What percentage of the bill is in the form of taxes? How much money was spent on directory assistance?
   b. Analyze your family's calling habits. What changes could your family make to save money each month?

CHAPTER 5

Computers

Objectives

After studying this chapter, you should be able to:
- describe how computers are used to communicate.
- identify the basic components of computer systems.
- tell what integrated circuits are and identify their functions.
- discuss the Internet, artificial intelligence, digital libraries, and virtual reality.

Terms

artificial intelligence
central processing unit (CPU)
computer
digital
e-mail (electronic mail)
fax (facsimile system)
integrated circuit (IC)
Internet
network
programming languages
virtual reality

Technology Focus

The Information Superhighway

The information superhighway has been a hot topic in the news. What is it, and is it really that important?

The term most commonly associated with the information superhighway is the Internet, but where did that term come from? A group of computers and printers may be hooked together somewhere in your school. That group of computers is called a **network**. That network can be connected to another network somewhere else in your school and form a larger network. Networks in one building can be connected to networks in other buildings and form even larger networks, and all the users have access to information on any computer on the larger network. The **Internet** is a huge interconnected, worldwide network of smaller networks. Its millions of connected users can get information from other computers all over the world. A portion of the Internet, called the World Wide Web (WWW), uses graphics to help users access information with only their computer's mouse as input.

Getting On-Line

Several commercial companies offer on-line (network) services to home and business computer users. Computer users can buy a membership to any of these services. The largest of these companies are America Online®, Microsoft Network®, CSi®(Compuserve), and Prodigy®. Each company offers similar services and enables its customers to hook into at least some parts of the Internet. Members pay a certain amount per hour or per month. Special software and a modem connect the member's computer to the service's powerful main computers. Subscribers download (bring from the large computers to their home computer) the information that they are looking for.

As with any new technology, there are problems. "Traffic jams" occur on the information superhighway. So many millions of people are now using the Internet that it can sometimes take a long time to get to the computer you are looking for and a long time to download information.

Take Action!

Familiarize yourself with the various search engines on the Internet. A search engine is a directory of available information. Try accessing some of these to find out what they offer.

Resources
- Chapter 5 Lesson Plan in the *Teacher's Annotated Edition* and in the *Teacher's Resource Binder.*
- Chapter 5 Review in the Teacher's Resource Binder.
  1. **Reinforcement.** Further discussion of the Internet can be found on pages 112-114.
  2. **Reinforcement.** Point out to the students the differences and similarities between a network and a system.
  3. **Extension.** Have students research and compare the features and costs of various Internet service providers.

CHAPTER 5: Computers

Computers in Communication

Computers are used in nearly every communication system today. They are the basis of the Internet and they make it possible for people to store and access vast amounts of information. Computer systems are found on desktops, but they are also hidden inside telephones, microwave ovens, automobiles, automated teller machines (ATMs), home burglar alarms, blood pressure monitors, and many other things that we rely upon. Computers are a part of every technological system, but they are essentially a communication device. As you learned in Chapter 4, communication technology sends and receives messages. A computer extends our ability to send and receive data in the form of images, sounds, words, numbers, and other information. It is a communication tool used in manufacturing, construction, transportation, and bio-related technology. Fig. 5-1.

Very simply, a **computer** is an electronic device that can store, retrieve, and process data. It can handle a great deal of information quickly. Computers are not smarter than people. They are simply much faster than people. A computer can do millions or billions of calculations in seconds.

FIG. 5-1 Computers aboard the Space Shuttle monitor and control its operations, enabling astronauts to work in space and maintain contact with people on earth.

Fascinating Facts

Computer giant Apple Computer began with two Steves in a garage. Steve Jobs and Steve Wozniak were the first people to focus on personal computers. Jobs went on to help design the Macintosh.

Parts of a Computer

You know what a computer does, but do you know the parts of a computer and what makes it work? A computer has four basic parts: input units, central processing unit, memory or storage units, and output units. Fig. 5-2.

4. **Extension.** Time students as they do a series of math calculations. Compare their times to that of a computer.

5. **Reinforcement.** Identify the four basic units of a computer in the classroom and show how they fit together.

SECTION 2: Communication Technology

FIG. 5-2 A computer system contains a central processing unit. Various input and output devices are connected to it.

Input Units

An input unit for a computer is any device that can feed information (data, images, sounds) into the computer. Examples of input units include keyboard, mouse, scanner, joystick, touch screen, sensor, and light pen. Fig. 5-3. A computer may have more than one input unit, but each input unit must encode (convert) the information into something that the computer can understand. Even though there are many types of input units, the keyboard and mouse are the most common ones. A mouse is an electronic device that fits in the palm of your hand. When you move it around on a flat surface, it allows you to move the cursor (flashing symbol on a computer screen that marks the current location) or other images on the computer screen.

Central Processing Unit

The **central processing unit (CPU)** is the heart of the computer. It is also known as the *microprocessor* or the "computer within the computer." This is where the computer fetches the instructions, analyzes the input data, executes the necessary operations, and sends the information to storage. It does all this work in a very orderly fashion, and it does it very quickly.

FIG. 5-3 This woman is using an input unit, called a digitizing tablet, along with a light pen to alter the design on the computer.

6. **Reinforcement.** Show the students some input units and list on the chalkboard their similarities and differences.

CHAPTER 5: Computers

It can make millions of mathematical calculations (such as multiplication, division, addition) every second! You'll learn more about the electronic components inside of the CPU later in this chapter.

Memory or Storage Units

The memory or storage units are an important part of most computers. They help the computer remember instructions and information. Every computer has both internal and external memory.

Some internal memory can be found within the CPU. The Read-Only Memory (ROM) in the CPU helps the computer remember which steps in the instructions come first. The Random Access Memory (RAM) in the CPU remembers what you tell the computer to do while the computer is on. An internal hard disk drive is also a common part of a typical computer system. Using the scientific principles of *electromagnetism* (magnetism caused by electricity), the hard disk drive can store data in a similar way to videotapes and audio tapes. It remembers instructions and information you store in it even after you turn the computer off.

You are probably most familiar with the external memory units of computer systems. The floppy disk drive allows you to store data or programs (special instructions for the computer) on floppy disks. Floppy disks are pliable pieces of plastic with a very thin coating of magnetic particles. They work on the principles of electromagnetism, too, and are most commonly found in a 3.5" size. Floppy disks have a much smaller storage capacity than the hard disk, and they work much more slowly. Electrical and magnetic fluctuations can erase the information stored on the floppy disks.

FIG. 5-4 A laser is used to burn pits into the surface of a **CD-ROM**. The pattern of these pits carries the coded data.

Another popular external storage unit is the compact disc, which uses CD-ROM (Compact Disc-Read Only Memory) technology. CD-ROMs can store large quantities of text, pictures, and sounds, which the computer can access. To store data in this manner, a *laser* (a narrow, high-energy beam of light) is used to burn tiny pits into the surface of the disc. The pits form a binary code. Fig. 5-4. (You'll learn more about binary codes in the next few pages.) When the disc is played, another laser reads the reflected light from these pits and sends that information to the computer.

Because light is used to write and read the data, CD-ROMs are called *optical storage media*. CD-ROM discs are durable and permanent. You cannot erase them, and you cannot add data to them as you can to a floppy disk. That's why many computer software programs are sold on CD-ROM.

7. **Reteaching.** Demonstrate for the students the principles of electromagnetism.

SECTION 2: Communication Technology

Output Units

Output units *decode* (convert into a language that humans can understand) information. There are many types of computer output units, but the video monitor and printer are the most common ones. The monitor displays the information on a screen. The printer puts the information onto paper. There are many types of monitors and printers.

Computers can be linked to other devices, too. For example, a computer that controls woodworking machines or medical devices has output units such as switches, fans, and pumps. A modem can be used as both an input unit and an output unit because it can be used for either task. As an input unit, it can feed information into your computer. As an output unit, it can send information out of your computer to other locations.

The Language of Computers

Communication can be very difficult if the people who are trying to exchange ideas don't speak the same language. Communicating with a computer can also be difficult. You have probably used a program to play a game or write a letter with a computer. The computer did not understand your keystrokes or mouse clicks. Your inputs had to be translated several times before the computer could understand what you wanted to do. Programs are written in programming languages. In turn, programming languages translate your inputs into a special machine language, which is finally changed into electrical impulses that the computer can understand.

Programming Language

Originally programmers tried to write programs in machine language. Later we will see that machine language is very difficult to work with. To make writing programs a little easier, they invented **programming languages**. The computers can translate these languages and then obey the commands. Some of the more popular programming languages include BASIC, FORTRAN, COBOL, Pascal, ADA, and C. Each of these languages is used to write different types of programs.

Fascinating Facts

The first computer programmer was a woman who lived during the 19th century. She invented a program for Charles Babbage's "analytical engine." Unfortunately, Babbage's machine never worked. In 1982, the U.S. Department of Defense named a computer language ADA, after the woman, Augusta Ada Lovelace.

Machine Language

Computers have their own language called *machine language*. It is made up of only two symbols because the computer knows only two signals. It knows on (1) and off (0). Every computer instruction, or bit, of data is written as a series of 1s and 0s. This sequence of 1s and 0s is the machine language known as a *binary code*. (Binary means having two parts.) Once

---

8. **Reinforcement.** Show the students some output units and list on the chalkboard their similarities and differences.
9. **Extension.** If possible, show the students examples of several different programming languages.
10. **Reinforcement.** Explain coding. Point out some types of coding with which students are familiar.
11. **Extension.** Play Twenty Questions to encourage students to think in terms of on and off, yes and no.

CHAPTER 5: Computers 103

FIG. 5-5 Every time you press a key on a computer keyboard, the symbol is automatically encoded into a form that the computer can understand, a series of on (1) and off (0) pulses.

something has been converted into binary code, it can be sorted, retrieved, sent, or altered. Pictures, sounds, numerals, words, and letters become series of 1s and 0s. Fig. 5-5.

Electrical Signals

All information must be changed electronically into this binary code so the computer can "understand" it. The code enters the computer as electric pulses. These turn tiny electronic switches on (1) and off (0), creating different combinations of paths for the electricity to follow. Electricity flows through these paths as the computer processes the information. Each on or off pulse of the code is called a *bit*. The term comes from *bi*nary dig*it*.

12. **Reteaching.** Using a panel with eight LEDs and switches, demonstrate how computers count in binary code.

Most machines combine eight bits into a *byte*. Then information can be handled in larger units.

The ability to convert analog data into digital information that can be used by the computer is the key to the electronic communication systems. Computers function using digital electrical signals that are really a series of on (1) and off (0) pulses of electricity. A digital signal is different from an analog signal which you may have heard about. Let's clarify the difference.

Most things in nature are *analog*. That means they have an infinite number of levels or variations; they are continuous. Voice levels, speed, time, pressure, and light intensity are good examples. **Digital** signals are discrete, which means they have a finite number of levels. Computers cannot understand analog signals because there are too many variations. Remember that computers know two signals: on (high or 1) and off (low or 0). An easy way to think about this is to consider two types of wristwatches. Fig. 5-6. An analog watch

FIG. 5-6 An analog watch (left) requires that you interpret the position of the hands in order to tell the time. On a digital watch (right), a specific time is clearly told by the numbers.

13. **Enrichment.** Computer capacity is often measured in bytes and megabytes (MB). A megabyte is one million bytes.

FIG. 5-7 An analog signal looks different from a digital signal. The analog signal is continuous, with many gradual changes. The digital signal is discrete. It is either completely on or completely off.

has hands that move around the dial. You determine the time by analyzing the location of the hands on the watch. When you look at a digital watch, you know the exact time because the numbers on the watch clearly display the exact time at that moment such as 11:32:09. If you looked at an analog signal and a digital signal side by side, you would see that the analog signal has many gradual changes whereas the digital signal is either completely on (high) or completely off (low). Fig. 5-7.

An analog system simply transfers the information from the input to the output. For example, a musician plays the guitar in front of a microphone. The microphone amplifies (increases) the sound level, but it doesn't electrically change the signal. The signal is still analog. You hear the guitar music as it is played, but it is louder. Fig. 5-8. Oftentimes you may also hear noise or distortion (such as static), too.

FIG. 5-8 The analog signal below has been amplified. Its wave pattern is the same, but the waves are larger and stronger.

CHAPTER 5: Computers **105**

SCIENCE CONNECTION
Semiconductors

A semiconductor is a solid material whose ability to conduct electricity lies between that of a conductor (such as a metal) and that of an insulator (such as rubber or glass). Silicon is the most widely used semiconductor material. Silicon is the brittle, crystalline element found in common sand.

Semiconductors are useful in electronics because, when impurities are added to them, they can be made to carry a limited amount of electric current. Depending upon the type of impurity used, the semiconductor will carry a negative or positive charge.

When semiconductors having different charges are sandwiched together in certain ways, they can act as switches that either stop the passage of electric current or allow it to flow. Layered semiconductors can be used as amplifiers that boost weak electric signals. They can also be used as rectifiers that convert alternating current from an outlet into the direct current needed by electronic devices.

Thousands of tiny semiconductor sandwiches—called microchips—can be built in fine layers of silicon smaller than a thumbprint. The main storage unit of a computer is made of these microchips, as are the workings of many other electronic devices.

SEMICONDUCTOR SANDWICH — POSITIVE CHARGES / NEGATIVE CHARGES

Try It!
Use an encyclopedia to find out more about silicon. What other products is it used for? What happens to it when it is heated?

---

14 If the guitar sounds were changed to digital signals, then you could take those signals and do other things with them. For example, you could combine the digital music signals with digital voice signals in your computer. You could filter out the distortion to make the music sound better. You could send the guitar signals through your computer modem to another computer in a South American recording studio. There the signals could be manipulated or enhanced by integrating them with sounds from the rain forest or the work of local Brazilian musicians.

14. Reinforcement. If possible, play for the students something that has been digitally recorded and the same piece that was not recorded in this way. Compare the two recordings.

SECTION 2: Communication Technology

The Amazing Integrated Circuit

You learned earlier that the CPU, or microprocessor, is central to the computer. It is composed of a complex device called an integrated circuit. Fig. 5-9. An **integrated circuit (IC)**, often called a microchip or chip, is a tiny piece of silicon that contains thousands of interconnected electrical circuits that work together. (A circuit is a path over which electric current or pulses flow.)

Transistors are the main electronic parts found in ICs. They are like miniature electronic switches. They turn on and off in order to process information using the binary code discussed earlier. However, transistors do not turn on and off like a light switch. Transistors have no moving parts. They allow current (the flow of electricity) to flow through them if a small amount of electricity is applied to one part of the transistor called the gate.

If electricity is applied to the gate, current is allowed to flow and the switch is closed (on). If the gate does not receive an electrical charge, current cannot flow and the switch is open (off). A typical CPU in a computer contains millions of these transistors. The amazing thing is that the CPU is only about the size of a dime! You would have to look through a powerful microscope to see the transistors. The CPU is one of many ICs found in a computer. However, it is the main one because it is the largest IC and it is the heart of the computer.

Integrated circuits are the basis of almost every modern communication tool you use today. Fig. 5-10. You can thank the IC for making it possible to carry a radio, telephone, or compact disc player in your pocket. Each of these communication tools contains several ICs, and each one performs a certain job. Some may process information. Others may be assigned for memory only. For example, you might have automatic redial on your telephone. An IC inside the phone "remembers" the last number dialed. You can simply press one button to redial the number.

FIG. 5-9 The picture on the left shows a tiny integrated circuit (IC) that can contain millions of microscopic electronic components. The packaged IC (on right) is in a plastic case to protect it. The "legs" of the case are made of metals that conduct electricity that can be connected to other computer parts.

15. **Reteaching.** Show students a real integrated circuit using a magnifying glass.
16. **Reteaching.** Identify the parts of a circuit and review the difference between a closed and open circuit.
17. **Reinforcement.** Other common devices use integrated circuits: irons with automatic shut-off feature, smoke alarms, VCRs, many watches, some of the talking toys, programmable thermostats, etc.

CHAPTER 5: Computers **107**

FIG. 5-10 In this camera, ICs control the amount of light reaching the film and the amount of time. They also control the focus.

How the Microprocessor Works

The microprocessor goes through a three-step sequence in order to make the computer work. The three steps are:

1. Fetch: Gets the instruction from the computer's memory or storage device.
2. Decode: Figures out what the instruction is.
3. Execute: Carries out the instruction.

The electronic circuitry in the microprocessor goes through the fetch-decode-execute process in order to make the computer do its job. It goes through this process tens of millions of times per second, and it does so accurately each time.

Integrated circuits carry a great deal of information, yet they take up very little space. Many can be fitted into a very small electronic device. This means that products can do more, yet be smaller and lighter in weight. Because the electric pulses travel so fast, the products work faster. Integrated circuits cost little to operate, and they are cheap to make. Because very little material is required, they can be produced cheaply in large quantities. Fig. 5-11. The use of ICs has made fast communication more affordable and available to more people.

FIG. 5-11 This silicon ingot (top) was grown in a vacuum oven. Then it was cut into thin slices, called wafers (middle). The third wafer has been chemically etched with integrated circuits. A finished wafer has been cut up into ICs (bottom) and packaged for easier handling.

108 SECTION 2: Communication Technology

FIG. 5-12 Police departments use computers to make sketches of criminals. Special programs allow the artist to try different noses and other features to try to match the description from witnesses.

All the instructions fetched by the microprocessor are in the machine language (binary code) discussed earlier. The instructions are actually very simple. For example, an instruction may be to add two numbers. Another instruction may be to compare two numbers to determine which is larger. The microprocessor can follow only about 150 different instructions.

Some of the instructions are in the computer's memory. Others are provided by the program. Depending on the way the instructions are written, they could make the computer operate as a word processor, a game, a graphic display, or many other things.

18. **Extension.** Show the students some of these instructions written in one of the programming languages.

Sending and Altering Information

There are many ways that computers alter, transfer, or move data around today. This ability to move information around quickly and change it as needed enables people and companies to do things that might otherwise be very difficult or time-consuming. It also enables us to develop and use a wide range of new and innovative communication systems.

Altering Information

When information has been reduced to binary code, it becomes fairly easy to manipulate. Pictures, for example, can be digitized and then altered. Fig. 5-12. Have you wondered how you would look with a different nose? There are computer programs that let you scan your photograph, erase your nose, select a new one from a menu of different noses, and print a picture of the new you.

Changing pictures in this way can be fun. It can also be useful. Plastic surgeons can show patients how they would look after surgery. Police departments can produce pictures of suspects based on the descriptions of witnesses. However, this type of alteration also poses ethical questions. Suppose a news photographer takes a picture of a senator surrounded by her political supporters. When the picture arrives at the news bureau, the editor notices that a tree behind the senator appears to be growing out of her head. He uses his computer to remove the tree. Is this ethical? Is the picture still true? What if it wasn't a tree behind the senator but instead a reputed gangster who appeared to be among the senator's supporters?

Sending Voices

Modern telephone systems use computers to change voices into pulses which are converted into microwave signals that are sent over long distances. When someone says "hello" to you on the telephone, you do not hear the actual voice. Instead, you hear the sounds that are re-created from the pulses that were sent. Sometimes these pulses are electric pulses. Other times they may be light pulses.

19. **Critical Thinking.** Ask the students to respond to these questions in a class discussion.

Sending Images

Computers can change words and pictures into digital information and send it by telephone. Pictures are sent as digital information every day. For example, a photographer for a New York magazine publisher may actually work overseas in a country such as Bosnia. He may mail his photographs to the magazine editor or send them electronically. The editor may choose certain photographs and combine them with text to create a magazine layout for a page. However, before sending the job to the printer, the magazine editor may want the photographer to approve of

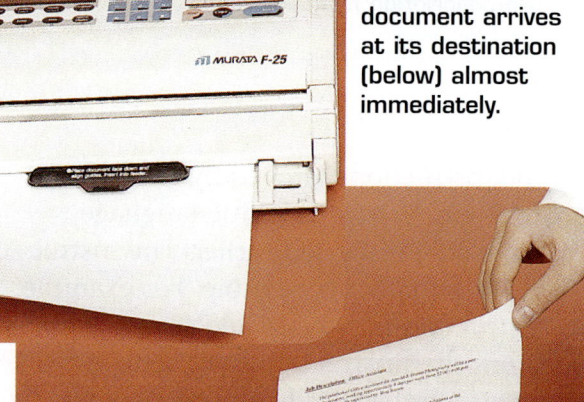

FIG. 5-13 A fax system can be used to send copies of documents from one location (above) to another. The document arrives at its destination (below) almost immediately.

SECTION 2: Communication Technology

FIG. 5-14 The laser scanner under the window reads the bar code on the product and sends the information to a computer. The computer checks its memory for the product's cost and sends that information back to the cash register.

how his photographs are being used. One easy way to send a copy of the layout (combination of text and pictures) to the photographer is to use a facsimile system. (Facsimile means a copy or duplicate.) The **fax (facsimile system)** turns the picture and words into a number code and sends this information over telephone lines to another fax machine. The fax machine at the receiving end converts this data back into pictures and words that the photographer can easily view to give his approval. Fig. 5-13.

Retrieving Information

Let's look at another example. Many large companies have offices in other parts of the world. Suppose the manager of a store in New York City wants to check the supplies of a store in Japan. She could do this by using her computer and the telephone lines because most major companies use data banks. A *data bank* is a central computer that stores the information from many smaller computers. The New York City manager can use her computer and modem to call the central data bank. The data bank could be located in yet another city, such as Dallas, Texas. The data bank sends the requested information back to the messenger over the telephone lines. She then sees on a computer screen the list of supplies in Japan. Amazingly, all this data transfer happens in less than a few seconds.

Using the Bar Code

A common data transfer method used today is the *bar code*. It is the striped code you see printed on most products. Each product has its own pattern of stripes. A computer is programmed to "read" the codes. The cashier at the store passes the code over a window. A laser scanner under the window reads light/no light from the stripes. The information is sent to a computer, which locates the price of the item in its memory. The computer sends that data to the cash register. Fig. 5-14. Then the computer writes the product's code number to a file that keeps track of how many units of that product have been sold.

20. **Extension.** Computers manipulate data quickly because the data is well organized. Have the students create a small database of addresses, using a separate field for each important piece of information.

21. **Critical Thinking.** Have the students bring in product boxes with bar codes. Try to discover in class what these numbers mean.

CHAPTER 5: Computers

The Internet and On-Line Services

As you know, you can find an immense amount of information on the Internet because it connects computers from all around the world. The Internet changes daily as new *nodes* (computer network sites) are added, deleted, and updated. When you access the Internet, you have the potential to reach every other computer network that is also tapped into the Internet. You move around the Internet electronically by jumping from one computer to another until you find the specific bit of information that you want. Fig. 5-15.

The Internet is a noncommercial (not owned by any one company or operated for profit) computer network that any person can access with the right hardware and software. The one part of the Internet called the World Wide Web is written in special language called *hypertext markup language (HTML)*. You can find information in almost any form on the Web—pictures, text, sound, and video—and on almost any topic. Many companies, organizations, educational institutions, and individuals have created web sites that you can visit with your computer. To understand how the World Wide Web is organized, think about it in terms of a massive library. A web site is like an individual book in that library. Most web sites have several web pages like pages in a book, 22

FIG. 5-15 When searching the Internet for information, you move through a maze of computers to the final destination.

DESTINATION

LOCAL NETWORK PROVIDER

YOU

22. **Extension.** Print out the HTML for a web site. Compare what you see on the screen to the printout.

112 SECTION 2: Communication Technology

and each web page has a specific address (like a page number) which is called a *universal resource locator (URL)*. You can tell your computer to go to a specific web page, or URL, by just typing in that address on your computer.

Before you can use the Internet, you must have a computer with a modem linked to a telephone line. In order to find and view information on the Internet, you must use a web browser. This is a software program that helps your computer to access and use the Internet. Netscape™ and Internet Explorer™ are two of the most popular web browsers today. Users frequently must pay a fee to use the Internet with these web browsers. Other commercial companies such as America Online®, Microsoft Network®, CSi®, and Prodigy® provide access to the Internet and other data bases for a fee.

There is so much information available on the Internet that finding what you want can be frustrating if you don't know the URL. It's like looking for a book in a large library without knowing the call number. The Internet does not have an index, card catalog, or accurate database for all the information in it. Many companies have created software programs that make it easier to search the Internet. These software programs are called search engines. Some of the more well-known ones are called Alta Vista™, Yahoo™, Infoseek™, and Lycos™.

Unlike traditional communication technologies, the Internet and commercial on-line services make it possible for individuals to communicate with large numbers of people instantaneously. "Chat rooms" are one example. Fig. 5-16. This

FIG. 5-16 Using an electronic bulletin board, a student can obtain information from other students all over the world.

23. **Extension.** Look at some search engines with the students. Decide as a class which one meets most of your needs.

CHAPTER 5: Computers **113**

type of setup is also convenient for distance learning. Here chat rooms are used to enable students and teachers who may be separated by hundreds or thousands of miles to discuss topics related to a course that they are enrolled in together. Imagine the impact that chat rooms will have in the future if more public schools begin using this technique. Would you prefer to attend school in a chat room instead of a traditional classroom?

You can also use on-line services and the Internet to send and receive electronic mail. **Electronic mail (e-mail)** refers to sending messages, letters, or documents using computers and telephone lines rather than traditional mail service. You can access library catalogs, weather services, college admission guides, and magazines around the world. You can correspond with a teenager in a foreign nation or question NASA scientists about the *Hubble Space Telescope*. You can even order flowers for a special friend. Moreover, you can do any of these things 24 hours a day.

The Internet is changing the way people can access information. Students once had to visit a library in order to find information about such things as volcanoes, trees, and other states or countries. Today, they can do a lot of their research while sitting at a computer linked to the Internet. Without leaving their laboratories, scientists can use the Web to instantly share their research findings with other experts from around the world. Consumers can purchase products like computers and televisions without ever visiting a store. The Internet is certain to affect your future if it has not already done so.

Artificial Intelligence

Artificial intelligence is the process computers use to solve problems and make decisions that are commonly solved or made by humans. Computers cannot think like humans. They are not intelligent. However, some people believe it is possible to "teach" computers to imitate human thought and decision-making processes.

Computers work by following a *program*. Basically, these instructions are logical steps to solving a problem. However, what about a problem that cannot be answered by logic?

Researchers are finding that computers can solve such problems when they have instructions in the form of reasoning processes that people use. Sometimes a person makes a decision based on a good guess, or a solution may be based on the way an expert might solve the problem. In fact, one application of artificial intelligence is the development of expert systems.

Expert Systems

When developing an expert system, developers prepare a program that has all the available facts about the topic. Experts in a particular area are interviewed. Based on the information that they provide and the facts that are available, a complex program is developed. The program instructs the computer how to solve a problem or reach a decision. This part of the program is referred to as the inference engine. It controls the order in which the experts' rules will be followed. It also infers new rules and facts when possible.

24. **Extension.** Have students respond to this question for a class discussion.
25. **Critical Thinking.** Ask students to describe situations in which using a computer with artificial intelligence would be preferable to having a person performing a task (e.g., working in outer space).

FIG. 5-17 Service engineers use expert systems to diagnose and repair problems with medical equipment. Here, a program is being used to diagnose problems in a high-frequency generator.

Various kinds of expert systems are in use today. Some repair shops use a system that figures out what is wrong with an engine or computer. Fig. 5-17. Another system can locate ore deposits in places where people could not.

Another expert system is used in the medical field to determine the right amount of treatment to give cancer patients who are undergoing *chemotherapy* (KEE-mo-THER-uh-pee). (Chemotherapy involves using chemicals to treat or control disease.) Since every person's body and disease is different, doctors can find it difficult to determine the appropriate amount of chemicals. The dose depends on the patient's body functions and on the present level of chemicals in the patient's body. Using an expert system that analyzes these factors based on the reasoning processes of medical experts, it is possible to give patients a more accurate treatment.

Research is being done to find new ways to use expert systems in communication industries. One possible use may be in the newspaper industry. Thousands of decisions are made in the production of every newspaper that is published. For example, which stories should appear on the front page? Which picture would be most appealing to readers? On what page should a certain advertisement be placed? Someday these decisions might all be made by using an expert system or some new form of artificial intelligence.

Speech Recognition

Another possible use for artificial intelligence is in the area of speech recognition. These systems can recognize human speech in all its variations. Then it can use that input to make decisions or do tasks. There are many potential applications of this technology in communication systems that rely on human input. For example, imagine being able to prepare your history paper by speaking the words to the computer instead of using a keyboard. Consider ways that speech recognition systems could be used to control robots or assist handicapped individuals.

26. **Enrichment.** Have an auto mechanic who uses an expert system explain to the class how the system works.
27. **Extension.** Another example in the medical field is the insulin pump, which can be programmed to meet the individual needs of a diabetic.
28. **Brainstorming.** Ask the students to list ways that speech recognition systems could be used to control robots or assist handicapped individuals.

CHAPTER 5: Computers

Digital Libraries

Imagine a time when all library materials—books, magazines, newspapers, recordings—would be available electronically. You would be able to access every written word, picture, graph, chart, recording, map, and other library item on your computer. This is the idea behind the digital library (or virtual library). Today's libraries are great warehouses of information. It is impossible for any one of them to carry copies of all print and nonprint media that have been made. Plus, rare books and special collections exist only in certain locations. Access to them is limited.

Some libraries are beginning to convert library items, including books, music recordings, photographs, and other media, into digital information in order to make them accessible electronically. This is most often done using a machine called a *scanner* which can change pictures and words into binary code. What's the advantage of this? Consider how electronic access would make the material accessible to more people throughout the world. Also, consider how rare documents and books could be preserved and viewed by millions instead of just a few individuals. Electronic access to information is convenient and efficient. Books that currently occupy thousands of shelves could be replaced with electronic copies that occupy a small space.

However, along with these benefits, there are some drawbacks. Scanning all this information into the computer is a long, tedious, and expensive process. Some people also believe that making some, but not all, material available on-line would cause people to ignore material that must still be accessed manually. There are also copyright issues to consider and the cost of keeping these electronic records updated as communication technologies improve and change.

FIG. 5-18 The virtual reality glasses shown here project a large-screen, full-color image on lenses in front of the viewer's eyes. These glasses eliminate all the former inconvenience of virtual reality equipment.

FIG. 5-19 In this virtual environment, the viewer watches the formation of a storm cloud.

Virtual Reality

Virtual reality (VR) is a multimedia application that uses 3-D graphics to create a realistic simulation. In VR, you are part of the computer's artificial environment, which is called *cyberspace*. You can interact with the computer by using special equipment (such as a viewer and a glove or handheld wand) that is connected to the computer. The computer senses your head, eye, hand, and finger movements and responds to them. Instead of viewing a flat computer screen, you are "surrounded" by the image. You can move through the cyberspace experience (the "virtual world") and control what is happening or what you are seeing by turning your head and moving your hand. Some scientists and engineers call this technology *telepresence* because it creates the feeling of really being there. Fig. 5-18.

Virtual reality shows promise for many applications. Some researchers are experimenting with VR in something called the *c*ave *a*utomatic *v*irtual *e*nvironment (CAVE). This is a special room like a small theater that is equipped with high-quality audio and video. Graphics are projected onto the walls and floor, and the VR user wears special glasses and sensors connected to the computer. Images surround the users and move with them. They get the correct perspectives and the environment seems real. Fig. 5-19. One day, surgeons may be trained using virtual reality. Pilots may practice their maneuvers in cyberspace before setting foot in the aircraft. You may "walk through" the home you want to build before construction even begins.

29

29. **Extension.** Much research is being done to develop human/machine interface to aid the disabled. Palo Alto VA Hospital (CA) and Wright State University (OH) are notable examples. Check institutions in your local area.

CHAPTER 5: Computers 117

Career File—Computer Programmer

EDUCATION AND TRAINING

A bachelor's degree in computer science or a related field is usually required. Courses in accounting, management, and engineering are helpful. Programmers must continually keep up to date.

AVERAGE STARTING SALARY

Starting salaries vary widely depending on a person's background and the size of the company. Earnings in the first year can range from $19,000 to $30,000.

OUTLOOK

Employment of computer programmers is expected to grow about as fast as average through the year 2005, according to the Bureau of Labor Statistics. Jobs will be the most plentiful in data processing, software, and computer consulting.

What does a computer programmer do?

Computer programmers write the detailed instructions that computers use to perform various functions. Business applications, such as accounting and inventory control, are examples. The instructions are called code, programs, or software. Some programmers maintain the software that controls the operation of the computer itself, such as how it communicates with a printer.

Does most of the work involve data, people, or things?

Other than the day-to-day interaction with other computer specialists, programmers work mostly with data and computers.

What is a typical day like?

Much of a programmer's day is spent coding and testing. Programmers break down each functional step into a logical series of instructions the computer can follow. They code the instructions in a programming language, such as C++ or COBOL. To test the program, they run it with sample data designed to test every possible processing condition of the program.

What are the working conditions?

Programmers work in offices in comfortable surroundings. They may work long hours or weekends to meet deadlines or fix problems that occur during off hours.

Which skills and aptitudes are needed for this career?

The job calls for patience, persistence, imagination, and the ability to think logically and pay close attention to detail. Much work is done under pressure. People skills are important, as programmers often work in teams.

What careers are related to this one?

Other occupations that require attention to detail include statisticians, engineers, accountants, financial analysts, and computer systems analysts.

One computer programmer talks about her job:

Each new application is a challenge. I get to use my creativity, logic, and problem-solving skills. I also like working as part of a team.

Chapter 5 REVIEW

 ## Reviewing Main Ideas

- Every computer consists of input units, central processing unit (CPU), storage units, and output units.
- Computers rely on a special machine language that consists of a series of 1s and 0s.
- Computers use many tiny integrated circuits in order to process digital information.
- The Internet and on-line services connect computers all over the world.
- Artificial intelligence is the process computers use to solve problems and make decisions normally made by humans.
- Digital libraries make library resources available to millions of people.
- Virtual reality uses computer graphics to simulate a real environment.

 ## Understanding Concepts
30
1. What does each of the four basic sections of every computer do?
2. What is an integrated circuit?
3. What is the purpose of programming languages?
4. How does a bar code work?
5. What makes the Internet a valuable communication tool?

 ## Thinking Critically

1. Suppose administrators in your school district decided that it is no longer necessary to have a school nurse. Instead, they have purchased a computer and an artificial intelligence system that will diagnose students' health problems. What are the advantages and disadvantages of this change?
2. How will on-line information systems affect businesses in your community?
3. Select a new computer application and tell what influence it might have on a group of employers and employees.
4. People are overwhelmed by the changes taking place in computers today. Why do you think this is true?
5. Computers are used in most classrooms today. How are they most useful? How are they least useful?

 ## Applying Concepts and Solving Problems

1. **Social Studies.** Investigate the role of the computer in your police or fire department. Interview the people who are involved and report your findings to the class.
2. **Language Arts.** Choose a topic related to communication, such as cell phones, e-mail, or digital television. Search the Internet for information about this topic. Look for different impacts. Then list all the URLs that you visit and write a summary of the information available at each site.
3. **Science.** Think about how your brain, nerves, and sensory organs work together to help your body function properly. Compare these inputs, processes, memory, and outputs to those of the computer. Create a diagram or chart to illustrate your findings.

30. **Resource.** Answers to "Understanding Concepts" and "Thinking Critically" can be found in the Teacher's Manual at the front of this *Teacher's Annotated Edition* and in the *Teacher's Resource Binder*.

CHAPTER 6

Telecommunication

Objectives

After studying this chapter, you should be able to:
- explain the importance of telecommunication.
- show how signals are transmitted.
- describe how telegraphs, telephones, radios, televisions, and satellites operate.
- identify the effects of digital technology.

Terms

amplitude
coaxial cable
communication satellite
earth station
electromagnetic waves
frequency
microwaves
optical fibers
personal communications services (PCS)
pixels
telecommunication
transceiver

Technology Focus

Telecommuting

Homework is taking on a new meaning. Today, thanks to communication technology that allows us to transfer information from one location to another easily and quickly, many people can work at home or at locations other than a typical office. Powerful personal computers, modems, and fax machines have made it easy for people working at home or in another remote location to communicate quickly with their companies, with each other, and with clients. Working from remote offices is called *telecommuting*.

American Express, J. C. Penney, Apple Computers, and General Electric now allow some employees to work at home. Many smaller businesses also employ telecommuters. Futurists say that there could be as many as seven million telecommuters by the end of the 1990s.

Why Telecommute?

Who are the people who telecommute? Some are skilled workers who choose to stay at home and raise a family. They can do this by telecommuting and still put their job-related talents and experience to good use and earn an income. Workers with physical disabilities can work out of a home office and avoid many of the difficulties they might encounter if they had to travel back and forth to work every day. President George Bush joined the ranks of telecommuters when he managed and directed the Persian Gulf War efforts from Washington, D.C., and other locations. President Bush was "telecommanding."

A home office worker usually has at least one computer and a modem. Many also have a fax machine. Some telecommuters work from mobile workstations. These workers may also use a cellular phone and a laptop computer.

Corporations find that telecommuters are very productive. Telecommuters also decrease the amount of office space and parking that is necessary, and they help keep a corporation's insurance rates lower.

Telecommuting workers save time by not having to drive back and forth to work every day. This also saves them the expense of fuel and of operating a car (oil, tires, maintenance, etc.). Telecommuters work on schedules that suit their needs, and they work in more pleasant surroundings.

The environment profits too. Because telecommuters do not drive to work daily, they use less fuel. In addition, telecommuting workers aren't crowding the highways and polluting the air with exhaust fumes.

Homework isn't just for school anymore.

Take Action!

As a class, write a letter to the human resources department at one of the companies mentioned in this article. Ask about the advantages and disadvantages of telecommuting.

Resources:
- Chapter 6 Lesson Plan in the *Teacher's Annotated Edition* and in the *Teacher's Resource Binder.*
- Chapter 6 Review in the *Teacher's Resource Binder.*
  1. **Brainstorming.** Ask the students to list the kinds of work that can be done at home.

What Is Telecommunication?

Telecommunication means communication over a long distance. When we examine telecommunication, we are exploring those methods used to transmit (send) and receive messages over a distance. Some of the most popular telecommunication devices today are the telephone, television, and radio.

Many inventions throughout history have led to the development of today's telecommunication system. How were long distance messages sent prior to the 20th century? People relied on humans or animals to move information over long distances. Families kept in touch through letters delivered by riders on the Pony Express. Native Americans and soldiers relied on smoke and fire signals to communicate with others far away. One inventor developed a visual telegraph during the late 18th century. Fig. 6-1. This device was mounted atop towers that were several miles apart. Using telescopes, messages could be seen from long distances, and the information could easily be transmitted to the next location by someone at each station.

Long distance communication has changed dramatically since these early days. No inventions stand out quite like the telegraph, telephone, radio, television, and satellite. In the next several pages, each of these technologies will be examined.

FIG. 6-1 The movable beams on this visual telegraph were positioned to represent certain words or phrases. The signal could be seen from a few miles away with the aid of a telescope.

Fascinating Facts

Telecommunication took place even in ancient times. Troy was defeated by the Greeks in the 11th century B.C. The Greeks were able to send word of this to Argos, over 500 miles away, in just a few hours by lighting signal fires on the tops of hills.

2. **Brainstorming.** Have the students think of ways to telecommute that have not been tried yet.

SECTION 2: Communication Technology

SCIENCE CONNECTION
Magnetism and Magnetic Media

At one time, all computer data was stored by means of magnetism. The hard disks, floppy disks, and tape we use today are all magnetic storage devices.

Magnetism is caused by the movement of electrons. In an atom, electrons orbit the nucleus. As the electrons make their circle, they also rotate. This orbiting and rotating create a magnetic force. In most materials, the magnetic force is weak. In some, such as iron, the force is strong.

To create a storage device, such as a floppy disk, a plastic base is coated with tiny particles of iron oxide. To record data on the device, a magnetized head is passed over the particles, arranging them into certain patterns. These patterns are in the form of the binary code the computer can interpret.

Unless the same process is used to arrange the oxides in a new pattern, they retain their data for a long time. After several years, however, the magnetic forces weaken and data can be lost. The data is also vulnerable to ordinary magnets, which can destroy it on contact by pulling the oxides out of position.

Try It!

Obtain a small magnet and some iron filings. Place the filings on a sheet of paper. Hold the magnet beneath the sheet of paper and move it around. What happens to the filings?

The Telegraph

One of the inventions that helped to make great strides in telecommunication was the telegraph. For the first time, people could send messages over long distances that could be received almost instantly. Samuel Morse has been given the most credit for inventing the telegraph. Even though others were also successful in their experiments with devices that could send messages over wire using electricity, Morse's model caught on. Morse's invention marks the beginning of telecommunication in America as we think of it today and has paved the way for future long distance communications.

How the Telegraph Works

Morse's telegraph was actually a very simple instrument that sends electric pulses over wire. Its operation was based on [3]

3. **Extension.** In class, demonstrate to the students how to make a simple electromagnet.

CHAPTER 6: Telecommunication

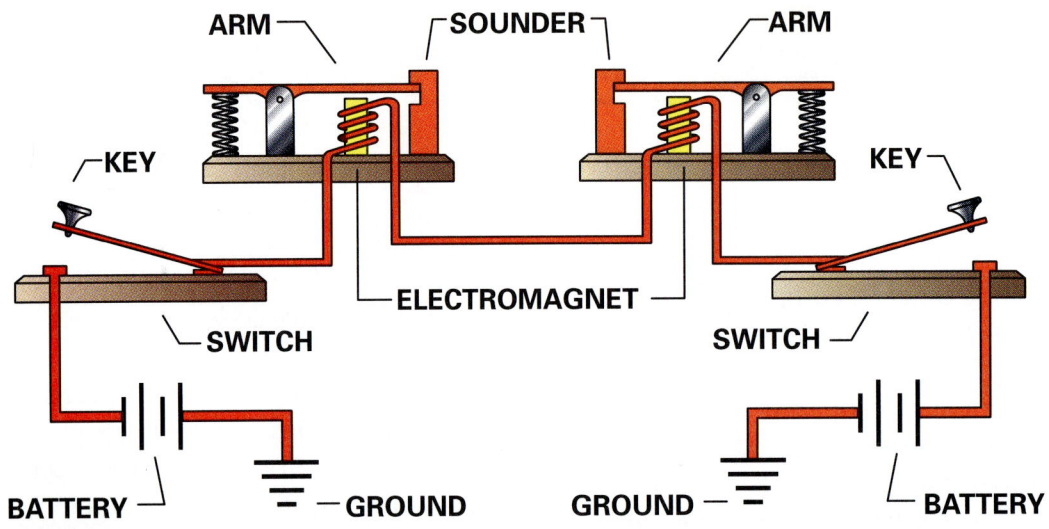

FIG. 6-2 When the key in Morse's first telegraph was pressed, current flowed through the circuit and the sounder at the other end clicked. The length of the clicks were varied to make a code.

the principles of electricity and magnetism. It consisted of little more than a battery, a sending device called a key, a wire, and a receiver called a sounder. Fig. 6-2.

The sounder consisted of an electromagnet and a metal arm that made a clicking noise. (An *electromagnet* is a soft iron core surrounded by a coil of wire that temporarily becomes a magnet when electric current flows through the wire.) The key acted as a switch that opened and closed the electrical circuit.

Morse developed a code to use with the telegraph based on varying the length of these clicks. The key is held down very briefly to make a short click, which represents a dot in the code. By holding the key down a little longer, a longer click is sounded. This represents a dash. Morse assigned a special sequence of dots and dashes to each letter of the alphabet, each number from one through ten, and each punctuation mark. Fig. 6-3. In 1844, using this code for

International Morse Code

A	.-	P	.--.	5	.....
B	-...	Q	--.-	6	-....
C	-.-.	R	.-.	7	--...
D	-..	S	...	8	---..
E	.	T	-	9	----.
F	..-.	U	..-	10	-----
G	--.	V	...-	'	.----. (apostrophe)
H	....	W	.--	:	---...
I	..	X	-..-	,	--..--
J	.---	Y	-.--	-	-....- (hyphen)
K	-.-	Z	--..	.	.-.-.-
L	.-..	1	.----	( )	-.--.- (parentheses)
M	--	2	..---	?	..--..
N	-.	3	...--	" "	.-..-.
O	---	4	....-	SOS	...---...

FIG. 6-3 The Morse Code made it possible for messages to be sent via (by way of) the telegraph using electric pulses. This system has been used for international communication.

4. **Enrichment.** Have students work in pairs to build model telegraphs. Point out features that involve electricity and magnetism.

5. **Extension.** Have students send messages in Morse code. If models were not made use flashlights to send light signals.

the first time, Morse sent the message, "What hath God wrought?" over 40 miles of wire, stretching from Washington, D.C., to Baltimore, Maryland.

The telegraph is no longer a common means of sending messages over long distances. It was, however, a significant step forward. It eventually led to the development of the telephone, a device that remains an important part of our modern telecommunications system.

The Telephone

6   A young inventor named Alexander Graham Bell was convinced that voices, too, could be sent over wire. Bell experimented with several different ideas. Through trial and error, Bell and his partner, Thomas Watson, developed a model telephone that became the basis of today's telephones. However, Bell was not the only one searching for a voice machine. When his patent for the telephone was
7 filed on February 14, 1876, Bell was only two hours ahead of another inventor, Elisha Gray. (A *patent* is a government document granting the exclusive right to produce or sell an invented object or process for a specified period of time. It ensures that the inventor's idea cannot be legally copied for that specified period.)

How the Telephone Works

Bell's telephone uses the principle that sound waves cause vibrations. In the transmitter, which is located in the mouthpiece, sound waves cause a parchment (paper) drum to vibrate. Then Granville Woods, another inventor, improved Bell's design. Woods invented

FIG. 6-4 When someone speaks into the mouthpiece of a carbon microphone, the diaphragm vibrates against the carbon granules. This creates an electrical signal that is transmitted over the telephone line to the receiver at the other end.

the carbon microphone for the telephone. 8 This microphone uses a flexible piece of metal called a diaphragm and carbon granules to produce a varying electrical current. When a person speaks loudly, the diaphragm presses against the carbon granules and produces a strong current. When the speaker pauses, the current stops. The current varies according to the sound waves that make the diaphragm vibrate. Fig. 6-4.

The telephone receiver is located in the earpiece. The receiver works much like the microphone (transmitter) in reverse. The receiver contains a wire-wrapped iron core. Connected to the iron core is a flexible metal diaphragm. Fig. 6-5. When the transmitted electrical signal enters the receiver, it travels through the coil, magnetizing the iron core. This forms an electromagnet. The magnetic field pulls on the metal diaphragm. The diaphragm vibrates and reproduces the sound.

6. **Extension.** When Bell invented the telephone, he also laid the groundwork for the development of the microphone and the loudspeaker.
7. **Extension.** Discuss patents—how they are obtained and why they are important.
8. **Reteaching.** Dismantle a telephone handset to show students the parts and wires inside the mouthpiece and earpiece.

CHAPTER 6: Telecommunication

FIG. 6-5 The transmitted signals pass through the coils causing the iron core to become magnetized. This causes the diaphragm to vibrate, and the vibration reproduces the sound.

FIG. 6-6 In an electret microphone, the space between two oppositely charged metallic plates varies according to the speaker's voice. This causes a varying current that is transmitted to the receiver electronically.

Electronic Telephones

Modern telephones are largely digital electronic devices. Integrated circuits (ICs) have replaced many of the parts in the older telephones, such as the bulky wire coils.

In the microphone, other devices that produce electrical currents in response to sound waves have replaced the carbon granules. For example, the electret microphone uses metallic plates with opposite (positive/negative) charges to produce the current. The varying sound waves created by the speaker's voice alter the space between the oppositely charged plates, and this produces a varying current. Fig. 6-6. Integrated circuitry has expanded the features available, such as last number redial, speed dialing, caller identification, voice mail, call forwarding, and more. It has also made portable phones possible and economical.

Cellular Telephones

The most significant development in telephone technology in the past few decades has been the trend toward wireless communications with lightweight cellular telephones. These devices use radio waves to transmit conversations, thereby omitting the need for connecting wires and cables. (You'll learn more about radio wave transmission later in this chapter.) You can carry your cellular phone in your car or on your boat, and you can easily slip it into your pocket or purse.

As you move about with your mobile telephone, your call is transferred from one operating area to another in order to maintain good signal transmission. Each operating area, or *cell* (hence, "cellular" phone), may range from a few hundred meters to several miles in diameter. Every cell has its own transmission tower which is linked to an electronic switching office. Fig. 6-7.

A cellular phone contains a device called a **transceiver** which is a transmitter and receiver combined into a single unit. The transmitter portion converts the signals into radio waves, which are transmitted to an antenna. The receiver portion of the transceiver receives radio waves from the antenna, changes them back into electrical signals, and sends them to the phone's control unit which makes it possible for you to both talk and listen.

Today cellular calls may use digital or analog technology, as discussed in the previous chapter. Analog cellular is very similar to sending messages by way of walkie-talkies because your voice is transmitted directly by radio waves. Digital cellular converts the voice into a computer language before transmitting it via radio waves. Digital cellular is fast becoming the preferred system because it offers better-quality voice transmission, more service options, and improved security. A common problem with analog cellular is the ease with which someone can eavesdrop on conversations. Digital technology makes it easy to *encrypt* (secretly code or scramble) electronic signals so that no one can listen in on the conversations. In addition, digital cellular technology is making available new telecommunications options called **personal communications services (PCS)**. PCS is a completely digital technology. It includes two-way paging, voice messaging, and many other services linked to your cellular phone. You need only one phone number to have access to all the services.

FIG. 6-7 With a cellular phone, as you move from one location to another, your conversation is bounced around from one cell to another. Each cell provides coverage to phones used within that cell. The switching office makes sure that signals reach the correct locations.

Transmission Channels

As you learned in Chapter 4, communication channels are the paths over which messages must travel to get from the sender to the receiver. They are also called transmission channels. Telephone messages may travel over wires, cable, or optical fibers. Some telephones, however, use microwaves or radio waves as a transmission channel.

Copper Wire

Many local telephone messages travel over twisted-pair wire. This consists of two thin, insulated copper wires twisted around each other. Twisted-pair wires may also be bundled together to form large cables that stretch thousands of miles. However, twisted-pair wire is not being used for most newly laid communications lines.

Coaxial Cable

Coaxial cable can carry many more messages all at once than twisted-pair wire. **Coaxial cable** consists of an outer tube made of a material that conducts electricity (usually copper). Inside the tube is an insulated central conductor (also copper). Usually several of these cables are combined into one bundle. Fig. 6-8. This bundle is then covered in lead and plastic for protection.

Optical Fibers

Optical fibers are being used increasingly to carry telephone transmissions including computer data. **Optical fibers** are thin, flexible fibers of pure glass that carry signals in the form of pulses of light. Fig. 6-9. Each optical fiber is surrounded by a reflective *cladding* (covering) and an outside protective coating. After the sound waves from the speaker's voice are converted to electrical signals, these signals activate a laser. The resulting laser light pulses enter the glass fiber and bounce rapidly back and forth off the reflective surface of the cladding as they pass through. Fig. 6-10. At the receiving end, the light pulses are changed back to an electrical signal, which is changed back into the sound of the sender's voice.

Optical fibers have many advantages over copper wire or coaxial cable. A single fiber is about the thickness of a human hair, yet it can carry thousands of voice or data circuits. (A pair of copper wires can carry only a few dozen circuits.) Optical

FIG. 6-8 This 22-tube coaxial cable can carry up to 108,000 two-way voice conversations.

9. **Reinforcement.** Obtain a section of coaxial cable from your local telephone company and allow students to examine it.

10. **Reinforcement.** Obtain a section of optical fiber and show students the fine fibers.

SECTION 2: Communication Technology

FIG. 6-9 Optical fibers are flexible strands of glass about as thin as a human hair.

fiber cables are much better for handling the high-speed data transmission needs of modern computer networks, including video data and the Internet. Because optical fiber cables are lighter and thinner than copper wires, they are also ideal for communication systems in large cities and other places where the space is limited. In addition, the signals that travel on optical fibers do not fade as quickly as electrical signals sent on copper wire. Many of the noises you sometimes hear on the telephone, like buzzing or clicking, are also eliminated when optical fibers are used.

Microwaves

Microwaves can be used to carry phone conversations over long distances. **Microwaves** are very short electromagnetic waves. (**Electromagnetic waves** are waves created by a magnetic field. These waves travel through the atmosphere and make communication without a connecting wire possible.) Billions of microwaves can pass a given point in less than a second.

In telephone communication, the sound waves are changed into microwaves. These are then transmitted using an antenna. The microwaves travel through the air like the ripples you see when you drop a stone into a pool of water. Another antenna is able to detect the signals and acts as the receiver. It boosts the signal to a usable strength and then converts it into the voice of the person you are talking with on the telephone. Fig. 6-11.

FIG. 6-10 The arrows show how light traveling through the glass fiber bounces off the reflective cladding, which keeps it inside. After leaving the optical fiber, the light is changed back into an electrical signal.

11. **Reinforcement.** Place a *clear* bowl of water on an overhead projector and dip your finger into the center to show the ripple effect.

CHAPTER 6: Telecommunication

of a cellular phone. The microphone changes the sound waves into electrical signals. Then the low-frequency sound signal is combined with a high-frequency carrier signal. Fig. 6-13(A and B). These waves are then *modulated* (altered).

Sometimes, the amplitude of the carrier wave is modulated. Fig. 6-13(C). This signal can be received only by AM (amplitude modulation) radio. Other times, the frequency is modulated. In frequency modulation, the waves are spread farther apart or crowded closer together. Fig. 6-13(D). These signals are received on FM (frequency modulation) radios and cellular phones. The modulated signal is then amplified (boosted) and sent to an antenna for transmission. When the antenna sends the radio waves, they travel through the air in all directions.

FIG. 6-11 Signals sent through the atmosphere are transmitted and received by antennas. The messages travel as electromagnetic waves.

Radio Waves

12   Radio waves are also electromagnetic waves, but they are longer than microwaves. All waves—sound waves, microwaves, radio waves—have both amplitude and frequency. **Amplitude** refers to the strength of the wave. **Frequency** is the number of waves that pass a given point in one second. Fig. 6-12.

The process of transmitting radio waves begins with the signal source, such as a person speaking into the microphone

12. **Reinforcement.** Use a rope in class to demonstrate amplitude and frequency.

FIG. 6-12 Radio signals are measured in cycles per second, or hertz. The signal shown here has a frequency of 3 hertz. Amplitude is measured from the wave's midpoint to its peak.

130   SECTION 2: Communication Technology

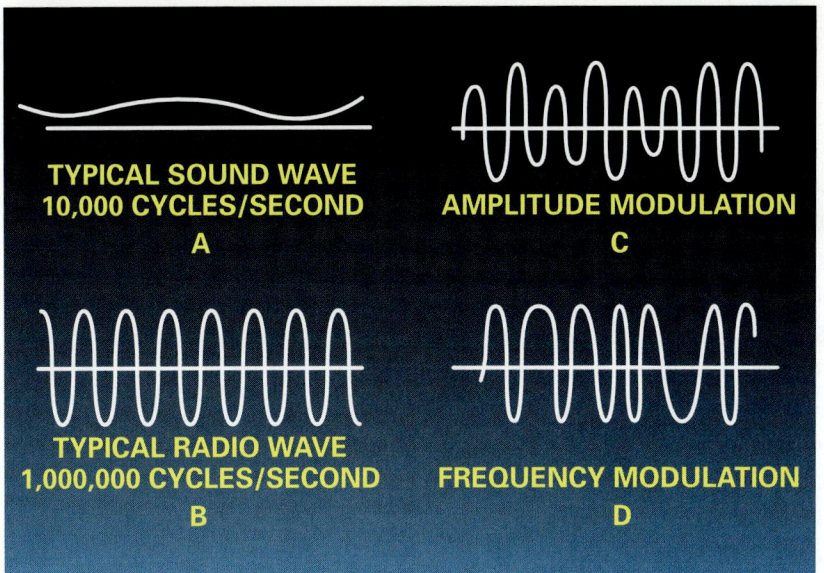

FIG. 6-13 A typical sound wave (A) is combined with a carrier wave (B). The carrier wave may receive amplitude modulation (C) or frequency modulation (D).

Getting a Message to the Right Place

No matter what transmission channel is used to send the message over the telephone, another major problem still exists. With millions of telephones in the world, how can you get a message to the right one? When telephones were first introduced, the number of connections possible was very limited. Today, however, almost all homes and other buildings have at least one phone. The number of possible connections is enormous! A system to handle this great flow of information had to be developed.

Early Switching Systems

The first simple telephone exchange system required human operators to connect the phone of the message sender with the phone of the expected receiver. These operators sat at stations and plugged the line of the person making the call into the right place. As the number of calls continued to increase, a number of new mechanical systems were developed. The first new system, invented by Almon B. Strowger, used electromagnetic switches to make telephone connections. The switching system received telephone signals by "counting" the number of pulses from the dialed number. The system held the signals until the entire number was registered. At that point, the system automatically created a "path" through all the switches to the receiver.

Automated Telephone Exchanges

Many other systems were used to make the line connection process possible. However, electronic switching systems did not become a reality until the late 1950s. Electronic switching makes connections faster and provides better signal transmission. These telephone exchange systems use computers to route a call through a series of telephone exchange centers to the intended destination.

---

13. **Reinforcement.** Have students draw a picture showing the number of connections possible between 6 or 8 stations.
14. **Enrichment.** Arrange for a class visit to the local switching center.
15. **Extension.** New area codes are being added on a regular basis to accommodate the growing number of telephone exchanges.

CHAPTER 6: Telecommunication **131**

Radio

The telephone greatly improved telecommunication for individuals. However, people still needed a method of *mass communication*—communication with large groups of people simultaneously. That meant they must find a way to send a message without using wires and cables.

Radio communication evolved slowly 16 from a series of discoveries in the field of electricity and electromagnetics. Finally, in 1895, Guglielmo Marconi used electromagnetism to send telegraph code signals a distance of more than 1 mile (1.6 kilometers) without wire. With this wireless telegraph, the forerunner of today's radio, the dream of mass communication was realized.

Improvements on Marconi's device followed rapidly. One such improvement was the vacuum tube. Vacuum tubes greatly strengthened a weak signal so that it traveled farther and more people received it. Soon everyone wanted a radio.

How Radio Works

As you just learned, the radio signal's amplitude or frequency is electronically modulated and amplified. Then an antenna transmits the signal through the air where other antennas can pick up the signal.

How does your antenna pick up the correct signal? Each radio station and device that works on radio waves (such as CB radios, military radios, and pagers) is assigned a certain frequency. The numbers on your radio dial indicate the frequency of the different radio stations. When you set your dial at a number, you are tuning in the frequency of a particular 17 station. The antenna or aerial on the radio picks up the waves that have

FIG. 6-14 Modern tactical radios use digital technology to get important voice, data, and position information to military troops, wherever they may be. Digital technology makes secure transmission possible. Plus, the radio can be kept small and lightweight.

16. **Extension.** Obtain audio tapes of narrative stories and comedy routines done on radio during the 1940s so students will understand that radio was not just music and news; it provided a variety of family entertainment features. Discuss what was *done* to create a "picture" in the listeners' minds.
17. **Reinforcement.** Have students create a simple diagram to show the process of sending and receiving radio waves.

that frequency. The circuitry in the radio separates the carrier wave from the sound signal. This process is called *demodulation*. The circuitry also boosts the electrical signal and sends it to the speaker, which changes the signal back into sounds.

Digital Technology in Radio

Digital technology is also influencing changes in the radio industry. For example, digital radio, also known as digital audio broadcasting (DAB) or digital audio radio (DAR), is the method of transmitting high-quality audio signals to radio receivers using digital instead of analog transmission. The technology currently being developed is called IBOC (in-band, on-channel) and will enable radio broadcasters to use their existing frequencies to transmit compact disc (CD)-quality sound through current FM radio bands. It will also allow high-quality sound through current AM bands. In order to receive this better-quality digital signal, you need a new digital receiver. You will not benefit from the improved digital broadcasting technology if you use your old analog radio receiver.

Digital technology has benefited our radio listening in other ways. Many modern radios rely upon digital components for improved tuning and operation. Fig. 6-14. In addition, new releases of recorded music can be sent to stations electronically so that they hit the airwaves in record time. That kind of speed isn't possible when recordings are shipped by way of regular shipping routes.

Television

Television is an electronic system of transmitting pictures and sounds over a wire or through the air. Today, television is a popular telecommunication tool with a variety of program options. Television sets come in many different sizes and are used in many different ways. Let's see how the system was developed and how it works today.

How Television Developed

Sending visual images over a long distance became the ambition of several inventors in the early 1920s. Many people experimented with various devices that could send and receive pictures. Fig. 6-15.

FIG. 6-15 One of the earliest televisions developed in America is this receiving device made by Charles Francis Jenkins around 1928. The motor and drum fit inside the wood case. When the motor runs, the drum spins. It projects an image up through the opening in the top of the case using a scanning motion. The mirror reflects the image through the magnifier. The viewer, looking through the magnifier, sees an orange and black image.

18. **Reinforcement.** Explain the difference between modulation and demodulation.

CHAPTER 6: Telecommunication **133**

Early televisions had poor picture quality. In addition, the screens were so small that you had to sit very close to and directly in front of them in order to see the picture. In spite of this, television had an enormous impact on the public. At first, most people thought television would be used solely for entertainment. However, it soon became apparent that this "picture machine" also had uses in education and information distribution. In 1939, only one or two programs were aired daily. During the early 1940s, the war effort slowed the development and spread of television. By 1946, however, the number of programs were growing rapidly.

Today, nearly every household in the United States has at least one television set—most have two, and some have one in almost every room. The picture is clear and sharp, and so is the sound. A complex broadcasting network is set up to transmit television programs around the world 24 hours a day.

How Television Works

A television set may seem like a very complicated device. Actually, it works according to some very simple principles.

The Television Camera

The images to be televised are focused onto a special target surface within a pickup tube inside the camera. This surface is divided into approximately 367,000 microscopic **pixels**, or picture elements. The light from the televised image gives each pixel an electrical charge. The charge varies according to the amount of light falling upon each one. Each pixel has a certain charge that equals a certain amount of light. The pattern of picture elements is scanned by an electron beam about the diameter of a pinhead. The beam moves from left to right over the entire target surface 525 times for each picture. Thirty pictures are scanned every second. The electrons pass through the target surface and strike a signal plate. When they hit the signal plate, an electrical current is created. This is the electrical video signal that will be transmitted.

If a color picture is to be broadcast, a more complex camera must be used. Color television is based on the principle that different mixtures of red, green, and blue light can produce every color in nature. Mixing light rays is not like mixing colors of paint. Paint simply reflects light. Color video cameras have three pickup tubes: one each for red, green, and blue. Light passing through the camera's lens strikes three mirrors. Each mirror is directed at a different pickup tube. In front of each tube is a filter that allows only one color to pass through. Fig. 6-16. The pickup tubes then process the images in the same way as the black-and-white pickup tube just described.

While cameras are picking up the video portion of a telecast and converting it to electrical signals, microphones are picking up the audio portion. The microphones are the same as those used in radio, and they convert the sound waves to an electrical signal just as they do in radio.

Transmission

The audio and video TV signals are amplified, modulated, combined, amplified again, and then sent to an antenna where they are transmitted. (In the case of cable television, however, these signals are sent over coaxial cable).

19. **Background.** Point out that all images reflect light differently, depending on the colors.

20. **Reteaching.** Demonstrate mixing colored light using a flashlight and thin, colored, transparent plastic sheets.

SECTION 2: Communication Technology

FIG. 6-16 A color video camera has three pickup tubes—one each for red, green, and blue. A black-and-white camera has only one pickup tube and does not need the mirrors to direct light. However, both cameras process the images in the same manner.

With the exception of cable television, signals are transmitted much as radio signals are transmitted. Your television set is like a radio with a picture attached. Just as we have AM and FM radio, we have VHF and UHF television. VHF stands for very high frequency, and UHF stands for ultra high frequency. As in radio, each television channel has a different frequency. The channel selector on the television receiver tunes in the frequency of the channel (VHF, UHF, or cable) you select, just as the dial does on the radio.

The television receives the transmitted signals and converts them to electrical signals. The video and audio signals are then separated. The audio signals are sent to the television's speaker, which converts the signals back into sound. The video signals are sent to the picture tube.

The Television Receiver

The flat end of the picture tube (the screen) is covered with phosphor salts. (*Phosphor* is a substance that emits light when given energy.) At the back of the picture tube is an electron gun. The electron gun projects a beam onto the phosphorescent (phosphor-covered) screen. The beam goes across the screen in a left-to-right direction 525 times. This occurs 30 times every second, just as when the camera scanned the picture elements. Because each scene is replaced 30 times every second, your brain interprets these rapid changes as a continuously moving image. The beam excites the phosphor salts and makes them glow. These glowing pixels create the image you see.

CHAPTER 6: Telecommunication

FIG. 6-17 Each of three electron guns inside the picture tube scans for one of the primary colors of light.

A color picture tube has three electron guns that sweep across the flat surface of the tube. One gun is used for each of the primary colors of light—red, green, and blue. The surface of the color picture tube is covered by groups of red, green, and blue phosphor pixels, which will make up the color image. Fig. 6-17. It is extremely important that the appropriate electron gun hits the correct phosphors. Most televisions have a masking guide that directs the beams to the correct phosphors. The beams excite the phosphors to create the color picture you see on the screen.

High-Definition Television

High-definition television (HDTV) is an innovation that makes cinema-quality video and CD-quality sound available on home television sets. Picture clarity is improved because the number of scanning lines are doubled. More scanning lines mean that the HDTV screen can be made larger without losing the sharpness of the picture. In the United States, the number of lines scanned will be increased from 525 to 1,050. In Europe, the number will be increased from 625 to 1,250.

HDTV viewers have remarked that the picture is so clear they can see wrinkles on actors' faces, textures in fabrics, and individual blades of grass on the ball field. In addition, the HDTV screen is wider in proportion to its height than present-day television. This makes the images look more like a movie screen. Fig. 6-18.

HDTV technology is expected to be ready for the market soon. However, the cost of these new sets is expected to be high during the first several years of production. This is common for new electronic products until the demand increases and the cost of producing them decreases.

Digital Television

Digital technology is having a tremendous influence on television broadcasting. Until now, broadcasters have relied on analog systems for television signals. However, new guidelines from the Federal Communications Commission

(FCC) will require that all signals be broadcast using digital technologies by the year 2010. This will bring significant changes in television broadcasting. During the transition phase from analog to digital, your analog transmission television set will still function. When the changeover is complete and analog transmission is discontinued altogether, you will need a digital television set in order to receive television broadcasts.

The initial expense of this technological change will be high for consumers and broadcasters, but the long-term advantages are great. In addition to offering better-quality transmissions, digital systems may be far more diverse than today's basic television set. They will be more like televisions combined with sophisticated computers. You will be able to manipulate video images, store images and sounds, electronically search for programs of interest, and more. Imagine zooming in on your favorite movie star or choosing a different camera angle as you watch a baseball game. You'll be able to modify colors and experiment with sounds. You will be able to divide your TV screen into sections and watch several television programs at the same time. The possibilities seem unlimited.

Satellite Communication Systems

A **communication satellite** is a device placed into orbit above the earth to receive messages from one location and transmit them to another. Fig. 6-19. The satellite acts as a relay station. That is, it simply reflects signals back to earth. For that reason, a satellite is often called a "mirror in the sky."

Communication satellites are placed in orbit approximately 22,300 miles above the earth. They travel at the same speed as the earth rotates. Thus, they remain above the same part of the earth at all times.

FIG. 6-18 The development of HDTV could revolutionize television in the next few years.

21. **Critical Thinking.** Ask students, "Someone once said that television is the most powerful communication tool known to man. Will the introduction of HDTV increase or decrease the importance of television?"
22. **Enrichment.** Electromagnetic waves travel at the speed of light—186,000 miles per second. Ask students how long it takes a signal to travel up to the satellite and back down to earth.

CHAPTER 6: Telecommunication **137**

FIG. 6-19 This satellite collects electromagnetic signals with one set of antennas, amplifies them, and then sends them using a different set of antennas. The solar panels collect energy from the sun. This energy is used to operate the satellite.

Satellites have greatly influenced our modern communication systems. For example, satellites help transmit numerous types of messages, including telephone and television signals. Satellites are used to transmit printed information as well. Stories and pictures for the *Wall Street Journal* and *USA Today* newspapers are transmitted by way of satellite to many printing locations throughout the country.

Satellites make it possible to communicate instantly. Live broadcasts depend on satellites to transmit messages as they are happening. Thanks to satellites, you can see things that happen far away while you sit in the comfort of your home. People all over the world can watch the World Series games while they are being played or witness the launch of a space shuttle. Millions of people from around the world were able to "attend" the funerals of Princess Diana of England and Mother Teresa of India by way of satellite television.

Since their development in the 1960s, satellites have changed a great deal in design and efficiency. This is a result of technological developments taking place in electronics and communication. Satellite systems are being developed today with greater capacity. They can carry more messages and provide better quality transmission.

23. **Extension.** Have students research satellite designs. Assign students to different generations (such as 1960s, 1970s, 1980s). Have them locate a picture of a satellite from their assigned time period, along with basic information.

How the System Works

Signals are sent to orbiting satellites through earth stations. An **earth station** (sometimes called a ground station) is a large, pie-shaped antenna. It receives signals and transmits them to the satellite. This is called the *uplink*. The satellite receives the signals and transmits them to another location back on earth. This is called the *downlink*. Receiving earth stations capture the signals and send them to the desired receivers. Fig. 6-20.

Imagine how many signals are passing through the air constantly! You may wonder how the right message gets to the right earth station. When a signal is transmitted, the sender puts a certain code at the beginning of the message. The code directs the signal to the intended receiver. In addition, messages can be "scrambled." Then, only certain earth stations can pick up the message and decode the information.

Satellites are also being used for navigation. You will learn more about these systems in Chapter 20.

Fascinating Facts

Radio telescopes work at picking up radio waves that come from outer space. The VLA (Very Large Array) in Socorro, New Mexico, consists of 27 radio telescopes. These antennas are spread out in a Y shape over 12 to 13 miles.

FIG. 6-20 This is how a satellite is used to transmit information from one place to another place far away.

24. **Extension.** Introduce the concept of cryptography.

CHAPTER 6: Telecommunication

Career File—Computer Repairer

EDUCATION AND TRAINING
Most employers prefer applicants with one to two years' training in electronics, which can be obtained from community colleges or technical institutes.

AVERAGE STARTING SALARY
Beginning computer repairers can earn about $22,000.

OUTLOOK
Employment of computer repairers is expected to grow much faster than average through the year 2005, according to the Bureau of Labor Statistics. Demand for repairers will increase as more computers are being used in homes and offices.

What does a computer repairer do?
Computer repairers install, maintain, and repair computers and devices used with computers, which includes printers, disk drives, modems, and monitors. They advise whether it is wise to buy a new part or machine or fix the old one. They may replace defective components, install such things as additional memory and circuit boards, and install network cables and wiring.

Does most of the work involve data, people, or things?
Repairers are most concerned with things. They use instruments such as voltmeters and oscilloscopes to install, test, repair, and adjust equipment. They also use hand tools such as pliers, screwdrivers, and soldering irons. Contact with people is limited.

What is a typical day like?
When equipment breaks down, repairers check for common causes of trouble, such as loose connections or defective components. For more serious problems, they may refer to technical diagrams and run special software designed to detect problems. Some repairers visit work sites on a regular basis to do preventive maintenance. They also visit whenever emergencies arise.

What are the working conditions?
Computer repairers generally work in clean, well-lighted, air-conditioned surroundings in a repair shop or at the site of installed equipment. The work may involve lifting, reaching, stooping, crouching, and crawling to make connections.

Which skills and aptitudes are needed for this career?
Repairers should have a mechanical aptitude, knowledge of electricity or electronics, and manual dexterity. Good eyesight and color vision are necessary to inspect and work on small, delicate parts. Good hearing is important to detect problems revealed by sound.

What careers are related to this one?
Workers in other occupations who repair and maintain electronic equipment include automotive electricians, broadcast technicians, electronics technicians, vending machine repairers, and appliance and power tool repairers.

One computer repairer talks about his job:
With this job I'm on the cutting edge of the new technologies. I run my own company, and that's very exciting, too!

Chapter 6 REVIEW

Reviewing Main Ideas

- Telecommunications is the ability to send and receive messages over distances.
- The telegraph uses the principles of electricity and magnetism to send coded messages over wires.
- In telephones, sound waves are converted to electrical signals that travel by way of wire, coaxial cable, optical fibers, radio waves, or microwaves.
- With radio, sounds are converted into signals that are sent through the air and picked up by receivers tuned in to the appropriate frequency.
- Pictures and sounds are converted into signals that are sent through the air to produce pictures and sounds on television.
- The development of radio and television as well as satellite communication has made mass communication a reality.
- Digital technology has greatly improved telecommunication services.

Understanding Concepts

1. How did the telegraph send a signal?
2. What are three advantages of using optical fibers to carry telephone messages?
3. How do amplitude and frequency differ?
4. What happens when you tune a radio to a particular station?
5. How are signals sent to orbiting satellites?

Thinking Critically

1. Compare the advantages and disadvantages of using radio versus TV to broadcast news information.

25. **Resource.** Answers to "Understanding Concepts," "Thinking Critically" can be found in the Teacher's Manual at the front of this *Teacher's Annotated Edition* and in the *Teacher's Resource Binder*.

2. The Federal Communications Commission (FCC) sets up rules and guidelines for broadcasting. What are the advantages and disadvantages of such regulation?
3. As more people can afford to buy a cellular telephone, new opportunities to communicate become possible. How do you think this affects their lives, communities, and businesses?
4. How do you think different groups, such as preschool children, seniors, and handicapped people, will be affected by HDTV and digital television?
5. How might satellites benefit national security?

Applying Concepts and Solving Problems

1. **Internet.** Discuss the effects of Internet news versus news in a newspaper or on TV.
2. **Science.** Images on a color television screen are composed of pixels. Use a magnifying glass to look at a patch of yellow on a color television screen to find out which receptors are stimulated so we see yellow.
3. **Mathematics.** The first cellular phones weighed close to 25 pounds, were mounted in the trunk of the car, and cost nearly $3,500. Today, you can purchase cellular phones that weigh about 8 ounces and cost less than $200. Calculate the percentage decrease in product weight and price.

CHAPTER 7

Graphic Communication

Objectives

After studying this chapter, you should be able to:
- discuss the importance of graphic communication in our society.
- identify what makes a good graphic design.
- explain how printing, photography, and drafting work and are used in communication.

Terms

computer-aided drafting (CAD)
drafting
electrostatic printing
graphic communication
gravure printing
holography
ink-jet printing
laser printing
photographic printing
planographic printing
porous printing processes
principles of design
relief printing processes

Technology Focus
Seeing Is Believing—or Is It?

In the past, a graphic artist often had to labor long and hard assembling photographs, artwork, and type for printed products. It was an important, but time-consuming, task. Today's new imaging systems make this work much easier and more fun.

The new graphic artist sits at a computer workstation. He or she can put together complete layouts (arangements of words and pictures) for magazines and posters, all on a computer screen. With the touch of a special electronic pen, a photo can be altered as needed. It can even be changed from a realistic picture into a special effect. A photograph can be moved from the top of the page to the bottom. Is there a tree or building in the background of a photo that is distracting? The graphic artist can electronically remove it. Would you like to combine several pictures? A photo can be altered as needed. It can even be changed from a realistic picture into a special effect. Parts of two or more photos can be combined. For example, Michael Jordan of the Chicago Bulls could appear to be playing basketball on your school playground!

The machine that makes this possible is called a *digital image processor*. It converts pictures into the thousands of tiny squares called *pixels*. The computer analyzes each pixel for color and brightness. The graphic artist chooses which pixels need to be changed and tells the computer how they should be changed. Thanks to systems like this one, printed products can be more clear-cut, more effective, and more exciting to create.

Wonderful as it is, however, this new process raises some questions of right and wrong that did not exist before. For example, the photographer who took Michael Jordan's picture on the opposite page was probably a professional who was paid for it. Taking pictures is how professional photographers make a living. Should the photographer also be paid for the new picture of Michael Jordan at the playground? Who deserves the credit for the playground picture, the original photographer or the artist who altered it?

How would you answer these questions? As technology creates easy and unusual ways to do things, we must give serious thought to the issues they raise.

Take Action!
Using a computer and photo-altering software, such as Adobe Photoshop™, try adding special effects to the sample photos provided with the program.

Resources
- Chapter 7 Lesson Plan in the *Teacher's Annotated Edition* and in the *Teacher's Resource Binder.*
- Chapter 7 Review in the *Teacher's Resource Binder.*
  1. **Enrichment.** Arrange for a visit to a nearby graphics firm for a demonstration of the latest techniques.
  2. **Critical Thinking.** Have the students respond to these questions for a class discussion.

CHAPTER 7: Graphic Communication

What Is Graphic Communication?

Graphic communication is the term given to methods of sending messages using primarily visual means. The basic methods are:
- printing
- photography
- drafting

Think about all the visual messages you receive every day. Photographs, pictures, product labels, signs, newspapers, magazines, books, and mail are just a few examples of graphic communication media. (Remember, media refers to the means of communication.) Can you name others?

Visual Design

To communicate a message effectively using graphic media, the visual design is extremely important. Visual design refers to how something looks.

FASCINATING FACTS

Even the best ideas are sometimes hard to sell. Chester Carlson printed the first photocopy in 1938 and patented his idea in 1940. When he tried to find a manufacturer for his photocopier, he was refused by twenty different companies. Finally, after ten years, a small company in New York bought the rights. That company is now Xerox Corporation.

What Makes a Design Effective?

Certain designs or pictures seem to catch our attention better than others. Many companies and organizations develop their own symbols with which they are quickly and easily identified. These symbols are called *logos*. Think about the logos (short for logotypes) that you know. Fig. 7-1. Why do you think they appeal to you and others? A good design captures the attention of the intended reader or viewer. It effectively communicates a message and leaves a lasting impression.

Designers must consider how a design will be used. They must think about the function of the design. Does it need to appeal to young people or older people?

FIG. 7-1 Logos have been developed for different companies, organizations, and services to communicate their identities quickly.

3. **Reinforcement.** Ask the students to name other examples of graphic communication media.

4. **Reinforcement.** Point out that symbols are an international language.

5. **Reinforcement.** Ask students to name additional identifiable logos.

Will it be used in a magazine or on a billboard? Will it be read and thrown away like a newspaper—or will people look at it for many years as they do a photograph?

Many factors can have an impact on the effectiveness of a design. Some of these will be discussed next.

Principles of Design

When designing a graphic message, the following **principles of design** should be considered: balance, proportion, contrast, variety, harmony, and unity. These factors help to determine the effectiveness of the design.

Balance

The visual weight of images in a design is referred to as *balance*. Some designs are formally balanced. Formal balance is achieved when a line drawn down the center of the design creates two halves that are very similar to one another, or symmetrical. Others are informally balanced. The objects in the design may look different, but they have equal weight to the eye. Fig. 7-2. The viewer must gain a sense of balance from the design—the space, type (words), and artwork should be positioned to give the viewer an impression of steadiness.

FORMAL BALANCE

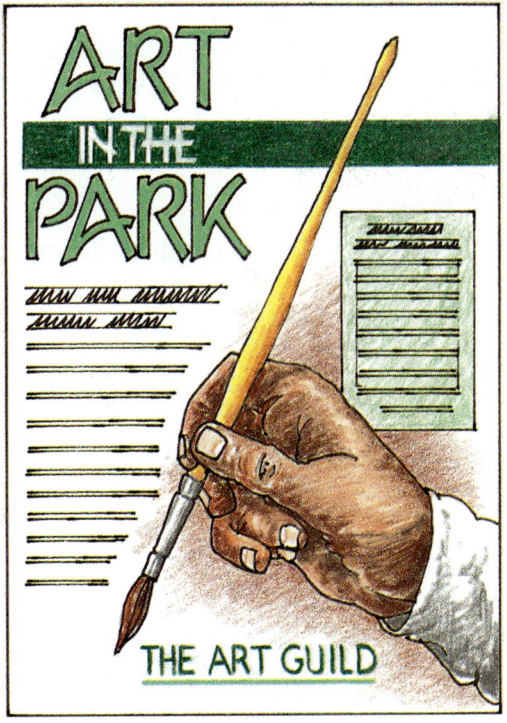
INFORMAL BALANCE

FIG. 7-2 A design with about the same amount of illustration and type on both sides of an imaginary centerline is formally balanced. If there is an unequal amount on either side, but both sides appear to have equal "weight," the design is informally balanced.

6. **Extension.** Explain that what an artist manipulates are the *elements* of design: line, color, shape, etc.

7. **Reteaching.** Have the students cut out magazine or newspaper ads that represent formal and informal balance.

Proportion

The size relationship of the various parts of a design is important. This relationship is referred to as *proportion*.

The sizes of type, drawings, and photographs are all considered carefully. The designer wants to give the right amount of attention and space to each part of the message. A design that is out of proportion will appear awkward and displeasing to the viewer.

Contrast

Contrast refers to techniques used to call attention to certain parts of the message. It is the emphasis in a design. The designer wants the viewer to especially notice important parts of the message. Contrast is often accomplished through simple techniques like underlining, adding color, changing the size of the type, or using arrows ➡➡.

Variety

Variety has to do with adding different things to a design. It's not all the same. It's not boring. Variety adds interest.

Harmony

Harmony refers to the pleasant aspects of a design that encourage the viewer to "look into" it. Harmony addresses eye movement and flow. Harmony is accomplished when the type and pictures are pleasing together. Fig. 7-3.

FIG. 7-3 The elements in this ad are placed in a pleasing way that encourages the reader to look at it. This is called harmony.

Unity

Unity refers to the overall effect of the design. If all aspects of the design appear to belong together and work well together, the design has unity. Unity is achieved when the designer plans the kinds of type, the colors, and the shapes used. If even one element doesn't "fit," the effect of the design can be ruined.

The Creative Design Process

All printed products (magazines, books, labels) are designed first. Designers go through a sequence of steps to create a design that is ready for reproduction. The creative design process may consist of four major steps:
- thumbnail sketches
- rough layouts
- comprehensive layouts
- camera-ready art

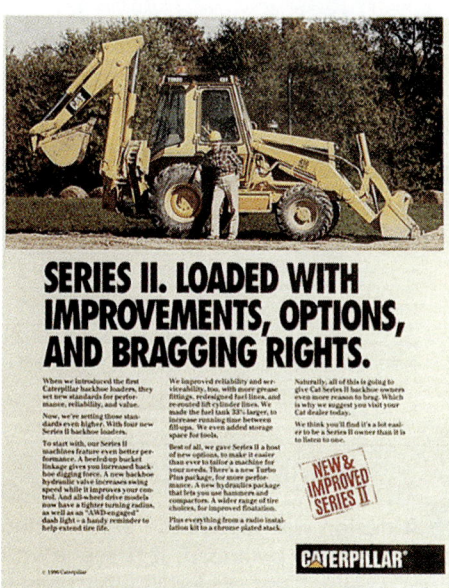

8. **Reteaching.** Display several advertisements and have students identify aspects of the designs that create contrast or emphasis.
9. **Critical Thinking.** Distribute copies of a newspaper ad. Have students cut the ad into separate components and redesign it, deleting components and adding art or line work as desired. Emphasize harmony.
10. **Activity.** Have students design a product package using the four steps.

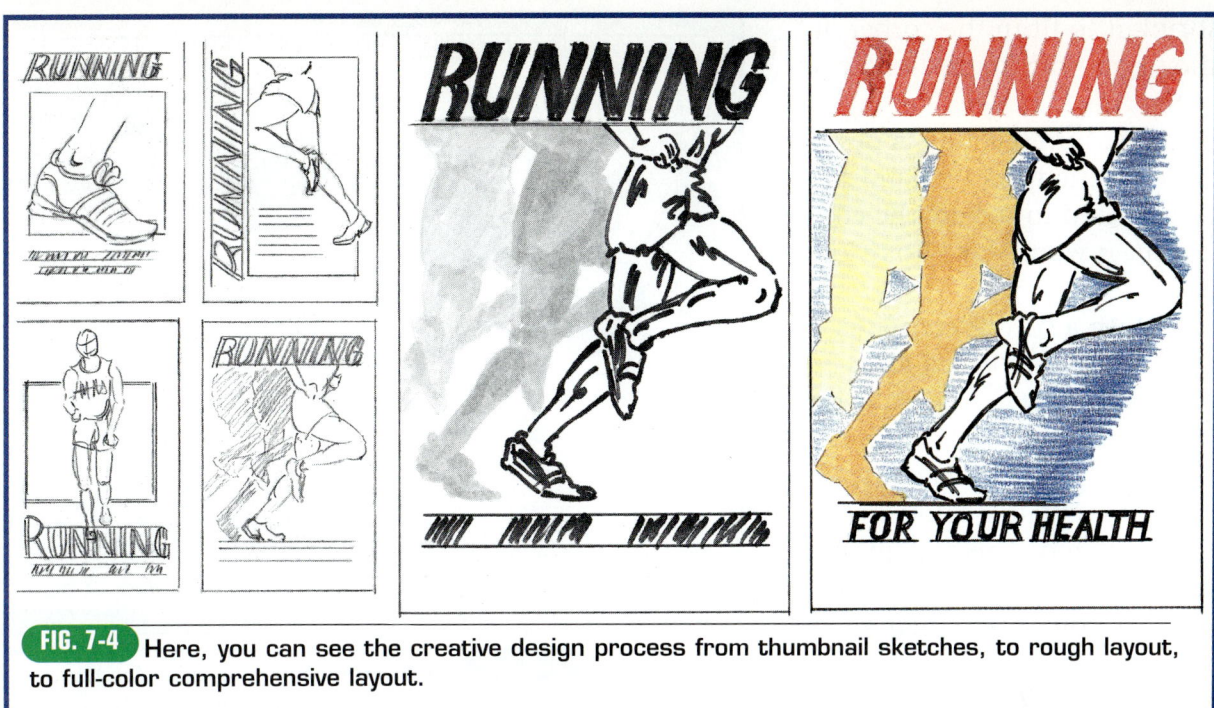

FIG. 7-4 Here, you can see the creative design process from thumbnail sketches, to rough layout, to full-color comprehensive layout.

Most designs start out as thumbnail sketches. Designers will often come up with many ideas before choosing the one they like best. A rough layout and a comprehensive layout are made of the design that a designer thinks is most appealing and functional. Camera-ready art (often called a paste-up) is prepared after the layout stages are complete. Computers and other special tools are used to make camera-ready art.

The first three steps of the creative design process are shown in Fig. 7-4. The last step of this process, camera-ready art, shows the type, art, photographs, and lines combined together. Everything is accurate and laid perfectly into place. The design is ready to be prepared for printing reproduction.

Designing with a Computer

Today, designers can use computers to create thumbnail sketches, rough layouts, comprehensive layouts, and camera-ready art. The computer makes it easy to combine text with illustrations. In fact, a person could use a small desktop computer to completely lay out a publication of his or her own. This is called *desktop publishing*. Many new software programs make it easy to do this work. Fig. 7-5.

Using a computer can make the creative design process much easier. This is because the computer makes changes quickly and accurately. Using a computer, a designer can turn 1-inch letters into 10-inch letters or move a picture from the top of the page to the bottom. The designer can fill a page with an illustration or

11. **Enrichment.** Invite a graphic artist to talk with the students about the creative design process.
12. **Critical Thinking.** Have students design a layout for a poster of an upcoming school event.
13. **Enrichment.** Have the students lay out a page using a desktop publishing program. They should keep in mind the principles of design.

CHAPTER 7: Graphic Communication **147**

shrink it to fit in the corner of the page. With the aid of a computer, these changes can be completed within seconds. A designer doing the work by hand would need to redraw the entire layout. It would take hours or days.

14 With a special computer graphics system called an *electronic pagination system,* a designer can assemble and prepare complex, full-color layouts. The designer can alter colors in photographs, too. This system uses scanners to input photographs into the computer's memory. The scanner turns the photographs into digital information that the computer can understand. Then the designer can alter that electronic information. Fig. 7-6. For example, a yellow dress can be changed to a bright red one. A black car can be changed to blue or white within seconds. The designer can also combine several photographs into one photograph.

The design process still relies on the creative and artistic skills of designers. However, digital technology has greatly affected how these designers turn their ideas into camera-ready art. Designers once spent most of their time using markers, paints, pencils, rulers, and adhesives in order to get layouts ready for printing. Today, many designers rely more on keyboards, monitors, mice, and color printers to do their jobs.

FIG. 7-5 Desktop publishing is easy for the home computer user. New computer graphic software packages make the task of designing and laying out signs, newsletters, reports, and other printed media much easier. Here you can see some of the options available on this drawing program.

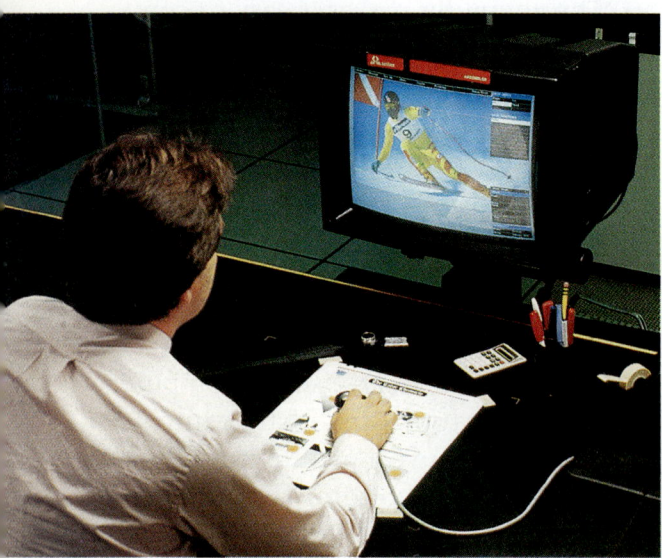

FIG. 7-6 Electronic pagination systems combine color scanners with computers. Designers use these systems to prepare camera-ready layouts for products to be printed.

14. **Enrichment.** Take the students to see an electronic pagination system at work.

SECTION 2: Communication Technology

Printed Communication

After the message is designed and camera-ready art is prepared, the message is converted into a form that can be printed. Many stages are involved in this process, but we will concentrate mostly on the printing.

The Importance of Printing

Most graphic messages sent and received daily involve some type of printing. Cereal boxes, candy wrappers, and wallpaper are printed. Printing is everywhere around us. Expressing messages without printed words and pictures would be difficult. Fig. 7-7.

Printing processes have made it possible for people to send many copies of a message in exact duplicate. In addition, printing has provided a way to preserve knowledge by recording information.

Before printing was invented, people recorded information by hand. If several copies were needed, someone would hand-write the information many times over. The invention of methods that made it possible to duplicate printed information were vitally important to the advancement of our culture.

Types of Printing Processes

Printing can be done in a number of ways. The type of printing process used depends on what is being printed, the quality of reproduction desired, costs, and the speed at which it must be printed.

Generally, printing techniques are grouped according to the type of printing surface used. Major groups of printing processes are: relief, porous, planographic, gravure, and electrostatic. Other printing processes include photography, ink-jet printing, and laser printing.

The processes used to get messages onto a surface have changed considerably over the years. Different processes are used for different reasons. However, the basic idea is still the same. No matter how the message is printed, it must be visually effective. As computers and other devices begin to change our printing machines and methods, keep in mind that the message being communicated is what's most important.

FIG. 7-7 How would you know what is inside a container that had no printed information on it? These containers could hold milk, orange juice, or eggnog.

15. **Reinforcement.** Discuss the impacts printing has had on the advancement of our culture.
16. **Reinforcement.** Show the students examples of the different types of printing processes.
17. **Enrichment.** Arrange a trip to a local printer or show a movie overviewing printing operations.

CHAPTER 7: Graphic Communication

MATHEMATICS CONNECTION
Measuring Type

Printers, typographers, and other people who work with material set in type have their own measuring system. This system accounts for sizes in the type itself and in the space between one line of type and another. The basic units of type measurement are the point and the pica.

A *point* is equal to 1/72 of an inch. Point size is measured from the top of a capital letter to the bottom of the descender on a lowercase letter. (The descender is the part of some lowercase letters—g, j, p, q, and y—that goes below the main body of the letter.) In general, capital letters are about two-thirds the height of the overall point size. The type shown below measures 18 points. To find its size you would measure from the top of the T to the bottom of the y, as shown.

18 POINTS Type

A *pica* is equal to 12 points (or 1/6 inch). Printers use picas to measure larger sizes, such as the length of a line or column of type. The main text in this book, for example, has a line length of 18.6 picas. The full columns are 49.6 picas deep.

A typographer's rule is used to measure points and picas. The rule has several other scales, including those for inches and rule widths.

Try It!

> Using a typographer's rule, measure the following lines of type. How many picas long are they? What type size is used in each case?
>
> **Myron liked to measure type.**
>
> By now the sun should be shining.

Answers to these questions can be found in the Lesson Plans in the Teacher's Manual at the beginning of this book.

SECTION 2: Communication Technology

Relief Printing Processes

Relief printing processes are methods that print from a raised surface. Fig. 7-8. Parts of the image are raised above the surface. These pick up ink and transfer it to paper. Printing processes that use raised printing surfaces include letterpress and flexography.

Letterpress is a very old form of printing. Johann Gutenberg is given credit for inventing the process upon which this technique is based. Gutenberg, the "Father of Printing," invented movable metal type. In *letterpress*, the reverse or mirror image of each character (letter of the alphabet, punctuation mark, or symbol) was made into a piece of metal type. Printers hand-set lines of metal type, letter by letter, to form sentences, paragraphs, and pages. Ink was rolled over the raised letters. Then, paper was pressed on the inked letters to transfer the message. This process was repeated until the desired number of pages were made.

Today, letterpress printing is almost entirely phased out of most companies. New machines can print messages more quickly and efficiently.

In *flexography*, raised letters are formed on plastic or rubber sheets. Letters do not have to be hand-set. Instead, they are produced using photographic and computer methods that are very fast.

Porous Printing Processes

A porous material has many small openings (pores) that allow liquids to pass through. In **porous printing processes**, ink or dye is passed through an image plate or stencil and transferred onto the material being printed. A *stencil* is a thin piece of material with holes cut out or etched in the shape of the desired design. Stencils can be made many different ways. The material being printed on (the receptor) may be paper, cloth, glass, or another material. The most common porous printing process is screen printing. It is sometimes referred to as silk-screen printing.

In screen printing a stencil is adhered (made to stick) to a porous screen. All areas of the screen are blocked out except for the area to be printed. A special tool, called a squeegee, is used to force the ink through the stencil and onto the product being printed. Fig. 7-9.

Screen printing is a simple and flexible printing process. This type of process can be used to print on almost any surface available. It is commonly used on fabrics, such as T-shirts and sweatshirts. Even three-dimensional products like drinking glasses, mugs, or shampoo bottles are commonly screen printed.

FIG. 7-8 Relief printing is printing from a raised surface.

18. **Reinforcement.** Have students make a linoleum block print of their initials or a logo.
19. **Extension.** Show the students movable metal type and type trays.
20. **Extension.** Demonstrate for the students silk-screen printing.
21. **Extension.** Show students some products that have been screen printed.

CHAPTER 7: Graphic Communication

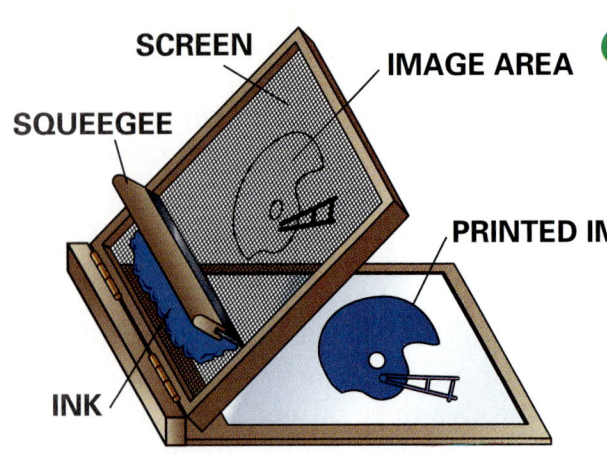

FIG. 7-9 Screen printing is a porous printing process. Ink passes through a stencil attached to a porous screen onto the surface below.

Planographic Printing Processes

Any process that involves the transfer of a message from a flat surface is called **planographic printing**. Offset lithography is the most popular planographic printing method. It is also the most common printing process used today.

Lithography is based on the principle that grease and water do not mix. An image is created on a flat plane using a material that has a grease base. The image attracts ink because ink, too, has a grease base. All areas of the printing surface that do not contain the image are covered with a water mixture, which repels the ink. Multiple copies of the image can be made by coating the image areas with ink and the remainder of the plate with the water mixture on a continuous basis with a printing press. Fig. 7-10.

Gravure Printing Processes

Gravure, or intaglio, printing is the exact opposite of relief printing. In **gravure printing**, images are transferred from plates that have sunken areas. The images are etched or carved into the surface. Each plate has many tiny holes, called cells, that combine to form the shape of the letters, symbols, and other design elements. The cells are filled with ink. Then, paper is forced against the plate. The paper absorbs the ink. The image transferred to the paper is the identical form of the image on the plate. Fig. 7-11.

FIG. 7-10 Lithography is based on the principle that grease and water do not mix. Because the printing is done from a flat (plane) surface, this is a planographic printing process.

22. **Reteaching.** Using a grease pencil and a stone, demonstrate this basic principle of lithography.

23. **Reteaching.** Demonstrate gravure printing by etching an image on a plexiglass plate.

152 SECTION 2: Communication Technology

FIG. 7-11 Gravure printing transfers images from a recessed surface.

Gravure printing is an expensive technique, but it provides very high-quality results. Gravure printing plates require precise equipment. The plates are large copper drums that are etched with the message or design. Etching is done using either chemical processes or a diamond etching tool. Because the plates are so expensive to create, this process is generally used for long press runs or jobs that require high quality. In fact, all U.S. paper money is printed using this technique. Some magazines, wallpaper, stamps, and newspaper supplements are also printing using gravure.

Electrostatic Printing Processes

Many quick printing companies use electrostatic printing processes. In fact, you have probably used this technique a few times yourself at the library or local copy center. Copier machines that you use to make quick duplicates of your original fall into this printing category.

Electrostatic printing is based on the principle that opposite electrical charges attract, whereas like charges repel. Basically, when your original is placed on the glass window, it is exposed to a plate inside the copier. The plate has been given a positive charge. A light removes the charge from the non-image areas of the plate. The image area remains positively charged. A toner material is given a negative charge. It sticks to the positive image areas of the plate. A paper is given a positive charge and passed over the toner plate. Because opposite charges attract and the positive charge is stronger than the negative charge, the toner is transferred to the paper. A heating element is used to set the toner on the paper permanently. Fig. 7-12.

The electrostatic printing process makes it possible to duplicate messages quickly. It is even possible to make color copies with this process. However, the reproductions are generally of poorer quality than those made using other printing techniques. For long runs, this process is too slow and expensive.

Photographic Printing

Photography is often considered to be a printing technique. Basically, in **photographic printing**, light is projected through a plate (usually called a negative) onto a light-sensitive material. After processing, the image appears. (Photography is discussed in greater detail later in this chapter.)

24. **Reteaching.** Demonstrate for the students that opposite electrical charges attract and like charges repel.
25. **Enrichment.** Demonstrate static electricity. Rub a plastic comb on a piece of nylon, fur, or wool and bring the charged comb near puffed rice, sawdust, or torn paper bits.

FIG. 7-12 This diagram shows the process involved in electrostatic printing.

Ink-Jet Printing

Another printing technique that does not easily fit into the other categories is called **ink-jet printing**. During this process, ink jets spray ink onto the surface to be printed. This process is computer-controlled. Digital data control the tiny nozzles that spray the ink droplets onto the receptor material. Because there is no contact between the image carrier and the material being printed upon, the process can be used to print on a wide variety of materials and in full color. In fact, some artists use large-format ink-jet printers to reproduce their work. These printers can print high-quality copies up to 40 inches wide on high-quality paper. Fig. 7-13.

FIG. 7-13 Large format inkjet printers can place more than four million droplets of ink per second on paper or canvas. This makes high-quality reproductions of artwork and layouts possible.

26. **Extension.** Show examples of ink-jet printing (bottom of soda can, some types of bulk mail, etc.).

Laser Printing

Laser printing is one of the newest forms of printing today. As you know, lasers are devices that strengthen and direct light to produce a narrow, high-energy beam. The most common laser printers work much like electrostatic copiers. They print onto regular paper. Many homes, schools, and businesses have these printers hooked up to their desktop computers. They provide good-quality output, and they operate much more quietly than typical computer printers. Other laser printers are designed to expose photographic paper. These laser printers are used to make very high-quality images and type to be used for camera-ready art. They are commonly found in art departments or printing companies.

27

Digital Technology and Printing

Digital technologies play a vital role in printing today. Integrated circuits and computers are an important part of nearly all modern printing processes. In addition to helping designers prepare camera-ready art, these devices can be found in every aspect of the industry. They ensure accurate ink coverage on your newspaper. They help to determine how black, cyan (blue), yellow, and magenta (red) inks can be combined to form whatever color the designer specified. They control printing operations and drive the machines that convert plain white paper into glossy color posters and vibrant magazines.

Computers also perform another important function through a technology called *printing on demand*. Here computers are used to print only the number of products needed at a specific time. Because everything is stored digitally, future printings can be made when needed and changes can easily be made, if desired. For example, a company may publish 100 copies of its product catalog in June. If quantities run low, another batch of 50 may be produced in July. When some prices and product lines change in August, new catalogs can be quickly updated and reproduced. This process helps to reduce costs because there is less waste from overproduction. Salespeople and customers benefit from having accurate materials on hand.

Communicating through Photography

Often, photographs can relate so many ideas and feelings that words are hardly necessary. Most graphic communication media depend on photography to send a message. Can you imagine magazines and newspapers that used only words? Photographs help us to capture feelings, expressions, concepts, informa-

Fascinating Facts

Chickens played a role in the history of photography. In the mid-1800s, photographic paper was coated with development chemicals and egg albumen. Some companies were using as many as 60,000 eggs a day for developing pictures.

27. **Extension.** See Chapter 15 for a Science Connection article on how lasers work.

tion, and images using a camera and film. In science and technology, photos enable researchers to study microscopic forms and to explore the failure or success of experiments. Teachers and authors use them to educate and to explain information. The rest of us use photos to record times and events in our lives.

The Photographic Process

The photographic process is similar to the way your eyes work. Everything reflects light. You see objects because your eyes pick up and focus the reflected light. In photography, light is reflected from a subject and focused through the lens of a camera onto film. When the film is developed, you see a reproduction of what the camera "saw." This image was captured when light rays passed through the camera lens and reached the film. Fig. 7-14.

Using a Camera

Many kinds of cameras are available today to help us record images. Regardless of the type or size, most have the same basic parts that work on the same basic principles. A basic camera requires six features or devices:
- light-tight space
- lens
- aperture with aperture diaphragm
- shutter
- filmholder
- viewfinder

It is necessary to control the light that reaches the film. Every camera must have a completely *light-tight space* in

FIG. 7-14 A camera works much like the human eye does but is much less sophisticated. Light passes through the cornea of the eye, which bends and focuses it. It then passes through a small hole called the pupil and reaches the lens. As light passes through the lens, the image is focused on the retina. In a camera, light passes through the lens, which bends and focuses it on light-sensitive film in the rear of the camera.

28. **Extension.** Have students make a pinhole camera.
29. **Reinforcement.** Use several different kinds of cameras to identify for the students the six basic parts of a camera.

SECTION 2: Communication Technology

FIG. 7-15 The size of the aperture is identified by numbers called f-stops. Notice that the larger the f-stop number, the smaller the opening. Small pictures near the settings like those shown here are sometimes used. These show that on cloudy days the opening must be large; on bright sunny days it should be small.

which to store the film in order to record an image.

A *lens* is needed to focus the image on the film. The lens directs the light rays so that they are correctly positioned and do not appear blurred.

An *aperture* (opening) lets light into the camera. An aperture diaphragm is required to control the amount of light that can reach the film. Just as the iris of your eye adjusts to changing light, the size of the aperture must be adjusted to varying light conditions. By adjusting the aperture diaphragm, the amount of light allowed to reach the film can be increased or decreased to get a good exposure. Fig. 7-15.

Every camera must have a shutter. A *shutter* is a device that controls the amount of time that light is allowed to reach the film. When you take a picture and hear a click, you are hearing the shutter open and close. It is like opening a door for a set amount of time. Usually, this set time is a fraction of a second, like 1/250th of a second or 1/30th of a second.

Next, a *filmholder* is needed. This holds the film in position so that it will not move during an exposure.

Lastly, the photographer looks through a *viewfinder* to locate the subject and position the subject or the camera for the picture. The viewfinder is like a window. It allows you to see the image being photographed.

Photographic Film

Photographic film comes in a variety of formats and sizes to fit different cameras and applications. Rolled film comes in canisters or cartridges that fit inside your camera. Sheet film is most often used by professional photographers and printers who will be duplicating photographs.

Despite the size or shape of films, they have one thing in common. They contain one or more thin layers of a light-sensitive material called an emulsion. The *emulsion* is the part of the film that captures the image. It must be chemically processed before the image can be seen.

30. **Reteaching.** Place a 35mm camera on an overhead projector with lens facing down on the glass and the camera back open (no film). Demonstrate size variations of the aperture.

31. **Reinforcement.** Show the students examples of rolled film and sheet film.

Films are rated according to how sensitive their emulsion is to light. This is often called the speed of the film. A high-speed film, such as ASA/ISO 400, is very sensitive to light. It does not need much light for a good exposure. A low-speed film, such as ASA/ISO 100, needs more light for a good exposure. Photographers choose films according to the lighting conditions where they will be shooting the film. Which type of film do you think would be better to use if you were taking photographs outside around the evening campfire?

Digital Photography

[32] Breakthroughs in photographic technology are improving the speed at which images can be recorded. New digital camera systems can record still and motion images on computer disks instead of film. This makes it possible to see the picture or motion instantly on a television screen or computer monitor. You don't need to take a roll of film to a store to be processed and have prints made. Fig. 7-16.

In digital photography, the image is stored as digital information. As such, it can be transmitted to other locations using telephone lines. You can send it to a friend using electronic mail or post your photographs on your web site for others to enjoy. Digital photographs are easily modified, too. You can crop out parts that you do not like, alter colors, or add special effects. Then you can integrate them with a letter or greeting card that you make on your computer. Photographs can also be output in full color on a special printer. The paper is laminated to give it a glossy finish like a real photograph.

People who do not have digital cameras can still reap the benefits of digital

FIG. 7-16 Digital cameras make it easy to get a perfect shot, because you can see the results instantly. You can view your shots on a TV or upload the images to your computer. If you don't like it, you can reshoot immediately. Color printers give quality output on paper or stickers.

photography. Some companies will process regular film and convert it to the digital format. They can supply regular photographic prints and a computer disk containing all the photographs.

Photographic Composition

Photography is a very effective communication medium. Certain factors make some photographs better than others. These factors are guides to composition. *Composition* is the way in which all the elements in a photograph are arranged. When you compose a photograph, you plan what will be in the picture and how it will be positioned. To communicate effectively and to create a visually pleasing photo, you will want to compose it carefully. Consider some of the following tips when composing a photo.

32. **Reteaching.** Demonstrate shutter speeds using the same setup you used to demonstrate the aperture.

Discuss the relationship between shutter speed and aperture.

Balance

Just as layouts should appear balanced, so too should the subjects in a photograph. When composing a photograph, consider the positions of the subjects. Ask yourself questions like, "Is there too much on one side of the picture?" or "Is there appropriate space at the top and bottom of the picture?"

Photographers often follow the rule of thirds to position a subject in a photograph. The *rule of thirds* divides the image area into thirds both vertically and horizontally. Composition is more interesting when the subject is located at the intersections of these imaginary lines, rather than the exact center. Fig. 7-17. By positioning the subject off-center at one of these locations, the viewer gains a better sense of balance. The picture is more visually pleasing.

Framing

A photo is sometimes more effective if the photographer uses a foreground image (part of the scene nearest the viewer) to frame a background image. This is a good technique to use when photographing scenic landscapes. Other times a background image can be used to frame a subject. For example, the trunk and branches of a tree can be used to frame a person standing in front of the tree.

Simplicity

A photographer should usually keep photos simple. This is done by concentrating on the center of interest in a picture. Background can be distracting to viewers. Sometimes, such as in portraits, no background is needed.

FIG. 7-17 This photo clearly shows the rule of thirds. Note that the main subject is placed off center where the lines intersect.

33. **Reinforcement.** Ask the students to bring in a photograph and apply the rule of thirds.

FIG. 7-18 The lines created by the rows of flowers and the sidewalk "lead" the viewer to the building.

Leading Lines

A photographer often needs to "lead" the viewer into a picture. By positioning a subject appropriately, leading lines can be used to pull the attention into the scene. This technique is especially important when photographing subjects that have apparent line structures. For example, the "lines" created by roads, bridges, fences, and buildings can be seen easily. Fig. 7-18.

Holography

Holography is the use of lasers to record realistic images of three-dimensional objects. In a photograph, you see only one side of the subject. In a hologram, which is the three-dimensional image produced by holography, you see the subject from different angles as you change your view. You can see the front, back, and sides. It is as if you were walking around it!

Holograms are a good way to record many details about an object. When viewed, they look more realistic than a photograph. It seems like the object is right there in front of you and that you can pick it up and hold it. Holograms are not used as often as photographs. They are more difficult and expensive to reproduce. For this reason, some companies use them for security purposes. Credit card companies and software makers have used holograms for several years in the hopes of deterring counterfeiters. Fig. 7-19. The technology to make and reproduce holograms is improving all the time. One day, holograms will be used in places where you now see photographs.

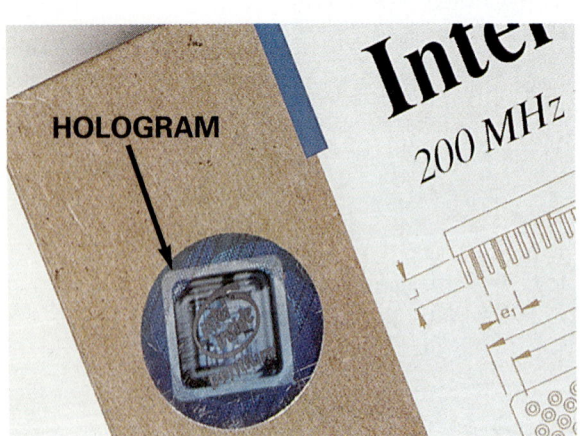

FIG. 7-19 The hologram shown on this product is of a computer chip.

34. **Extension.** Watch for newspaper and magazine articles detailing new developments in photography. Share these ideas with students.

160 SECTION 2: Communication Technology

Drafting: The Universal Language

Drafting is the process of accurately representing three-dimensional objects and structures on a two-dimensional surface, usually paper. It is an accurate drawing process used for nearly every product or structure made today—large or small. Integrated circuits, shoes, tools, cars, bridges, and skyscrapers are just a few examples of things that start as drafted drawings.

Drawings are also used to communicate ideas effectively and accurately. For example, all the design details of a building could not be verbally described to construction workers. Fig. 7-20. To describe the details of a planned product or structure in a way that others can create an exact model of the planned idea, drafting is required.

FIG. 7-20 This house floor plan is an example of a drafted drawing. It accurately shows the placement and size of each room and other structural features, such as fireplaces, stairs, windows, doors, and closets.

CHAPTER 7: Graphic Communication

Drafting is often referred to as the universal language. A person who understands the basic symbols, lines, and rules can understand the message regardless of who communicates it. You don't need to understand the Japanese language in order to understand a drawing made by a Japanese-speaking person. People who are skilled at recording and understanding drafted messages are called drafters. Many other careers also require a thorough understanding of drafting. Architects, engineers, electricians, and plumbers, for example, regularly make or use drafted drawings.

The Process of Drafting

Drafters use many different techniques and tools to create a message. They can create messages using simple tools like pencils, rulers, T squares, triangles, and compasses. However, today drafters often use computers in much the same way that graphic artists use them to design printed matter. Regardless of the materials and tools used to communicate the message, the thinking process is the same.

The first step in creating a drafted drawing is sketching. Sketches are often created freehand and are drawn quickly. They may also be done using the computer. Sketching allows the drafter to experiment with design ideas. Fig. 7-21. Sketches show the ideas roughly but neatly. Sketches should also show the proper size relationship between parts.

After the drafter chooses the sketch that best represents the idea, detailed drawings must be made. The goal of drafting is to describe an object or structure so accurately that someone else can use the drawings to create it. To achieve this goal, the drawings must accurately describe the shape, size, dimensions, and details of the object and all its parts.

Drafters use many different types of drawings to describe the shapes of things accurately. These drawings show the object from different views and angles. Each view communicates a different message. Each drawing has a different purpose and use.

Multiview Drawings

One common type of drawing is called a *multiview drawing*. It shows two or more different views of an object drawn at right angles, or perpendicular, to one another. The technique used is called *orthographic projection*. The best way to think about this process is to picture the object that you want to draw inside of a glass box. Each view of the

FIG. 7-21 A designer often creates many sketches of possible designs before creating accurate, technical drawings of the best design.

35. **Extension.** Let the students examine a set of blueprints. Draw their attention to the symbols and shapes rather than to the words.
36. **Extension.** Demonstrate for the students how to use these drafting tools.
37. **Extension.** Have students draw a picture of an object from your verbal description of it. Compare different interpretations.

SECTION 2: Communication Technology

FIG. 7-22 Picture the bookend inside a glass box. When the box is unfolded, all six views of the bookend appear on a flat surface.

object is projected a different side of the imaginary box. When the box is unfolded, you can see each view as it would appear on a piece of paper. Fig. 7-22.

The position of each of these views is a standard drafting practice. Note that the rear view is to the left of the left-side view. It is often possible to completely describe the shape of the object using only three views: top, front, and right side. The object's height can be shown in the front and right-side views. The object's width can be shown in the front and top views. The object's depth is seen in the top and right-side views. Fig. 7-23.

In order to create accurate multiview drawings, drafters must understand and follow standard drafting practices. For example, every detail of the object must

FIG. 7-23 This is an orthographic drawing of a cabinet. Most orthographic drawings show the top, front, and right side. However, the more complex the object, the more views the drafter must provide.

CHAPTER 7: Graphic Communication   **163**

FIG. 7-24 Note the various types of lines used to describe the shape of this product shown in three views. Hidden lines show surfaces that are not visible in that view. Centerlines show centers of holes. Object lines define the visible shape of the object.

be shown in each view. If a certain feature cannot be seen from a particular view, it must still be shown with special lines or symbols. Study Fig. 7-24. Notice that the holes in this object are shown as dashed lines in some views. This means that the holes are hidden in that view. Notice that some lines have curves on the end to show the curved surfaces of the object. Centerlines are shown as alternating long and short dashed lines. They show the centers of holes and arched shapes. Object lines are the most prominent lines on the drawing. They represent all surfaces and details that are visible from that view.

Many multiview (or three-view) drawings are called *working drawings*. This is because they are a very common type of drawing used in manufacturing and construction industries. Products and structures are made from these working drawings. Working drawings are fully dimensioned to show the exact size, shape, and location of every detail. Drafters must indicate height, width (or length), and depth of every feature. Different symbols may be used to show the material from which something is made. Thousands of symbols are used to represent specific types of fasteners, electrical components, plumbing fixtures, and structural parts.

Pictorial Drawings

Other drawings are used to show an object in depth. The object is tilted or rotated so that three sides are visible in one view. *Pictorial drawings* show objects as they appear to the human eye. Most people can easily understand a pictorial drawing because it looks quite realistic.

Pictorial drawings are a useful communication medium. They clearly communicate how something looks, and they help people to visualize an object's appearance. For example, architects and builders rely on pictorial drawings of houses in order to communicate with people who want to build new homes. Homeowners rely on pictorial drawings in books and manuals in order to fix leaky faucets, assemble new lawn mowers, and repair broken appliances.

Pictorial drawings generally fall into one of three categories: isometric, oblique, or perspective.

- *Isometric*. In an isometric drawing, the object is tilted so that its edges form equal angles. (Isometric means of equal measure.) In isometric drawings, one corner always appears closest to you. Fig. 7-25.
- *Oblique*. An oblique drawing is made from the front view, with the top and side views lying back at any angle other than 90°. The front of the object appears closest. Fig. 7-26. This method is often used in drawing irregularly shaped objects.
- *Perspective*. Perspective drawings resemble the way something would appear in real life. Think about when you look down a railroad track or long, straight highway. The track or road seems to get narrower and come together at some distant point, doesn't it? In perspective drawings, receding parallel lines appear to come together in the distance. Fig. 7-27.

Technical illustrations may be based on any type of pictorial drawing (isometric, oblique, perspective). Technical illustrations provide technical information in a visual way. They often include shading to

FIG. 7-25 This is an isometric drawing of the cabinet shown in Fig. 7-23.

FIG. 7-26 This is an oblique drawing of the cabinet shown in Fig. 7-23.

FIG. 7-27 A perspective drawing often looks more realistic. The lines appear to meet at a distant vanishing point. Compare the actual scene in the photograph (above) with the drawing.

CHAPTER 7: Graphic Communication

FIG. 7-28 This type of pictorial drawing is known as an exploded view because it shows the object as if all the parts have been pulled apart. Here you can see how the parts of the wrench handle fit together.

help improve clarity or add emphasis. Technical illustrations are commonly found in repair manuals and assembly instructions. They are often accompanied by graphs, charts, or diagrams. Exploded views are one type of technical illustration. They show how parts of a product fit together. Fig. 7-28.

Computer-Aided Drafting

For hundreds of years, drafters have used hand tools. Today **computer-aided drafting (CAD)** systems are used more often in businesses, industry, and homes.

A CAD system includes special computer graphics software that makes it easier to create accurate technical drawings using a computer. CAD software provides the electronic tools needed to create multiview and pictorial drawings according to the specifications of the CAD operator. CAD programs usually have large data banks of common drafting symbols that the operator can simply pick and place into the drawing. Using various computer commands, the CAD operator can create any shape imaginable to any scale desired. Dimensions can be added and color can be used to enhance details and add clarity. Using commands such as "copy" or "move," the CAD operator is able to modify drawings quickly and accurately.

When drafting was done by hand, sometimes an entire drawing would need to be redrawn in order to change details. A digital drawing can be changed quickly and a new copy can be output to a printer or plotter.

All CAD drawings are saved as digital data that can be easily sent electronically to other locations using electronic mail, fax machines, or the Internet. They also may be electronically integrated within reports or other written documents.

One area of CAD that is increasingly used in industry is called *modeling*. Modeling enables the drafter to create a three-dimensional representation of an object on the computer. Models can be made as wireframe models or solid models. They both look three-dimensional, but the wireframe model does not have solid surfaces. Fig. 7-29. Engineers often use solid modeling to assess mass properties, stress, and thermal analysis of products before they are ever built.

New devices are being developed and marketed that also allow you to *digitize* (change into electronic data) real objects. A special stylus, or arm, passes over the object, "tracing" its shape. As it moves, the computer creates a model on the screen. Fig. 7-30. Once digitized, the computer model can be edited and enhanced using various software packages.

38. **Enrichment.** Invite a CAD operator to talk to the class.

SECTION 2: Communication Technology

FIG. 7-29 The impeller shown on this computer screen was created with CAD software that is capable of modeling. The larger impeller on the main screen is an example of solid modeling. A wireframe model can be seen in the upper right corner.

FIG. 7-30 This digitizing arm can be used to trace a three-dimensional object in order to create an electronic computer model. The model can be modified using various software packages.

CHAPTER 7: Graphic Communication **167**

Career File—Technical Writer

EDUCATION AND TRAINING
Employers prefer college graduates with a degree in journalism and some knowledge about a specialized field—engineering, business, computers, or one of the sciences.

AVERAGE STARTING SALARY
Beginning salaries for technical writers average about $18,000 per year.

OUTLOOK
Through the year 2005, technical writing positions are expected to be keenly competitive, according to the Bureau of Labor Statistics. However, employment is expected to increase faster than average because of the continuing increase in scientific and technical information.

What does a technical writer do?
Technical writers make scientific and technical information understandable to a nontechnical reader. They prepare instruction manuals, catalogs, parts lists, assembly instructions, sales promotion materials, and project proposals. They also oversee preparation of illustrations, photographs, diagrams, and charts used in published materials.

Does most of the work involve data, people, or things?
Technical writers work with all three. They gather information from technical personnel and specifications. Most writers use computers to prepare drafts, and many use desktop or electronic publishing systems.

What is a typical day like?
During or after gathering information about a subject, technical writers organize the material in a logical way. Depending on the purpose of the written piece, the writer may develop step-by-step instructions that readers must follow. Initial drafts are reviewed for accuracy by technical personnel, for usability by non-technical people, and for grammar and punctuation by an editor. The writer changes the material based on these reviews.

What are the working conditions?
Most technical writers work in comfortable offices. The search for information may require visits to different workplaces, such as factories, offices, or laboratories. Writers may work overtime to meet deadlines. Many technical writers work for computer software firms or manufacturers of aircraft, chemicals, pharmaceuticals, and electronic equipment.

Which skills and aptitudes are needed for this career?
People who love to write and are able to express ideas clearly and logically are well-suited to technical writing. Creativity, curiosity, a broad range of knowledge, self-motivation, and perseverance are also valuable.

What careers are related to this one?
Other occupations whose primary purpose is to communicate information and ideas include newspaper reporters, radio and television announcers, advertising and public relations workers, and teachers.

One technical writer talks about his job:
I like making technology easy to understand. It's also great to see your own words in print.

SECTION 2: Communication Technology

Chapter 7 REVIEW

 ## Reviewing Main Ideas

- Balance, proportion, contrast, variety, harmony, and unity are the principles of design.
- Printing processes are divided into categories according to the surface from which the image is printed.
- In photography, light is reflected from a subject and focused through the lens of a camera onto film.
- Drafters create many types of drawings that can accurately record information about the size, shape, and dimension of objects and structures.

 ## Understanding Concepts

39
1. What is a logo? Why do companies and organizations develop logos?
2. Name and briefly describe the four steps in the creative design process used by graphic designers.
3. What is the main difference between relief printing and gravure printing?
4. Name the six basic features every camera must have and briefly tell the function of each.
5. What is the goal of drafting?

 ## Thinking Critically

1. Analyze the visual design of this textbook's cover. Consider its appeal and function. Evaluate how each of the principles of design were applied.
2. Computers have altered the creative design process. Discuss the advantages and disadvantages of such changes.

39. **Resource.** Answers to "Understanding Concepts" and "Thinking Critically" can be found in the Teacher's Manual at the front of this *Teacher's Annotated Edition* and in the *Teacher's Resource Binder.*

3. Discuss ways that graphic communications media sometimes give us too much information.
4. How could holograms be used in schools to help students learn?
5. How would one type of drawing be better for the builder of a house and another better for the buyer?

 ## Applying Concepts and Solving Problems

1. **Language Arts.** Select a product and develop an advertising campaign. Organize your thoughts and outline your strategy before you begin. Present your final promotional campaign to the class.
2. **Science.** Use an eyedropper to drop water on various surfaces to check porosity. Test natural materials, such as different kinds of paper, fabric, and ceramics. When would porosity be important in choosing materials for a printing process?
3. **Mathematics.** Drafting requires an understanding of scale measurement. Make a scale drawing of one room in your home. Use a scale of 1/2" = 1'. Do a second drawing of the same room using a scale of 1/4" = 1'.
4. **Technology.** Select a simple square or rectangular object, such as a pencil box, and make a multiview drawing of it. Show at least three views.

CHAPTER 7: Graphic Communication   **169**

1 DIRECTED ACTIVITY

Finding Information on the Internet

Context

Because there is so much information available on the Internet, it can be difficult to find what you need. Most often people use an Internet "browser" or search engine to locate a topic. You simply type in a word or statement related to your topic and the browser searches the Web for related words. A *universal resource locator* (URL) is the exact "address" of a website. Although it takes longer to type the URL, you reach the site faster. Here is an example of a URL:

http://www.test.org:1962/test/files/file.html

- **This tells you the type of connection needed; "http:" indicates the World Wide Web.**
- **A colon and number after the name of the site usually means that you will not be accessing the site directly but through a "side" port, which in this case is labeled "1962."**
- **Another slash means you"ll be accessing a directory or file within a file.**
- **Double slashes mean that the following information is the name for the computer site you are looking for on the Internet.**
- **This slash indicates that you're going to a specific directory or file at the site.**
- **This is the name of the actual file that will appear on your screen; "html: stands for HyperText Markup Language, which indicates the file has been formatted for reading with Web software.**

SECTION 2: Communication Technology

The different parts of a URL mean different things. The following abbreviations appear in URLs. They identify the type of organization sponsoring the site:

.com = commercial organization (for profit)
.org = nonprofit organization
.edu = educational institution
.gov = government institution
.mil = military institution
.web = web activities
.infor = information services
.arts = arts and cultural activities
.net = network
.rec = recreation/entertainment
.firm = business

Equipment and Materials
- Computer with Internet capabilities
- Your copy of a URL table like the one shown below

Goal

For this activity you will find sites on the Internet. You will keep a record of the sites you visit.

Procedure

1. On a piece of paper, make a table like the one shown below. (*Do not write in this book.*)

2. Access the Internet on your computer. (You may be working with a partner or teammate if there aren't enough computers for everyone.)

3. Type in one of the URLs listed.

URL Listing	Description of Site
http://www.nrel.gov/	
http://www.sendit.nodak.edu/	
http://www.latimes.com/	
http://www.radiohistory.org/	
http://www.usmc.mil/	
http://www.art.net/	

SECTION 2: Communication Technology

4. Explore the site. On the table you made, describe what you found.

5. Now, try to find information without having a URL. Access an Internet search engine, such as Yahoo or Infoseek.

6. Under "Search," type the word "technology."

7. On a separate sheet of paper, write the number of categories and site matches for "technology" the search engine locates.

8. Review the first 20 matches. What percentage of these seem related to your subject?

9. Access one of the seemingly related sites. Describe what you find.

Evaluation

1. Did you have any difficulty using a URL? If so, what caused the problem?

2. Have you discovered any drawbacks to using the Internet? If so, what are they?

Useful Skills across the Curriculum

1. Language Arts. Access the following site on the Internet:

www.mcgraw-hill.com

Leave the site. Then try to access it again, but this time misspell the URL. For example, type "hall" instead of "hill." What happens? What conclusion can you draw from this experiment?

2. Mathematics. Do some research to find out how many people use the Internet regularly, both worldwide and in the U.S. What percentages of both populations do those numbers represent?

Credit: Peter A. Tucker

DESIGN AND PROBLEM-SOLVING ACTIVITY 2

Designing a Cellular Telephone Network

Context

Cellular phones operate within a wireless communication network that is divided into small sectors called *cells*. Each cell has its own tower with three antennas that cover the area serviced by that cell. The size of the cells varies, depending on the number of calls that need to be handled. Each tower can support only a certain number of calls at one time. In a large city many cells will be close together, while in rural areas the cells will cover a larger area and be more spread out.

> **Health and Safety Notes**
>
> This activity may involve the use of electricity. If so, observe all safety precautions for working with electrical circuits. Ask your teacher to review your setup before you connect the power.

Goal

For this activity, you will design and build a working model of a cellular telephone network. Your model should show how the network is divided into cells and how a typical call is bounced around from cell to cell as the caller moves from one area to another.

1. State the Problem

Design a cellular phone network and create a model.

On a clean sheet of paper, type or write neatly a design brief (a short statement of your problem or task), including some of the major points to be considered in solving the problem.

Specifications and Limits
- The model must show how the network is divided into cells.
- It must demonstrate how a typical call is bounced from cell to cell as the caller moves from one area to another.

SECTION 2: Communication Technology

2. Collect Information

Several questions need to be answered before you build your model. For example, what materials are available? How much time do you have? What age group will you design it for— young children? adults? all ages?

You can collect information from many different sources to help you design your model. Here are some suggestions to get you started:

- Visit your school or community library and get additional information on cellular telephones and how they work.
- Research ways you can use electronics and basic circuits in your design.
- Visit a local cellular telephone company. Many will give you a tour and let you see the switching circuits they use.
- Locate and find website addresses of major cellular telephone companies. Visit the websites to locate information about how these networks operate.

3. Develop Alternative Solutions

There are many ways you can model a cellular network. Develop several possibilities that demonstrate how a call is transferred to different cells as the caller moves within the network. For example, a model car could be moved from cell to cell and lights or LEDs could indicate an active cell. Make sketches of your ideas. Include notes that explain how things work.

4. Select the Best Solution

Remember that your solution should do an effective job of modeling the network. It must also be appropriate for the age group chosen.

SECTION 2: Communication Technology

5. Implement the Solution

Now you are ready to build your model. Remember that neatness and accuracy are important and that first impressions count. Always keep safety in mind, and, if you are using electricity, let your teacher check your work before you apply power.

6. Evaluate the Solution

Ask your classmates to make suggestions on your design. After getting their reactions, write a paragraph describing its strong points and weak points. What would you do differently if you were to repeat this activity?

Useful Skills across the Curriculum

1 **1. Mathematics.** To keep radio and phone signals organized, the Federal Communications Commission (FCC) assigns each user a different frequency. Find out what frequencies are used by cellular networks. What unit of measurement is used to measure frequencies?

2. Science. Some people are concerned that electromagnetic waves may cause cancer. Do some research on this subject and report your findings to the class.

Credit: Jeff Colvin

1. **Answers.** 46.610 to 46.970 MHz; measurement is in megahertz (millions of cycles per second)

3 DESIGN AND PROBLEM-SOLVING ACTIVITY

Using a Digital Editor

Context

Editing videotape used to be a difficult, time-consuming process. Recorded segments were copied or deleted using a manual start-and-stop method. Today, digitally recorded tape can be easily edited on a computer using digital editing software.

Goal

For this activity, your team will be given digital video and audio clips—short segments of prerecorded images and sounds. Using a computer and digital editing software, you will combine these clips into a final video that conveys a message to your audience. Your video should be no less than 30 seconds or more than 90 seconds long.

1. State the Problem

You must use a computer and digital editing software to create a message from prerecorded audio and video clips.

On a clean sheet of paper, type or write neatly a design brief (a short statement of your problem or task), including some of the major points to be considered in solving the problem. (In this case, those might include the time constraint, the order of the clips, or how to make a transition from one clip to the next.)

2. Collect Information

Before you get started, there are several important questions that must be answered.
- Who is the audience for your message?
- What is the intended purpose of the message?
- Which clips should be used?
- Should the entire clip be shown or just a portion of it?
- How can the audio clips be used to add "punch" to the message created by the video images?
- How long will the video be?

SECTION 2: Communication Technology

You can collect information to help you develop your video from many sources. Here are some suggestions to get you started:
- Consult the manual for the digital editing software you will be using (or information given to you by your teacher) to determine what features are available.
- Study videos, such as movie or television program segments or samples provided by your teacher. Determine the qualities that distinguish an interesting and entertaining video. Discuss your ideas with teammates.
- Contact professional video editors in your area, such as those who make commercials for your local TV stations. Ask their advice.
- Locate video editing and planning resources in your school or community library.

3. Develop Alternative Solutions

Most digital editors offer several methods for changing from one clip to another. This is known as a *transition*. You may choose to use an immediate transition, or you may wish to slowly "fade out" from one clip and "fade in" to another. Experiment with the various transitions available on the software you are using.

Develop several possible solutions by organizing your ideas on 3" x 5" cards. You may want to use one card for each clip you plan to include. This will help you determine the order of the clips, how much of the clip you want to include, what audio clips to insert, and what transitions you plan to use.

4. Select the Best Solution

Determine the best solution by discussing the alternatives with your teammates. Keep the following questions in mind as you do this:
- Do the clips fit together in an appropriate way?
- Does the audio add to the video or detract from it?
- Are transitions used effectively?
- Does the video meet the time requirement?
- Does the video communicate the intended message in a clear way?
- Is the message appropriate for the audience?

5. Implement the Solution

Now you are ready to use the digital editing software to create your message. Import (load) the clips you have decided to use into the editing program. Edit them to meet your team's specifications. For example, you may want to use only the first 10 seconds of a 15-second clip. The digital editing software will let you cut or delete the last 5 seconds. Now add the transitions. Transitions can help make your video flow smoothly as well as give it a professional touch. Finally, add the audio clips.

6. Evaluate the Solution

Present your video to your classmates and have them critique it by using an evaluation form like the one shown below. (Do not write in this book.)

Category	Rating—from 1-10, 10 being the best
Proper length (Subtract 1 point for each second of variation)	
Transitions (Did the video flow smoothly from clip to clip?)	
Impact on viewer (Held your attention; message was clear)	
General effectiveness (Were clips good and used effectively?)	

What did your classmates think of your final video? What were its strong points? What areas could have been improved?

Was the digital editing software easy to use? How could it be improved?

Useful Skills across the Curriculum

1. Mathematics. Suppose you were given 15 video segments. One of them is 14 seconds long, 4 of them are 10 seconds long, 3 are 24 seconds long, and the remainder are 12 seconds long. If you were to use them all in a video, how many minutes of tape would you have?

2. Language Arts. Write a script for a voice-over narration for your videotape. Time your reading of the script to be sure it matches the length of the video.

Credit: Jeff Colvin

1. **Answers.** 3.5 minutes

SECTION 2: Communication Technology

DIRECTED ACTIVITY 4

Making an Internet Home Page

Context

Having a *home page* on the World Wide Web is like having your own interactive bulletin board that people surfing the Internet can visit. You can design it to tell visitors about a favorite subject, such as baseball, and refer them to other sites where they can get more good stuff about that subject. You can also add graphics to make it fun and interesting.

Web pages are created using Hypertext Markup Language, or HTML. This language allows you to write text, add graphics, and link your page to other sites. In HTML, everything is programmed using commands enclosed between the "<" and the ">" symbols. These commands are called "tags."

Goal

For this activity, you will make your own Internet home page.

Procedure

1. Decide what the theme, or subject, of your home page will be.

2. Make a list of three or four websites related to your theme that you like to visit when you're using the Internet. As you create your home page, you will add links to these sites. By clicking on a link, visitors will be able to go directly to the sites from your home page.

3. To make creating your home page easier, use a tutorial to learn about Hypertext Markup Language. Try one of these:

   ix.urz.uni-heidelberg.de/~il1/steve/HTML.html

   www.maui.net/~lcoleon/Usinghtml.html

4. Obtain a free home page somewhere on the Internet. Both of the following sites offer free space. You may discover others.

   www.geocities.com

   www.angelfire.com

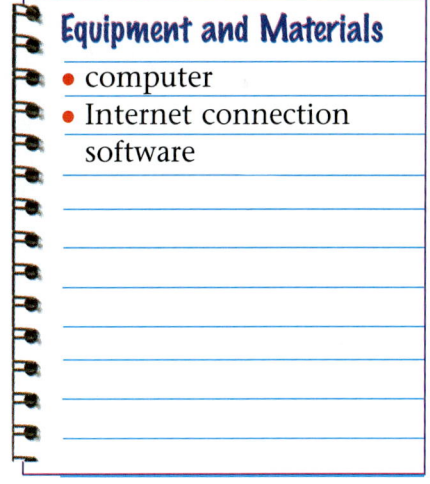

Equipment and Materials
- computer
- Internet connection software

SECTION 2: Communication Technology

GeoCities organizes home pages by "neighborhood." Neighborhoods are subject areas. For example, home pages about sports are located in the "Colosseum." Angelfire organizes home pages by state, such as Florida.

Although space for home pages is free at these sites, keep in mind that they are commercial sites. Although items may be offered for sale, you do not have to buy anything.

5. Use the editor provided by the site to create headings, graphics, links, and so on. GeoCities has both a basic and an advanced editor.

6. After you have created your home page, register it with search engines like Yahoo! or Infoseek. Registration allows people to find you on the Web. Someone who asks for sites about baseball, for example, would receive a list that included the address of your home page. Other places to register include:

www.submit-it.com

www.addme.com

Again, keep in mind that while registration is free, you should watch out for offers that cost money.

Evaluation

Access some of your classmates' home pages. Compare theirs to yours. Look for ways in which to improve your page.

Useful Skills across the Curriculum

1. Mathematics. Include a counter in your home page. A counter will keep track of how many "hits" (visitors) your page receives. Keep track of the hits for one month. What was the average number of hits per day?

2. Science. Use the Net to learn if your home page subject could be related to science in some way. For example, baseball might be linked to medicine and injuries received by players.

Basic HTML Commands

Try creating a sample page of HTML commands to help you learn the language. Then print the page to see what the tags look like. Then you can select the ones you like best for your home page. Basic commands are shown in the table. Notice that a slash (/) indicates the end of the command.

BASIC HTML COMMANDS

Command	What It Means
<A HRef="Uniform Resource Locator (URL)> </A>	Places a hypertext link to another URL
<B> </B>	Sets text in boldface
<Body> </Body>	Indicates the body of the document
 	Inserts a line break
<Center> </Center>	Centers the enclosed text or image
<H1> </H1>	Indicates text for a heading (there are 6 possible levels)
<Head> </Head>	Indicates the header section
<HTML> </HTML>	Defines the document
<HR>	Inserts a horizontal rule
<I> </I>	Sets text in italic
<Img Src=ImageFile" Align="alignment">	Inserts an image; right or left alignment should be specified
<P>	Inserts a paragraph break
<Title> </Title>	Indicates the title of the document

SECTION 3

CHAPTERS

8 Manufacturing Systems

9 Product Development

10 Production Planning

11 Production

12 Marketing

13 Computer-Integrated Manufacturing

TECHNOLOGY TIME LINE

8000 B.C. Bowls and other items are made from ceramics in Nubia (Sudan).

3600 B.C. Important agricultural tools are perfected.

1400 B.C. Metal smelted and cast in Egypt.

550 The silk industry becomes a state-run monopoly in the Byzantine Empire.

1793 Eli Whitney credited with invention of cotton gin.

1830 Labor unions seek to reform the workplace.

SECTION 3: Manufacturing Technology

Manufacturing Technology

People have always manufactured the things they needed. As technology progresses, the way things are manufactured changes. In this section, you will learn how products are manufactured, from the planning stages to the time the product is marketed. You will also learn about the role computers play in designing and assembling products. You will discover that all manufactured products, no matter how complex, begin with an idea.

1846
Elias Howe patents a practical sewing machine.

1901
Chrysler Corporation uses the assembly line in factories.

1930
First analog computer is invented by Vannevar Bush.

1961
First industrial robot used in a General Motors factory.

1970
Japan becomes a world leader in manufacturing.

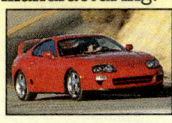

1996
The Internet emerges as a new world marketplace.

SECTION 3: Manufacturing Technology

CHAPTER 8

Manufacturing Systems

Objectives

After studying this chapter, you should be able to:
- define manufacturing.
- describe how manufacturing developed.
- name the three basic types of production systems.
- explain how manufacturing and the economy influence each other.
- describe the four parts of a managed production system.

Terms

assembly line
assets
continuous production
custom production
global market
Industrial Revolution
intermittent production
just-in-time (JIT)
manufacturing
materials processing
profit
total quality management

Technology Focus

A New Strategy

During the 1980s, economic pressures caused companies to recognize that keeping large inventories of materials on hand cost money, but did not contribute immediately to profit. The *time* that unused inventory spent on the shelf was equal to *money* invested without return.

Then Bose Corporation led the way in developing a no-inventory approach. Items from suppliers arrived **just in time (JIT)** to be used on the assembly line to make the product. It took courage to try this new strategy, because, if a supplier failed to deliver a needed item, the assembly line had to be shut down. To make JIT work, the manufacturer had to choose suppliers who could be trusted. Suppliers got contracts promising long-term business, and manufacturers got reliable deliveries of parts in small batches.

The Next Step

The success of JIT led to JIT II, an even closer relationship with suppliers, introduced by Bose in the mid-1990s. In JIT II the supplier puts a representative on the manufacturer's site full time. The supplier's representative works as part of the manufacturer's purchasing team. The manufacturer relies on the supplier's representative to place orders with the supplier. The supplier and manufacturer make compatible equipment purchases and plans for growth. JIT II is fine-tuned cooperation.

Is Cooperation Better than Competition?

As with any trust-based relationship, JIT II imposes certain limits. For instance, the supplier is not free to pursue business with competing manufacturers. The supplier may be dependent on one significant customer.

Moreover, there is an assumption in JIT II that the supplier will manage to reduce costs. Locked into a long-term contract, with pressure to reduce costs, a supplier might wonder if it would be more profitable to compete in the open market. To compete would involve negotiating the best possible price with a variety of customers.

Which is better, cooperation or competition? The answer varies for different companies in different markets. Either way, JIT and JIT II have added a new dimension to the world of manufacturing.

Take Action!

Suppose you own a small trucking company. How might your company contribute to a JIT II relationship between manufacturer and supplier? Write a brief proposal (offer) to the other companies.

Resources
- Chapter 8 Lesson Plan in the *Teacher's Annotated Edition* and in the *Teacher's Resource Binder.*
- Chapter 8 Review in the *Teacher's Resource Binder.*

What Is Manufacturing?

Imagine your life without manufacturing. You would have no bicycle to ride, no television to watch, and no athletic shoes to wear. There would be no clothes, no furniture, no airplanes. Lifesaving devices such as artificial hearts could not even be imagined. We are all very dependent on **manufacturing**, which is the making of parts and putting the parts together to make a product. The products can be large or small, simple or complex.

Anytime you make parts and put them together to make a product, you are manufacturing. However, today, when most people think of manufacturing, they think of the manufacturing industry.

The manufacturing industry is important to our society. It's important to our economy. Many people work in manufacturing to help produce products. They also buy products with the money they earn. The more products people buy, the more products are manufactured. This allows more people to work.

Manufacturing is also important to the economy in another way. A piece of material is worth more after it has been changed into a useful product. That's *value added*. The value is increased by the manufacturing process.

The Development of Manufacturing

Manufacturing began long ago when a person first changed a material into a useful item. Through the years, it has developed into a complex system of production.

In earlier times, each person or family made the products that were needed. Any extra products were traded for other items. Cloth, brooms, and utensils were common products. All items were handmade. There were no power tools.

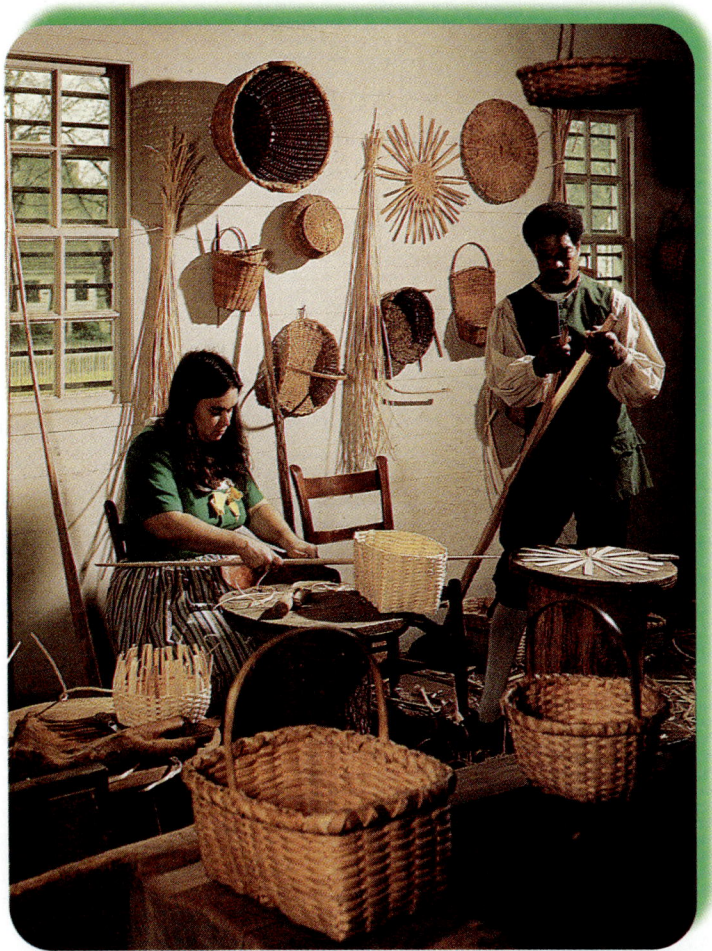

FIG. 8-1 Cottage industries grew as families began to specialize in producing certain products.

1. **Discussion.** Ask students to describe an item they have made.
2. **Reinforcement.** Discuss contributions of local or nearby manufacturing industries to your area's economy.
3. **Reteaching.** Give examples to ensure that students understand the "value added" concept.

SECTION 3: Manufacturing Technology

FIG. 8-2 Large buildings called factories were built to house manufacturing machines. Workers went to the factory to work instead of staying in their homes.

Cottage Industries

Later, families began to specialize. One family specialized in baking, another in weaving, and so forth. These families were not just making their own things and selling the rest. Now they were actually making goods to sell. Commercial manufacturing had begun. Everything was still handmade. Manufacturing took place in the home. Workers were members of the family. This was called *cottage industry*. Fig. 8-1.

The Factory System

In the 1800s, the factory system came into being. Machines were developed. A steam engine or waterwheel provided power. People went to work in a special building that housed these machines. This building was called a factory. Fig. 8-2. A company owned the factory and paid the workers. These changes were the beginning of the Industrial Revolution. The **Industrial Revolution** refers to the great changes in society and the economy caused by the switch from products being made by hand at home to being made by machines in factories.

At first, the factory system had many faults. Workers were paid very little. Many worked 16 hours a day. Working conditions were usually poor. Often, there was no fresh air, no concern for safety, and no breaks in the work routine. Even young children were made to work under these conditions. Gradually, concerned people caused changes to be made. *Labor unions* were formed by workers to present their demands to the company. Gradually the unions forced factory owners to improve wages and working conditions. Laws were passed to protect workers.

4. **Critical Thinking.** Discuss how the Industrial Revolution changed the way of life of the people at that time. Ask: Is the "revolution" continuing yet today?
5. **Extension.** Discuss the effect of safety on productivity and worker morale.
6. **Extension.** Discuss with students the situation some celebrities encountered recently when foreign companies they had invested in were found to use poor labor practices.

CHAPTER 8: Manufacturing Systems

Fascinating Facts

One of the first automated factories was a flour mill that had a system of conveyors.

Modern Manufacturing

Today manufacturing is still done in factories. However, modern factories are much safer and more efficient than old-time factories. Fig. 8-3. Production systems have become very technical and specialized. Much knowledge has been gained over the years.

The demand for products determines what will be manufactured. If fewer consumers need or want audio-cassette players, fewer will be manufactured. If fewer people take bus transportation, the demand for buses will decrease and fewer buses will be manufactured. On the other hand, if more consumers want CD players or four-wheel drive vehicles, more will be manufactured to meet the increase in demand.

Consumer products that are widely used, such as candy bars and compact discs, are produced for *stock*. The company makes large amounts of the product and keeps them on hand. When orders come in, the products are shipped to customers.

When a product is not widely used, the manufacturer may not start producing it until the customers have placed orders. In this case, the product is produced *on demand*. For example, aircraft manufacturing companies make aircraft only when they have a firm order. The order may be for a single plane or for 100 planes, but the company doesn't start production until the order is placed.

A manufacturer generally chooses from three basic types of modern produc-

FIG. 8-3 Today's factories are clean and neat and emphasize safety.

FIG. 8-4 Products such as race cars are produced one at a time. Each one is unique.

tion systems to meet the demand for products. These are custom, intermittent (job-lot), and continuous production. As you read about the three basic types of production systems, you will be able to see why each has its own advantages.

Custom Production

In **custom production**, products are made one at a time according to the customer's specifications. Each one is different. This type of production is usually the most expensive per number of parts made. Custom-made products may be large or small, simple or complex. Many ships and pieces of jewelry are custom-made. Fig. 8-4.

At one time, manufacturers believed that making products one at a time by hand would become a thing of the past. Instead, handcrafting industries have survived and grown. There are many small businesses producing specialized and/or custom-made, handcrafted products. For example, while furniture manufacturing is becoming more automated, the most valued pieces of furniture are handcrafted pieces, carefully designed and constructed.

Intermittent Production

In **intermittent production**, a limited quantity of a product is made. Then production of that product is stopped. Any necessary retooling or changeovers are made so a different part or product can begin to be produced. Many seasonal items, such as lawn mowers and snow blowers, are manufactured this way. Per part, this type of production, also called job lot production, is less expensive than custom production. The cost can be spread over more products.

Continuous Production

Continuous production is the system used for mass-producing products. This means a large quantity of the same product is made in one steady process using an assembly line. In an **assembly line**, the product moves from one workstation to the next while parts are added. This type of production is also called *line production* or *mass production*. A production line is

7. **Internet.** Have students look for custom-made products on the Internet.
8. **Reteaching.** Show food or health product containers that have batch or lot numbers printed on them. Discuss the concepts of batches and lots.

9. **Background.** A "production line" may also be called an "assembly line."

CHAPTER 8: Manufacturing Systems **189**

set up and the products are continuously produced. Thousands, maybe millions, of the same product are made. Because changing a production line setup is expensive, continuous production is the most economical type of manufacturing system. Cars and electronic products, such as radios and computers, are made this way.

Many manufacturers today are combining the ideas of continuous and intermittent production. They are making large quantities of the same product with slight variations. Automobiles are an example. Two models may seem the same at first glance, but, on close inspection, they may have many differences. The engine, seats or other interior features, and trim may vary. However, by producing many basic models at the same time, production is more efficient. Very close coordination of parts and assembly procedures are required to make this happen.

Who Does Manufacturing?

Anyone can manufacture a product. If you built a bookcase for your room, you would be custom-producing a product. Usually, however, manufacturing is done by companies that specialize in making products.

Companies and Corporations

A *company* is an organization formed by a group of people for the purpose of doing business. A *corporation* is a company that is owned by many people who have bought shares in it.

FIG. 8-5 One company may manufacture many different products. All of these products were made by the same company.

You probably know many manufacturing companies and corporations by name, like Apple Computer, IBM, Ford Motor Company, and General Electric. You might not recognize other companies by name, but you know their products. Parker Brothers (games), Levi Strauss & Co. (jeans), and Sony Corporation of America (electronic products) are examples.

Some small companies have only a few workers. They manufacture such products as computer graphic designs, ceramic vases, specialized newsletters, or stained-glass windows. Much industrial manufacturing is also done by small, locally owned companies employing a few hundred workers. These companies usually make parts that go into other larger products. For example, a company may make metal brackets that hold computer parts in place,

10. **Reinforcement.** Show how spreading costs over a long run results in lower unit costs.
11. **Reinforcement.** Have students identify nearby manufacturing plants. Discuss what is manufactured in each plant.

SECTION 3: Manufacturing Technology

seat covers and door panels for cars, or molded wires and plugs for radios.

The largest companies may manufacture many different products. The General Motors Corporation manufactures over 50 different types of cars and trucks. The Procter & Gamble Company makes many different types of products. Soaps, shampoos, cake mixes, and peanut butter are a few examples. Fig. 8-5.

Sometimes large companies own smaller companies. The smaller companies are called subsidiaries. They are separate, but controlled by the same "parent" company. Sometimes a company may be *diversified*. That means the subsidiaries don't necessarily make the same parts or products. For example, United Technologies has more than 100 subsidiaries. Otis makes elevators and escalators. Carrier makes air conditioners. Essex makes wire and cable. Pratt & Whitney makes jet engines. Hamilton Standard makes helicopters. All of these companies, however, are owned by United Technologies Corporation.

Sometimes a company has subsidiaries that are suppliers for the main product they manufacture. A company that makes telephones may also own a company that creates plastic molding, another company that makes electronic chips, and still another company that manufactures wire.

The Global Market

Many U.S. companies have manufacturing plants in different parts of the world. Fig. 8-6. Sometimes these locations are chosen to take advantage of lower labor costs and/or lower cost of materials.

FIG. 8-6 This map shows Caterpillar's® manufacturing plants, licensees, and contract manufacturers around the world. Caterpillar® manufactures large construction equipment.

12. **Critical Thinking.** Many corporations diversify. Ask students: What advantages would this practice have? What might be some disadvantages?

CHAPTER 8: Manufacturing Systems

FIG. 8-7 This automotive plant in California is a joint venture between Toyota, the largest automobile manufacturer in Japan, and General Motors, the largest automobile manufacturer in the United States. The plant has about 3.7 million square feet of manufacturing space and employs 3,500 workers.

Other times, these locations are chosen so that products can be manufactured in the countries in which they are to be sold. This helps eliminate the high cost of transporting finished products.

Some companies from different countries have entered into manufacturing agreements. In some cases, manufacturing plants are managed and operated in a cooperative manner. Toyota Company of Japan and General Motors Corporation of the United States have a joint agreement to cooperatively manufacture automobiles in California. Fig. 8-7. Increased cooperation and mergers across national boundaries have resulted in multinational companies.

Rapid transportation and communications via satellite have helped products become popular throughout the world rather than in just one nation. Automobiles are manufactured and sold worldwide. Many products, such as clothes, appliances, and medicines, are in demand around the world. This has created a worldwide or **global market**.

The need for many services, such as health care and transportation, keeps companies busy manufacturing new products. These needs and demands for products and services are what drive our *global economy*. 13

This growing need for products is being helped by a gradual and steady reduction of trade barriers. For example, the European Common Market permits the countries of Europe to sell goods throughout Europe without import taxes or restrictions. A similar trade agreement, NAFTA, the North American Free Trade Agreement, also permits free trade between countries of North America. This decrease in trade barriers helps manufacturers find larger, worldwide markets for their products.

13. **Reteaching.** Emphasize that in a global economy what happens in one country often affects what happens in others.

International Competition

One definite effect of the development of the global economy is the increase in competition for a share of the world marketplace. With goods available from all over the world, consumers have more products from which to choose. Fig. 8-8.

Throughout the world, everyone wants a good buy for his or her money. Worldwide competition has made quality an important issue. Often, the major difference between products produced by different manufacturers is the reliability and potential life of the product. To make consumers want to buy their product, manufacturers must make a quality product at a good price.

To stay in business, a manufacturer must also make a profit. A **profit** is the amount of money a business makes after all expenses have been paid. However, the selling price of a product must be competitive with similar products. Global markets make selling more competitive. Manufacturers must try harder to keep costs down without sacrificing quality.

One method for staying competitive is to increase productivity. *Productivity* is the comparison of the amount of goods produced (output) to the amount of resources (input) that produced them. If you increase your output without increasing input, you increase your productivity. High productivity helps keep costs down. This not only helps keep the manufacturers more competitive, it also makes the market larger. When a product is more affordable, more people are able to buy it.

14

FIG. 8-8 Competition from foreign manufacturers, like Honda and Toyota, first gained public attention when Japanese cars became popular in the U.S.

14. **Enrichment.** Discuss how many companies have moved operations overseas to take advantage of cheaper labor.

CHAPTER 8: Manufacturing Systems

MATHEMATICS CONNECTION

How Earnings Become a Paycheck

You might be surprised when you receive your first paycheck and find that the check is for less than you thought you were earning. There is quite a difference between your gross salary—the actual amount you earn before taxes and other deductions—and your net salary—your actual "take-home" pay.

The kinds of deductions that are taken from gross salary vary. The two major deductions are for income tax (federal, state, and sometimes city) withheld to pay your taxes and for Social Security tax. Employees may also request other deductions, such as union dues, medical insurance, life and/or disability insurance, credit union payments, savings plans and stock plans, savings bonds, the employee's portion of unemployment insurance, and charitable contributions.

The amount taken out for federal income withholding tax is usually the largest payroll deduction. This amount usually does not change from pay period to pay period.

Social Security taxes are also a major paycheck deduction. These deductions are usually labeled FICA, which stands for Federal Insurance Contributions Act. This FICA deduction contributes to a Social Security pension account and a Medicare (hospital insurance) account for your retirement.

In 1997, the pension portion of the Social Security tax was about 6.2% of the first $65,400 of gross annual income. The Medicare portion of the Social Security tax was about 1.45%. Together the two components of the FICA tax were about 7.65% of the first $65,400 earning base.

On the paycheck stub shown here, federal taxes amounted to $49.70. State taxes were $13.22, and FICA was $38.25. This check also shows deductions for regular medical insurance, a company pension fund, and charitable contributions.

Take-home pay may also reflect overtime worked or lost work time. Always review your paycheck for accuracy and make sure that your take-home pay is correct.

Try It!

1. Calculate the take-home pay for the employee whose paycheck stub is shown here.
2. If a person earns gross pay of $400 per week, how much will the Social Security portion of the FICA tax be?

Answers to these questions can be found in the Lesson Plans in the Teacher's Manual at the beginning of this book

SECTION 3: Manufacturing Technology

A Managed Production System

As you have learned, a system is a group of parts that work together to achieve a goal. A system is needed to produce manufactured goods. To work efficiently, a manufacturing system needs to be well managed. Like other systems, a managed production system has inputs, processes, outputs, and feedback. Fig. 8-9.

Inputs

Input includes anything that is put into the system. The seven inputs in manufacturing include people, materials, tools and machines, energy, capital, information, and time.

People

People make the system work. They are the most important input. People in manufacturing may design products, purchase materials, run machines, assemble parts, inspect products, or sweep floors. No matter what their job is, they all contribute to a team effort.

Many companies organize people by departments. In one company you might find an engineering department, a production department, a marketing department, and a research and development department. Employees usually have a job title that describes the kind of work they do. A manufacturing engineer, for example, plans the sequence of operations and determines which machines will be used. A materials manager makes sure that

FIG. 8-9 A managed production system is the result of inputs, processes, outputs, and feedback. Here, the chef takes the order (input), mixes ingredients (processes), completes the meal (output), and the customer expresses enjoyment (feedback).

15. **Discussion.** Ask students if they know people who work for manufacturing companies. Discuss the various types of work these people do.

CHAPTER 8: Manufacturing Systems

FIG. 8-10 Some raw materials, such as coal, are mined from the earth.

commonly used in manufactured products. There are two categories of materials: raw materials and industrial materials.

Raw materials are materials found in nature. Iron ore, plant fibers, and petroleum are all examples of raw materials. Fig. 8-10. All industrial materials are made from raw materials.

Most raw materials need some refining or processing to convert them into industrial materials. Industrial materials are in a form that can be used to make products. For example, metal ore must be heated to a very high temperature to remove its impurities before it can be used to make a product.

Fascinating Facts
Aluminum was as costly as silver until a cheap way of extracting it from ore was discovered in 1886.

enough of the right materials are available, so the parts can be made. Following a Japanese practice, some companies call all their employees "associates" rather than giving them specific titles. Other companies are reorganizing their production workers into "cross-functional teams," where workers perform multiple jobs.

Materials

Products are made from one or more materials. Metals, plastics, wood, glass and other ceramics, textiles, and rubber are all

Industrial materials are usually made as *standard stock*. This means that the material is formed or packaged in a widely used (standard) size, shape, or amount that is easy to ship and to use. Fig. 8-11. Standard stock includes sheets of plywood, steel, and aluminum. Bolts of cloth and barrels of liquid chemicals are also standard stock.

Scientists continue to develop new materials capable of providing specific properties not available in conventional materials. An *alloy* is made by combining

16

16. Enrichment. Show examples of standard stock.

196 SECTION 3: Manufacturing Technology

FIG. 8-11 Metal is available in a variety of standard shapes and sizes.

two or more metals or a metal and a nonmetal. The new material has improved properties. Fig. 8-12. A *composite* is a material made by combining two or more materials. A composite is usually much stronger and more durable than the materials from which it was made. *Ceramics* are materials made from nonmetallic minerals that have been heated to very high temperatures. Ceramic materials are strong, hard, and resistant to corrosion and heat.

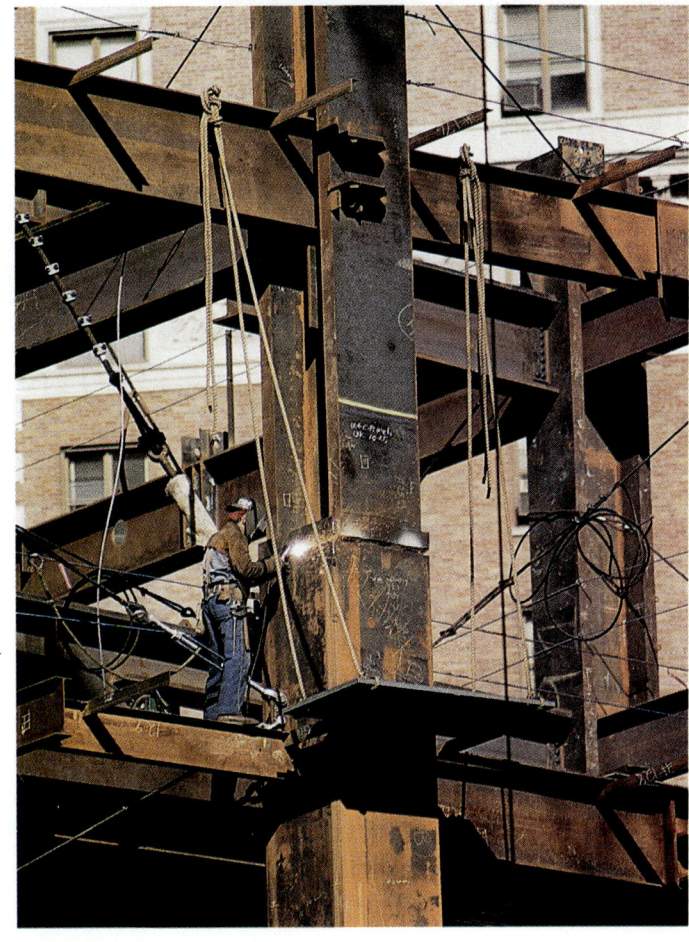

FIG. 8-12 New steel alloys are manufactured from raw materials and can be formed into structural beams used in high-rise construction. The alloys are stronger than ordinary steel.

CHAPTER 8: Manufacturing Systems

FIG. 8-13A These pliers are an example of a hand tool.

Tools and Machines

17 Tools and machines are used to help change materials into a finished product. The tools and machines used in manufacturing can be classified into three main categories:

- hand tools
- portable power tools
- machines and equipment

Simple tools powered by humans are called *hand tools*. Hammers, screwdrivers, and wrenches are all hand tools. Special hand tools are often used in manufacturing. Fig. 8-13A.

If a tool is powered by electricity or air and is small enough to carry, then it is a *portable power tool*. Portable electric- or air-powered tools include drills, grinders, and wrenches. Fig. 8-13B.

FIG. 8-13B This battery-powered electric drill is a commonly used portable power tool.

The *machines and equipment* category includes the large, powerful machines used in manufacturing. They are usually installed permanently. This means they are not movable. Saws, milling machines, and drill presses are examples of manufacturing machines. Fig. 8-13C. Equipment includes ovens, paint booths, and welding outfits.

18

17. **Extension.** During the study of this section, you may wish to introduce students to tools, machines, and equipment available in your lab.

18. **Enrichment.** Discuss the six basic machines and how tools and machines function.

198 SECTION 3: Manufacturing Technology

FIG. 8-13C This vertical milling machine is used to cut and shape metal parts.

Energy

Energy is used to provide power and light for manufacturing. For example, electricity powers motors that run machines. Gas is burned to provide heat for furnaces. Fluid power from air and water pressure is also used in manufacturing.

At the present time, oil is the principal source of usable energy, followed by coal and natural gas. The supply of these fossil fuels is limited. They are nonrenewable sources of energy, and the world is rapidly using the remaining supply. Alternate sources of energy will continue to be sought and developed. Manufactured products, such as solar cells and wind turbines, will be needed to capture and convert energy to usable power. Fig. 8-14.

19

New materials, with their special properties, permit the manufacture of many new and different products. However, they also create manufacturing problems. Some new materials are very hard and brittle and require new methods to form and shape them. This means that new tools and machines need to be developed to keep up with the continual development of new materials.

FIG. 8-14 Solar energy can be collected using photovoltaic cells (cells designed to convert solar energy into electrical energy) arranged in large banks. These "solar farms" may become major sources of electricity in the future.

19. **Review.** Review the sources of energy discussed in Chapter 3.

CHAPTER 8: Manufacturing Systems

Research continues in the development of ways to conserve energy. A *superconductor* is a material that will carry electrical current with virtually no loss of energy. Using superconductors, small generators can be made very powerful and efficient. Electricity can be transmitted great distances with minimal energy loss when superconductors are used.

Capital

Capital is another term for money or financial resources. The two basic types of capital are fixed capital and working capital.

Fixed capital refers to buildings and equipment owned by the company. It's called fixed capital because it represents money spent to buy things that belong to the company and will be there permanently. Tools and machines, though often listed separately as an input, are also fixed capital. Another term used to describe anything the company owns that has value is **assets**.

Working capital refers to cash. It's the money the company uses to buy materials and supplies, pay workers, buy advertising, and pay taxes.

In order to raise capital, some owners sell shares in the company. These shares are also called stock. Fig. 8-15.

Information

Information is an important input to manufacturing. A toy manufacturer, for example, must know the characteristics of different materials so that no toys will be made from a toxic plastic. Manufacturers need to know the characteristics and capabilities of the machines in the factory. They will also want to know about consumers who may buy the product. Such information can help as products are being designed. Information is often kept in a computer data bank.

FIG. 8-15 This certificate represents a share of ownership in a corporation.

FIG. 8-16 Manufacturing must be managed. Today, managers use computers to help them make decisions about planning, organizing, and controlling the manufacturing system.

Time

Time is needed to order materials, produce parts, and assemble products. Manufacturing companies try to figure out the best ways to use production time in order to produce as many quality products as possible in the least amount of time. This is called *efficiency*. Companies that are efficient make more money than those that are inefficient.

Processes

Processes are all the activities that need to take place to make the product. There are two basic kinds of processes: management and production.

Management Processes

In order to work properly, every part of the manufacturing system must be managed. Basically, managers make decisions in three areas: planning, organizing, and controlling. *Planning* means deciding how something should be done before you do it. You make a plan of action so that things will work smoothly. *Organizing* is gathering and arranging everything needed to do a job. *Controlling* means keeping track of things. Correcting an error is an example of controlling. Fig. 8-16.

Several new strategies are being used in management. **Total quality management** means that employees are expected to meet a performance standard for their job. Their job description tells how they will be measured to see if they're doing a good job. *Empowerment* means to allow employees to make decisions without asking the boss every time.

Reengineering is organizing the company so all of the employees focus on the customer, in order to provide good service and quality products.

Production Processes

All the processes used to actually produce the product are production processes. These processes can be classified as preprocessing, materials processing, and postprocessing.

20. **Reteaching.** Use an example to show planning, organizing, and controlling.

21. **Language Arts.** Note the root word "process" and how adding different prefixes changes its meaning.

CHAPTER 8: Manufacturing Systems

"Pre" means before. *Preprocessing* happens before any work is done on the material. When the raw material or standard stock is received at the factory, it must be unloaded, stored and protected until used. These activities are examples of preprocessing. Note that preprocessing activities don't actually change the material.

Materials processing means changing the size or shape of a material in order to increase its usability or value. Industrial materials are turned into components or parts. Material can be changed using several basic processes: forming, separating, conditioning, and combining. Fig. 8-17.

Forming is a way of changing the shape of a material without adding or taking anything away. An example is hammering a round piece of steel into a flat shape. You still have the same amount of steel, but it is flat instead of round. Other forming methods include forging, extruding, and molding.

Separating is a way of changing the shape of a material by taking some away. If you saw a two-by-four in half, you have changed its shape by separating. Cutting a piece of fabric from a large roll to make a seat cover is another example. The amount of material has been changed; there is less than when you started. Common methods of separating include sawing, drilling, shearing, and punching. In one special type of separating, called non-contact cutting, there is no contact between the tool and the material. An example is cutting by laser beam. The laser melts a small amount of material along the cutting path. Other examples are water jet cutting and electrical discharge machining.

FIG. 8-17 Industrial materials are converted into manufactured products by materials processing.

22. **Reinforcement.** Show examples of products made by each process.

SECTION 3: Manufacturing Technology

MATERIALS PROCESSING	
Process	**Example**
FORMING Casting or molding Compressing and stretching	Injection molding of CDs Forging tractor axles
SEPARATING Shearing Chip removal Non-contact separating	Punching holes in sheet metal Drilling holes in metal Laser cutting saw of blades
CONDITIONING Thermal conditioning Chemical conditioning	Hardening steel for beams Hardening fiberglass shovel handles
COMBINING Mixing Bonding Mechanical fastening Coating	Blending cola drinks Welding bicycle frames Bolting a wooden desk together Painting car bodies

FIG. 8-18 This table shows the main categories and sub-categories of materials processing.

Conditioning means changing the structural properties of a material. Heat treating is done to the teeth on a gear to make them extra hard so they won't wear out as fast. Adding a catalyst to fiberglass resin produces a chemical reaction that causes the resin to harden and take shape.

Combining means putting two or more materials or parts together. Using bolts, rivets and screws to hold parts together is a type of combining called mechanical fastening. Gluing is another form of combining. Welding two pieces of metal together is also a combining process, as is putting a coat of paint on a part. Fig. 8-18.

"Post" means after. *Postprocessing* activities include things done to the product after the materials have been changed in form or shape. Handling, protecting, and storing products are all examples of postprocessing activities. Once again, there is no value added at this stage.

Postprocessing may include recycling activities. Collecting glass or plastic bottles, sorting them and grinding them into chips is an example of recycling. The chips can then be used as raw material or standard stock for manufacturing other products. Perhaps your community has an organized recycling program. Fig. 8-19.

Outputs

The result of inputs and processes is called output. In manufacturing, as in any

FIG. 8-19 Some products can be recycled after their functional life. This plastic will be ground up and remelted to be made into other products.

system, there are many outputs. One output, of course, is the manufactured product. This output is expected and desirable. Other outputs, such as waste and pollution, may not be expected, and they certainly are not wanted.

The outputs of a system affect us and our world. In other words, they have an impact. For example, cars are an output of a manufacturing system. They have become so numerous and widely used that they have an impact on nearly every aspect of our lives. Cars affect our

- *Economy.* Many people's jobs depend on the automobile industry.
- *Society.* Cars are more than just a way to travel. They are part of our culture. We use cars as a way to express who we are or who we would like to be.
- *Politics.* Who should pay for road-building? Should the government protect the American car industry from foreign competition? Should the speed limit be raised or lowered? All these are political issues that result from our use of automobiles.
- *Environment.* To build and drive cars, we use nonrenewable resources, such as metals and oil. Car exhaust pollutes the air. Automobile junkyards are an eyesore, and they can pollute the land and water. Fig. 8-20.

By recognizing impacts, their effect can be made to work for us. For example, one way to meet environmental goals is to design products to be recyclable. Ford Motor Company has developed a pilot program in Germany to improve techniques for dismantling cars. Developers at Ford have collected information about materials, the amount of time it takes to remove parts, and what parts have the most value. This information helps design engineers develop new cars and trucks that are easier to disassemble and recycle. Currently, about 70% of an automobile can be recycled.

Other companies have found that garbage can be valuable industrial material. For example, some park benches and picnic tables are manufactured from recycled plastic and soda bottles. Ford uses plastic soda bottles to make grill reinforcements, door padding, luggage rack side rails and other parts for automobiles. Also, one company's waste may be another company's raw material. A company that punches sheet metal blanks for bottle caps may sell their scrap metal to another company to make furnace filters.

23. **Extension.** Ask students to name as many products related to cars as they can, such as traffic lights, car wax, etc.

SECTION 3: Manufacturing Technology

FIG. 8-20 Salvage yards are important to solving environmental problems. Here, automobile tires are cut up and sent to recycling centers.

Feedback

The feedback loop of a system links the output back to the input. It is the information about the outputs that is sent back to the system to determine whether the desired results are being achieved. Consider the car example. We know that car exhaust pollutes the air. This is an undesirable output. People let the government and the car manufacturers know that they want cars to pollute less. That's an example of feedback. That feedback has led to changes in the way cars are designed.

Other industries are responding to feedback in ways that affect how many products are produced. With the growing concern for the environment, companies are actually setting environmental goals. More money is being spent to address environmental issues. Current research and new technology will enable manufacturers to make products without harming the environment.

Changing manufacturing processes and modifying raw materials are two other responses to feedback about environmental concerns. Fig. 8-21. Many utilities are reducing harmful carbon dioxide emissions by using solar and geothermal technologies. New methods of packaging are another way to help reduce waste problems.

Manufacturers are aware we must all strive to protect our most valuable resource—our environment. Feedback is an important part of making changes.

FIG. 8-21 Many manufacturers are trying to find ways to reduce or eliminate environmental hazards caused by their manufacturing processes.

CHAPTER 8: Manufacturing Systems   **205**

Career File—Industrial Engineer

EDUCATION AND TRAINING
A bachelor's degree in engineering is required.

AVERAGE STARTING SALARY
The average salary for a beginning industrial engineer is around $33,500 per year.

OUTLOOK
According to the Bureau of Labor Statistics, the demand for industrial engineers will grow at an average rate through 2005. Manufacturers will be making new products and improving existing products, always seeking to reduce costs and increase productivity. Industrial engineers will help them meet these goals.

What does an industrial engineer do?
Industrial engineers are productivity experts. They determine the best way to use resources—people, machines, materials, information, and energy—to make and distribute a product faster, better, cheaper, or safer. They may design machinery, such as industrial robots, or the product the robots will assemble. They are usually involved in quality control.

Does most of the work involve data, people, or things?
Industrial engineers work with all three. They study the procedures that people follow, the machines they operate, and the tools and materials they use. Engineers often use computers to develop and analyze alternate solutions.

What is a typical day like?
At the beginning of a project, engineers may interview managers and workers for in-depth understanding of what is needed. Depending on the purpose of the project, they may survey potential sites for a new manufacturing plant or determine the most economical way to ship a product. Toward the end of the project, they analyze the information they gathered and prepare written recommendations. Many industrial engineers are in management positions, which means they also supervise the work of others.

What are the working conditions?
Most industrial engineers work in an office, with frequent visits to the plant.

Which skills and aptitudes are needed for this career?
Industrial engineers should be creative, analytical, and detail-oriented. They should work well with people and be able to explain their ideas orally and in writing.

What careers are related to this one?
Other workers who use scientific principles in their jobs include mathematicians, all types of engineers, and all types of scientists and science technicians.

One industrial engineer talks about her job:
My job is to solve problems and make our products better and more cost effective.

Chapter 8 REVIEW

Reviewing Main Ideas

- Manufacturing is the making of parts and putting those parts together to make a product.
- The history of manufacturing includes cottage industries, the factory system, and modern manufacturing systems.
- Three types of modern production systems are custom, intermittent, and continuous.
- International competition for products in the global market influences quality, price, profit, and productivity for consumers as well as manufacturers.
- A production system of inputs, processes, outputs, and feedback needs to be managed to work efficiently.

Understanding Concepts

1. Name two or three products that you have manufactured yourself. Describe the procedures you used.
2. Briefly describe each of the general developments in the history of manufacturing that are discussed in this chapter.
3. Give an example of a product that would best be manufactured by each of the three types of modern production systems.
4. In what ways do people's needs and demands around the world affect manufacturing?
5. Give two examples of each of the four parts of a system that are important for manufacturing.

Thinking Critically

1. Why is it important for manufacturers to have information about consumers who want to buy their product?
2. Why is efficiency so important to a manufacturing system?
3. How would the construction of a new factory in your area affect your community?
4. Name five manufactured products. Tell what materials each item is made from and which methods you think were used for processing the materials.
5. Tell what you would do about waste and pollution if you were the president of a manufacturing company.

Applying Concepts and Solving Problems

1. **Language Arts.** The early days of clothing manufacturing produced workplaces known as "sweat shops." Do some research, then write an essay describing these shops.
2. **Social Studies.** Search the World Wide Web to find information about recycled products or recycled manufacturing materials. Try to find at least three companies who use recycled materials to make new products. List the company name and the products they make and what recycled materials they use.

24. **Resource.** Answers to "Understanding Concepts," "Thinking Critically," and "Applying Concepts and Solving Problems" can be found in the Teacher's Manual at the front of this *Teacher's Annotated Edition* and in the *Teacher's Resource Binder*.

CHAPTER 9

Product Development

Objectives

After studying this chapter, you should be able to:
- explain what product engineers do to develop a product.
- explain design for manufacturability.
- name three types of drawings used in manufacturing.
- explain the importance of product testing and analysis.

Terms

computer-aided engineering (CAE)
concurrent engineering
design for manufacturability
interchangeable parts
mock-up
modular design
product engineering
prototype
standardization
tolerance
working drawings

Technology Focus

CAD and CAM, Together at Last

CAD (computer-aided design) and CAM (computer-aided manufacturing) are often linked in print as "CAD/CAM," in spite of the fact that they have been worlds apart in practice. Finally, that is changing. Boeing led the way by designing its 777 jumbo jetliner entirely on computer, using the CAD software called CATIA.

Chrysler took the next step, using CATIA and another program called DMAPS (Digital Manufacturing Process System) to design not only a car, the Concorde, but also the processes to manufacture the car. DMAPS is CAM software. By linking it to the CAD program, the time required to develop a new car from concept to full production was reduced from over four to just two years.

Clay vs. Electronic Models

While design engineers focus on the product itself, process engineers are concerned with the tools and dies that make its parts. They also solve problems in delivering materials and setting up the steps in assembly. Because of their different functions, design and production departments have traditionally used different software tools—designers using CAD, production using CAM.

Then manufacturers realized that the two could be linked. Instead of creating clay models of products, designers made computer models using CAD. The CAD data was then transferred to CAM programs. Working from a computer model rather than from clay yielded more precise specifications. Such precision extended the life of tools and reduced waste. The software also allowed users to try out such things as various arrangements of robots.

Concurrent Engineering

Most important, access to the computer model of a new product means that all departments can be involved in product development from the beginning. Production no longer has to wait until the design is completely finished. This strategy of involving all departments is called concurrent (simultaneous) engineering.

Chrysler got there first with CATIA and DMAPS, but other major manufacturers are also committed to this new, unified way of developing products. Shorter times to market have become important to profits. In a competitive economy, it helps to get there first.

Take Action!
Make a list of several characteristics of different products that cannot be designed on computer. Tell why you think each is computer-resistant.

Resources
- Chapter 9 Lesson Plan in the *Teacher's Annotated Edition* and in the *Teacher's Resource Binder*.
- Chapter 9 Review in the *Teacher's Resource Binder*.

CHAPTER 9: Product Development

New Products, New Ideas

Hardly a day goes by without a new product being introduced. Sometimes the product isn't really new. It's just "new and improved." Why do we see such a constant parade of new or improved products? Where do manufacturers get all these ideas for new products or improved products?

The Product Life Cycle

The product life cycle starts when a new product is introduced. When it's new, the product is usually expensive. It may also be unreliable. However, as more people begin to buy the product, it enters the growth period. During the growth period, product improvements are made. The "bugs are worked out." More of the products are made, and the price usually gets lower. For example, when cellular phones were first introduced, they cost over a hundred dollars. Now, some can be bought for just a few dollars, and they're becoming quite common. More people can afford to buy the product.

Because of advertising and general acceptance by the public, the product enters the maturity stage. During this period, the price stabilizes (stops changing). Also, the product is usually very reliable.

After a while, sales start to taper off. Nearly everyone who wanted the product now has one. This period in the product life cycle is called saturation. The demand has been satisfied fully. More products are available to sell than can be sold. Finally, sales start declining, and the product may be taken off the market. Fig. 9-1.

The length of a product's life cycle varies, depending on the product. It may last a few months, a few years, or many, many years.

To be successful, manufacturing companies must continue to produce and sell products. Because of the product life cycle, each company tries to keep creating new product ideas. That way, as one product is declining, a new product can be entering the growth stage. This is what keeps a successful manufacturing company going and growing. It's a continual process.

FIG. 9-1 Most products go through the product life cycle. The length of time for each stage may vary, but this chart shows average times.

1. **Extension.** VCRs and clothing products are other good examples of products that go through a life cycle.

SECTION 3: Manufacturing Technology

SCIENCE CONNECTION
Designing a Scientific Experiment

Product research involves testing new products and the materials and methods used to make those products to see how they perform under certain conditions. The most reliable testing is done by changing all of the conditions, called *variables*, one at a time, in a controlled environment. The results are then carefully measured and compared.

In order to design your own scientific experiment, you must first decide what its purpose is and what you wish to demonstrate. Next, you must plan how you will perform the experiment, what materials and equipment will be needed, and what conditions must be controlled. Unless you consider all of these factors, your experiment will not be a success, because the results will not be reproducible and meaningful.

For example, suppose the company you work for wants to use steel wool in a product but is concerned that it may burn under certain atmospheric conditions. For your experiment, you will need a sample of fine steel wool, a closed system so you can control the atmosphere (a container such as a flask with a tight stopper will do), and gases such as oxygen and nitrogen. In this experiment, the variable will be the atmosphere in the flask. Changing the atmosphere by replacing the room air with oxygen and then with nitrogen will allow you to compare what happens in each case. (When you use oxygen, the burning will occur quickly. With a nitrogen atmosphere, however, the flame will be extinguished.)

In many cases, you will have to repeat an experiment in order to understand what the results mean. When you repeat it, you may wish to change one of the test conditions to see what effect this has on your results. Research may require hundreds of similar tests before determining the best materials or methods or the best product.

STOPPER

STEEL WOOL

Try It!

1. Explain how you could set up an experiment to show whether or not various solid materials dissolve in water or other liquids. Do you think the temperature of the liquid would affect your results?
2. If you were an inventor or a manufacturer testing a new toy, what kinds of experiments or tests might be used before the toy was put into production?

Answers to these questions can be found in the Lesson Plans in the Teacher's Manual at the beginning of this book.

CHAPTER 9: Product Development

FASCINATING FACTS

Thomas Edison, inventor of the light bulb and countless other items, was granted 1,093 patents during his life, more than any other individual.

Research and Development

Research and development are processes used to turn ideas into new products. *Research* is done to gather information. How can we make a car that gets better gas mileage? How can we make lighter bathtubs? What is the best design for a computer keyboard? Fig. 9-2. These are just a few of the questions that could be answered by research. *Development* is using research information to solve a problem.

New Product Ideas

Every new product starts as an idea. Generally, all product ideas are the result of one or two occurrences. Sometimes product ideas come from someone who thinks of a new product to solve a particular problem. The problem creates the demand for the product. Other times, someone suggests an idea that would make a useful or desirable product. The company may then, through advertising, create the demand for the product. Fig. 9-3.

Sometimes the idea for a new product comes from an individual, who may then sell that idea to a manufacturer. However, most product ideas come from research and development (R&D) groups. The research and development department may be a part of the company or an independent R&D organization hired to create workable ideas for new products. People working in R&D departments use brainstorming, market research, and problem-solving techniques to come up with

FIG. 9-2 Research into the medical condition known as carpal tunnel syndrome (pain in hands and wrists) has led to the development of more user-friendly keyboards.

2. **Critical Thinking.** Encourage students to think creatively. They could have good ideas for new products or think of ways to improve existing products.

3. **Extension.** Collect articles and ads showing new inventions and new product ideas. Share these with students and suggest they bring in examples they find.

SECTION 3: Manufacturing Technology

FIG. 9-3 The demand for children's toys is usually created by advertising.

FASCINATING FACTS

In the 18th century, when someone violated a patent, it was the patent holder's responsibility to defend it. Eli Whitney, inventor of the cotton gin, spent his entire fortune on legal fees to protect his patent from those who tried to steal it.

new product ideas. Fig. 9-4. It is the responsibility of the R&D group to develop ideas and then refine them.

Actually, not every idea turns out to be a good product. From about sixty new product ideas, only one may finally make it as a new product. That's another reason a company is always looking for new ideas.

FIG. 9-4 Brainstorming is one way engineers can come up with new product ideas. All kinds of ideas are suggested, even those that may sound silly at the time.

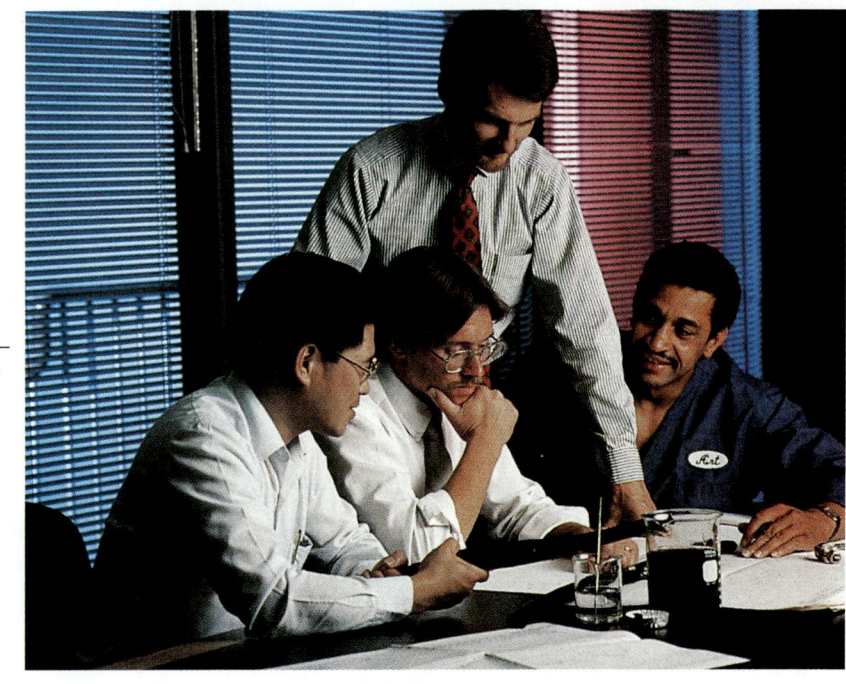

CHAPTER 9: Product Development

Types of Research

There are two types of research: basic research and applied research.

Basic research is done to learn more about the world around us. This kind of investigation helps us understand things. Often new information can be put to use immediately. Sometimes, however, it can't. Scientists discovered the transistor as they investigated electronics. It wasn't until later that they found uses for it, such as making smaller radios and TVs.

Most research done in manufacturing is *applied research*. This type of research is done to solve a problem. The new information can be applied immediately. After transistors were in use for a while, researchers looked for a way to reduce the size of a radio still further. They tried to find ways to make electronic components even smaller. The result was the development of the microchip. Fig. 9-5.

The same methods are used for both basic and applied research: retrieving, describing, and experimenting.

FIG. 9-5 Technology research has led to the miniaturization of electronic products.
A. Early radios used large vacuum tubes.
B. Then transistors replaced vacuum tubes. Transistors were much smaller.
C. Most radios now contain microchips, usually tiny bits of silicon treated to carry and control the flow of electricity.

4. **Critical Thinking.** Ask students to name other products and identify how they were changed.

SECTION 3: Manufacturing Technology

Retrieving is getting information that is already known. This method identifies what already exists. That way we don't "reinvent the wheel."

Describing is telling about present conditions. If you wanted to manufacture stereo headphones, you'd need to know the sizes of human ears. If you measured some ears to find out, you would then have the information to describe what size of headphones would be best.

Finally, research is done by *experimenting*. When experimenting, researchers try things out to see what happens. This provides information about how things will work. If necessary, they can make corrections and try again. They do this until they find the answers they need.

> **FASCINATING FACTS**
>
> The search for a cheaper billiard ball led to the discovery of plastic. The balls were originally made of ivory. Looking for a new material, an inventor mixed treated cotton, camphor, and alcohol and heated it. The result? Celluloid plastic!

Product Design

After the product idea is established, the product itself must be designed. Designing the product's appearance is an important part of the development process. The design should be eye-catching and appealing.

Drawings

Product designers start by thinking about the product idea. They try to picture in their minds what the product should look like. Then they begin making simple drawings. First, they make lots of *thumbnail sketches*. Fig. 9-6. As you learned in Chapter 7, these are small drawings done quickly. They are called "thumbnail" because of their small size. They don't include much detail. The purpose of making thumbnails is to capture many different ideas.

FIG. 9-6 A designer begins by making thumbnail sketches.

5. **Extension.** Explain how any experiment is conducted: controlling some factors and varying others, measuring and recording results, drawing conclusions.

6. **Critical Thinking.** Ask students to identify products that they think have a good appearance. Discuss factors that make designs pleasing or attractive.

CHAPTER 9: Product Development   **215**

The next step is to make some *rough sketches*. Fig. 9-7. The product designer reviews the thumbnails. Then he or she picks the best parts of each and combines them into a larger, more detailed drawing. Features like color and surface texture may be shown. As each new drawing is made, the product's final "look" becomes more and more apparent.

Renderings are made next. These realistic drawings are done very carefully. A rendering shows the designer's idea of the final appearance of the product. It is usually shown to potential consumers to see if they like the design. Fig. 9-7.

Today, many designers use CAD software for product sketches and drawings. The design can be easily changed on the computer as the product goes through the development process. CAD is discussed in more detail in Chapter 13.

Mock-ups

Designers also make mock-ups of the finished drawing. A **mock-up** is a three-dimensional model of the proposed product. It looks real but has no working parts. People who see it can understand what the final product will look like.

7

Most of the time a mock-up is built full size. However, sometimes a scale model is made. Fig. 9-8. In a scale model, one size is used to represent another size. For example, one inch on the scale model might be equal to one foot on the actual product. Mock-ups may be made of wood, plaster, plastic, or even cardboard. The type of material used does not matter as long as the mock-up looks like the real product.

CAD software can also be used to create models. Some models can even be analyzed using different software. Analysis will be discussed later in this chapter.

FIG. 9-7 Rough sketches (left) and renderings (right) show a product in more detail.

7. **Extension.** The "Technology Focus" in Chapter 13 discusses making mock-ups and models of designs using stereo lithography.

SECTION 3: Manufacturing Technology

FIG. 9-8 This is a scale model of a new aircraft. The model is much smaller than the actual product, but size relationships are constant.

Management Approval

During the product design process, various people make comments or suggest improvements. For example, managers from different departments in the company are asked to give their opinions about the design. The marketing manager is concerned with how the product will sell. The manufacturing manager is concerned with how easy or difficult it will be to manufacture the product. The financial manager is concerned with costs. All of these managers and more must agree that the product design is a good one.

Many companies have combined managers from different departments into management teams. Fig. 9-9. These management teams work together to oversee all of the product design, development, production, and marketing of one or several products.

When everyone finally agrees on the product design, an official approval is given. Much developmental work remains to be done, but this approval is the signal that development should proceed.

FIG. 9-9 This design team is approving the prototype for a new automobile intake manifold.

8. **Reinforcement.** Explain why so many managers are involved in approving new products.

Product Engineering

Engineering involves predicting the future behavior of a material or system. Engineers base predictions on known facts, mathematics, and scientific principles. **Product engineering** means planning and designing to make sure the product will work properly, will withstand extensive use, and can be manufactured with a minimum of problems.

In a process called **concurrent engineering,** groups of specialists from all areas of design and production work as teams. Concurrent engineering is also known as simultaneous engineering.

In **computer-aided engineering** (**CAE**), computers help engineers plan, design, and test products. They are valuable in all stages of product development. Engineering work involves lots of mathematics. A computer can solve math problems much faster than a human can. In computer-aided engineering, computers are not only used to perform needed math calculations, but also to produce working drawings and to analyze parts.

Many things must be considered when engineering a product. Two of the most important involve functional design and manufacturability.

Functional Design

Functional design ensures that a new product will work properly. Proper function often depends upon the materials used and the structure of the product.

The designers must answer questions such as: Will the clips designed for the straps of backpacks, fanny-packs, parachutes, or cameras hold? Will they be too difficult to hook and release? What is the best material to use—plastic or metal? Will the clips be uncomfortable to wear? How durable are they? Will they malfunction or break easily?

If power is needed, engineers must select the best type of power for the product. For example, what would be best to power a calculator, pocket computer game, or portable CD player? Would a combination of power sources be best? Should the power sources be replaceable or renewable?

The success of all product designs depends upon the materials, the physical design, and, if needed, the appropriate power source. All must be carefully planned to ensure that the new product will function properly.

Materials

A very important part of engineering any product is specifying the exact materials to be used. Should the product be made of plastic or metal? Does the material need to be fireproof? Maybe it has to be able to withstand physical impacts or extreme temperatures. The product engineer has to know exactly what is required. Then he or she can select just the right material.

The composition of a material has an effect on its characteristics, such as strength. Steel, for example, is available in several degrees of strength, referred to as grades. The engineer must determine which grade of steel is best for a particular product. A tractor axle and a door hinge are both steel products, but they have different strength requirements. Thus, different steels would need to be specified.

---

9. **Reteaching.** Review the material covered in the Technology Focus feature at the beginning of the chapter.

10. **Science.** Demonstrate how two "identical" materials can have different strengths. Use two pieces of ¼" steel rod, one annealed, and one quench-hardened. Have a student bend them in a vise. Discuss the results.

FIG. 9-10 The U. S. cycling team rode this type of bicycle in the Olympics. All of the structural and mechanical elements were carefully planned.

Structural and Mechanical Elements

Structural elements are frameworks and supporting members. *Mechanical elements* are the working parts. Such things as fasteners, the shape of a piece of aluminum, and the meshing of gears and levers can all affect product performance. What holds a car body together? Why does the cap on a pen snap on and stay there? Why does the door open when you turn a doorknob? It's because the structural and mechanical elements of these products were engineered properly. Fig. 9-10. The engineer is responsible for how each part will fit and work with the other parts.

Power Sources

Not all products are powered. Dishes, hats, and baseballs aren't powered. However, toasters, computers, and lamps are. If a product requires any kind of power to make it work, then those details must be engineered.

Many products use electricity. Fig. 9-11, part A. Some, like a tractor, are powered by the force of a moving liquid. Fig. 9-11, part B. This is *hydraulic* power. Still others may be mechanically powered, such as by one or more springs. Even compressed air is a power source. Fig. 9-11, part C. The size of each piece of wire, the horsepower of each motor, and the stroke of each cylinder are all determined by the engineer.

Design for Manufacturability

What effect will the design of the product have on the ability to actually manufacture it? How will the design affect the production of parts? Engineers try to design parts that will be easy to manufacture and assemble. This process is called **design for manufacturability**. For example, a part may be made with rounded corners instead of square corners if it is easier to make it that way. Factors engineers must consider include interchangeability of parts, standardization, simplification, modular design, and design for assembly.

11. **Reinforcement.** Point out the structural and mechanical elements of a folding chair.
12. **Brainstorming.** Ask students to name examples of powered products—electrical, airpowered, hydraulic, and mechanical. Make a list on the chalkboard according to category.

CHAPTER 9: Product Development

A. This power saw uses electricity.

B. This farm tractor requires a combination of electrical, mechanical, and hydraulic power.

C. A pneumatic jackhammer is powered by compressed air.

FIG. 9-11 Products may be powered in different ways.

Interchangeability of Parts

Mass production involves producing large quantities of a particular product. In order to do this efficiently, **interchangeable parts** are required. These parts are identical, and any one of them will fit the product. Assembly is thus speeded up.

13. **Reinforcement.** Demonstrate this concept using a box of 1/4-20 nuts.

However, because of human error, or differences in machines, or tool wear, there are always tiny differences. To allow for this, engineers plan for a certain amount of variation called tolerance. **Tolerance** is the amount that a part can be larger or smaller than the specified design size and still be used.

14. **Reteaching.** Use a micrometer to measure 10 "identical" parts. Record your results on the chalkboard and discuss them with students.

Standardization

Manufacturers who make similar products often use the same parts. Makers of bicycle tires, for example, can sell them to more than one bike manufacturer. In order to do this, however, the sizes must be the same. Agreement on common sizes of parts is called **standardization**. Fig. 9-12.

Standard sizes make manufacturing easier. For example, bolt sizes are standardized. Instead of every manufacturing company making their own sizes of bolts, they can buy standard-size bolts. It's cheaper and faster. Threads on a light bulb, tire sizes for cars, and electrical plugs and outlets are other examples of standardized products. Standard parts in standard sizes reduce the amount of new engineering work that must be done on a new product.

Simplification

Reducing the number of different sized parts in a product also makes the product easier to manufacture. Suppose that a design for a typewriter specified sixteen screws of different sizes. Then, sixteen different screws would have to be ordered and assembled correctly. However, if the same size screw fits in each place, just one size would be required. Assembly would be simpler, too. Any one of the screws could be placed in any one of the sixteen locations.

Modular Design

A *module* is a basic unit. In manufacturing, using preassembled units is called **modular design**. For example, in automobiles, units such as engines, wheels, and radiators are modules. It's possible to

FIG. 9-12 The bicycle hub shown here can be used in combination with any of the 5-speed, 6-speed, or 7-speed rear freewheels.

15. **Reinforcement.** Show two different brands of a standard-sized product.

FIG. 9-13 This printer has been designed so that parts can be clipped into place. This makes it easier for robots to assemble.

have a variety of car models by starting with basic parts and adding different modules. The same basic car could be made with a four-cylinder engine, a six-cylinder engine, or even an eight-cylinder engine.

FIG. 9-14 This is a detail drawing for the control handle of an electric motor limit switch. It gives all the information needed to process the part.

Design for Assembly

Design for assembly means planning so that product parts can be put together more efficiently. For example, if parts can be held in place with clips that are molded into the product instead of with nuts and bolts, the nuts and bolts can be eliminated. Assembly can also be done much faster, since using clips is quicker than threading nuts onto bolts with wrenches.

Many products, such as computer printers, automobile dashboards, and refrigerators, use clips instead of screws or nuts and bolts to hold parts together. Fig. 9-13.

Working Drawings

As the details for production are determined, they must be recorded. This is done by preparing a set of **working drawings**. These drawings show exact sizes, shapes, and other details. Products are manufactured according to the information these drawings provide, so they must be accurate and readable.

There are three main types of drawings in a set of working drawings: detail drawings, assembly drawings, and schematic drawings.

A *detail drawing* specifies the details of a particular part. All the necessary information is given so that the part can be manufactured. The detail drawing shows shape, dimensions (sizes), and locations of features like holes and bends. Tolerances are indicated. Special information about materials and surface finish may also be provided. Fig. 9-14.

Generally, a detail drawing is needed for every part of a product, except standardized parts. Just imagine how many detail drawings are needed for a large or complex product like an airplane!

Assembly drawings show parts in their proper places and how they fit together. Fig. 9-15. There are several different types of assembly drawings, but they all serve the same purpose.

FIG. 9-15 Exploded assembly drawings, such as this one, are needed for repair and service manuals. They also help workers assemble products.

ASSEMBLY DRAWING

KEY NO.	PART NO.	DESCRIPTION
1	34750C AAC	FRAME
2	34781	GRIPS
3	34771C	HANDLEBAR ASSEMBLY
4	302108	BINDER BOLT
5	302107	WEDGE
4-6	32752	HANDLEBAR STEM
7	303384	HEAD BEARING SET
8	34784C	FORK
9	303005 CCD	FRONT WHEEL - LESS TIRE (TIRE SIZE 20X1.5)
10	303155	FRONT SPOKE & NIPPLE SET (6 EACH)
11	303520	AXLE BEARING SET
12	32375Z	FRONT WHEEL RETAINER
13	27679Z	KICKSTAND ASSEMBLY
14	14590C	SPROCKET
15	12300C	CRANK
16	32809	PEDALS
17	34779C	CHAIN GUARD
18	12834	CHAIN & LINK
19	303333	CRANK HANGER BEARING SET
20	303013 CCD	REAR WHEEL - LESS TIRE (TIRE SIZE 20X1.75)
21	303155	REAR SPOKE & NIPPLE SET (6 EACH)
22	98X250	SADDLE
23	32942Z	SEAT POST
24	303404Z	SEAT POST CLAMP ASSEMBLY
25	34350	REFLECTOR PACKAGE
26	34792Z	FRONT REFLECTOR BRACKET
27	34800Z	REAR REFLECTOR BRACKET
28	34696	HANDLEBAR PAD
29	32618	TIE STRAPS
30	34764	NUMBER PLATE
31	34783PA	CALIPER BRAKE
*	64X286	FRONT PLATE DECAL
*	F-4661	OWNER'S MANUAL

FIG. 9-16 This schematic diagram is used to build an automobile wiring harness.

Also given on an assembly drawing is a parts list. This is a listing of all parts needed to assemble the product. Usually, the quantity needed of each part is also shown.

Sometimes a special drawing called a *schematic drawing* is used to show the position of parts in a system—that is, the "scheme" of things. Parts are not shown as they actually look. Instead symbols are used to represent the parts. Electrical and hydraulic systems are usually shown in schematic form. Fig. 9-16.

Analyzing the Design

A product engineer's job includes making sure that everything about the new product is right. One very important aspect of product engineering is reliability. It's one thing for a product to work properly, but will it continue to work? The reliability of a product can seriously affect its success. Multiple reliability problems can quickly give a manufacturer a bad reputation. Consumers will then choose a competitor's product instead, and sales will go down. On the other hand, when a manufacturer's products have a high reliability rating, consumers are more likely to buy them. This increases demand, and manufacturers can then ask a higher price for their product.

Engineers use the results of testing to make predictions about the working life of the product. For example, based on their design data and on test results, engineers can calculate and predict the life of a car battery, a motorcycle tire, or a computer keyboard. Some products are designed to last only a certain period of time or under certain conditions. This is called *planned obsolescence*. For example, a paper facial tissue is designed for a single use. It does not last as long as a washable cotton handkerchief.

Each part, as well as the completed product, is carefully analyzed during the design process. Various methods are used to assure that the product will serve its intended purpose, including computer simulation and testing and analyzing prototypes.

16. **Enrichment.** Discuss how the quality of cars caused a shift in the industry in the 1970s.

Prototypes

A **prototype** is a full-size working model of the actual product. It is the first of its kind, and is usually built by hand, part by part. Only one may be built or maybe a few hundred. The number depends on the size and complexity of the product. Because prototypes are hand-built, they are usually very expensive. A prototype for a new automobile can cost a million dollars or more! Fig. 9-17.

Prototypes can be used to check manufacturing procedures. As the prototype is being built, manufacturing engineers keep track of the progress. They note any difficulties. This provides feedback for planning the processes that will be used to mass-produce the product.

Types of Analysis

Prototypes serve as experimental models for analyzing product design and engineering. Each prototype is subjected to a variety of tests. Fig. 9-18. The tests are designed to try out the product under extreme and difficult conditions. Some tests actually destroy the prototype. This is called *destructive testing*. Crashing a prototype car into a brick wall in order to test its "crashworthiness" is an example. Most tests, however, do not actually destroy the prototype. *Nondestructive tests* are used to analyze the prototype and its performance.

Functional analysis means analyzing the prototype to see if it works as predicted. During the early stages of product engineering, engineers predict (mathematically) how the product will work. After

FIG. 9-17 Each part for this prototype car was custom made.

17. **Extension.** Mention that computers can be used to make prototypes, which will be discussed in Chapter 13.

CHAPTER 9: Product Development

FIG. 9-18 If the prototype can perform well in extensive tests, then the manufacturer knows the product will be a good one. Here, a life-size dummy is used to test a reclining chair.

the prototype is built, it can be tested to see if the predictions were correct. For example, a prototype of a new flashlight might be subjected to a temperature test to make sure it works within a wide range of temperature. The prototype would be tested in freezing cold and boiling hot temperatures for various amounts of time.

Failure analysis is used by engineers if the prototype fails. The prototype is then analyzed to see why it didn't work as it should.

Usually every prototype is tested under certain conditions until it breaks down or fails. Then engineers tear it apart and analyze the parts very carefully. They look for clues that would show why the product failed. For the flashlight example, they might use a drop test and a switch test. In the drop test, the prototype would be dropped repeatedly from different heights. The engineers want to see how well it can stand shock. The switch test would show how many cycles (times switched on and off) the switch could stand before it failed. Depending on test results, the product may be redesigned to make it better.

Value analysis is an attempt to reduce the cost of materials and purchased parts without sacrificing appearance or function. Every part of the new product is studied carefully to see if the right material has been selected. A certain part, for example, might be made of steel in the original design. A value analysis might reveal that the part could be made of plastic. Plastic would cost less, and yet the plastic part would still work as well as the steel part. Then the plastic part would have to undergo the same kind of product testing as the steel part. The engineers specify the most functional, yet lowest cost, material for every part.

As you know, a computer simulation can help analyze a product's design even before any parts are made. Because it works so fast and can handle large amounts of information, the computer is also valuable for functional, failure, and value analysis. Fig. 9-19. (More about computer technology will be discussed in Chapter 13.)

FIG. 9-19 A product drawn with CAD software can also be analyzed using the computer. Here, the structural integrity of a stereo speaker cabinet is being analyzed.

18. **Critical Thinking.** Have students analyze the parts of a product and list alternative materials for certain parts.

CHAPTER 9: Product Development

Career File—CAD Drafter/Technician

EDUCATION AND TRAINING

Employers prefer applicants who have had at least two years of drafting training after high school. College courses in engineering, architecture, and mathematics are useful. Some companies have their own training programs. Because the technology changes so rapidly, drafters must expect to continue their education even after they've found jobs.

AVERAGE STARTING SALARY

Depending on background, beginning drafters earn about $16,500 per year.

OUTLOOK

Drafting jobs are expected to grow slower than average through the year 2005, according to the Bureau of Labor Statistics. Although the demand for drafting services will increase, the increasing use of CAD software means that fewer people will be needed to do the work.

What does a CAD drafter or technician do?
CAD drafters or technicians use computers and CAD software to prepare technical drawings. The drawings are then used by other workers to build everything from spacecraft to industrial machinery to products sold in stores. The drawings show exact dimensions, specific materials to be used, and procedures to be followed. Some CAD drafters have a knowledge of engineering and manufacturing and may use technical handbooks and tables in their work. Many specialize in one design area.

Does most of the work involve data, people, or things?
Although CAD drafters work closely with engineers and other professionals, most of their work is done with the tools of their trade—a computer and CAD software.

What is a typical day like?
CAD drafters create drawings using rough sketches, specifications, codes, and calculations previously made by engineers and other professionals. The drawing is developed on the computer and can be printed at any stage. It is stored electronically so that revisions and design variations can be made easily. Drafters may meet with other workers to determine the final design of a product and how it will be manufactured.

What are the working conditions?
Drafters usually work in offices. They often sit at a computer for long periods of time doing detailed work, which may cause eyestrain and back discomfort.

Which skills and aptitudes are needed for this career?
Some artistic ability is helpful. Final drawings require accurate, detailed work. Drafters should be able to communicate well with other professionals.

What careers are related to this one?
Other workers who prepare or analyze detailed drawings and make precise calculations and measurements include manual drafters, architects, engineers, science technicians, map makers, and surveyors.

One CAD drafter/technician talks about his job:
With CAD I don't have to worry about being neat! Everything always comes out looking good.

19. **Review.** You may wish to review the discussion of drafting in Chapter 7.

Chapter 9 REVIEW

 ## Reviewing Main Ideas

- A product's life cycle consists of its introduction, growth period, maturity, saturation, and decline.
- New ideas are researched, developed, drawn, and approved before becoming products.
- Product engineering means planning and designing to ensure the product will work properly and be manufactured with a minimum of problems.
- Designing for manufacturability involves the use of interchangeable and standardized parts, simplification, modular design, and design for assembly.
- Working drawings include: detail drawings, assembly drawings, and schematic drawings.

 ## Understanding Concepts

1. List the stages of a product's life cycle.
2. How does an idea become a product?
3. What is the general role of product engineers?
4. What five factors play a part in design for manufacturability?
5. Name and define three types of working drawings.

 ## Thinking Critically

1. Discuss some ways consumers benefit from mass production of goods.
2. How is product testing beneficial to manufacturers? How is it beneficial to consumers?
3. Why do you think some products go quickly through the life cycle?
4. Think of a product that has a poor appearance. How would you improve its appearance without harming its usability?
5. Suppose you had an idea for a product. How would you go about communicating your ideas to a manufacturer?

 ## Applying Concepts and Solving Problems

1. **Mathematics.** Determine the following tolerance ranges:

   a.) The dimension given for the length of a flange is 13.12 in. (+ or - .06). What are its maximum and minimum acceptable lengths?

   b.) A dowel rod is dimensioned 18 in. (+ 1/16, - 1/8). What is the range of acceptable lengths?

2. **Science.** Obtain a plastic bag used to hold produce from a grocery store. Weigh a number of items and put them, one at a time, into the bag. How much weight did the bag hold before it failed? Should the bag be redesigned? Why or why not?

3. **Technology.** Take apart a product that no longer works. What went wrong? Was the product poorly designed, or was it planned obsolescence?

20. **Resource.** Answers to "Understanding Concepts," "Thinking Critically," and "Applying Concepts and Solving Problems" can be found in the Teacher's Manual at the front of this *Teacher's Annotated Edition* and in the *Teacher's Resource Binder*.

CHAPTER 10

Production Planning

Objectives

After studying this chapter, you should be able to:
- describe planning procedures used to achieve production efficiency.
- describe how the layout of a manufacturing facility is developed.
- explain how machines and tools are changed for a new process.
- explain why a pilot run is important

Terms

bill of materials
die
fixture
group technology
jig
mold
part print analysis
pilot run
plant layout
process chart
tooling-up

Technology Focus

Awards for Innovation

What do these products all have in common—Band-Aids®, Tylenol®, Pampers®, Pringles®, Scotch Magic Tape®, Post-It® Notes, Teflon®, personal computers (PCs), and Microsoft Windows®? The answer: all were developed by winners of the National Medal of Technology (NMT) award.

The President of the United States awards the National Medal of Technology to America's leading *innovators*. (Innovators are people who come up with new ideas.) Beginning in 1985, the first NMTs were awarded to encourage the development of technologically innovative and commercially successful products, processes, or services. Between 1985 and 1996, 98 individuals (65 working alone; 33 working in teams) and eight companies received the award.

Famous Winners

You may recognize the names of some of the winners. Steven Jobs and Stephen Wozniak, founders of Apple Computer, Inc., were among the first winners in 1985 "for their development and introduction of the personal computer . . . extending the power of the computer to individual users." Microsoft Corporation founder Bill Gates was one of the 1992 winners "for his early vision of universal computing at home and in the office . . . and for his contribution to the development of the personal computer industry." Proctor & Gamble Company won in 1995 "for creating, developing, and applying advanced technologies to consumer products that have strengthened the American economy while helping to improve the quality of life for millions of consumers worldwide." 3M also won the NMT in 1995 "for its many innovations over decades, producing thousands of successfully commercialized products." Johnson & Johnson was a 1996 winner "for a century of continuous innovation in research, development, and commercialization of products that are critical in the management of disease, improvement of quality of life, reduction of health care costs, and fostering of U.S. global competitiveness."

Just imagine—if you're a U.S. citizen, one day you may receive this prestigious award for a technologically innovative product, process, or service you develop!

Take Action!

How innovative can you be? Take 10 minutes to list as many different uses for a brick as you can. Don't be afraid to list silly uses. They may inspire you to think of more constructive ones.

Resources
- Chapter 10 Lesson Plan in the *Teacher's Annotated Edition* and in the *Teacher's Resource Binder.*
- Chapter 10 Review in the *Teacher's Resource Binder.*

Process Planning

Changing the form or shape of an industrial material requires several processes. The processes are chosen carefully. The company wants to make high-quality products in the least costly way.

Management Approval

As you read in Chapter 9, management gives approval to go ahead with the design of the product. Management also decides whether the company should proceed with production. Managers review information about the product design. They also consider future sales. If these factors look promising, then the managers allow production planning to begin.

Product Requirements

Manufacturing engineers plan production. To do so, they must know all the details about the product. They study the bill of materials and the working drawings.

A complete **bill of materials** is a list of the materials or parts needed to make one product. The quantity of each part or material is given. Items are listed in the order in which they are used. Figure 10-1 shows a sample bill of materials.

BILL OF MATERIALS

Part Number	Part Name	Quantity
256100	Wheel	1
256110	Rim	1
256120	Hub	1
256121	Axle	1
256122	Cone	2
256123	Bearing	16
256124	Locknut	2
256125	Nut	2
256126	Washer	2
256130	Spoke	36
256131	Nipple	36

FIG. 10-1 A bill of materials lists all the parts needed to make a product.

Working drawings provide valuable information for the process planner. By carefully studying the drawing for a part, a process planner can begin to get ideas about how the part could be made. This is called **part print analysis**. Fig. 10-2. Specified features and dimensions are identified on the drawing. Information about the finished shape and size of the part, the materials needed, and the tolerances is also collected during part print analysis.

Make-or-Buy Decisions

Should the company make or buy the parts needed? Three questions must be answered for every part in the product.

- Can the item be purchased? (Availability)

FASCINATING FACTS
Industry depends on about 80 different minerals that are obtained through mining. Some mines can yield as much as 300,000 tons of ore and waste rock every day.

1. **Reinforcement.** Careful planning makes any activity proceed more smoothly. Give examples.

2. **Reteaching.** Disassemble a ball-point pen. Make a list of the parts and the material from which each part is made.

SECTION 3: Manufacturing Technology

- Can the company make the part? (Manufacturability)
- Is it cheaper to make the part or to buy it? (Cost)

Information is gathered about parts that are available for purchase. Quality is important. Also, suppliers must be able to provide the proper quantities at the time they are needed.

Engineers also consider whether the company is able to manufacture the part. Can the tools and equipment now owned by the company be used? Do workers have the knowledge and skills needed to make the part?

Parts that cannot be purchased, must be made by the company. However, "make-or-buy" decisions are usually based on cost. Which way will be least expensive?

Selecting the Process

Manufacturing engineers make production plans for each part that will be made in the plant. They identify the processes needed, select the equipment for each process, and arrange all the processes into a logical order.

One common procedure for organizing this information is to make a **process chart**. Fig. 10-3. There are several types of these charts. The amount of detail varies, but all process charts show the sequence of manufacturing steps. These steps reflect the decisions manufacturing engineers have made.

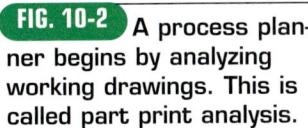 **FIG. 10-2** A process planner begins by analyzing working drawings. This is called part print analysis.

3. **Extension.** Making this type of decision is a never-ending process. Companies continually look for ways to reduce costs. Conditions change. At a later time, another decision might be more economical.

4. **Critical Thinking.** Select a sample product. Show students each part and have them discuss what process(es) might have been used to make the part.

CHAPTER 10: Production Planning **233**

PROCESS CHART - MEMO HOLDER

HOLDER

- CB-1   JOINT EDGES OF STOCK SMOOTH
- CB-2   RIP BOARDS TO WIDTH
- CB-3   CROSSCUT STOCK TO ROUGH LENGTH
- I-1   INSPECT FOR SIZE
- TS   STORE UNTIL GLUE DRIES
- CB-6   CLEAN ALL EXCESS GLUE
- CB-7   SURFACE TO THICKNESS
- CB-8   JOINT OUTSIDE EDGES
- CB-9   CROSSCUT TO FINAL LENGTH
- D   JOINT EDGES OF STOCK SMOOTH
- I-3   INSPECT FOR SIZE
- CB-10   ROUTE PENCIL GROOVE
- CB-11   ROUTE EDGES, ROUND ON ALL SIDES
- CB-12   SAND WITH ORBITAL SANDER
- I-4   INSPECT FOR MILL MARKS
- CB-13   SPRAY ONE-STEP FINISH
- CB-14   RUB DOWN FINISH
- I-5   FINAL INSPECTION

PACKAGE

- P-1   CUT CARDBOARD BOX TO SIZE USING TEMPLATE
- P-2   LAY OUT BOX USING TEMPLATE
- P-3   CUT OUT WINDOW IN LID PORTION OF BOX
- P-4   SCORE BOX ALONG DOTTED LINES AND FOLD FLAPS UP 90 DEGREES
- P-5   TAPE BOX WITH CARDBOARD TAPE
- P-6   INSERT CELLOPHANE WINDOW AND TAPE IN PLACE ON INSIDE
- I-2   INSPECT BOX
- TS   HOLD FOR PACKAGING

- CB-17   INSERT HOLDER INTO PACKAGE
- I-6   INSPECT PACKAGED PRODUCT
- TS   HOLD FOR DISTRIBUTION

FIG. 10-3   A process chart shows the sequence of operations required to make a product.

SECTION 3: Manufacturing Technology

FIG. 10-4 In what ways are these two parts the same? In what ways are they different? In group technology, one set of production plans can be used for both as long as changes in dimensions are made.

Production planning can be done more quickly when the manufacturing engineers don't have to start all over each time. The plans for any similar part can be used again, as long as dimensions are changed. Fig. 10-4. Computers have made this type of planning very efficient.

Facilities Planning

Well-planned facilities are important in manufacturing. Buildings must be planned. Everything that goes into the buildings must be placed according to the plan. Then products can be produced efficiently.

Plant Location

Where should a factory be located? Which part of the world? Which part of the country? Which state? Which city? Which part of the city?

Manufacturing facilities are located in various places for various reasons. A plant may be located near a large city because many workers are available there. Sometimes the best location is near a source of raw materials. Other times, a location near the marketplace is better. Each company must decide which is best for producing and selling its product. Fig. 10-5.

Group Technology

Another interesting procedure used in production planning is called **group technology**. This is a method of keeping track of similar parts that a company manufactures. Using a classification scheme, all parts are coded according to their shape and features. When the company has to manufacture a new part, the engineers first check the file of existing parts to find similar parts.

Suppose the company manufactures a certain cylindrical part with a hole through the center. For a different product, they may need to make a similar but different size cylindrical part. Since they already have the production plans for one cylindrical part, all they have to do is to retrieve the information about the first part and change the dimensions to make the part needed for the new product.

5. **Critical Thinking.** Ask students why they think local manufacturing plants chose their particular location. What environmental impacts should be considered?

CHAPTER 10: Production Planning **235**

FIG. 10-5 The location of a manufacturing plant is important. Transportation services must be available to bring in materials and to ship out finished goods.

FIG. 10-6 Some layout planners build a model of the plant from sketches to help them visualize production flow and space requirements.

SECTION 3: Manufacturing Technology

FIG. 10-7 This diagram shows a machine that requires a space seven feet by ten feet. How much space would ten of these machines occupy?

Plant Layout

A plan is developed for the interior (inside) of the plant. A drawing of this plan shows the arrangement of machinery, equipment, materials, and traffic flow. This is called the **plant layout**. This layout is determined by production flow, related activities, and space requirements.

Using the process charts, a layout planner develops a plan for production flow. The plan shows how materials and parts are moved into, through, and out of the plant. The production flow can be a kind of "road map." Sometimes lines are drawn on a factory floor plan to show where things go.

To make the production flow more efficient, related activities are usually located near each other. For example, a drying room would be located near the painting room. However, some activities that occur in the factory are not directly involved with production. Every factory has a shipping and receiving area. There are also offices, rest rooms, and maybe even a cafeteria. These areas support the manufacturing process and are planned in relation to the production area.

The amount of space for each activity has to be determined. Fig. 10-6. Usually this is done mathematically. Consider cafeteria space, for example. Suppose 50 people will be using the space and each person needs 12 square feet. Then 600 square feet of space will be needed altogether. (50 x 12 = 600) In the same way, if you know how much space is needed for each machine and you know how many machines will be used, then by multiplication you can determine how much space will be needed. Fig. 10-7.

6. **Reteaching.** Show a flow diagram superimposed on a floor plan. For example, show the flow of students through the school.
7. **Reinforcement.** Using the same floor plan, show how activities are related.
8. **Mathematics.** Have students solve the problem in Fig. 10-7. Discuss workspace requirements. Answer: 600 sq. ft.

CHAPTER 10: Production Planning

FIG. 10-8 Roller conveyors are used to move boxes from one place to another.

Materials-Handling Systems

Materials handling is moving and storing parts and materials. Many different and automated types of equipment are used. The equipment and its automated systems are known as the *materials-handling system*.

Conveyors are used to move materials, parts, and products from one place to another along a fixed path. Fig. 10-8. There are many different types of conveyors. Each one is used for a specific purpose. Belt conveyors consist of a rolling belt on which items are placed. Many supermarkets have belt conveyors at checkout counters. Roller conveyors are good for carrying heavy loads. Skate-wheel conveyors are good for transporting boxes.

A *hoist* has one or more pulleys and is used to lift heavy loads. A *crane* is actually a hoist that can be moved about in a limited area. One kind of crane travels on overhead rails. Fig. 10-9. The parts move along the track suspended from the hoist.

Trucks used for materials handling are different from the trucks you see on the road. The proper name for them is industrial trucks, but most people just call them *forklifts*. They have forks on the front that can be raised and lowered. Forklifts are used for lifting and carrying loads from one place in the plant to another. Fig. 10-10. Forklifts are often used to move *pallets*, which are special platforms on which large parts or materials are often stored. Loaded pallets can be stored on pallet racks to save floor space.

Today, computers are also used to control materials-handling systems. These computerized systems will be discussed in Chapter 13.

9. **Reinforcement.** Have students name places where they have seen forklifts used

238 SECTION 3: Manufacturing Technology

FIG. 10-9 An overhead crane is used to lift large, heavy items and move them to another spot. The crane shown here can also transfer its load to another crane.

FIG. 10-10 Industrial trucks, like this forklift, are used to lift and move materials from one place to another.

CHAPTER 10: Production Planning

Putting the Plan into Action

After planning is complete, further preparations are made. Workers are chosen and organized. Machines and other equipment are obtained and prepared for production. The system is tested and refined. Production can begin!

Organizing People

People are needed to work in production. The *human resources department* employs and trains workers. However, specifying the number and types of workers needed is part of production planning. Fig. 10-11.

Production planners have carefully studied all the possible methods for manufacturing the product. They decide which methods and machines to use. Then they know how many workers will be needed and exactly what each worker will be required to do.

Tooling Up

Getting the tools and equipment ready for production is called **tooling-up**. All the machines and tools are prepared for certain jobs. If necessary, the company buys new tools, machines, and equipment. Tooling-up often includes adapting (changing) existing machines. Special tools may be needed. These are made or purchased. Special tools include jigs, fixtures, molds, and dies. Gages are instruments also used as an aid to tooling-up. They will be discussed in Chapter 11.

Jigs and fixtures are used to adapt general purpose machines to do certain operations. They are frequently used during the drilling and cutting of metal parts. A **fixture** clamps and holds the part in place during processing. Fig. 10-12. A **jig** is like a fixture in that it holds the part, but it also *guides* the tool. For example, a jig might hold a piece of wood in place while it also serves as a guide for the saw being used to cut the wood.

Some machines require molds or dies to shape materials. A **mold** is a hollow form. Liquid material is poured, squirted, or forced into the mold cavity (space inside the

FIG. 10-11 Production planners determine how many workers are needed and what kind. These people are assembling hard disk drives for computers. Why do you think they are wearing special clothes?

10. **Reteaching.** Use sports as an example of the need for organizing people (coach, manager, various playing positions).
11. **Enrichment.** Demonstrate examples of jigs and fixtures. Discuss the design and usefulness of jigs and fixtures in the school lab and in manufacturing.
12. **Enrichment.** Show a sample mold and explain how it is used to shape a product.

240 SECTION 3: Manufacturing Technology

FIG. 10-12 Most fixtures are custom-made to hold a specific type of part.

FIG. 10-13 Molds are used to shape parts or products. This picture shows a robot removing a finished part from an injection molding machine. In injection molding, a heated liquid is forced (injected) into a mold where it is allowed to cool and harden.

mold). As the material hardens, it takes the shape of the cavity. Fig. 10-13.

A **die** is a piece of metal with a cut-out or raised area having the desired finished shape. The material is then forced through or against the die to take on its shape. A punch and die set is used to punch or cut shapes from materials. This is called *die punching*. It works like an ordinary paper punch. Coins are made this way. Fig. 10-14. Sheet metal and plastics are frequently cut to shape using a punch and die.

FIG. 10-14 Coins are made using a die-punching process.

CHAPTER 10: Production Planning

MATHEMATICS CONNECTION

Working with Fractions on Product Drawings

Sometimes working drawings contain fractions. Working with them is no different than working with fractions in other problems.

For example, suppose you wanted to enlarge the diameter of the top of the wastebasket in this drawing by ⅝". To find the new diameter, you would add: 12 ¼" + ⅝". Because the denominators are different, you must find a common denominator. In this case, ¼ can be changed to 2/8. You can then add the fractions to find the new diameter:

12 2/8" + ⅝" = 12 ⅞"

Suppose you want to cut the bottom of the wastebasket from a piece of wood that measures 11" x 22". You want to use the leftover wood for another project, and for that you need a piece that measures 11" x 12". Will you have enough? To subtract 10 ⅛ from 22, you must first borrow from the whole number in order to have another fraction to work with, then subtract:

WASTEBASKET

DIMENSIONS ARE IN INCHES

$$22 = 21\ 8/8$$
$$-\ 10\ 1/8$$
$$\overline{11\ 7/8}$$

You will not have enough wood for the second project.

Try It!

> The outside diameter of the top of the wastebasket is 12 ¼". If its walls are ⅜" thick, what is the inside diameter?

Another kind of die is used for forming sheet material. The material is stretched over the die or the die is pushed into the material. This is *die stamping*. The body of a car is shaped this way. Many times die stamping and die punching are done at the same time with the same machine.

One machine can be used to make different parts by simply changing the mold or die. For example, an injection molding machine might be used to make plastic hairbrushes. By changing the mold, the same machine could be used to make toothbrushes.

242 SECTION 3: Manufacturing Technology

FIG. 10-15 An engineer times the assembly process during a pilot run.

The Pilot Run

Will the production plan work? One way to know for sure is to conduct a **pilot run**. A pilot run is like a practice where all parts of the system are operated together before production really starts. The main purpose is to find and correct production problems, but it also gives workers a chance to learn new tasks.

During the pilot run, engineers watch closely and keep records. As problems are identified, corrections are suggested and tried. This process of finding and correcting problems in the operation of a system is called *debugging*. Because this is a pilot run, the system can be stopped while corrections are made. Actual production is not delayed.

The pilot run also allows engineers to check and adjust the timing along the production and assembly lines. They may speed up or slow down equipment to see the effects on production. Fig. 10-15. Speeding up may increase the production or assembly rate, but the quality of the product may be affected. Workers must have enough time to do tasks properly.

The pilot run also serves as a training time for workers. They can learn new tasks or see how their jobs have been changed. Workers need to know exactly how to make the new product. They also need to understand how their work fits into the system.

Fascinating Facts

Henry Ford revolutionized car manufacturing by using a *moving* assembly line. He streamlined his process so well that for a while in 1925 a car was completed every 10 seconds.

CHAPTER 10: Production Planning

Career File—Employee Training Specialist

EDUCATION
Employee training specialists must have at least a bachelor's degree in human resources. A master's degree is preferred.

AVERAGE STARTING SALARY
Beginning salaries for employee training specialists with a bachelor's degree are about $26,000 per year. Specialists with a master's degree start at about $38,500 per year.

OUTLOOK
Through the year 2005, job prospects for employee training specialists are expected to grow faster than average, according to the Bureau of Labor Statistics. Training needs will increase as the work environment grows more complex, technology brings changes, and employees' skills become obsolete.

What does an employee training specialist do?
Employee training specialists plan, organize, and direct various programs to improve the skills, productivity, and morale of a company's employees. They determine what training is needed and make arrangements for the training to take place. Training may be in the form of classroom instruction, on-the-job training, workshops and conferences, or computer-assisted instruction (CAI).

Does most of the work involve data, people, or things?
An employee training specialist works mostly with people and ways to improve their skills. They confer with managers and supervisors to identify training needs.

What is a typical day like?
Training specialists perform a variety of activities depending on the size, goals, and nature of the organization. They may conduct orientation sessions for new employees, arrange on-the-job training and apprenticeships, develop special classroom programs, prepare schedules, or conduct training sessions themselves.

What are the working conditions?
Training specialists normally work in clean, pleasant, and comfortable office settings. Most work a standard 35- to 40-hour week.

Which skills and aptitudes are needed for this career?
Training specialists should have a desire to help people improve and succeed. They should be able to communicate well, as they may be instructors themselves.

What careers are related to this one?
Other occupations that require skills in interpersonal relations include career counselors, employment recruiters, lawyers, psychologists, social workers, and teachers.

One employee training specialist talks about her job:
The best part of teaching is when shy students start to catch on. With my help they begin to think they can do the job after all!

Chapter 10 REVIEW

Reviewing Main Ideas

- Product analysis involves planning of materials and an analysis of parts for the finished product.
- Manufacturing engineers consider availability, manufacturability, and cost.
- Process charts and group technology are ways to plan efficient manufacturing.
- Facilities planning includes decisions about plant location and plant layout.
- Equipment, such as conveyers, hoists, cranes, forklifts, and pallets are part of a materials-handling system.
- Tools and equipment can be adapted to do different tasks.
- A pilot run is a test of the system to correct problems and to check or adjust production and assembly time needed.

Understanding Concepts

1. What do manufacturing engineers study to analyze a product before they plan the actual production?
2. What are some planning procedures engineers use to achieve efficiency?
3. Why do manufacturing engineers need to know about the flow of production?
4. How are tools used or prepared for particular jobs?
5. What process takes place to test a system before actual production begins?

Thinking Critically

1. What might happen if a supplier cannot supply enough parts to a manufacturer by the time they are needed?

13. **Resource.** Answers to "Understanding Concepts," "Thinking Critically," and "Applying Concepts and Solving Problems" can be found in the Teacher's

2. Name two manufacturing plants and tell where they are located. Tell why you think the facilities are located there.
3. How does having a machine that can use different molds or dies affect costs?
4. Discuss several consequences of *not* making a pilot run.
5. Suppose that a new computer does not have one part of the keyboard in place. As the manufacturer, how would you go about trying to fix the problem?

Applying Concepts and Solving Problems

1. **Science.** Substances used in molds are those that are solids at normal environmental temperatures but that can be melted. Use a reference work to find the melting points of the following substances: silicon dioxide (glass), tin, lead, and copper. Which has the lowest melting point? The highest?

2. **Mathematics.** On a trip to a builder's supply store, Carlo bought:
- one 20' extension ladder at $112.00.
- two boxes of 8" x 1" FH wood screws at $3.19/box.
- five sheets of 100c abrasive paper at $.49/sheet.
   a. What was the total cost?
   b. If the tax on the items was 6 1/2%, what was Carlo's total bill?

Manual at the front of this *Teacher's Annotated Edition* and in the *Teacher's Resource Binder*.

CHAPTER 10: Production Planning **245**

CHAPTER 11

Production

Objectives

After studying this chapter, you should be able to:
- explain the difference between components and assemblies.
- explain the purposes of packaging.
- describe how production and product quality are controlled.
- explain how inventory is controlled.

Terms

acceptance sampling
assemblies
component
gages
International Organization for Standardization (ISO)
inventory
inventory control
production control
quality assurance
specifications
statistical process control
subassembly

Technology Focus
Quality in Electric Guitars

Construction of a solid-body electric guitar begins with putting the body and neck together. Both are cut from wood, such as mahogany (which favors mellower tones) or maple (higher, brighter tones). Because of the decreasing availability of these woods, poplar, ash, and others are also used. The guitar body may be cut from a block or built up from laminated pieces. [1]

The neck has to resist warping, so it is often an assembly of glued pieces. The grain of one piece is laid crosswise to the grain of the next piece. The neck may also be made of a single piece reinforced with an internal rod. The fingerboard is glued onto the neck, and frets (metal strips) are pressed into the fingerboard by machine or tapped in by hammer. Attaching the neck to the body may be accomplished by elaborate joinery or by using hardware.

On the face of the guitar, a router cuts grooves for pickups. Pickups are coils that generate an electromagnetic field which responds to the vibrations of the strings. Control knobs, in combination with amplifier controls, are added to adjust volume, bass/treble, tremolo, and echo.

Automation Changes the Product

In the past, quality in electric guitars always depended on handwork. Fine guitars were made by skilled woodworkers.

Eventually, automation found its way into guitar manufacturing. The results were disappointing, and mass-produced electric guitars reached a low point in the 1970s. In the 1980s, however, Japanese manufacturers steadily introduced improvements. Japanese guitars became popular in the mid- to low-priced part of the market.

A lot of effort is invested in sanding, painting, and finishing. Japanese innovation in these steps reduced processing time and improved consistency of results. However, not all parts of a guitar can be reached easily by sanding and buffing machines. Musical instruments have oddities of shape that make complete automation impractical. Some work is still done by hand.

In mid-priced guitars today, quality is better than ever. The high quality of Japanese products has forced everyone else to meet high standards. Presently, the leader in production is Korea, where low wages keep labor-intensive guitars affordable.

Take Action!
Visit a music store to compare a high-priced guitar to a low-priced guitar. List all the quality differences you can find in materials and workmanship.

Resources
- Chapter 11 Lesson Plan in the *Teacher's Annotated Edition* and in the *Teacher's Resource Binder.*
- Chapter 11 Review in the *Teacher's Resource Binder.*
  1. **Reteaching.** Ask students to define "laminated."

Producing Products

Producing products is what manufacturing is all about. *Production* is the multi-step process of making parts and assembling them into products.

Components

Each individual part of a product is called a **component**. Some components are simple, like a wire. Other components, such as a casting for an automobile engine, may be very complicated. Fig. 11-1.

Before production actually begins, decisions are made whether to make or buy each part. A computer manufacturing company, for example, may buy many electronic components. However, the company would probably make some of the special parts.

FIG. 11-1 Castings, like these parts for an auto engine, can be complex.

Assemblies

Components are assembled with other components. This means they are put together in a planned way. Assembled components are called **assemblies**. If an assembly will be used as a component in another product, it is called a **subassembly**. The handlebars and brakes of a bicycle are examples of subassemblies. Fig. 11-2.

FIG. 11-2 Even a fairly small product like a portable electric saw can contain many components and subassemblies.

2. **Extension.** If there is a company in your area that manufactures components for other products, use it as an example.

When assembly operations begin, all the necessary parts must be available in the right quantities. Sometimes assembly work is done by hand. That is, workers pick up parts and put them together. They may glue the parts together or perform other tasks. Sometimes automatic assembly machines are used.

Components may be assembled into subassemblies, and subassemblies may then be assembled with additional components or other subassemblies. The point at which all the parts are combined to form the product is called *final assembly*.

Packaging

After final assembly, many products are packaged. There are many reasons for packaging products. Fig. 11-3. Fragile or easily broken items such as glassware or electronics must be packaged to protect them. Items that could spoil or deteriorate, such as food or beverages, must be packaged to maintain freshness. Some products require more than one package. Think of chewing gum. Many companies first place each stick of gum in its own foil wrapper. Then several sticks are put in another package. These packages may in turn be placed

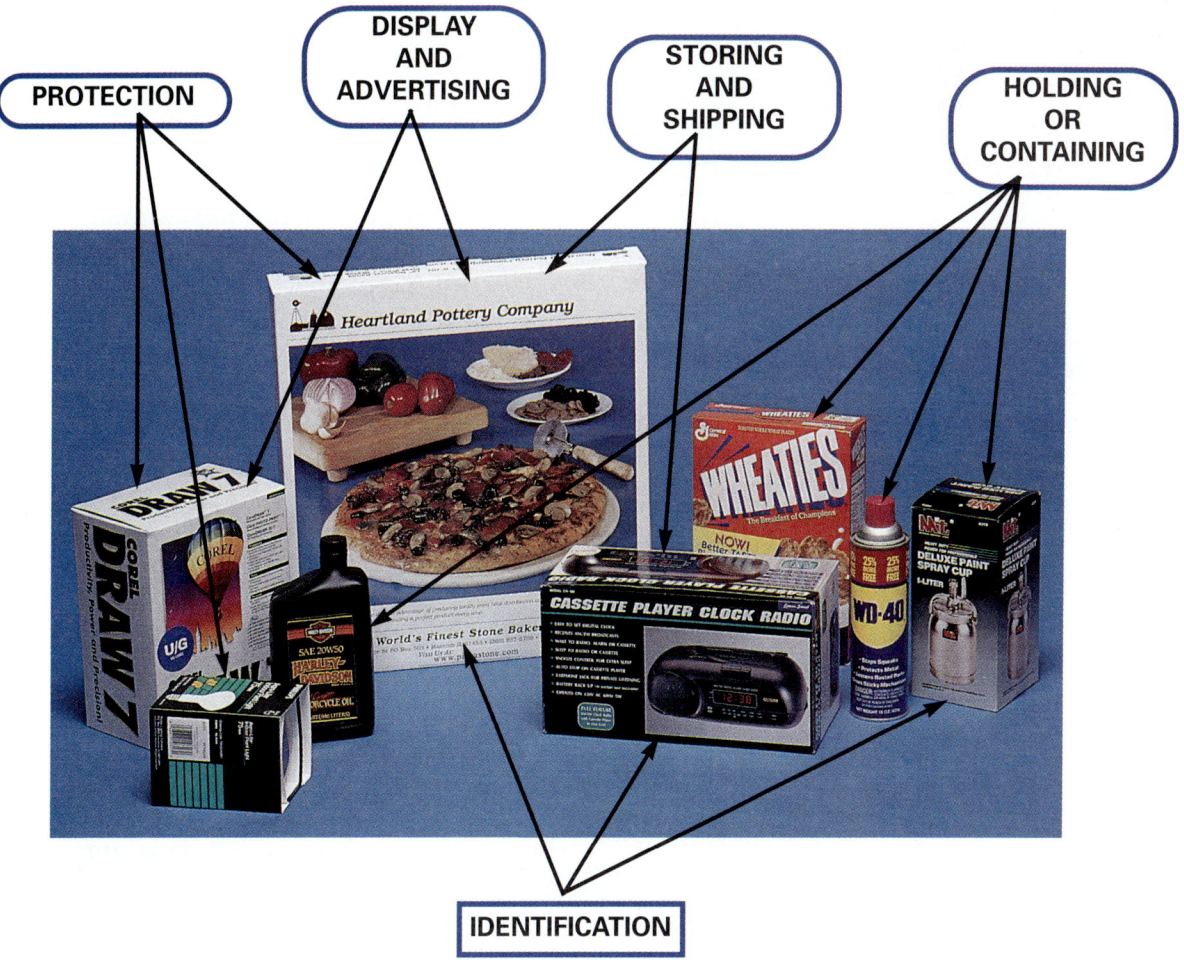

FIG. 11-3 Products are packaged for many different reasons.

CHAPTER 11: Production 249

in a bag to be sold as a multi-pack. For shipping, the packages are placed in a large fiberboard or cardboard carton.

Packaging is extremely important in marketing (selling to the consumer). You will learn more about marketing in Chapter 12. The package must be designed to attract the consumer. It must be pleasing to look at. Can you think of a time when you bought one product instead of a similar one because the package made it look better?

Most manufacturing companies buy packages from a package manufacturer. The packages may already be printed. If the packages are not preprinted, they can be printed or labeled after a product is put inside. A printed bar code is usually applied after packaging. This code contains information that identifies a product. That information can then be used for such things as updating inventory records.

Production and Inventory Control

During the actual production process, someone must make sure there's enough material on hand. Someone also needs to see that the right number of parts are being made. These tasks are part of production and inventory control.

Production Control

Production control is controlling what is made and when. How do workers know when to start on a certain product? How many parts should be made at one time? When will products be ready to ship to the customer? A plan for controlling production provides answers to these and other questions.

The master *production schedule* is very important to production control. This is a time chart that lists all the parts and shows how many of each one are to be made in a certain period of time. Fig. 11-4.

Usually a schedule is prepared several months in advance. It gives start and stop dates as well as the number of machines to be used. The schedule is a plan for what is supposed to happen. The actual time spent may be longer or shorter than planned. If so, the schedule is changed.

Controlling production also involves keeping track of what work has been done, when it was done, and who did it. After raw material is released to the production

FIG. 11-4 A production schedule shows what work is to be done and the dates for starting and ending production.

**Auto Parts, Inc.
Chicago, Illinois
MASTER PRODUCTION SCHEDULE**

PRODUCT	Week Beginning					
	1/7	1/14	1/21	1/28	2/4	2/11
serpentine belts	100	100	100	100	100	100
wheel covers	—	600	1200	600	—	600
hub caps	—	375	—	—	375	—

3. **Reinforcement.** Discuss packages. What kind of information is printed on them? Of what materials are they made?

department, production control must know what is happening to it at all times. All material that is being worked on is called *work in process* (WIP). A system called *shop floor control* is often used to keep track of work that has been done. The needed information is collected "on the shop floor." Workers record information about the work they've done on computer cards or special note pads. They may also enter data into the computer at a terminal near their workstation. The computer records are then updated. Fig. 11-5.

FIG. 11-5 Production control requires that workers keep records of tasks completed.

Inventory Control

4 **Inventory** is the quantity of items on hand. In a manufacturing plant, **inventory control** means keeping track of:
- raw materials
- purchased parts
- supplies
- finished goods

For purposes of inventory control, a *raw material* is any material before it enters processing. For example, potatoes, corn, and cheese may be raw materials at a snack food factory.

Purchased parts are ready-made parts that the company buys. A lawn mower factory might buy engines and wheels.

Supplies are different kinds of items needed to keep the plant running smoothly. These items do not become part of the product, but they are needed to support the production process. Supplies include staples, computer paper, oil, and lightbulbs.

Products that are completed but not yet sold are called *finished goods*. In a furniture factory, tables and chairs might be the finished goods.

Keeping good records of all inventories is important. When the inventory for materials gets low, more must be ordered. Without records the company would not know how much was on hand. Inventory records are usually kept in computer files. 5

The purchasing department helps make sure proper inventory levels are maintained. Buyers, or purchasing agents, buy things a company needs, such as materials, parts, equipment, and supplies. They make sure the items ordered are the proper quality and are reasonably priced. They also make sure the items are delivered at the right time. Timing is very important.

4. **Reinforcement.** Give examples of inventories stockpiled outside local factories. Can students cite others?
5. **Reinforcement.** Develop with students an inventory report of items in your classroom or lab.

CHAPTER 11: Production

MATHEMATICS CONNECTION
Using a Balance Sheet

A *balance sheet* is a financial statement. It shows the financial status of a company at the end of a certain period. The first part of a balance sheet shows the company's assets. *Assets* are resources owned by the company that benefit its operation. Cash, inventory, land, buildings, equipment, and accounts receivable are all assets. (Accounts receivable are amounts customers owe the company.)

Look at the balance sheet for Klassy Klothes Corporation. The value of each asset is written in a column to the right. Its total assets amount to $350,000.

The second part of the balance sheet shows liabilities and equity. Amounts of money the company owes others are called *liabilities*. If the company has borrowed money from a bank, the bank has a claim on the assets of the company. Employee salaries accrued since the last pay period are also liabilities. So are all the accounts payable (bills the company must pay). Look at the balance sheet for Klassy Klothes. The total value of all its liabilities is $80,000.

The owners of the company, because they have invested their funds in the business, also have a claim. The claims of owners are called owners' equity. Owners' equity in Klassy Klothes is $270,000.

Added together, the value of the liabilities and the value of the owners' equity should be in balance with the assets. This is why the statement is called a balance sheet.

KLASSY KLOTHES CORPORATION
Balance Sheet
December 31, 19XX

Assets		Liabilities	
Cash	$40,500	Bank loan	$26,000
Accounts receivable	50,000	Accounts payable	36,000
Inventory	62,500	Salaries payable	18,000
Office equipment	5,000	Total Liabilities	80,000
Supplies	2,000		
Land	100,000	Owners' equity	270,000
Buildings	90,000		
Total Assets	$350,000	Total liabilities and equity	$350,000

Try It!

1. Do the accounts for Klassy Klothes balance? Explain your answer.
2. If you were a banker, would you loan money to a company whose liabilities were greater than its assets? Explain your answer.

Answers to these questions can be found in the Lesson Plans in the Teacher's Manual at the beginning of this book.

SECTION 3: Manufacturing Technology

Quality Assurance

6   The quality of a product is how well it is made. Manufacturing companies want to produce high-quality products. They want each product to fulfill its purpose in the best way possible. They also want to satisfy consumers. **Quality assurance** means making sure the product is produced according to plans and meets all specifications. Sometimes this is also called *quality control*.

The level of quality must be set in advance. This level is called the *quality standard*. However, it is impossible to do something perfectly over and over again. There will always be slight changes, or variations, from one part or piece to another. Variation occurs because of differences in workers, materials, machines, and processes. Controlling this variation to keep the best quality possible is the goal of quality assurance. Fig. 11-6.

There are two basic ways to approach quality assurance: prevention and detection. Prevention involves doing everything possible to prevent variation in materials or processes before parts are made. Detection means inspecting parts or products after they have been produced to find any variations. However, preventing mistakes is better than finding mistakes. It costs less to correct errors before or during production instead of afterwards.

Standardization

One important step toward quality assurance in today's global market is standardization. Because the world's economy is so interrelated, there is a need for international standardization. The **International Organization for Standardization (ISO)** promotes and coordinates worldwide standards for many things. One important ISO standard is known as ISO 9000. This is a set of standards for manufacturing companies to use in establishing and maintaining documentation of their manufacturing practices. A company can have their procedures certified or approved and then can say that they meet ISO 9000 standards. This implies that they produce quality products because they consistently follow procedures.

FIG. 11-6 Products must be checked to be sure those that are defective are caught. How would you react if you bought a keyboard that had missing keys?

6. **Discussion.** Discuss the concept of quality. Point out that quality may be relative or subjective.

CHAPTER 11: Production 253

QS 9000 is an effort led by automobile manufacturers. This standard adds additional requirements to the ISO standards. The major auto manufacturers require their suppliers to be certified under QS 9000. That improves the possibility that the vendor will supply quality parts. The goal is to help companies develop and sustain good quality control practices at every step in their operation. If the standards are followed, the company's products and services should be of good quality. Fig. 11-7.

In the United States there is an organization called the American National Standards Institute (ANSI). ANSI does not make the standards, but coordinates and organizes qualified groups who then agree on the standards.

Process Improvement

Process improvement involves continually working to improve the processes by which things are made. This could involve changes in machines, for example, or in the ways workers do their jobs. Some variation is normal in any process. With careful checking and analysis, quality assurance workers can discover the type and amount of variation that is normal for a process. Then they can monitor the process and determine when it is either *in control* or *out of control*.

Statistical process control (**SPC**) is one technique used in process improvement. It is based on a special type of mathematics called *statistics*, which involves collecting and arranging facts in the form of numbers to show certain information.

Computers and other recording devices keep a record of what a particular machine is doing. This information is recorded on a *control chart*. Fig. 11-8. The chart shows a mean and the upper and lower acceptable limits of variation from that mean. (As you learned in Chapter 9, this is called *tolerance*.) Quality assurance workers use the control charts to analyze the process. Let's look at an example.

Suppose a certain machine automatically fills empty cereal boxes with cornflakes. The label on the box says that there are 16 ounces of cereal in the box, but the boxes are not

FIG. 11-7 ISO sets standards for quality for components like these machined parts.

FIG. 11-8 A control chart is used to track the process. At what point did the process go out of control?

all filled exactly the same. There is a tolerance of plus one-half ounce. That means the box can actually have any amount between 16 and 16.5 ounces. By checking the weight of every 100th box and plotting that information on a control chart, it's possible to monitor the machine's correctness. As long as the machine is running within the upper and lower limits, the product is OK. If the control chart shows that the boxes are being over- or underfilled, workers can stop the machine and make the necessary adjustments.

Fascinating Facts
Before electricity, waterwheels, windmills, and huge steam engines provided the power to run machines used in manufacturing.

Inspection

To *inspect* something means to look at it and compare it to some standard. Inspectors examine a part or product to see if it meets the specifications. **Specifications** are the detailed descriptions of the design standards for a part or product, and they may include drawings. These standards include rules about its size, shape, function, performance, and the type and amount of materials used.

Inspectors check on materials, parts, and processes. Sometimes they visually inspect the part. Most often they use some kind of measuring device.

Inspections are made when materials are delivered, when the work is in process, and when the goods are finished. If delivered materials don't meet the standards, they are rejected and returned to the supplier. Work in process is inspected to make sure it is being done properly and that the parts are correct. After the product is

7. **Reteaching.** Compare final inspection at a factory with how students inspect a product before buying it.

CHAPTER 11: Production

made, it is given a final inspection. Everything is checked to make sure it works and looks right. Fig. 11-9.

Inspection Tools

A variety of inspection tools and devices are used to check materials, parts, and products. Some are used for measuring and others for comparing.

Various **gages** are used to compare or measure sizes of parts and depths of holes. One simple gage is a go/no-go gage. Fig. 11-10. By slipping a part into this gage, the inspector can tell at a glance whether or not the part is the right size.

Not all inspection tools are as simple as gages. Computer-controlled devices can make very precise measurements. They will be discussed in Chapter 13.

FIG. 11-9 Final inspections are done on products before they leave the factory to make sure that everything works properly. In this photo, an inspector is checking a refrigerator.

GO/NO-GO GAGE

FIG. 11-10 A go/no-go gage can be used to quickly check if the part is within tolerance limits.

TOO BIG   WITHIN LIMITS   TOO SMALL

8. **Enrichment.** Demonstrate various precision measuring devices (micrometers, calipers, etc.).

Acceptance Sampling

Many products are made in large quantities, so it is not possible or practical to inspect each product. An inspection procedure called acceptance sampling is used in these cases. **Acceptance sampling** means randomly selecting a few typical products from a production run, or lot, and inspecting them to see whether they meet the standards. If they do, then the whole lot is approved. If the samples are rejected, the other products from that lot are also rejected.

The size of the sample depends on the lot size. Generally speaking, the smaller the lot size, the greater the percentage that should be inspected. The table in Fig. 11-11 shows one company's sampling plan. You can see that for a lot size of two, all products are tested. However, for a lot size of 10,000 the sample size is only 125.

The *acceptance level* is how many of the sample must pass if the lot is to be accepted. In a lot size of 280, the sample size is 20. A minimum of 16 must be acceptable. So, of the 20 pieces inspected, you would reject the entire lot if more than four of the sample pieces did not meet acceptable standards.

Burn In

One special quality assurance measure is called *burn in*. It is done to electronics products like computers. Electronic products that fail tend to do so in the first few hours of operation. Because of this, a computer manufacturer actually runs every computer for the first few hours. Those that fail are repaired, if possible. If the computer passes the burn-in test, then it will probably last a long time.

XYZ COMPANY SAMPLING PLAN				
Lot Size	Sample Size	Sample Percentage	Acceptance Level	Acceptance Percentage
2-8	2	100%-25%	1 out of 2	50%
9-15	2	22%-13%	1 out of 2	50%
16-25	3	19%-12%	2 out of 3	67%
26-50	5	19%-10%	3 out of 5	60%
51-90	8	16%-9%	6 out of 8	75%
91-150	13	14%-9%	10 out of 13	77%
151-280	20	13%-7%	16 out of 20	80%
281-500	32	11%-6%	27 out of 32	84%
501-1,200	50	10%-4%	43 out of 50	86%
1,201-3,200	80	7%-3%	71 out of 80	89%
3,201-10,000	125	4%-1%	114 out of 125	91%

FIG. 11-11 A sampling plan establishes how many items must pass if the lot is to be accepted.

9. **Extension.** Sampling is based on statistics. Discuss sample population, normal distribution (curve), etc. Relate these to students' standardized test scores.

10. **Critical Thinking.** Have students discuss what could be done with rejected parts.

CHAPTER 11: Production

Career File—Painting and Coating Machine Operator

EDUCATION AND TRAINING
Completion of high school is generally not required but is advantageous. Most operators learn their skills on the job, usually by helping experienced operators.

AVERAGE SALARY
Beginning apprentices usually start at about half the salary of a fully qualified machine operator—around $8,000 a year. The income for experienced operators can range from $14,500 to $29,000 per year.

OUTLOOK
Employment of painting and coating machine operators is expected to remain about the same through 2005, according to the Bureau of Labor Statistics. Jobs in manufacturing are expected to decline because robots can apply paint. Automotive painters, on the other hand, can expect steady work repairing and refinishing automobiles damaged in accidents.

What does a painting and coating machine operator do?
Painting and coating machine operators control the machinery and equipment that apply paints and coatings to a wide range of manufactured products. They apply the paint or coating using different methods. The most common methods include dipping in vats, tumbling in special barrels, or spraying.

Does most of the work involve data, people, or things?
Painting and coating machine operators work mostly with things—machines, coating materials, and the articles that get painted or coated.

What is a typical day like?
Paints and coatings are commonly applied using spray guns. Spray machine operators fill the spray tanks with a mixture of paints or chemicals. They adjust valves and nozzles to get the proper spray pressure, and they position the guns to direct the spray onto the article. They visually inspect the quality of the coating. They may also regulate drying ovens.

What are the working conditions?
Painting and coating machine operators work indoors and may be exposed to fumes from paint and coating solutions. Spray-gun operators may have to bend, stoop, or crouch in uncomfortable positions to reach all parts of the article.

Which skills and aptitudes are needed for this career?
Painters should have keen eyesight and a good sense of color. Operators should have physical stamina because they have to stand for long periods of time.

What careers are related to this one?
Other occupations in which workers apply paints and coatings using machines include construction painters and electrolyte metal platers.

One painting and coating machine operator talks about his job:
I like best the way the product looks when it's finished—all shiny and new.

SECTION 3: Manufacturing Technology

Chapter 11 REVIEW

Reviewing Main Ideas

- Components are simple parts that can be put together to create assemblies. An assembly used as a component in another product is called a subassembly.
- Packaging is important to protect the product and appeal to customers.
- Production control is usually accomplished with production schedules and shop floor control.
- Inventory control means keeping track of raw materials, purchased parts, supplies, and finished goods.
- Quality assurance, or quality control, means making sure that products meet all specifications.
- National and international organizations promote standardization of products and good quality control practices.
- Statistical process control is a technique to check machines and improve the process of production.

Understanding Concepts

1. Tell the difference between components, assemblies, and subassemblies.
2. In what ways is packaging important to the sales of a product?
3. Describe a production schedule. How is it related to shop floor control?
4. Name the two standards organizations and describe how they influence the quality of produced goods.
5. Name several methods used by quality assurance workers to improve products.

Thinking Critically

1. Think of a product you or your family has recently bought. How was it packaged? What purposes did the packaging serve? Can you think of any way the packaging might have been improved?
2. Name a product that you use every day. Then describe what raw materials you think were used; which parts, if any, were purchased; and what components were assembled.
3. What would happen if a company didn't have a certain part for a product to be delivered in two weeks?
4. What are some consequences if an inspection is not performed at the beginning, middle, or end of production?
5. Think of how you would go about solving this production problem: Production has stopped because several components are too large to complete the assembly.

Applying Concepts and Solving Problems

1. **Language Arts.** Production is a multi-step process that should run smoothly. Write a sequence chain to analyze a production process, such as painting a chair.
2. **Mathematics.** Use a ruler to measure the width of this book. Next, use a metric ruler to measure this book. Which measurement is more precise? Why?
3. **Technology.** Analyze each side of a cereal package by categorizing what appears on it.

11. **Resource.** Answers to "Understanding Concepts," "Thinking Critically," and "Applying Concepts and Solving Problems" can be found in the Teacher's Manual at the front of this *Teacher's Annotated Edition* and in the *Teacher's Resource Binder*.

CHAPTER 11: Production

CHAPTER 12

Marketing

Objectives

After studying this chapter, you should be able to:
- define market and marketing.
- describe how market research is done.
- explain why advertising is important.
- name three major ways of selling products.
- explain how products are distributed.

Terms

advertising
chain of distribution
commission
consumers
direct sales
distribution
market
marketing
market research
retailers
sales forecast
test marketing
warehouse
wholesalers

Technology Focus

Snowboarding: A Case of Uphill Marketing

Marketing is tough enough when you're a big company with money to spend. Using opinion polls and other methods, you survey potential buyers. You refine your product, packaging, and advertising to match what consumers want. You test the product. Marketing is even tougher if you're not a big company and must risk everything on an idea nobody else believes in . . . yet. Such was the case with snowboarding.

Carving out a Market

The first snowboarders got their start in the 1970s with a snowboard they called the Snurfer (snow + surfer). An inspired few went on to develop snowboarding into a sport.

At first, there were no spectators or customers to support a snowboarding career. The early enthusiasts had to create the market. They put on events, drew crowds, and encouraged new customers.

When Jake Burton set up shop as a snowboard manufacturer, he took his product to trade shows but soon found that wholesale buyers in established markets were reluctant to put a new product on their shelves. Burton told *Forbes* magazine that business was so slow in those days he didn't dare leave his exhibit space, even to use the bathroom, for fear of missing his only sale for the day. Winter resort operators, too, had to be persuaded to accept snowboarding.

However, snowboarders spread the word when a resort offered good rides and a friendly attitude. They came out to the slopes both early and late in the season, and their business was worth too much to ignore. Snowboarding customers and the needed distribution chain gradually came together.

Resorts Join the Effort

In the 1990s snowboarding boomed, as the share of ski-lift tickets purchased by snowboarders rose from 5 percent to 25 percent. At some resorts, snowboarders accounted for as much as 40 percent of business in the off-season.

Hoping to attract some of the more than two million snowboarders, resorts began marketing efforts on their own. Some sponsored professional riders and hosted events. They built jumps, half-pipes, and trails. A few resorts tried advertising, especially on radio and MTV. They hired consultants to make sure their messages included snowboarding slang.

The rest is history. Today snowboarding is a nationally recognized sport. Snowboarders competed in the 1998 Winter Olympics. From now on, marketing, as they say, is all downhill.

Take Action!

Suppose you're promoting a new sport but have only $250 to publicize an upcoming event. How would you spend the money to draw a crowd?

Resources
- Chapter 12 Lesson Plan in the *Teacher's Annotated Edition* and in the *Teacher's Resource Binder.*
- Chapter 12 Review in the *Teacher's Resource Binder.*

Markets

The goal of manufacturing is to produce and *sell* the product. Therefore, a manufacturing job isn't complete until the product is sold. **Marketing** includes all the activities involved in selling the product.

Manufacturers often try to make products for a certain market. A **market** is a specific group of people who might buy a product. A market might be teenagers, senior citizens, young married couples, mechanics, the military, or even a school district. For example, a watch manufacturer may make colorful plastic watches aimed at the teenage market. Fig. 12-1.

Markets change as population, incomes, and lifestyles change. For example, in more and more households all adults have jobs. This has increased the demand for prepared foods. People in charge of marketing must be aware of these changes in the market.

There are two major types of markets: industrial and consumer.

Fascinating Facts

Marketing promotions can be *too* successful. When McDonald's Restaurants decided to start giving out Teenie Beanie Babies® in 1997 with each Happy Meal®, they had no idea that they would start a collection frenzy. Within a few weeks, many of the Teenie Beanies were worth as much as $15.

The Industrial Market

Businesses and industries make up the industrial market. They buy products to use in their own companies. They also buy products to use as parts in the products they make.

Car dealerships, hospitals, and magazine publishers all buy paper goods, lightbulbs, and pencils. The purchased products are used in their businesses.

Often, manufacturers buy parts for products they manufacture. Automobile manufacturers may buy tires from one company, windshield wipers from another, and so on. Buying parts to use in their products also makes them part of the industrial market.

FIG. 12-1 Marketing techniques are often aimed at particular groups of people. Who is the market for these products?

1. **Brainstorming.** Have students identify various markets and the kinds of products manufactured for sale to those markets.
2. **Reinforcement.** Point out how the school is part of the industrial market.
3. **Critical Thinking.** Have students identify industrial markets for products shown in Fig. 12-1.

The Consumer Market

Consumers are ordinary people (like you and your family) who buy products for their own personal use—things like toothpaste, videotapes, and sneakers.

The consumer market is a big one consisting of many millions of people. Manufacturers spend a lot of time and money making sure their product is one consumers will want to buy and then persuading those consumers to buy it.

The Marketing Plan

Before a company begins to sell a product, the marketing department develops a marketing plan. This plan includes a **sales forecast**, which is a prediction of how many products the company will sell. Marketing arrives at this estimated figure by looking at the market potential (amount of possible sales). A percentage of this number is figured as the company's expected *market share*. (Each manufacturer competes with other manufacturers of similar products for its portion, or share, of the total sales of that type of product.) The plan also includes ideas for advertising and sales.

Market Research

Market research includes all activities used to determine what people want to buy and how much they will pay for it. The results indicate how well the company can expect the product to sell. If the market research shows that lots of people will buy the product, then the company can plan a large-scale production. Suppose, however, the research is inaccurate. The company can lose a lot of

FIG. 12-2 A company often test markets an item to see if customers like it. Have you ever tried a sample of a food item in a grocery store?

money. Imagine making thousands of items and selling only a few hundred! Market research must be done carefully and accurately.

One way of researching the market is to *interview* people. Potential consumers are asked their opinions about a product. If it's a food product, people may be asked whether they like its appearance or taste. The most important question asked is "Would you buy this product?"

Another way of doing market research is **test marketing**. A company produces a small number of products and sells them in a very limited area, such as in one city or region of the country. If sales are good in the test market, then the company can usually expect that the product will also sell elsewhere. Fig. 12-2.

4. **Mathematics.** Using population figures for your community, make up examples of market share and have students figure percentages.
5. **Reteaching.** Bring to class magazines, such as *Advertising Age* that illustrate the link between research and marketing.
6. **Critical Thinking.** Ask: "Does it really matter what city or region is chosen for a test market?" Help students see that cities and even regions have different market characteristics.

CHAPTER 12: Marketing

FIG. 12-3 Advertising is everywhere.

Advertising

Advertising tells people about a product. It is the method or methods the company uses to persuade, inform, or influence consumers to buy a certain product. The whole goal of advertising is to convince consumers that they need the product. Market research is used to determine the best type of advertising to reach the target market. The *target market* is the group of consumers most likely to use and want the product.

Most of us are familiar with TV commercials. They cost a lot of money, but they reach a lot of people. Advertising is also done in magazines and newspapers, and on radio and billboards. Fig. 12-3.

One interesting method of advertising is to give away free samples. Sometimes a company will send you a sample of their product in the mail. If you like the free sample, you will probably buy the product.

Sales

If a company can't sell the product, all is lost. Sales personnel are trained to know about the company's products and services. They are also taught selling techniques. Fig. 12-4. Many people in sales work on **commission**. That is, they receive a certain percentage of the amount paid for the product. The more they sell, the more money they make. This encourages them to sell more products.

There are three main ways of selling products: direct sales, wholesale sales, and retail sales.

Fascinating Facts
In the movies, advertising can be big business. Mercedes-Benz paid $5 million to supply the jeep in the movie, *The Lost World: Jurassic Park*. For its money, it got a mini-commercial.

7. **Resource.** You may want to refer to the activity at the end of Section II, "Using a Digital Editor."
8. **Extension.** Coupons are a subtle form of advertising; they encourage people to buy a certain product.
9. **Critical Thinking.** Ask students how, as sales people, they might determine a customer's needs.

SECTION 3: Manufacturing Technology

Direct Sales

Sometimes a manufacturing company sells its product directly to the customer. This method of sales is called **direct sales**. Manufacturers commonly sell direct to purchasers in the industrial market. For example, an aluminum producer sells aluminum tubing directly to an air conditioner manufacturer. A manufacturer who sells products directly to another manufacturer is called an *original equipment manufacturer* (OEM).

Direct sales are also made to consumers, but this is less common. Perhaps you've gotten advertising and an order blank for a product in the mail. Maybe a TV commercial invited you to send for a CD or jewelry. These products are usually sold directly from the manufacturer.

A type of direct selling that has grown very popular is the factory outlet store. An outlet store carries products made by one manufacturer. These products are usually last season's merchandise that did not sell in retail stores. The items are usually available at reduced prices.

Wholesale Sales

10   **Wholesalers** are people or companies who act as intermediaries (go-betweens). They buy large quantities of products from manufacturers. Then they sell the products to commercial, professional, retail, or other types of institutions that also purchase in quantity. Wholesalers may also be involved in such activities as financing, storing, and transporting products.

Retail Sales

Retailers buy products from manufacturers or wholesalers. Then they sell the products to consumers who will actually use the product.

There are millions of retail stores in the United States. Retail stores include department stores, chain stores, discount stores, and many others. They all have the same function—selling products to consumers.

FIG. 12-4 Knowledgeable salespeople help customers locate the products they're looking for.

10. **Reinforcement.** Give examples of wholesalers in your local area.

CHAPTER 12: Marketing   **265**

Distribution

Distribution refers to methods used to get goods to the purchaser. As the result of distribution activities, we have many well-stocked stores from which to choose.

The path that goods take in moving from the manufacturer to the consumer is called the **chain of distribution**. The chain of distribution can be long. It may involve both a wholesaler and a retailer, as well as the manufacturer and the consumer. A short distribution chain includes only the manufacturer and the consumer, as in direct sales.

Managing distribution is important. The product must be available at the right time and in the right place. A distribution manager has to make certain that products move smoothly through the entire chain of distribution.

Warehousing

A **warehouse** is a building where products are temporarily stored until the next part of the chain of distribution is ready for them. Having products already made and stored ahead ensures that there will be enough on hand for when they are needed. Fig. 12-5.

The number and location of storage facilities are very important. One company may have many warehouses. A large manufacturer of snack foods, for example, may have warehouses located in 20 states. This makes distributing the product easier.

Transportation

Regardless of the chain of distribution used, some form of transportation is involved. All products need to be shipped from the factory and must eventually reach the industrial or consumer market. Trucks and trains are commonly used. Depending on the product, air freight may be used. For products that are exported (sent out) to other countries, ships are the most common method of transportation.

FIG. 12-5 Products may be temporarily stored in a warehouse until they are ready to be distributed to wholesalers and retailers.

11. **Reinforcement.** Help students see that warehousing keeps a steady supply of products available to consumers.

12. **Critical Thinking.** Ask: Why is transportation important to the manufacturing industry? Might it affect the number of products sold or the price of the product?

SCIENCE CONNECTION
Effects of Storage on Product Quality

Foods and other manufactured items can lose quality when they are stored for any length of time. This loss of quality can cost manufacturers millions of dollars every year. Four enemies of stored items are microorganisms, enzymes, pests, and the chemical process known as oxidation.

Microorganisms. These are tiny forms of life, such as bacteria, molds, and yeasts. They primarily affect foods, although some also work on wood and petroleum products.

Some microorganisms can cause decay or even make food toxic (poisonous). Molds, which look fuzzy, can cause a change in flavor. Yeasts cause fermentation, which turns food to alcohol. Bacteria can turn foods sour or give them the odor of rotten eggs. In canned goods, a resulting gas can accumulate and burst the containers.

If water leaks into kerosene, fungi grow in it, producing a gel-like contaminant. Water-soluble lubricants used in manufacturing also attract fungi and must be filtered before use.

Enzymes. An enzyme is a protein molecule that exists in all living things and speeds up chemical reactions. Enzymes can make fruits overripe and change the way they taste.

Pests. Pests include rodents and insects. Rodents can contaminate foods and chew other items. Termites can attack wood and wood by-products, such as cardboard. Moths and beetles are also hazards, especially to textiles. Moths lay their eggs on the textile and the larvae eat holes in it. Many clothing manufacturers try to prevent infestation by storing clothing items in dark areas where moths and beetles are less likely to invade.

Oxidation. Oxidation occurs when oxygen in the air combines with other substances. When food oxidizes, the fat or oil in it changes flavor, often becoming rancid tasting.

Oxidation also damages metals. For example, iron forms rust, which weakens the metal and makes its surface unattractive, as anyone who owns an older car knows. Surface coatings and paint slow down the process but do not stop it. Paints themselves are subject to oxidation, which makes them dull. A clear finish coating helps protect the surface.

Try It!

In 1997, the Hudson Food Company had to recall millions of pounds of hamburger patties, which may have been contaminated with a dangerous bacteria. Do some research about this incident. What other companies were affected by the recall? Did Hudson survive the loss?

Career File—Manufacturer's Sales Representative

EDUCATION AND TRAINING
A college degree is preferred. On-the-job training is usually given so that sales representatives can become familiar with the products they are to sell.

AVERAGE STARTING SALARY
Compensation is usually a combination of salary plus commission. A new sales representative may earn $15,500 the first year, although this can be higher or lower depending on the person's sales volume. The median income is $32,600.

OUTLOOK
The availability of jobs is expected to grow faster than average through 2005, according to the Bureau of Labor Statistics. This growth will be due to the continued increase in the number of new products.

What does a manufacturer's sales representative do?
Sales representatives in the manufacturing industry market a company's products to wholesalers, retailers, government agencies, other institutions, and even other manufacturing firms. They look for new clients, visit client sites, demonstrate or explain how their company's products can fill the client's needs, and take orders for merchandise. After the sale, they follow up and resolve any problems or complaints involving the merchandise.

Does most of the work involve data, people, or things?
Manufacturer's sales representatives deal with people most of the time, encouraging them to purchase a company's products.

What is a typical day like?
Much of a sales representative's time is spent traveling to prospective and current client sites. During a sales call, they discuss the client's needs and how their products can fill those needs. They may show product samples, and they emphasize the qualities that make their products superior to products from competitors. The sales representative's goal is to negotiate a sale.

What are the working conditions?
Manufacturer's sales representatives work in an office when they are not traveling. However, they may be away from home for several weeks at a time. Although the hours are long and often irregular, most sales representatives have the freedom to set their own schedules.

Which skills and aptitudes are needed for this career?
Sales representatives should have a pleasant personality and appearance, and they should be able to communicate well with people. Patience and perseverance are needed because completing a sale can take several months. The ability to understand the way the product works is desirable.

What careers are related to this one?
Other occupations that require selling are real estate agents, insurance agents, wholesale and retail buyers, and retail sales associates.

One manufacturer's sales representative talks about his job:
I really believe in the products I sell. I'd never try to sell something I wouldn't buy myself.

268 SECTION 3: Manufacturing Technology

Chapter 12 REVIEW

Reviewing Main Ideas

- Two types of markets are: industrial (businesses) and consumer (individual customers).
- Marketing plans include research to predict what and how much of a product the manufacturer can expect to sell.
- Advertising is intended to persuade, inform, or influence customers to buy a product.
- Three ways of selling products are: direct sales, wholesale sales, and retail sales.
- Distribution of goods includes moving the goods from the manufacturer to a consumer. This often requires managing transportation and warehousing.

Understanding Concepts

1. What is the difference between a market and marketing? Give an example of each.
2. Name two activities that may take place during market research.
3. What is the goal of advertising?
4. What is the difference between direct sales and retail sales?
5. Describe the chain of distribution. Name the activities that take place in this chain.

Thinking Critically

1. When do you think it is necessary to begin a marketing plan? Explain your answer.
2. How can advertising benefit the consumer?
13. **Resource.** Answers to "Understanding Concepts," "Thinking Critically," and "Applying Concepts and Solving Problems" can be found in the Teacher's Manual at the front of this *Teacher's Annotated Edition* and in the *Teacher's Resource Binder*.

3. Many businesses pay sales personnel a commission rather than a salary. What are the advantages and disadvantages of working for a commission?
4. Select three advertisements. Then describe the product and the market at which each ad is aimed. Are there differences in the style of advertising?
5. What effect do you think the chain of distribution has on the final price of a product?

Applying Concepts and Solving Problems

1. **Language Arts.** Select a product and write one marketing campaign that would appeal to different age groups.
2. **Social Studies.** Try to think of a product that was introduced but failed and was taken off the market. (If you have trouble coming up with one, ask family members.) What consumer needs were not fulfilled by this product? What improvement might have made the product more successful?
3. **Science.** The symbol pictured here is printed on recyclable plastic. The number inside the triangle varies to indicate different chemical structures and is used to help sort items. Compare the symbols of as many different kinds of plastics as you can.
4. **Mathematics.** Daniel is paid a straight commission of 5% of the sale price of each used car he sells. What is his commission if he sells an auto for $6,400?

CHAPTER 13

Computer-Integrated Manufacturing

Objectives

After studying this chapter, you should be able to:
- describe how different computer programs help product designers and production planners.
- describe computerized production methods.
- explain the differences between an automated factory and an automatic factory.

Terms

computer-aided manufacturing (CAM)
computer-integrated manufacturing (CIM)
computer numerical control (CNC)
finite element analysis
flexible machining center
manufacturing resource planning (MRP II)
programmable controller
rapid prototyping
robots
statistical quality control

Technology Focus

Stereo Lithography—Turning Ideas into Plastic Reality

We've all heard the old saying that a picture is worth a thousand words. After all, words can only tell about something, while a picture can *show* it. Sometimes, however, even a picture is not enough. Sometimes only a full-size model, or *prototype*, will do the job.

Manufacturers have always built prototypes of certain new products, even though they are expensive and can take from a week to several months to create. By means of a new process called stereo lithography, however, designers can now use their computers to design and build three-dimensional prototypes in a fraction of the time and at far less cost.

Terry Kreplin works in the Advanced Engineering Design Center for Baxter Health Care Corporation. Like other engineers, Kreplin can see his ideas in his imagination before they are constructed. "However, I have to be able to communicate the design of a product to other people," he says. A prototype must then be made of the small valves and other medical equipment his group designs.

Kreplin is manager of Baxter's "rapid prototyping" organization. His group uses a 3D Systems Stereo Lithography Apparatus (SLA), which produces full-size, 3D plastic models directly from CAD drawings. The technique is called photopolymerization and is based on the principle that some liquid polymers, when exposed to light, become solid. ("Photo" means light; a polymer is a plastic.) Stereo lithography is photography and chemistry combined.

The SLA system uses an ultraviolet laser beam to "draw" the first thin layer, or cross-section, of the object in liquid polymer. The ultraviolet light causes that polymer layer to harden. As soon as one layer hardens, the laser beam draws another layer. As layers build up, complex objects having precise details take form.

Kreplin recently designed a helmet. A full-size model of the shell, bill, and internal supporting mechanisms was constructed in one operation from his CAD drawing. It took nearly 24 hours for the SLA to complete the helmet in polymer. "That's a long time," says Kreplin, "until you realize that if someone were to cut it out of solid plastic or build a mold, it would take weeks to accomplish the same thing."

With stereo lithography, ideas can be quickly turned into the next best thing to reality.

Take Action!

1A

In a few sentences, describe the handle shown in Fig. 9-14 on page 222. Give your description to three different people and ask them to sketch it. How do their drawings compare to the picture?

1A. **Critical Thinking.** Ask students if a drawing or model coveys an idea better than words.

Resources
- Chapter 13 Lesson Plan in the *Teacher's Annotated Edition* and in the *Teacher's Resource Binder*.
- Chapter 13 Review in the *Teacher's Resource Binder*.

CHAPTER 13: Computer-Integrated Manufacturing

What Is Computer-Integrated Manufacturing?

In **computer-integrated manufacturing (CIM)**, computers are used to help tie all the phases of manufacturing together to make a unified whole. Computer-integrated manufacturing helps make all areas of manufacturing as efficient as possible. In CIM, computers are an essential part of planning, production, and control.

Computerized Planning

Computers can be used for planning products and production. Product development is made easier with the use of computer-aided design (CAD) and computer-aided engineering (CAE) programs. Fig. 13-1. The designer creates a plan or design right on the computer screen instead of on paper. The plan or design can be changed for improvement or analysis. There are many CAD and CAE programs developed by different companies. One function of a program may be to add geometric elements (parts) to a model. Another function may be to manage the database. A database contains specific information about the product, such as material, strength, and dimensions.

As helpful as the CAD/CAE programs are, people are still needed for their creative abilities and complex reasoning skills. The computer programs are a valuable tool for the designers and engineers to use. However, they cannot replace people.

FIG. 13-1 Computer-aided engineering (CAE) puts advanced technologies to work to produce products faster and less expensively. When combined with computer-aided design, changes can be made quickly.

1. **Reinforcement.** Point out that computers can be used in one or more of these ways without CIM. The key difference with CIM is that all aspects of planning, production, and control are linked together.

272 SECTION 3: Manufacturing Technology

SCIENCE CONNECTION
Can Computers Really Think?

As computers grow "smarter," they become able to do more of the things people can do. When, in 1997, an IBM computer called Deeper Blue* finally beat Garry Kasparov, the best chess player in history, some people were worried. Could machines not only think, but think better than humans could?

Machines have always been better than humans at some things. Cars can move faster than a human can run. Forklifts can carry heavier objects than the strongest man. Computers are sometimes better at analytical problem solving than people because they work so fast. They can review enormous amounts of data in a fraction of the time it would take a human being. Deeper Blue, which was really a series of 32 general-purpose computers, each having eight special chess-playing processors, was an analytical giant. It could scan between 200 and 300 million chess positions per second. No human can match that. However, does this mean that Deeper Blue could actually think?

Analytical problem solving—the ability to sort through remembered facts to find likely combinations that fit certain criteria—is only one small part of real human thinking.

---
*Second generation of Deep Blue, the computer Kasparov beat in 1996.

After just a few weeks, a human baby can recognize its parents; feel emotions, such as love or sorrow; laugh at things that strike it funny; and begin storing information that may one day make it not only smart, but wise. Computers will never be able to imitate human thinking until they can imitate these other aspects as well.

Unfortunately for those of us who love science fiction, computers are just machines. They do not win at chess because they want to. They don't care about chess. They don't care about anything. They play chess because they are designed to play it just as vacuum cleaners are designed to suck up dirt.

In truth, Garry Kasparov was not beaten by a machine. He was beaten by a team of five computer scientists and another chess grandmaster who worked with them for over a year. Without *their* ingenius, human thinking, Deeper Blue would just be a useless heap of metal and silicon instead of the remarkable achievement that it was.

Try It!

Use the Internet to learn more about chess at http://www.chesschampions.com/

CHAPTER 13: Computer-Integrated Manufacturing

Product Design

In designing a new product, using computer-aided design (CAD), the designer can "draw" simple or complex shapes. These drawings can be enlarged, rotated, or reduced. If the designer wants to make changes in a product's appearance, it can be done rapidly. This is called *interactive computer graphics*. It's interactive because the designer interacts with the computer. The computer reacts to what the designer does, and the designer reacts to what happens on the screen. When satisfied with the design, a hard copy (drawing on paper or film) can be printed.

Product Analysis

Most CAD/CAE programs are capable of *simulation*. Information about the new product is entered into a computer. The computer runs the information through the program. The program simulates (imitates) the conditions under which the product will be used to see if there are any problems in its design.

Simulation can also show how any moving parts will work. Other elements can be added to show how they will interact with the product and its actions. Other elements could include humans, robots, and other machines. These simulations allow the designers to analyze and make improvements in the product's features before it is actually produced.

Finite element analysis (FEA) is one type of computer simulation. FEA predicts how a specific component or assembly will react to environmental factors such as force, heat, or vibration. It is called analysis, but it is used in the design phase to predict what will happen when the product is used.

For example, the wheel assembly on a car can be "road" tested using a computer. When the car crosses railroad tracks, hits bumps, encounters water and other road hazards, how will it work? Will parts rub against one another, or will vibrations loosen a component? These types of analyses can be simulated and checked before any parts are made. This allows for any needed changes before producing real parts. FEA predictions save the company time and money.

Design for Assembly and Manufacturability

A company in Massachusetts has used a relative of the CAD/CAE programs to develop their *design for assembly/design for manufacturability (DFA/DFM)* plans. By analyzing their product in terms of how it was assembled and made, they were able to reduce the number of parts by 30%. Three hours of assembly time were saved. They eliminated $500 worth of parts per item, made the product easier to fix, and put into effect other cost-saving and product-improvement measures. Fig. 13-2.

Fascinating Facts

The first machines to move automatically were clockworks created in the Middle Ages to keep track of time. Systems of gears and pulleys enabled the workings and figures attached to the clocks to move.

FIG. 13-2 Computer simulations help planners to anticipate the amount of time required to complete each process. Shown here is a simulation of robots placing and gluing a windshield in a car.

Virtual Reality

A computer, special glasses, and other devices such as gloves or special keyboards can be used to create virtual reality. Fig. 13-3. As the user's head moves, the scene changes to adjust to the new position, just as the view changes when you move your head in real life. Virtual reality permits a designer to work in three dimensions when designing automobiles, airplanes, or other products. Virtual reality can also be used to plan the layout and movement in a new factory.

FIG. 13-3 Virtual reality simulates the environment surrounding the user. Here, an engineer works with a virtual environment in order to develop safer airplanes.

2. **Extension.** Boeing has created a virtual airplane that flies in virtual space. VPL Research has a virtual car. You may wish to find out more about these to share with the class, or select students to do short reports.

CHAPTER 13: Computer-Integrated Manufacturing **275**

Product Engineering

Engineering a new product requires many mathematical calculations. A computer can solve math problems much faster than a human can. In computer-aided engineering, computers are not only used to perform math calculations, but also to produce working drawings and to analyze parts.

Computers can be used when drafting detail, assembly, and schematic drawings. After the drawing is created on the computer screen, it can be turned into a hard copy using a plotter. A plotter is a specific kind of drawing machine. The computer causes the plotter to move a pen point or ink jets across a piece of paper or film to make the printed drawing. Fig. 13-4.

Using CAD and CAE computer programs, an actual three-dimensional model can be created for evaluation. This is called **rapid prototyping.** After a CAD/CAE drawing is made, the computer "tells" a machine the product's dimensions and the machine constructs it. There are different types of machines that make the prototypes in a variety of materials and processes.

The difference between a normal prototype and a rapid prototype is that the latter does not work. It is only a detailed physical model. It shows the designers what the product will look like and how it handles. For example, if the product is a telephone, the rapid prototype is checked to be sure the design is attractive and the receiver is comfortable to hold and use.

Production Planning

Process planners use computer software designed for *computer-aided production planning (CAPP)*. With it they can quickly determine the best processes for obtaining materials and scheduling production.

A system used to help in both production and inventory control is *material requirements planning (MRP)*. A computer analyzes information from the bill of materials, the master production schedule, and inventory records. Based upon this information, the computer can make calculations and recommendations about resources and costs.

Another computer program called **manufacturing resource planning (MRP II)** helps plan for materials, people, time, and money. The computer calculates when, how much, and where materials are needed during production. It also analyzes costs, the skills needed to perform a job,

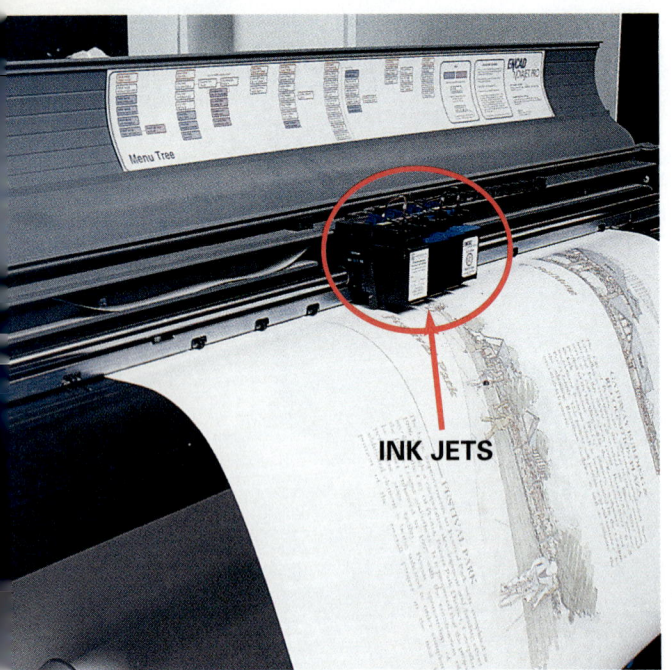

FIG. 13-4 A plotter produces a "hard" copy of the drawing made with the computer.

INK JETS

3. **Reinforcement.** Continue to emphasize the importance of planning.

SECTION 3: Manufacturing Technology

how long the job will take, and how much it will cost to pay the workers and run the equipment.

Companies can save time and money if changes and adjustments can be made *before* production begins. By using manufacturing resource planning programs, production planners can check their plans. They may discover that additional resources are necessary. They may find potential problems.

Still another program, *enterprise resource planning (ERP),* performs all these tasks plus an evaluation. It answers questions as to which resources are best to use, where additional resources can be obtained, and how to get the most from them.

Computerized Production

Parts and products can be made more efficiently using computerized production methods. Computers are used to tell the machines what to do. This is called **computer-aided manufacturing (CAM)** or *computer-aided machining*.

FIG. 13-5 This computer numerical control (CNC) machine performs many operations quickly and automatically. Here, a laser cutter is used to cut precision parts.

Computer Numerical Control

In the past all machines were controlled by human operators. They turned handles and pressed levers to make the machine work. Then *numerical control (NC)* was developed. An electrical controller was connected to the machine by wires. Then a special number code was entered into the controller by a reader that "read" a punched paper tape. The holes in the tape represented the series of numbers. These were the directions to the machine.

Now **computer numerical control (CNC)** is commonly used. Fig. 13-5.

Fascinating Facts
Computer chips are pretty small already, but *nanotechnology* will be able to make them even smaller. By building computer chips atom by atom, no space will be wasted and nanotechnology will revolutionize manufacturing.

CHAPTER 13: Computer-Integrated Manufacturing **277**

FIG. 13-6 This waterjet cutter can cut metals up to six inches thick, leaving a smooth, finished edge.

This is a form of NC, but it doesn't need paper tape or a special reader. The numerical directions are contained in a computer program. Not only can the computer give directions to machines, but it can also detect what is happening. For example, if a cutting tool breaks, the computer receives this information and stops the machine.

Cutting Technology

For many years materials were cut using a saw, knife blade, or cutter. Much cutting is still done that way, but computers have made several new processes possible.

Waterjet cutting is the use of a highly pressurized jet of water to cut a material. The water pressure at your house is about 50 pounds per square inch (PSI). At that pressure, water won't cut anything. Squirting water through a very tiny hole at 50,000 PSI, however, turns the water into a knife blade! Fig. 13-6. Sheet materials like cloth, plywood, rubber, and plastic can be cut easily and quickly this way. The waterjet follows a path guided by a computer program.

Lasers strengthen and direct light to produce a narrow, high-energy beam. Whatever the beam strikes becomes so hot that the material vaporizes (turns into a gas). In *laser cutting*, this concentrated, high-energy beam of light is used to cut materials. Another use of laser cutting is to engrave the molds used to make compact discs (CDs). Like waterjet cutting, laser cutting is computer-controlled.

Flexible Machining Centers

A **flexible machining center (FMC)** is a combination machine tool. It's capable of drilling, turning, milling, and other processing. Fig. 13-7. A computer controls the various functions.

One place a flexible machining center might be used is in an engine factory. The FMC might work first on an eight-cylinder engine. Next, it might work on a four-cylinder engine and then on a six-cylinder engine. Each different engine would otherwise require a change in tool setup, but an FMC can handle these differences easily.

4. **Enrichment.** Ask students to bring in articles and pictures from newspapers and magazines that show lasers being used. Point out that there are different types of lasers—low power lasers for measuring (supermarket checkout); high-power lasers for medical or manufacturing purposes.
5. **Reinforcement.** Review the meaning of tooling-up, as explained in Chapter 10.
6. **Extension.** Discuss how new technology has affected the tooling-up process.

Automated Assembly

Most assembly machines are specially built to assemble a certain product. They are often automated. For example, automated assembly is common in the electronics industry. Small components, such as resistors or diodes, are packaged on a role of tape and loaded into an automatic insertion machine. As circuit boards travel past the insertion machines, the parts are rapidly and accurately inserted into the right places. Computers, video recorders, and stereos are examples of products assembled in this manner.

Robotics

Robotics means using robots to perform tasks. **Robots** are special machines programmed to automatically do tasks

FIG. 13-8 Robots have movable joints, much like a human does. This allows them to reach any point inside their work envelope.

FIG. 13-7 A flexible machining center is capable of performing several processing operations.

that people usually do, such as moving objects from one place to another, assembling parts, welding, or spray painting.

There are several types of robots, but they have common features. Fig. 13-8. They all have a "hand," usually called an *end effector*. The end effector may be a gripper for holding things or it may be a built-in tool like a drill or a welder. Robots also have joints just as a human has joints. Robots typically have a wrist, an elbow, a shoulder, and a waist. These joints allow a robot to stretch, turn, raise, and lower within a limited work area. This work area is called its *work envelope*. A robot can be programmed to manipulate its end effector anywhere inside the work envelope.

One main advantage of a robot is that it can be programmed to repeat a task over

7. **Critical Thinking.** Ask students how they would feel if they found out their job was going to be taken over by a robot. Is there anything they might be able to do to make sure they remained employed?

CHAPTER 13: Computer-Integrated Manufacturing

and over. It doesn't need a coffee break or rest period. Also, it's very accurate. Robots are good for doing work that is hazardous to humans, such as in paint shops filled with fumes, in high or low temperature conditions, or where very heavy loads must be lifted repeatedly. Fig. 13-9.

Robots have even been introduced to the health-care industry to help provide hospital care. A mobile "hospital helpmate" is already being used in at least one hospital. It is designed to bring patients food and medicine, carry medical records and blood samples, and dispose of contagious wastes. Fig. 13-10. If it proves successful, there will be increased demand for these service robots in other hospitals and health-care facilities, as well as in homes to care for the physically challenged.

FIG. 13-10 This four-foot-tall robot is designed to perform many necessary hospital procedures. It uses 23 sensors, including devices that measure distance using sound waves.

FIG. 13-9 Robots are being used in place of human workers for hazardous tasks, such as spray-painting.

SECTION 3: Manufacturing Technology

Computerized Control

A very important part of CIM is *control*. Computerized production not only uses computers to automate the various machines that handle materials and make products but also monitors and controls the flow and quality of work. For example, if one workstation develops a production problem, the other workstations adjust their schedules. Many methods and devices are used.

Inspection Devices and Methods

Computers are often used to inspect products for quality. One kind of quality control device is a *coordinate measuring machine (CMM)*. Fig. 13-11. A CMM, as it is usually called, is a very accurate computer-controlled measuring device. The main advantage of a CMM is its high degree of accuracy and its consistency. A CMM is usually used to measure hard-to-measure parts, like rounded or spherical items. It measures the part and compares the measurements with the specifications for the part's dimensions, which have been stored in its memory.

Another quality-control machine is a *laser curtain*. This machine uses a moving laser beam to record the measurements of a part. The measurements are automatically entered into the computer for comparison. For example, to check whether a part's surface is flat, the laser beam is projected across the surface. The beam hits anything that sticks up.

Other instruments used for making comparisons include optical comparators and special scanning microscopes which magnify small parts. X-ray machines are used to see inside welded metal parts. Other devices emit sound waves to check product characteristics.

A computer is used to perform **statistical quality control (SQC)**. The computer gathers information from a small sample of parts to determine how well the parts are being made. Then, by using statistics, it can predict the percentage of parts not meeting specifications in the entire batch.

FIG. 13-11 Parts can be measured very accurately using a coordinate measuring machine (CMM). This machine is inspecting 70 different features of an engine block.

FIG. 13-12 A bar code contains information. The information can be entered into the computer by scanning the code with a special reader device.

JIT Manufacturing

In CIM, computers are used to keep track of inventory, to order materials, and to schedule deliveries. Computer control helps make just-in-time (JIT) manufacturing practical. In JIT manufacturing, materials are delivered as they are needed—just in time. This reduces the need for warehouse space. It also means fewer workers are needed to organize and keep track of the materials. However, the system must be set up carefully. If the right materials, in the right amount, do not arrive at the right time, production is halted. (See the article about JIT on page 185.)

Automatic Identification

Controlling requires current information. One way that current information is entered into a computer is by a process called *automatic identification*. A special tag is attached to a part or product. The tag contains an identification code. A machine can read the code to identify the part.

The best example of "auto ID" is *bar coding*. You've seen bar codes on products you buy in the store. The code is a series of black and white lines. The computer can read the code and thus identify the product. Fig. 13-12. In manufacturing, the information in a bar code can be used to keep track of inventory, to direct a part to the right workstation, and to otherwise control what happens to it.

Speech Recognition

Another method of control is speech recognition. A worker speaks commands into a microphone. The computer recognizes the words, and it carries out the instructions. That way, the worker's hands are free to do other work. Speech recognition is not yet widely used, but technological advancements will soon make talking with computers an everyday experience.

Programmable Controllers

A **programmable controller** is a small, self-contained computer used to run machines and equipment. It's housed in a heavy-duty case. The case protects it from the factory environment. Fig. 13-13. The fact that it is programmable means that workers can change the way it functions. This makes it more useful than controllers that are built to do only one thing.

One programmable controller system is called manufacturing executive systems (MES). It operates the MRP and MRP II systems discussed earlier in this chapter. It helps companies to make sure all the materials, workers, and other resources are available for use during the production stage. MES also helps companies to plan so resources are paid for within the budget.

8. **Reinforcement.** Help students understand how JIT can help keep inventories low and save money.

FIG. 13-13 This programmable controller is controlling a manufacturing process.

Computer Networks

Some computer-controlled machines have their own built-in computers. Others require a direct link to a larger computer. A company's main computer may be in one central location. Each of its factories or plants would then have its own computer. In addition, there would be many small computers located throughout the various departments of each plant. Fig. 13-14. For CIM to work, all the computers must be connected so that they can all communicate. This is done using computer networks.

A *network* is a way of hooking together several computers. The computers are linked by wires or fiber optic cables. Instructions and information are transmitted from one device to another over the network. Inside a plant there is a *local area network (LAN)*, while between plant locations there is a *wide area network (WAN)*. Because of networking, all of a company's computers (and the machines they control) can be linked together for greater efficiency.

Some companies use their computer systems to talk to their employees all over the world. They may use *video-conferencing*. In video-conferencing, everyone can see everyone else by means of video cameras. Images are transmitted via satellite. Their CAD/CAM/CAE systems may also be connected so that design and production engineers can share ideas. Their messages are all e-mail (electronic mail) that can be sent

CHAPTER 13: Computer-Integrated Manufacturing

FIG. 13-14 This diagram shows how various types of computers are linked together into one large system.

to the office next door or across an ocean. Computer networks provide instant worldwide communication and have become an important part of global manufacturing.

Automated Materials Handling

Computer control can also make sure that the right material is in the right place at the right time within the factory.

Automatic Guided Vehicle System

Popular for materials handling is the use of an *automatic guided vehicle system (AGVS)*. Specially built driverless carts (AGVs) follow a wire "path" installed in the floor. Fig. 13-15. Movement is controlled by a central computer. The computer keeps track of each vehicle's location. It directs starting, stopping, and speed, and causes the vehicle to switch from one path to another. Some AGVs can carry heavy loads like car engines. Others are made to carry lighter loads, such as computer chips and circuit boards.

Automated Storage and Retrieval System

Another type of system used in manufacturing today is an *automated storage and retrieval system (AS/RS)*. This is a special set of tall racks with a computer-controlled crane that travels between them. Fig. 13-16. Loads are usually on pallets.

9. **Critical Thinking.** Have students consider the possibility of using an AGV in the school. How might it be useful?

SECTION 3: Manufacturing Technology

FIG. 13-15 These automatic guided vehicles (AGVs) are part of a system that moves materials and products around a factory.

The computer causes the crane to pick up a load. Then it selects an empty slot in one of the racks and directs the crane to travel to that spot and store the load. The crane can also be directed to retrieve or pick up loads from storage.

One kind of AS/RS is called a *miniload*. It is designed to handle items small enough to fit in drawers or tote pans. The miniload is like an automatic set of cabinet or dresser drawers. Imagine if your dresser drawers were unloaded and loaded automatically!

FIG. 13-16 Computer-controlled stacker cranes store and retrieve parts.

10. **Critical Thinking.** Ask students how an AS/RS could be used to save space in the media resource center.

CHAPTER 13: Computer-Integrated Manufacturing

Computer-Integrated Manufacturing in Action

The combination of computerized planning, production, and control is an efficient way to manufacture parts and products. However, buying all the hardware and software is expensive. For this reason there are two stages of development in CIM: the automated factory and the automatic factory.

The Automated Factory

An *automated factory* is one in which many of the processes are directed and controlled by computers. An automated factory includes:

- *Manufacturing cells.* Manufacturing cells are groups of machines working together as directed by a computer. Fig. 13-17.
- *Islands of automation.* Manufacturing cells that aren't connected in any way are called islands of automation. They're like islands in an ocean. In manufacturing, each island of automation is independent of the others.

FIG. 13-17 In a manufacturing cell, several machines are grouped and operated together. A computer controls the operation.

11. **Enrichment.** Collect articles about automated and automatic factories for student reference.

FIG. 13-18 A flexible manufacturing system is computer controlled and can produce a variety of parts or products in any order. No time-consuming tasks of changing machine setups are necessary.

- *Flexible manufacturing systems.* Flexible manufacturing systems (FMS) are groups of manufacturing cells and flexible machining centers. The cells and FMCs are tied together by an automated materials-handling system and by computer control. Fig. 13-18.

  There are several advantages to automated manufacturing. Most of these machines can be adjusted and changed as product needs change. Their use can help improve quality control and reduce production cost per item.

12. **Critical Thinking/Language Arts.** Have students write a scenario about an automatic factory producing a product they'd like to manufacture.
13. **Reinforcement.** One of the most important parts of CIM is that all records are kept in a common

The Automatic Factory

A factory in which almost everything is done by machines—automatically—is an *automatic factory.* There are no people making products. All the parts are made by automatic machines. All materials are moved by automatic materials-handling equipment. All assembly work is done by automatic assembly machines. The quality control checking is also all done automatically. The factory may be dark inside. Most machines do not need lights to operate.

In an automatic factory, all the various systems work together. This integration of the various systems into one giant system makes the factory automatic. All the information collected from all the subsys-

database. Information does not need to be recreated. This saves paperwork and time.

CHAPTER 13: Computer-Integrated Manufacturing   287

tems is kept in a main computer. The main computer directs the other computers.

Since all records are kept in computer files, there is much less need for file cabinets and similar types of office equipment. A copy of a report or drawing can easily be printed at any time.

One manufacturing plant that is completely automatic already exists. It is a Japanese manufacturer of heavy machinery. Their machinery is used throughout the world. The plant operates in the dark, except when humans are needed for repair or changes in manufacturing procedures.

This type of automated plant may gain greater use in the future for certain types of products. The cost of such plants is very high, and they can be used only to manufacture products that will require few design and tooling changes. The products will need to be in demand for many years to make the investment in an automated plant worthwhile.

Will people be needed at all in such a factory? Yes; they will be needed to control, service, and repair the computers and other equipment. However, far fewer workers will be needed than in today's factories.

Factories in Space

Manufacturing in space is an area of growing interest. As you may hear or read in the news, the astronauts in the space shuttle spend most of their time working on experiments. Why is this? In space, natural conditions like gravity are different. In fact, in space, objects are almost weightless because the forces of gravity acting upon them are very small. This condition is called *microgravity*. Some products can be improved when they are made under microgravity. For example, microgravity permits droplets of metal to be formed into nearly perfect spheres, making superior ball bearings. This reduced gravity also makes it possible to make mixtures from materials of different densities, which is not possible on Earth.

FIG. 13-19 Crystals grown in space are used in electronics and other precision applications. The crystals grow free of defects since there is very little gravity acting upon them.

SECTION 3: Manufacturing TECHNOLOGY

FIG. 13-20 Shown here is an artist's conception of a future space factory. This factory will provide new materials and products for us to use here on earth.

On Earth, if you tried to make a mixture from materials with different densities, the materials would separate—just like when you try to mix oil and water. In space, new metal alloys can be made because heavier materials do not settle to the bottom. (An *alloy* is a combination of two or more metals or a metal and a nonmetal.) Other materials and products that can be produced better in space than on earth include crystals for electronic applications, ceramics, glass, and new medicines. Fig. 13-19.

Manufacturing in space will involve the use of both manned and unmanned factories in which materials will be processed automatically. These materials may be finished products or used to make a finished product. Fig. 13-20. Additional applications are being developed as many nations of the world look to the commercial uses of space.

CHAPTER 13: Computer-Integrated Manufacturing

Career File—Industrial Machinery Repairer

EDUCATION AND TRAINING

Most industrial machinery repairers are high school graduates. They learn their jobs through a four-year apprentice program, which is usually sponsored by a local trade union.

AVERAGE STARTING SALARY

Industrial machinery repairers earn about $16,000 the first year. Those with more experience earn between $21,000 and $37,000.

OUTLOOK

Through the year 2005, the available jobs for industrial machinery repairers are expected to grow at a slower rate than other jobs, according to the Bureau of Labor Statistics. Fewer repairs will be necessary on modern machines.

What does an industrial machinery repairer do?

Industrial machinery repairers keep machines, motors, and engines in good working order. They regularly inspect and lubricate the equipment to prevent breakdowns that would delay production, damage the machine, or injure an operator. They adjust and calibrate automated equipment, such as robots. Industrial machinery repairers are also called maintenance mechanics.

Does most of the work involve data, people, or things?

Industrial machinery repairers spend most of their time with tools and machines. They may also refer to technical instructions and drawings to diagnose and fix a problem. They work with electronic repairers or electricians.

What is a typical day like?

Much of the work is preventive maintenance—regularly inspecting, adjusting, and lubricating the machines to prevent breakdowns. When an operator reports a problem with a machine, or when the output from a machine is below standard, the repairer is called in to correct the problem as quickly as possible. A broken machine delays the production schedule. Repairers may be called to the plant at night or on weekends for emergency repairs.

What are the working conditions?

Machinery repairers often work underneath or above large machinery in cramped conditions or on the top of a ladder. They are subject to common shop injuries such as cuts and bruises, and they use protective equipment such as hard hats and ear plugs.

Which skills and aptitudes are needed for this career?

Industrial machinery repairers like to figure out how things work, and they like to work with their hands. They should be agile and in good physical condition to get into hard-to-reach parts of machinery. They should have an interest in and willingness to learn about electronics.

What careers are related to this one?

Other occupations that require similar skills and knowledge about how machines work include aircraft and automobile mechanics and heating and air conditioning mechanics.

One industrial machinery repairer talks about his job:

I like working on a lot of different machines, and this job has variety.

Chapter 13 REVIEW

Reviewing Main Ideas

- Computer-integrated manufacturing (CIM) is the use of computers to link the phases of planning, production, and control of manufacturing together.
- Advances in technology result in new tools and machinery used for manufacturing, such as CNC machines, lasers, waterjets, flexible machining centers, and robots.
- Coordinate measuring machines (CMM), just-in-time manufacturing, automatic identification, speech recognition, and programmable controllers are all used for computerized control.
- Two systems are used for materials handling—the automatic guided vehicle system (AGVS) and the automated storage and retrieval system (AS/RS).
- In an automated factory, many of the processes are computer controlled. In an automatic factory, almost everything is done by machines.
- Manufacturing in space, under microgravity, enables new materials and products to be produced.

Understanding Concepts

1. What are the advantages of computer-integrated manufacturing? Can you think of any disadvantages?
2. What are CAE programs used for?
3. What kinds of information can production planners obtain from computer programs such as CAPP and MRP?
4. Which advanced technology machine would you use to weld under dangerous conditions?
5. Give the full name for and describe the following: CMM, SQC, JIT, LAN, and WAN.

Thinking Critically

1. Name some products that would be good choices for rapid prototyping. Explain your selections.
2. If you were responsible for planning for resources, which computer program—MRP, MRP II, or ERP—would you choose and why?
3. Which of the following companies would benefit from a LAN network and why?—a plant with 150 workers, a company with plants in two states, a company with many plants in several countries.
4. Suppose an automatic factory advertised for a human worker. What types of jobs might that worker be needed for?
5. How could virtual reality be useful in designing an automobile?

Applying Concepts and Solving Problems

1. **Science.** List at least five environments where robots could be used in place of humans for scientific work.
2. **Internet.** Access the following web sites and write a paragraph about each, describing what you find:
http://www.piglet.cs.umass.edu:4321/robotics-mpegs.html
http://www.cs.washington.edu/research/jair/home.html

14. **Resource.** Answers to "Understanding Concepts," "Thinking Critically," and "Applying Concepts and Solving Problems" can be found in the Teacher's Manual at the front of this *Teacher's Annotated Edition* and in the *Teacher's Resource Binder*.

DESIGN AND PROBLEM-SOLVING ACTIVITY

Researching and Developing a Composite Material and Product

> **Health and Safety Notes**: As you work with materials, wear safety glasses, handle cutting tools carefully, and read and observe any warnings on adhesives or resins.

Context

Composites are new materials created by combining two or more different materials. Materials such as wood, paper, glass fibers, or plastics can be mixed or layered with fillers or plastic resins to make a composite. Plywood, fiberglass, particleboard, and Foamcore® board are examples. Composites are created to improve strength, dimensional stability, and temperature and shock resistance. Products, such as furniture, doors, shelving, electronic circuit boards, fishing rods, skis, and bicycles, are made with composite materials.

Goal

For this activity you will design a product to be made from one or more composite materials. You will then create a composite material to be used in manufacturing the product.

1. State the Problem

Design a product and create a composite material from which it will be made.

Specifications and Limits

- The product's size must not exceed 1 ft. x 1 ft. x 1 ft.
- It must be made entirely from a composite material.
- A set of sketches and technical drawings that describe the product must be turned in to your instructor.
- Three samples of composites you are considering must be created; their size should not exceed 1 in. x 6 in. x 6 in.

2. Collect Information

With your research and development team, brainstorm product ideas. Then determine what characteristics the material from which the product is made should have. Consider such things as strength and dimensional stability. A chart, such as the one shown here, can be used to evaluate each product based on several characteristics.

PRODUCT EVALUATION CHART				
Key A = Important B = Desirable C = Unimportant	Must be durable	Must be temperature resistant	Must be shock stable	Must be dimensionally stable
Product 1				
Product 2				
Product 3				

3. Develop Alternative Solutions

Sketch several product ideas and try several composite combinations. Keep in mind ease of manufacturability and assembly. To the right are some items you might want to have on hand.

4. Select the Best Solution

Select one product design and three possible composite combinations for development.

5. Implement the Solution

Produce samples of your composite materials, and decide how to build your product. For example, can the product be mass produced using an assembly line?

6. Evaluate the Solution

Review your experiences. For example, as you created your composites, did you have trouble controlling the adhesive? What could you do during actual production to remedy the problem?

Equipment and Materials
- sketching paper and pencils
- computer and CAD software, such as AutoSketch®
- materials for making the composite, such as paper, cardboard, wood shavings, fabrics, fabric mesh, and screening
- adhesives for joining materials, such as polyvinyl acetate (white) glue, resorcinol glue, or contact cement
- cutting tools, such as scissors, scroll saws, and hand saws
- clamps, vises, and other holding devices
- shaping tools, such as files, surform tools, sanders, and Dremel® tools

Useful Skills across the Curriculum

1. Mathematics. What is the formula for finding how many square feet are in an object? Approximately how many square feet are in your finished product?

2. Science. Review the characteristics of at least three adhesives to determine which will be best for your product. Consider such factors as drying time, appearance when dry, and water resistance.

Credit: Brigitte Valesey

1. **Answers.** Area = length x width.

DESIGN AND PROBLEM-SOLVING ACTIVITY 2

Testing Composite Materials and Prototypes

Health and Safety Notes: Before doing this activity, make sure you understand how to use the tools and materials safely. Have your teacher demonstrate their proper use. Follow all safety rules.

Context

The characteristics of a product depend upon the materials used to manufacture it. Materials can be tested in a variety of ways. For example, they can be stretched, torn, compressed, punctured, and exposed to hot or cold temperatures. Material testing can involve the use of power equipment or hand tools and other simple devices.

Goal

For this activity, you and your team members will develop a method and a simple device for testing the composite materials you created in Activity 1. Then you will manufacture a prototype and test it as well.

1. State the Problem

Design a device for testing a composite material, then create a prototype using the material and test it as well.

Specifications and Limits

- The testing device should have no more than eight parts.
- The device should be designed to hold material no larger than 1 in. x 6 in. x 6 in.
- The device should be designed with safety in mind.
- The prototype should be built using the specifications and technical drawings developed for Activity 1 on p. 292-294..

2. Collect Information

What properties did you design your composite to have? These are the properties you should test for. Collect the drawings you made for Activity 1 to build the prototype.

SECTION 3: Manufacturing Technology

3. Develop Alternative Solutions

Brainstorm with your team ways in which the composite and prototype could be tested. For example, if you are creating a locker memo board, you may want to test the composite by trying to tear it or pull it apart. Consider what equipment and materials you might use. To the right are some suggestions.

4. Select the Best Solution

Select the design for the testing device that you think will do the job best.

5. Implement the Solution

Construct the testing device and test your composite samples. You may want to use a chart like the one shown here to record your results. Select the composite having the most desirable characteristics. Then prepare a sufficient amount of it and build the prototype.

Equipment and Materials
- sketching paper and pencils
- computer and CAD software, such as AutoSketch
- composite material samples developed in Activity 1
- materials for constructing a testing device, such as wood, plastics, mechanical and chemical fasteners, and assorted hardware

COMPOSITE TEST RESULTS

Composite	Weak	Fairly strong	Strong
1			
2			
3			

6. Evaluate the Solution

Test the prototype in use. Is it strong and stable? Will it withstand repeated use? Will the product or the composite material have to be redesigned?

Useful Skills across the Curriculum

1. Science. What are several key properties of the composite material you selected? Which components give it those characteristics?

2. Science. Determine which forces acted upon your composites during the test.

3. Language Arts. Suppose you were responsible for testing these materials for a manufacturing company. Write a report directed to the president of the company describing the results of your test and outlining your recommendations.

Credit: Brigitte Valesey

DESIGN AND PROBLEM-SOLVING ACTIVITY 3

Creating an Inventory Management System for the Manufacture of a Product

Context

Manufacturing companies continually seek efficient ways to manage their inventory of materials, preassembled parts, and tools. Having too large an inventory results in wasted storage space or extra materials that may not be used. Too small an inventory may result in insufficient materials for production.

Goal

For this activity you will develop an inventory management system for a class manufacturing enterprise, such as production of the product created for Activities 1 and 2 on pages 292-296. You will use computer spreadsheet software, such as FoxPro® or EXCEL®.

1. State the Problem

Develop an inventory management system for production of a particular product, such as the one you developed for Activities 1 and 2.

Specifications and Limits
- The system must be designed to log in and check out materials, preassembled parts, and tools.
- The data collected should be used to determine inventory needs for production of the product.

2. Collect Information

Determine which materials and tools need to be inventoried, how much storage is needed, and what storage is available. Prioritize inventory needs and locations. Where do materials and tools need to be stored in relation to the manufacturing activity?

3. Develop Alternative Solutions

Experiment with several ways to set up your database. Have your instructor help with creating rows and columns, as well as how to input and save data.

4. Select the Best Solution

Choose the method that seems to work best.

5. Implement the Solution

Use the database for each day of the manufacturing activity. Keep a log of materials and tools signed out at the beginning of the session. Log in returned items. Use the spreadsheet to keep track of partially assembled products. Post information concerning completed products, subassemblies produced, and materials left for remaining products.

6. Evaluate the Solution

Did you discover any benefits to keeping an inventory? What key information did your database provide? If a manufacturing company were to produce this product, would you recommend using JIT methods or not? What improvements could you make in your database?

Useful Skills across the Curriculum

1. Mathematics. Estimate the square footage required to store your finished products. How much space would be required for a manufacturing company to store 15,000 of the items?

2. Science. Review the Science Connection regarding stored items on page 267 of this text. Does your product contain materials that could deteriorate? Explain.

Credit: Brigitte Valesey

DESIGN AND PROBLEM-SOLVING ACTIVITY 4

Creating a Multimedia Advertisement

Context

Marketing a product by means of advertising is an important part of the manufacturing process. Through advertising, companies compete for the buyer's attention and communicate information about a product's strong points. Product advertising must be attention getting, have eye appeal, and convey a strong message. Today, many advertisers use multimedia (more than one medium) to sell their products.

Goal

In this activity, you will use a computer and multimedia or presentation software, such as Power Point® or Astound®, to develop an advertisement for the product you manufactured in Activities 1, 2, and 3, pp. 292-298.

1. State the Problem

Create a multimedia advertisement for a particular product.

Specifications and Limits
- The advertisement must run for 30 seconds.
- The advertisement must make use of at least two different media, such as written or spoken words, music, visuals, and so on.

2. Collect Information

With your team, decide what information about your product you wish to communicate. Then determine who your potential customers are.

3. Develop Alternative Solutions

Brainstorm strategies for advertising the product. Think of slogans or story lines. Sketch possible logo designs.

4. Select the Best Solution

Select the strategy that will communicate best what your potential customers need to know.

5. Implement the Solution

Prepare a storyboard or script, showing key visuals in your advertisement. Keep words to a minimum; use color and strong symbols to convey information; incorporate action, movement, and flow of visuals. Present your advertisement to the class.

6. Evaluate the Solution

In what ways does your advertisement influence buyers to purchase your product? What types of advertising make you want to buy? Is this advertisement of that type?

Useful Skills across the Curriculum

1. Science. Plan some kind of packaging for the product that takes the environment and waste management into account. Explain the benefits of your ideas to the class.

2. Mathematics. Contact a local TV station to find out their charges for airing a 30-second commercial during these time periods: after midnight, late morning, 5:00-8:00 p.m. Calculate how much it would cost to run your advertisement once a day during each of those time periods for one week.

3. Language Arts. Create another advertisement for the same product for radio. Write a 30-second script and perform it for the class.

Credit: Brigitte Valesey

DIRECTED ACTIVITY 5

Selecting and Buying Standard Stock

Context

After a product has been researched and developed, the production stage of manufacturing is put into action. Before a product can be manufactured, the tools, equipment, and consumable supplies needed for production must be purchased.

The supplies are usually purchased in the form of standard stock. *Standard stock* is material that has been formed or separated into widely accepted sizes, shapes, or amounts. Figures A and B show some standard types and sizes of fasteners and wood.

Goal

In this activity, you will play the role of a buyer. You will research catalogs from at least two vendors for standard-stock and find out the costs of standard-stock wood, metal, and fasteners.

Procedure

1. Obtain at least two catalogs for standard-stock materials.

2. Determine the cost of each of the following from two different sources.

Dimensional lumber
—Eight 2" x 4" boards, each 16 ft. long
—One 1" x 10" board, 12 ft. long
Plywood
—3/4", one 4' x 8' sheet, exterior grade
Galvanized sheet metal
—22 gauge, one sheet, 3' x 8'

Equipment and Materials
- notebook paper
- pencils and ball-point pens
- catalogs from companies that sell standard stock materials (for example, structural steel companies and lumberyards)

3. What is the price per pound (from each source) for each of the following types of nails:

16d, common
10d, common
6d, finishing

SECTION 3: Manufacturing Technology

Fig. A. Examples of standard fasteners.

Fig. B. Wood is cut into standard sizes, such as 2" x 4". After the wood is surfaced (smoothed), it is smaller. The actual size of a 2 x 4 that you buy is 1½" x 3½".

Fig. C. Use this formula to figure board feet:
$$\frac{\text{length (ft.)} \times \text{width (in.)} \times \text{thickness (in.)}}{12}$$

Evaluation

1. For each type of standard stock that you researched, how much did the different sources' costs differ? Share your findings with the class.

2. Why do you think these cost differences exist between different companies' standard stock?

Useful Skills across the Curriculum

1. Mathematics. Hardwood is purchased in board feet. See Fig. C. Compute the amount of board feet in four 2" x 8" pieces of oak 3 ft. long. Figure the costs (from two sources) of this wood if it is surfaced on two sides.

2. Science. To order the best materials for the job, you need to know something about the properties of materials. Find out about the properties of wood, such as strength, heat conductivity, and hardness. How do these properties affect wood's usefulness as a construction material?

1. **Answer.** 16 board feet.

SECTION 4

CHAPTERS

14 Construction Systems

15 Planning Construction

16 Managing Construction

17 Constructing Homes and Other Buildings

18 Other Construction Projects

TECHNOLOGY TIME LINE

2750 B.C. Stonehenge begun in England.

214 B.C. Great Wall of China is begun.

50 Glass is first used for windows.

880 Angkor Wat, a vast temple complex, is begun in Cambodia.

1100 Anasazi people begin cliff dwellings at Mesa Verde.

1400 Machu Picchu is built by the Incas in Peru.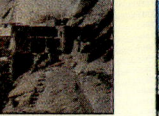

1883 Brooklyn Bridge is completed.

SECTION 4: Construction Technology

Construction Technology

Construction technology involves the building of structures, such as houses, bridges, highways, dams and canals. In this section, you will learn how construction takes place, from the planning stage, through actual building, to finishing the site.

1884 The Home Insurance Building in Chicago is world's first skyscraper.

1914 Panama Canal is completed in Central America, connecting the Atlantic and Pacific Oceans.

1947 R. Buckminster Fuller perfects the geodesic dome.

1976 Canadian National Tower becomes tallest freestanding structure in the world.

1988 In Japan, the world's deepest underwater tunnel is built between two islands.

1996 Petronas Towers in Malaysia is world's tallest building.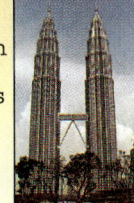

SECTION 4: Construction Technology

CHAPTER 14

Construction Systems

Objectives

After studying this chapter, you should be able to:
- explain the importance of construction.
- define and give examples of the four basic types of construction.
- describe how the seven resources are used in construction systems.
- detail some advantages and disadvantages of several types of structural materials.

Terms

adhesives
architects
building codes
commercial construction
engineered wood materials
engineers
geotextiles
industrial construction
public works construction
residential construction
site
structural materials

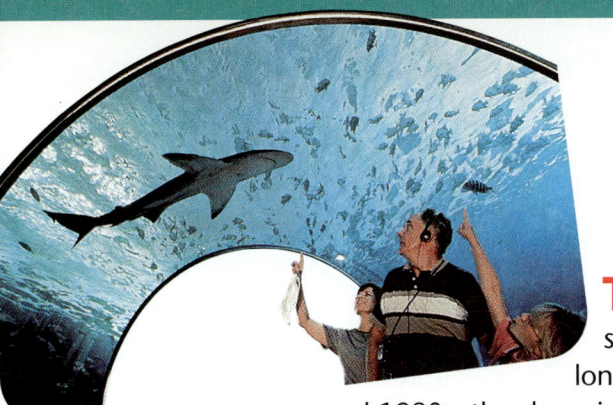

Technology Focus
Shopping Malls Invest in New Business

The regional shopping mall, with its 50 to 150 stores, often attracts customers who travel as long as 40 minutes to shop there. In the 1970s and 1980s, the shopping mall was a main source of retail business. This changed in the 1990s, however. Discount stores and factory-outlet centers competed for price-conscious shoppers, while big department and specialty stores lured customers seeking more glamorous products.

Faced with a decrease in sales, mall developers have taken action to bring more business into aging malls. Strategies may vary, but they nearly always involve new construction. New construction is intended to give the old mall a new look and is usually worth the investment. Shoppers respond to new facades (fah SAHDZ), trim, and attractive features. Some malls add special attractions, like amusement parks!

Making Changes

One of the ways that designers achieve a new look that shoppers may not notice is the lighting. Lighting designers can dramatize entrances and gathering points. A good mix of accent lighting and general lighting can make walking distances appear shorter.

Skylights have strong appeal but also present a challenge in glare-control. One solution is to surround the source of natural light with intense electrical lighting, which eliminates contrasting dark areas. Photocells ("electric eyes") can be used to adjust the lighting for twilight and nighttime. Mercury vapor lights, with ample blue-green in their spectrum, make indoor plants look more green.

Plants can subtly enhance or, if neglected, detract from the quality of an interior. Potting systems are available that can ensure proper growing conditions. In some cases, artificial plants make a more reliable choice.

Flooring is another design element that influences an interior. Any floor finish must be strong to withstand wear and tear, because people do notice when a floor is worn and shabby. Tile, sometimes with colored grout, makes possible an endless variety of colors and patterns. A terrazzo floor, one that is made of chips of stone, can be inlaid with images or designs that match a fountain or other focal point.

Take Action!
At a local mall, make a map to show its lighting, landscape, or flooring details.

Resources
- Chapter 14 Lesson Plan in the *Teacher's Annotated Edition* and in the *Teacher's Resource Binder.*
- Chapter 14 Review in the *Teacher's Resource Binder.*

CHAPTER 14: Construction Systems

What Is Construction?

Construction is the process of producing structures. These structures can be as varied as houses, skyscrapers, athletic stadiums, highways, bridges, or dams. The construction industry is the production system that produces these constructed products. It offers many opportunities for different types of work. The skills required to work in construction are varied, but they all have something to do with building. That, in a word, is exactly what construction is—building.

Our Constructed World

When people think about construction, they usually think about the building of houses, the type of constructed product that has helped us meet one of our basic human needs—shelter. Construction has also helped us meet many other needs as well. For example, constructed products include buildings and other structures where we can eat and sleep, work, play, study and learn, shop, receive health care, and worship.

Because of construction, we have a network of highways, bridges, airports, and tunnels. These structures enable us to travel freely about the country.

Certain buildings have been so beautifully designed and constructed that they are considered great works of art. Other structures have become important historical or cultural symbols. Fig. 14-1.

FIG. 14-1 Structures do more than meet our basic needs. The Lincoln Memorial in Washington DC is a symbol of respect for our sixteenth president. The Cathedral of Notre Dame in Paris is considered a work of art.

1. **Extension.** Have students bring pictures to class of structures (not just buildings) that are historically or culturally significant or that they think are beautiful.

308 SECTION 4: Construction Technology

FASCINATING FACTS

The Great Wall of China stretches 1,400 miles across northern China and has an average height of 23 feet. Built entirely by hand to keep out invaders, it took hundreds of years to complete.

Types of Construction

Look at Fig. 14-2. Each structure shown falls into one of the four major types of construction:
- residential
- industrial
- commercial
- public works

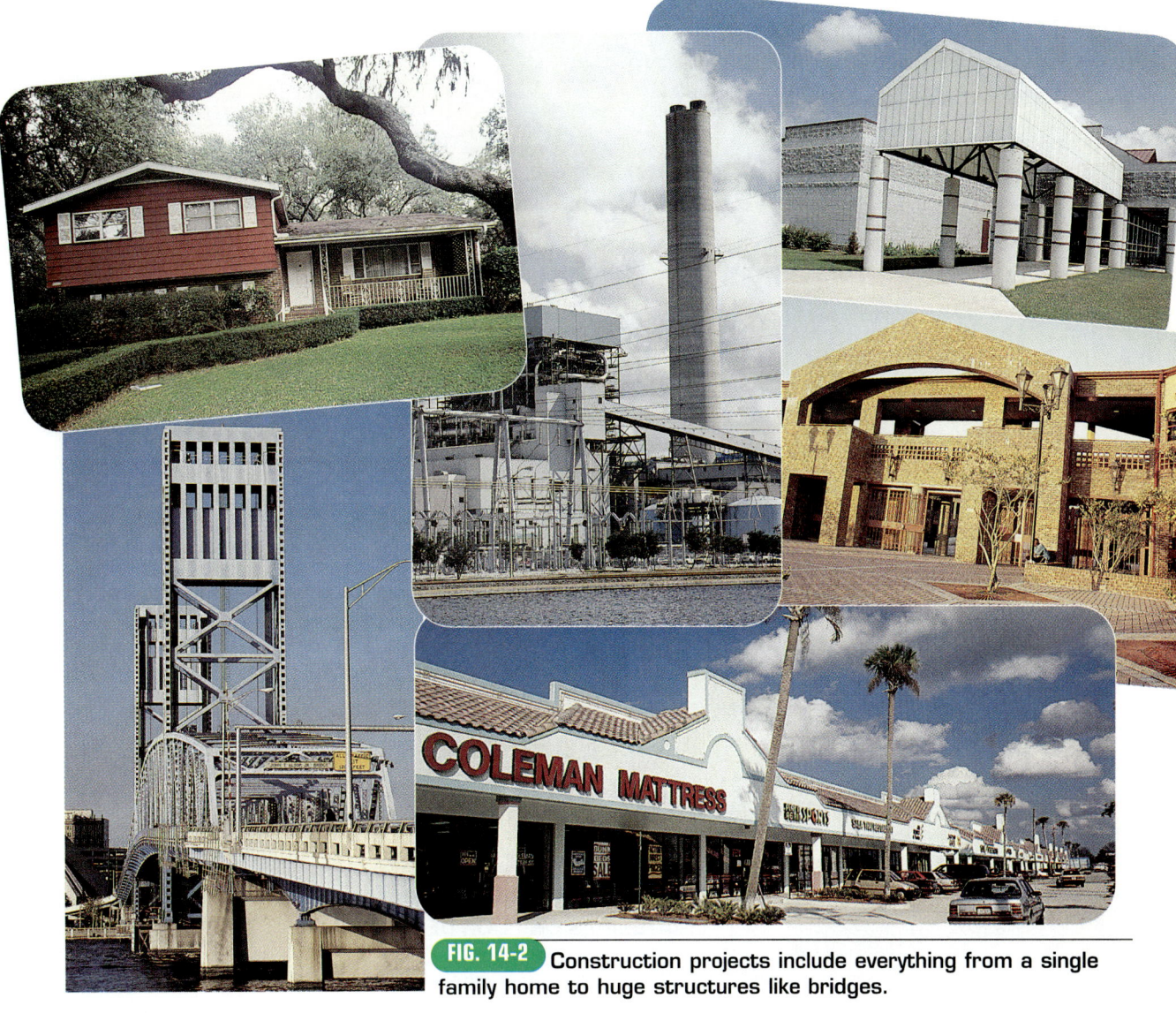

FIG. 14-2 Construction projects include everything from a single family home to huge structures like bridges.

CHAPTER 14: Construction Systems

Residential construction refers to building structures in which people live. Most residential structures are single-family or private homes. However, residential construction also includes the building of small multifamily units, which are residences that have two or more apartments or dwelling areas.

Duplexes and town houses are multi-family units. A *duplex* consists of two apartments, which are usually side by side under a single roof. A *town house* is a single two-story unit, but several of these are built side by side, sharing adjoining side walls, to form a single long building. Town houses are sometimes called *row houses*.

Most residential construction is done by fairly small construction companies. Usually, common materials (such as wood and brick) and basic building techniques are used.

Industrial construction includes the building and remodeling of factories and other industrial structures. This type of construction is usually planned by specialized engineering firms. Industrial structures are usually built by large construction firms that have many employees.

Commercial construction consists of building structures used for business.

FIG. 14-3 The parts of a construction system are input, processes, output, and feedback.

2. **Critical Thinking.** How would the environment (climate, land contours, soil composition, water supplies, etc.) affect residential construction?

3. **Critical Thinking.** Ask students: Why might industrial plants need to be planned by specialized engineering firms?

SECTION 4: Construction Technology

Supermarkets, shopping malls, restaurants, and office buildings are examples of commercial construction. Commercial projects are usually large-scale construction projects that involve millions of dollars and many workers.

The building materials and techniques used in commercial and industrial construction are somewhat different from those used in residential construction. These structures often have steel frames and many concrete parts.

Public works construction is building structures intended for public use or benefit. This type of construction includes large projects such as dams, highways, bridges, tunnels, sewer systems, airports, hospitals, schools, and parks. The projects are usually funded by federal, state, or local taxes.

Public works construction projects are built by large construction firms. Most of these have hundreds of employees and much heavy equipment.

Inputs

Like other technological systems, a construction system has inputs, processes, outputs, and feedback. These all work together to achieve the goal of constructing a finished structure. Fig. 14-3.

The inputs include the seven technological resources that will be used in building the structure: people, information, materials, tools and machines, energy, capital, and time.

People

The building of structures requires the skills, hard work, and cooperation of many people. Fig 14-4. Some people are directly involved in the building process. **Architects** design structures and develop the plans for building them. **Engineers** make sure the design is structurally sound and determine how the structure will be built and what structural materials should be used. Skilled trades workers carry out the actual construction process. For example, heavy equipment operators use earth-moving machines such as bulldozers and operate cranes; electricians install wiring; plumbers install fresh and waste water systems; masons lay bricks and blocks or pour concrete; and carpenters measure, saw, and fasten boards.

There are other people directly involved in the process who oversee the workers, monitor the overall job progress, order materials, schedule work and deliveries, and inspect the structure.

Other workers are more indirectly involved in construction projects by providing needed materials and equipment. They may be nearby in the local hardware store or thousands of miles away cutting lumber in a forest.

Information

Many kinds of information are needed for construction systems. For example, when architects decide what kind of structure to design, they collect information about the needs of the people who will be using that structure. If it is a school, for example, how many students will attend it, what are their ages, and do they have any special needs? Will physical education facilities be needed, and if so, what kind? Is an auditorium needed? A cafeteria? A lab or shop? All of this information is necessary for the architect to design a good school for these particular students.

---

4. **Reinforcement.** Stress the importance of cooperation. For example, plumbers and carpenters must work in the same area without interfering with each other's work. Carpenters must complete some of their work to enable plumbers to do theirs, etc.

5. **Brainstorming.** Ask students to name some other jobs related to construction. Ask them which jobs they might be interested in pursuing.

FIG. 14-4 People are the most important resource in all construction systems. Many people in a wide variety of jobs contribute to the process.

SCIENCE CONNECTION

The Chemistry of Plastics

Although a few occur in nature, most of the plastics we use today are manufactured from chemicals. These chemicals come from such things as petroleum, coal, limestone, salt, and water.

The raw, unfinished material from which a plastic is made is called a *resin*. Although resins may be used by themselves to create plastics, other ingredients, called *additives*, are usually combined with them. These additives change the properties of the resin and make it usable for different jobs in construction.

Certain *fillers* may be added to resins to extend them and reduce manufacturing costs. Other fillers are used as reinforcements. For example, fiberglass improves a plastic product's rigidity, strength, and resistance to cracking.

Plasticizers added to resins loosen the chains of plastic molecules, which softens them. This makes the resulting plastic more flexible. Plasticizers can also make resins flow into molds more readily.

Additives that keep plastics from breaking down are called *stabilizers*. Some stabilizers are effective against heat. Others stabilize the plastic in the presence of oxygen or ultraviolet light.

Colorants can be added to resins to make them a certain color. Unlike paint, which can be added only to the surface of a product, colorants color the plastic throughout.

Many plastics burn easily. Some give off poisonous smoke. This makes them dangerous to use in carpeting, clothing, and airplanes. Flame retardants can be added to the resins to make the plastic less flammable and toxic.

Chemists are learning more control over certain additives. Soon they will be able to produce plastics with new properties, such as toughness, stretchiness, and the ability to withstand extreme temperatures.

Try It!

Do some research on one of the following: catalysts, biodegradable plastics, celluloid. Report your findings to the class.

Information must also be gathered about the construction site itself. The **site** is the land on which a project will be constructed. Information must be gathered about whether the land is flat or hilly, whether the soil provides good drainage, and whether the soil can support the structure. For buildings, information must be gathered about lot size and the exact boundaries of the lot.

Information about any laws or codes regulating construction must also be gathered so the structure can meet these requirements. For example, **building codes** specify the methods and materials that can or must be used for each aspect of construction.

Materials

Many different kinds of building materials are available. The size and nature of the project under construction determine what is needed.

Structural materials are those used to support heavy loads or to hold the structure rigid. They are chosen for strength and stiffness. Wood, steel, aluminum, concrete, masonry, plastics, and adhesives are often used as structural materials.

Wood

Wood has many practical advantages as a structural material. Wood is fairly durable to varying weather conditions, and it can be a renewable resource. It can readily be cut, shaped, and fastened together with nails, staples, screws, bolts, or adhesives. Wood can be used as it has been for thousands of years—as conventional solid-sawn lumber, or it can be used as the main ingredient in modern engineered wood materials.

Conventional solid-sawn lumber is wood that is sawed and planed to standard sizes. The two types of conventional solid-sawn lumber are dimension lumber and board lumber. *Dimension lumber* is lumber that measures between 2 and 5 inches thick. It is classified by its dimensions (as it is cut from the log, before it is planed and dried); for example, a board

NOMINAL SIZE (before drying and planing)	ACTUAL SIZE (dried and planed)
For dimension lumber	
2 x 4	1 1/2" x 3 1/2"
2 x 6	1 1/2" x 5 1/2"
2 x 8	1 1/2" x 7 1/4"
2 x 10	1 1/2" x 9 1/4"
2 x 12	1 1/2" x 11 1/4"
For board lumber	
1 x 4	3/4" x 3 1/2"
1 x 6	3/4" x 5 1/2"
1 x 8	3/4" x 7 1/4"
1 x 10	3/4" x 9 1/4"
1 x 12	3/4" x 11 1/4"

FIG. 14-5 Common sizes of conventional softwood (fir, pine, and spruce) lumber used in construction. It is important to realize that the stated size of lumber is not its actual finished size. The stated *nominal* size is the size of the lumber when it is cut from the log. After it is cut, the lumber is dried and planed on all four sides for smoothness. The finished, or *actual*, size is therefore smaller than the nominal size.

6. **Reteaching.** Collect samples of construction materials for students to examine.

7. **Reinforcement.** Bring in samples of solid-sawn lumber (both dimension and board lumber) to show students.

SECTION 4: *Construction Technology*

FIG. 14-6 In *plywood*, each ply or layer is glued with its grain at right angles to the layer above and below it, making the plywood strong and resistant to warping. In *waferboard*, the chips and flakes of wood have nondirectional, or random alignment. In *oriented-strand board*, the reconstituted wood strands are directionally layered, and each of the 3 to 5 layers is perpendicular to the layer above and/or below it.

that is 2 inches thick and 4 inches wide is called a two-by-four (2x4). *Board lumber* is lumber that measures less than 1/2 inch thick and 4 or more inches wide. Both types of lumber are commonly available in even-numbered lengths, such as 6, 8, 10, or 12 feet. Common sizes are shown in Fig 14-5. Solid-sawn lumber is strong for its weight and still relatively cost-effective, especially in smaller sizes like 2x4's and 2x6's, which are commonly used for framework of walls, floors, and roofs.

In spite of the fact that most lumber companies plant seedlings to replace trees they cut down for lumber, our timber supply is gradually being used up. It is becoming increasingly important to stretch the supply of our valuable wood resources, especially large, older trees. **Engineered wood materials** have been developed to help make the best use of our wood supply.

Plywood, particleboard, waferboard, and oriented-strand board (OSB) are engineered wood panels made by mixing woods from small, crooked trees that would otherwise be unprofitable to harvest. Fig 14-6. Glulam and laminated veneer lumber beams are engineered dimension lumber made of smaller pieces of wood glued together with waterproof adhesive and bonded under heat and pressure. Fig 14-7. These engineered materials are up to 30 percent stronger than conventional lumber of similar sizes. In addition, they are a more efficient way of using our wood resources. About 80 percent of a tree can be used when making

8. **Extension.** Explain to students the difference between softwoods and hardwoods. Explain that softwoods are used for framework, while more expensive hardwoods (e.g., oak, cherry, maple) are used for their beauty and durability in cabinets, floors, and trim.

9. **Extension.** Discuss with students ways in which the forests can be made to produce wood products for generations to come.

FIG. 14-7 Laminated engineered veneers were used to make this joist, which is used in the part of the framework that supports a house's floor.

these wood products. Only about 50 percent of a tree is used when it is harvested for conventional lumber.

Wood has some disadvantages as a construction material. Depending upon the conditions to which it is exposed, wood may burn, warp, split, or rot. Untreated, it can be attacked by termites and other insects.

Metals

Steel is an outstanding structural metal material made by combining iron with small amounts of carbon. Its major advantages are that it's very strong and rigid (stiff). Fig. 14-8. Steel can be formed into various structural shapes, such as tees (shaped like the letter T), channels (shaped like the letter U), beams or girders, sheets, and rods. Steel can also be formed into wire, which can be woven into extremely strong rope or cable. Steel units can be welded together or fastened with bolts or rivets.

Steel also has disadvantages as a structural material. For example, ordinary steels may rust or corrode. These must be protected by either being painted or embedded (enclosed) in concrete. Also, when steel is exposed to temperatures above 700°F (371°C), it will rapidly lose its strength. Therefore, it is usually covered with a fireproof material, especially in high-rise buildings.

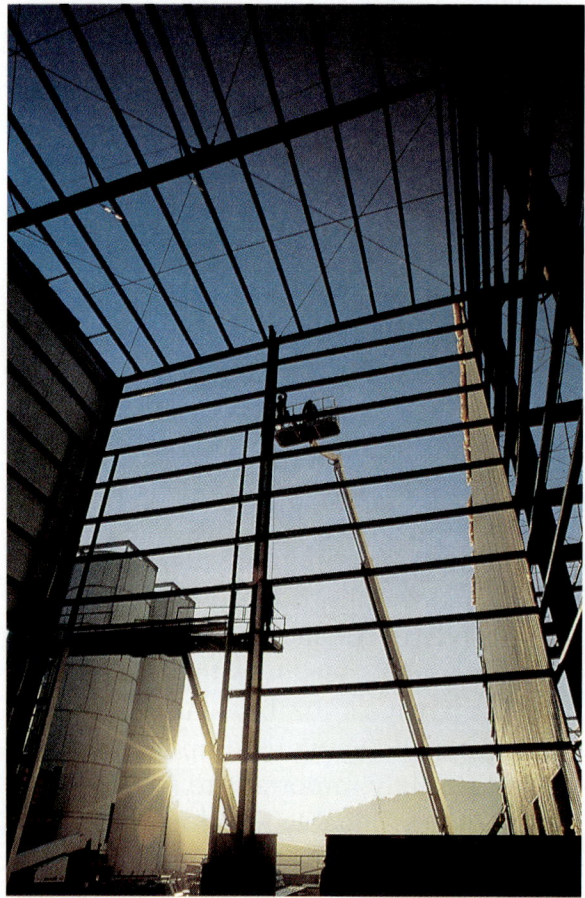

FIG. 14-8 A steel frame is used when constructing most large buildings.

10. **Extension.** If possible, locate a picture of the remains of a burned building that had been constructed with steel beams. Show students what the beams look like after a fire.

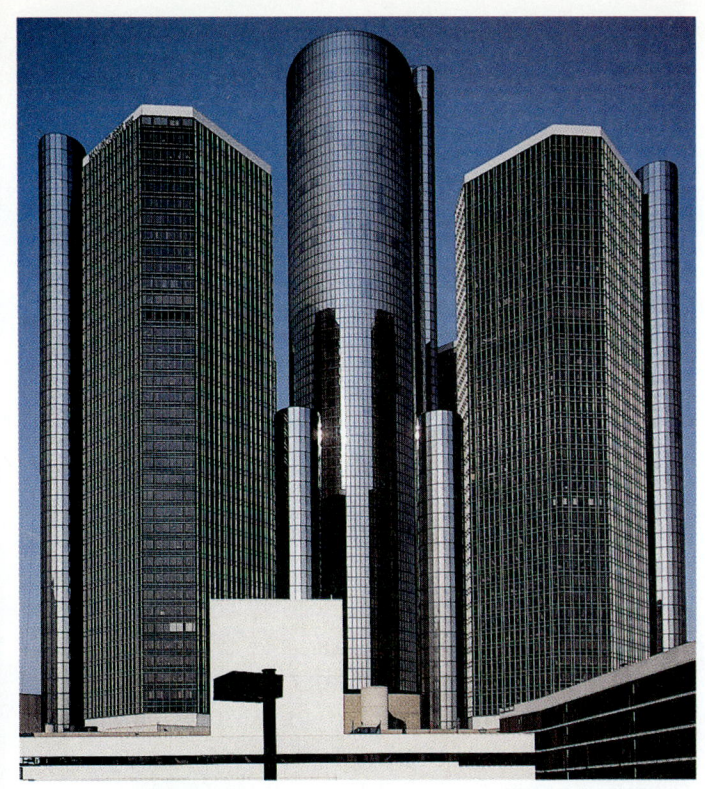

FIG. 14-9 New lightweight, yet strong, materials are often used in building skyscrapers.

Aluminum is a metal used when light weight and resistance to corrosion are important. Like steel, aluminum can be formed into a variety of shapes. For example, it can be formed into beams, rods, and wire. It can also be rolled to form plates (flat sheets 1/8" or thicker), and foil. Aluminum parts can be welded or fastened together with rivets or bolts.

A variety of new, lightweight, ultra-strong steel and aluminum materials have been developed. These unique materials give architects and design engineers greater freedom to experiment with new design and construction techniques. Examples can be seen in the steel and glass towers in many major cities. Fig. 14-9. Other examples of new techniques can be seen in the increasing number of clear-span structures. Fig. 14-10.

FIG. 14-10 Lightweight materials and new design techniques are creatively combined in this clear-span structure. The curved roof of the structure requires no vertical support members inside the structure.

Concrete

Concrete is a mixture of sand, gravel, water, and cement. When properly cured, it is a very strong material. *Curing* is not simply "drying." It is a chemical reaction that makes the concrete hard and strong. It can be used to build such structures as bridges, tunnels, highways, and large buildings.

Concrete has many advantages as a structural material. Besides being strong, it is moldable. It can be poured into molds to form almost any kind of shape. It may be made into blocks. Huge panels or entire walls can be made of concrete. Separate concrete units can be joined to form structures, such as high-rise buildings.

Concrete also has some disadvantages as a structural material. It is heavy. In addition, it must be prepared properly, poured carefully, and finished and cured correctly. If even one of these processes is not done correctly, the concrete may crack or crumble. Fig. 14-11.

Masonry

Masonry is a broad term that includes both natural materials (such as stone) and manufactured products (such as bricks and concrete blocks). The material selected depends on the type of project. Most masonry structures are held together by some kind of cement mortar or other bonding material.

Masonry has the advantage of being fireproof and is also a sturdy structural material. It is usually more costly than wood, however, and the bricks or stones must be laid well. This means a lot of skilled labor is involved. In addition, the mortar that holds the brick or stone together tends to crumble over time and must be replaced periodically.

FIG. 14-11 When placing concrete, all processes must be done carefully in order for the concrete to become hard, strong, and durable.

SECTION 4: Construction Technology

Plastics

Plastics have been used in many areas of construction for several years. Plastic pipe is now used in most plumbing systems. Plastic is used to make many insulating materials. Some bathtubs and sinks are made from fiberglass, a composite made from plastic resin and spun glass fibers. New, strong, lightweight plastics and plastic composites are beginning to be used in place of other heavier structural materials. Because they are lightweight, strong, waterproof, economical, and resistant to corrosion and rust and can be formed into any desired shape, new plastics might someday replace many of the materials with which we now build.

Plastics are now being used for roofing materials, liquid storage tanks, exterior siding, protective coatings, and fasteners. A new type of "lumber" made from recycled plastics is being used to make outdoor items such as picnic tables and decks. Fig. 14-12.

Another new application of plastics is in geotextiles. **Geotextiles**, also called engineered fabrics, are like large pieces of plastic cloth. Geotextile materials are spread on the ground as an underlayment, or bottom layer, for roadbeds. They are also used on slopes along highways to help keep soil in place and prevent erosion. Fig. 14-13. Entire roofs of some new structures are made from these specialized plastics. The Denver International Airport terminal building has a textile roof. Fig 14-14.

FIG. 14-13 Dirt is being moved onto this geotextile, which is being used to help reinforce the levee (embankment) being constructed to prevent a flood.

FIG. 14-12 Plastic is increasingly being used as a construction material. This park bench is made of plastic recycled from milk bottles.

11. **Critical Thinking.** The text discusses many advantages of plastics as structural materials. Ask the class about some possible *dis*advantages. Develop a list on the chalkboard.

12. **Extension.** Provide samples of geotextiles and other plastic construction materials.

CHAPTER 14: Construction Systems

FIG. 14-14 Geotextiles help allow architects to create new architectural designs such as the roof of the new Denver Airport terminal building.

Adhesives

Adhesives are materials that are used to bond together, or adhere, two objects. Glue is a common type of adhesive. New, stronger adhesives that can bond almost any kind of material to almost any other kind of material are being used and perfected. Fig. 14-15. They are used to make engineered lumber and to bond many of the new plastic construction materials.

Strong adhesives are being used more and more in place of nails in today's construction to help save time on the job. For example, plywood or drywall can be fastened to studs (boards used to frame walls) more quickly with adhesives than with nails.

13. **Reinforcement/Critical Thinking.** Ask students to list those places in the school building where adhesives might be used for fastening.

Other Materials

Other materials used in construction include roofing materials, vinyl siding, insulation, drywall or gypsum, electrical wiring and lighting, plumbing supplies and fixtures, and heating and cooling systems.

Fascinating Facts

Safety glass is "safer" than other glass because it breaks into many, small, dull-edged pieces. Regular glass breaks into bigger pieces and the edges are much sharper.

SECTION 4: Construction Technology

FIG. 14-15 These concrete bridge segments are bonded together with an adhesive.

Light equipment is equipment that can be moved about fairly easily. Ladders, sawhorses, and wheelbarrows are commonly used. Some light equipment, such as surveying equipment, is fairly complex. Laser equipment is often used in surveying, leveling construction sites, and leveling interior walls. Fig. 14-17.

FIG. 14-16 Some portable power tools such as this drill are powered by rechargeable battery packs.

Tools and Machines

There are four basic types of tools and machines used in construction: hand tools, portable tools, light equipment, and heavy equipment.

Hand tools are simple tools powered by humans. Hand tools commonly used in construction include hammers, saws, screwdrivers, shovels, and wrenches.

Portable power tools are tools that are powered by electricity or air and are small enough to carry. Examples include electrically powered saws and drills and air powered nailers and staplers. Fig. 14-16.

14. **Reinforcement.** Ask students to name tools they have in their homes that are the same type that a construction worker uses on a construction project. Make a collective list of these on the chalkboard.

CHAPTER 14: Construction Systems **321**

FIG. 14-17 Construction workers will use this laser level to accurately align interior walls.

Large and powerful equipment is referred to as heavy equipment. Cranes, bulldozers, and trucks are examples of heavy equipment. Fig. 14-18.

Energy

Energy is needed to build a structure. Petroleum and electrical power run the many machines and tools that construction teams use, such as bulldozers, power drills, and cranes. Human energy is used every time a worker drives a nail with a hammer or an electrician installs wiring.

For most types of structures, once the structure is completed, energy will still be needed to run the internal systems. Can you imagine your house without a furnace or electrical power? It would be difficult to drive over the Brooklyn Bridge or Golden Gate Bridge at night without all the lights marking the outline of the bridge. Energy is an input to construction systems both when structures are being built and when they are being used.

FIG. 14-18 This bulldozer is being used to clear away rocks and trees and to move the earth to desired levels.

15. **Background.** Moving heavy equipment to a construction site is expensive. Typically, it costs as much to move a large crane to a construction site as it costs to buy a Corvette.
16. **Reinforcement.** Have students identify the energy systems currently being used in their own school.

322 SECTION 4: Construction Technology

Capital

Basically, the term *capital* refers to all the money used in construction processes. The land on which a structure is built is a very important part of capital in construction. Land is very expensive to buy and a valuable asset to own. In addition, many hundreds of millions of dollars are spent on equipment, tools, materials, and other supplies. Large sums are also spent on workers' wages.

Time

Time is a critical resource in construction technology. Scheduling is of the utmost importance. Materials must be delivered so that they are available as they are needed. The jobs that need to be done are dependent on one another, so if one job is not done on time, it will affect the schedule of many other jobs. For example, the ground must be cleared and prepared before construction can begin. Bad weather can cause serious delays in this process, as well as other outdoor construction processes. If the ground is not prepared, workers cannot build the foundation, which is the part of the structure that rests on the earth and supports the body of the structure from the first floor on up. The frame cannot be constructed if the foundation has not been completed. Electricians cannot install wiring and roofers cannot install roofs until the framing is complete.

Construction companies do not want to pay employees who are waiting for their phase of work to begin. In addition, construction contracts often include financial penalties if construction is not complete by a certain date or bonuses if it is completed before a certain date.

17. **Brainstorming.** Ask students if they can think of any reasons other than bad weather that might delay construction one it has begun.
18. **Brainstorming.** Have students think of all the processes of construction. Make a list on the chalkboard.

Processes

The processes of construction include everything from choosing and preparing a site to turning over a finished building to its new owner. Fig. 14-19. Perhaps you have watched the gradual progress of a construction project, such as a new home or a shopping mall, in your own community. If so, you have seen a number of construction processes going on. The next chapters in this section will examine some of the construction processes in more detail.

Output

The main output of a construction system is, of course, the finished structure. Depending on the purpose of the particular system, this can range anywhere from a backyard treehouse, to a new dam or bridge, to a whole new development of residential or commercial buildings.

Output also includes any impacts on society and the environment. Suppose a city builds a new skating arena. Perhaps people will begin to use the arena rather than nearby ski areas for recreation. The arena may encourage skating professionals to move to the community and set up a skating school. These are some of the possible social impacts.

Construction systems also have an impact on the environment. This is one of the reasons construction projects need to be monitored so closely from their very beginning. Sometimes a proposed project may present possible threats to natural features and wildlife. The benefits of the project may not be worth the potential harm. Even if the project doesn't damage

19. **Extension.** Wetlands are often a protected environment. Have students research the reasons construction is often prohibited on wetlands.

CHAPTER 14: Construction Systems

FIG. 14-19 Planning a project, putting up a frame, and finishing the interior of a structure are all processes of a construction system.

the environment, it may change it in undesirable ways. For example, people might not want to build a factory in the middle of a residential area.

Controversies often arise over proposed construction projects and the impacts they might have on the surrounding environment and community. As in any other technological system, the advantages and disadvantages of a proposed project need to be carefully weighed.

Feedback

Feedback on construction systems is important. Even though people try to plan the best possible structures, they sometimes overlook important factors. People using a building, for example, might realize there is not enough ventilation, and the ventilation system will need to be improved.

A dramatic kind of feedback was provided during the Tacoma Narrows Bridge accident. The Tacoma Narrows Bridge,

SECTION 4: Construction Technology

built in Washington state, was a suspension bridge that opened on July 1, 1940. On November 7, 1940, the bridge began to sway and twist violently. At 11:00 a.m., a 600-foot length of the main span broke off and collapsed. (Fortunately, the bridge had been closed to traffic earlier in the day.)

One of the main reasons for the disaster turned out to be the proportion of the vertical depth of the span to the length of the span. Fig. 14-20. The thin span was too flexible to withstand the motions set in action by the wind.

The collapse of the bridge was feedback that resulted in major structural changes. When the bridge was rebuilt, the vertical depth of the span was increased from 8 to 33 feet.

FIG. 14-20 The small vertical depth of the bridge span is one of the factors that led to the Tacoma Narrows Bridge catastrophe.

20. **Resource.** A more recent bridge disaster that provided a dramatic kind of feedback involved the Schoharie Thruway Bridge. There's a handout in the *Teacher's Resource Binder* that describes this disaster.

CHAPTER 14: Construction Systems

Career File—Architect

EDUCATION AND TRAINING

Architects must hold a bachelor's degree in architecture, spend at least three years as an intern, and pass a licensing examination.

AVERAGE STARTING SALARY

Intern architects earn an average of $24,500 per year.

What does an architect do?

Architects design buildings and other structures, such as schools, hospitals, industrial parks, and entire communities. The structures must be functional, safe, and economical. They must suit the needs of the people who use them.

Does most of the work involve data, people, or things?

Architects work mostly with people and drafting tools or computers. They often interact with clients, engineers, urban planners, interior designers, landscape architects, and other people involved in construction. Most architects use computer-aided design and drafting (CADD) technology.

What is a typical day like?

A new project starts with defining needs with the client. At the drawing board, which is usually a computer, architects create their designs. The design must follow building codes, fire regulations, and other requirements. Their drawings include such things as the heating and cooling systems, electrical systems, and plumbing systems. They may also assist the client in getting construction bids and selecting a contractor.

What are the working conditions?

Architects generally work in a comfortable office advising clients, developing reports and drawings, and working with other architects and engineers. Visits to the construction site may require a hard hat. They may occasionally be under great stress, working nights and weekends to meet deadlines.

OUTLOOK

Employment of architects is expected to increase about as fast as average through the year 2005, according to the Bureau of Labor Statistics. Even with a slowdown in new construction, existing buildings will need to be repaired and renovated, particularly in urban areas.

Which skills and aptitudes are needed for this career?

Artistic and drawing ability is helpful, but not always essential. More important are a good visual imagination and an understanding of the relationships among objects in three-dimensional space. Good communication skills, creativity, computer skills, and knowledge of CADD are also very important.

What careers are related to this one?

Others who engage in similar design and construction work include landscape architects, building contractors, civil engineers, urban planners, interior designers, and graphic designers.

One architect talks about her job.

When I was a kid I enjoyed designing and building things. Now I just make the design and somebody else gets to build the structure.

SECTION 4: Construction Technology

Chapter 14 REVIEW

Reviewing Main Ideas

- The four main types of construction include residential, industrial, public works, and commercial construction.
- The inputs include people, information, materials, tools and machines, energy, capital, and time needed to design, plan, and complete the structure.
- Processes include activities ranging from choosing and preparing a site to finishing the walls with paint and the grounds with walkways and plants.
- The output includes the finished structure, plus any impacts that structure has on society and the environment.
- Feedback provides the construction system with information about any problems with the structure itself and about any negative impacts on individuals, the community, or the environment.

Understanding Concepts

1. Define the four types of construction and give examples of each.
2. Provide job descriptions for at least three different types of workers directly involved in the building process.
3. What are structural materials and what qualities do they need to have?
4. Name the six structural materials commonly used in construction and list any advantages or disadvantages they have.
5. Briefly discuss how important scheduling is to construction systems.

Thinking Critically

1. Why might some people object to certain construction projects?
2. Choose a building or other structure in your community or nearby that interests you. What features or qualities of this structure interest you and why?
3. Describe an ecosystem in the area of the country where you live. Has construction or the harvesting of construction materials had any effects on this ecosystem? If so, what is being done or can be done to reverse these effects?
4. How do building codes benefit consumers?
5. If you could choose a new construction project for your community, what would it be? Explain your choice.

Applying Concepts and Solving Problems

1. **Social Studies.** Look up Louis Sullivan in an encyclopedia or other resource. What did he develop and why?
2. **Science.** Using a plastic container, mix plaster of Paris with water according to label directions. Observe that it first forms a *colloidal gel*. Heat is given off by the reaction between gypsum and water. Next, crystals grow and interlock, forming a hard mass, just as they do in concrete.
3. **Technology.** Design and construct a scale model of a residential structure using nontraditional construction materials, such as straw bales or recycled plastics.

21. **Resource.** Answers to "Understanding Concepts" and "Thinking Critically" can be found in the Teacher's Manual at the front of this *Teacher's Annotated Edition* and in the *Teacher's Resource Binder*.

22. **Critical Thinking.** Inform students that, in western states, straw bales covered with stucco really are used for constructing houses.

CHAPTER 14: Construction Systems

CHAPTER 15

Planning Construction

Objectives

After studying this chapter, you should be able to:
- explain how construction projects are started.
- discuss regulations that apply to construction.
- identify factors to consider when selecting a site.
- list the steps in the design process.
- describe architectural working drawings.

Terms

city planners
detail drawings
elevations
floor plan
power of eminent domain
private sector
public sector
scale drawings
site plan
survey
topography
utilities
zoning laws

Technology Focus

The New Old Ball Parks

In the late 1960s, sports stadiums were multi-purpose. The huge, crater-shaped arenas were designed to host almost any event, from baseball and football to rodeos. Inside, the grass was artificial. Outside, a vast prairie of scorching blacktop set the places apart from everything around them.

Then in 1992 Baltimore opened Oriole Park at Camden Yards. It had the look of ball parks built before the 60s, like Wrigley Field in Chicago and Fenway Park in Boston, which fans loved for their charm and design quirks.

Squeezing into the City

The quirks in beloved ball parks were usually the result of site problems. Fenway Park has a scrunched left field because of space limitations. Its towering wall, called the Green Monster, was erected because shopkeepers on the other side didn't want baseballs crashing through their windows. The ball park's individuality comes from its relationship to the neighborhood around it.

The same thing is true of Wrigley Field. If you've ever noticed fans perched on rooftops during a televised Cubs game, you've seen another example of what happens when a ball park must fit into the tight spaces of a city.

New Design Solutions

In the new ball parks, structures are still big, but that 60s ugliness is gone. Oriole Park, for example, was built to offer good views of the city beyond its walls. Designers helped connect the park to its surroundings by using brick construction, which is featured in the nearby Camden yards development.

The acclaim for Oriole Park encouraged construction of new sports facilities with an old-time feeling. The same designers, HOK Sports Facilities Group, used similar ideas to produce a different look at Cleveland's Jacobs Field. Exposed steel lattices mirror the open-work design of the many steel bridges that define Cleveland's heritage and character.

Smaller Is Better

With seating for 42,000, Jacobs Field looks cozy compared to its predecessor. The old stadium, known as the "Mistake by the Lake," could seat more than 70,000 fans. Jacobs Field is more inviting.

The smaller overall look of these parks makes the heroics of the game seem larger. A home run looks all the more exciting when you see it fly over a roof, and more people want to be part of the crowd at the old ball game.

Take Action!

Sketch an idea for redesigning your school's sports facility. Highlight features that relate the facility to its surroundings.

Resources
- Chapter 15 Lesson Plan in the *Teacher's Annotated Edition* and in the *Teacher's Resource Binder*.
- Chapter 15 Review in the *Teacher's Resource Binder*.

Initiating Construction

Most buildings and other structures are built for ordinary people. These people make up the **private sector** (part) of our economy. As you learned in Chapter 14, there are four types of construction—residential, industrial, commercial, and public works. The private sector is directly responsible for three of those types. For example, a family may need a home (residential). A person in business may need a store or a warehouse (commercial). A company may need a factory or other facility (industrial). Private funds are used to pay for the design and construction of these projects.

The public sector of our economy is responsible for public works construction. This **public sector** includes municipal (city), county, state, and federal governments. People are appointed to or hired by the government to serve on boards or in agencies, bureaus, departments, or commissions. These people are responsible for initiating (beginning) construction projects such as highways, post offices, and fire stations. Tax money is used to pay design and construction costs. Fig. 15-1.

Constructing new buildings is very expensive. For most families, buying a home is the most expensive purchase they will ever make. Most nonresidential construction projects are even more expensive. Careful construction planning can make the difference between a business making a profit or going bankrupt. Poorly conceived public works projects can waste millions of dollars collected from taxpayers.

Community Planning

Like the building of a house or other structure, the building of a community must be carefully planned so that it can be designed to best meet the community's needs. This planning is usually done by

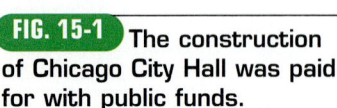 **The construction of Chicago City Hall was paid for with public funds.**

1. **Extension.** Identify boards, committees, etc. that are responsible for public works construction in your community.
2. **Reinforcement.** Have students identify local structures that were paid for with tax money.
3. **Critical Thinking.** Ask students to speculate about what effects large-scale public works construction projects might have on elections.
4. **Background.** Community planning is increasingly becoming known as urban planning.

city planners. These people have studied all aspects of community development. Many larger communities have planners on their permanent staffs. Smaller communities usually hire planners on a temporary basis. These consultants work closely with city, county, and state officials and various governmental agencies. In the course of their work, they study:

- the size and character of the population
- the economy of the community
- the nature and quantity of natural resources such as oil, gas, water, timber, and farmland
- transportation facilities
- educational facilities
- the history and culture of the area

After learning all they can about a community, planners identify areas of potential growth. They also identify potential problems that might limit future growth and development and work to find solutions to these. Finally, planners make recommendations for future community development. Fig. 15-2.

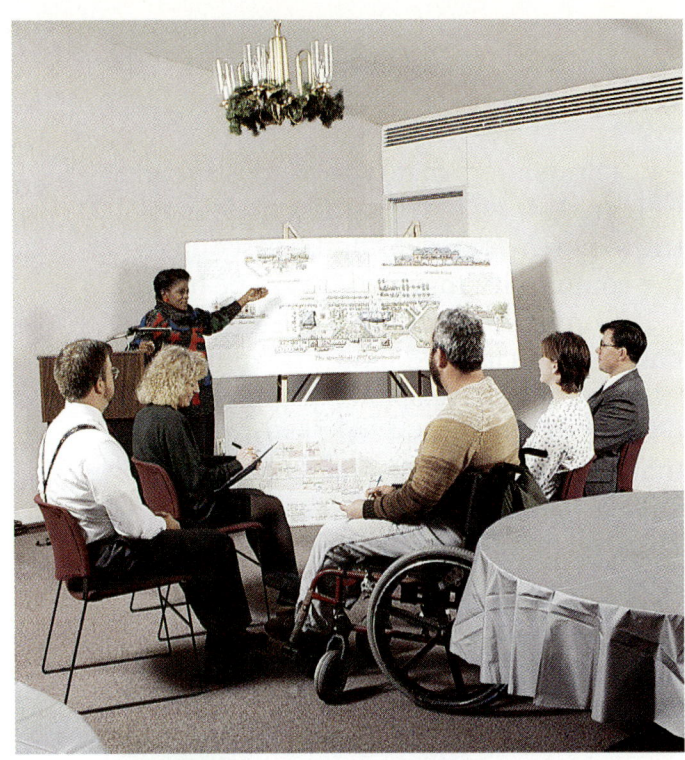

FIG. 15-2 A planner presents development ideas to the planning board. Interested citizens may attend these sessions and make their wishes and opinions known. All of us must work together to build good communities.

What will be best for the community? Citizen representatives and elected officials sit on planning commissions and boards. These people study all recommendations or plans carefully. Before deciding whether to initiate any construction, they consider the potential impacts on:

- the lives and property of the people living in the area
- the local economy
- the level of employment
- property values
- taxes
- the health, safety, and general welfare of the people in the area

Fascinating Facts

Designed by Joseph Strauss and finished in 1937, San Francisco's Golden Gate Bridge is crossed by 40,715,000 cars each year. People pay about $55,000,000 in tolls each year to cross it.

5. **Enrichment.** Have a city or area planner speak to the class about his or her responsibilities and about possible plans for the future of your community.

6. **Extension.** What commissions and boards are in your community? Discuss their responsibilities.

Controlling Construction

No matter what type of construction is desired, or who desires it, it must meet the requirements and standards set up by the community in which it is to be built.

Communities are divided into residential, commercial, and industrial zones. Communities have boards of appointed officials that set up special **zoning laws** that tell what kinds of structures can be built in specific parts of the community. These laws may also specify such things as:

- maximum property size
- maximum height of a building
- the number of families that can occupy a house
- the number of parking spaces a commercial building must provide
- the distance structures must be from the property's boundary lines

These laws are designed to protect homeowners from traffic, noise, and other environmental problems. Zoning regulations also attempt to limit noise and traffic congestion near hospitals.

In addition to zoning requirements, all structures must meet certain building codes. As you learned in Chapter 14, building codes are local and state laws that specify the methods and materials that can or must be used for every aspect of construction. Fig. 15-3. To make sure each structure is constructed according to these building codes, the structure must be inspected throughout the construction process and when construction is completed. You will learn more about building inspection in Chapter 16.

Site Selection and Acquisition

Two basic decisions must be made before construction can begin. One is choosing the best site. The other is choosing the best design. These two decisions are not usually made independently. The design may be influenced by the nature of the site. The site choice must be suitable for the design. Once the best site has been chosen, it must be acquired.

FIG. 15-3 Building codes establish the rules by which a building is made.

---

7. **Enrichment.** Obtain a planning map from your area planning commission. Discuss it with students.
8. **Extension.** Explain that each zone is divided into various subzones. For example, there are size or type limitations for businesses that can develop in certain commercial subzones. If possible, obtain a local zoning map to show students how their area is zoned.
9. **Enrichment.** Have a government official speak to the class about zoning laws in your area.

SECTION 4: Construction Technology

Selecting a Site

In addition to being suitable for the basic design of the structure, the site must also meet a number of other important criteria. Although specific criteria will vary from structure to structure, the same basic factors must be considered:

- *Location.* Is the site in the city or the country? Is it near roads or highways? For stores and restaurants, is it an area where there will be a lot of potential customers?
- *Size.* Is the site large enough? If not, can an adjoining site be acquired?
- *Shape.* Is the site long and narrow, short and wide, pie-shaped, or L-shaped? Will the planned structure fit well on the site?
- *Topography.* **Topography** refers to the site's surface features. Is there a lake or stream? Are there hills, gullies, large rocks, or trees? What is the nature of the soil? Is it dry sand or wet clay? Fig. 15-4.
- *Utilities.* **Utilities** are services such as electricity, natural gas, and telephone. Are these available at the site? How much will it cost to have them installed?

FIG. 15-4 Falling Water, a home designed by the famous architect Frank Lloyd Wright, makes use of the natural topography of its site.

10. **Reinforcement.** Explain that the Leaning Tower of Pisa leans because it was built on soft soil that began to settle even before the tower was finished.
11. **Critical Thinking.** Ask students if they can think of any ways natural features can be used to complement the design of a structure.
12. **Enrichment.** Explain that sites may have utility easements that can affect where buildings can be located.

CHAPTER 15: Planning Construction

FIG. 15-5 Are utilities included with the lots that are for sale for this new subdivision?

Also, if the property is outside the city limits, what kind of water and waste disposal systems—if any—are available? Fig. 15-5.
- *Zoning.* Will zoning laws permit the type of planned structure to be built there?
- *Cost.* Is the price of the site reasonable and affordable?

The Survey

Sometimes during the site selection process, prospective buyers may consult a survey or may have a survey made of a piece of land being strongly considered. A **survey** is a drawing that shows the exact size and shape of the piece of property, the position of the property in relation to other properties and to roads and streets, the levels or elevations (heights) of the property, and any special land features (rivers, streams, hills, gullies, trees, etc.). The survey also includes a written description of the property.

Surveyors have traditionally used specialized telescopic instruments to measure elevations and distances. Fig. 15-6. To determine elevations, the surveyor looks through the instrument while a helper holds a level stick at a specific spot. The surveyor can read the numbers on the level stick and record the exact elevation of that spot.

Laser equipment is now frequently used to determine elevations on construction sites. Instead of the small telescope, a laser is precisely located. A worker can then observe where the laser beam strikes the level rod to determine the elevation at that spot. Fig. 15-7.

Surveyors can also use the laser instrument to measure distances by aiming it at a sophisticated mirror device called a prism. The laser's light will bounce right back to an instrument called an *electronic distance meter* or EDM. The EDM translates the time difference into a precise distance, whether the prism is 100 feet away or one mile away. Fig. 15-8.

Once all measurements have been made and recorded, a stake is driven into the ground at each corner of the property to mark the boundaries. These stakes will later be used as points of reference when workers must identify and mark the exact location of the structure on the property. (You will learn more about this process, called *laying out the site*, in Chapter 17.)

13. **Background.** Make sure students understand that consulting a survey or having one made is not necessarily part of every site selection process. Also, surveys may be initiated for various purposes and by various people—sellers, buyers, developers, architects, etc.
14. **Enrichment.** If possible, obtain a copy of a survey for students to examine.

SECTION 4: Construction Technology

FIG. 15-6 A *level* is a small telescope that only rotates horizontally (side to side); it can be used to find exact elevations. A *transit* can be rotated vertically (up and down) as well as horizontally; it can be used for determining elevations, angles, and distances. A *theodolite* is an extremely accurate transit used for precise surveying and larger distances. *Level sticks* are marked with measurements along their length.

FIG. 15-7 Laser light shining onto the level rod allows this worker to determine heights accurately and rapidly.

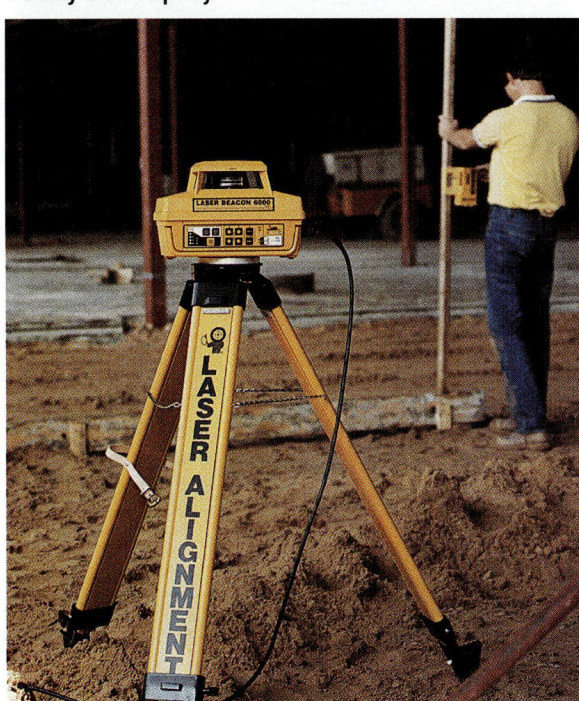

FIG. 15-8 This surveyor will use this EDM to precisely locate property boundaries.

15. **Enrichment.** Demonstrate how a transit is set up. Show how it is used to lay out angles, line up stakes, measure elevations, and plumb corners.
16. **Enrichment.** Ask a local surveyor to bring in survey equipment and demonstrate it to students.
17. **Background.** EDM is rapidly replacing older optical equipment because of time savings and increased accuracy.

CHAPTER 15: Planning Construction

SCIENCE CONNECTION
How a Laser Works

Lasers have found a place, not only in construction, but also in many other technologies. Do you know how they work? As you learned in your science classes, all things are made of atoms. At the center of an atom is its nucleus. Electrons whirl around the nucleus in different orbits. Sometimes an electron jumps from one orbit to another. When this happens, the electron loses energy in the form of light.

When light energy travels, it does so as waves radiating in all directions. As the light waves move through the air, they bump into one another, which scatters them. This is why light gets dimmer the farther it moves from its source.

Light waves can be prevented from scattering. They can also be made to travel in only one direction. When controlled in this way, they are very powerful and able to span long distances without growing dimmer. The result is laser light. Laser light waves march in harmony with one another in the form of a narrow beam.

The Ruby Laser

Laser light can be produced from a crystal, a semiconductor, gas, chemicals, or dyes. A ruby laser consists of a crystal inside a tube similar to a fluorescent lightbulb. A

photographic flash is beamed inside the tube. As the light strikes the crystal, it excites the crystal atoms. Electrons in the crystal atoms leap to other orbits and give off additional light. The ends of the crystal are like mirrors, which reflect this light back and forth until the light waves are all moving "in step." Finally, a shutter is opened and the light waves leave the tube as a powerful laser beam.

Try It!

Do some research to learn how a gas or semiconductor laser works. Report your findings to the class.

SECTION 4: Construction Technology

Acquiring the Site

Once the site has been selected, the land or property must be acquired.

The simplest way to acquire land is to make a direct purchase, meaning to buy it from the owner. Doing this may not be as easy as it seems. For example, the land may be prime (the best) farmland; owners may consider it too fertile (rich in nutrients) to be used for buildings or other structures. The land may be located in or near a major city; land there is scarce, and is usually very expensive. Quite often, the owner of a piece of land may simply not want to sell. He or she may ask for more money than the buyer is able or willing to pay.

The right to own property is precious. However, sometimes land may be needed for public purposes. For example, federal, state, or local government officials may decide to make certain public improvements. Perhaps a new road, school, dam, or public housing is needed. A landowner may refuse to sell. If there are no acceptable alternatives, the government can take legal steps to force the owner to sell by exercising its power of eminent domain. The **power of eminent domain** is a law that states the government has the right to buy private property for public use. (The rights or needs of all come before the rights or needs of one.)

The legal process for taking over land that an owner has refused to sell is called *condemnation*. The government must prove that the property is needed. If it does, then the owner has to sell. However, he or she must be paid a fair price for the land. Individual rights must be protected.

The Design Process

In this section, we will look briefly at the design process used for residential and commercial structures. Similar steps are followed for all types of construction.

Identifying Specific Needs

Suppose a family needs a new home. They may decide to have one *custom-built*. This means the house will be planned and built especially for them. To begin the construction process, they hire an architect to design the structure and develop the plans for building it. The family chooses someone who is experienced in designing custom-built houses.

A meeting is held between the architect and the family. At the meeting, the architect asks questions to identify specific needs or problems:

- How large is the family? What is the age and sex of each member? This may determine how many bedrooms and bathrooms the house should contain.
- Does either spouse work at home? If so, what kind of work does he or she do? This may determine whether there is a need for an office or den. Perhaps there should be a separate entrance, sound-proofing, or special wiring for telephones, computers, or other equipment.
- What are the family's special interests or hobbies? For example, if a family member enjoys woodworking, there may be a need for a workshop.
- Does the family entertain a lot? There may be a need for a large family room, a formal dining room, or a guest room.

---

18. **Review.** Review the problem-solving process discussed in Chapter 2, and relate its similarities to the design process.
19. **Background.** Architects refer to this as "programming."
20. **Extension.** Have each student write a description of his or her ideal house.
21. **Background.** Architects refer to this as "schematic design."

- How much money can the family afford to spend? Money is the single most important factor. The amount available influences nearly all other decisions.
- Where is the family going to live? Suppose they plan to live in town. Land there may be scarce and lot sizes rather small. The architect may recommend a two- or three-story house. If, on the other hand, the family plans to live in the country, a sprawling, one-story house may be preferred. The family may have already looked at some home designs and have an idea of what they want. The architect might show the clients plans of other homes that he or she has designed. Fig 15-9.

A similar procedure is followed in other types of construction. For example, suppose a company needed a new office building. Then company representatives would meet with the architect. Questions would be asked to identify company needs. For example:

- How many people will be working in the new building
- What types of work will be done?
- Is the company planning to expand in the near future?

Developing Preliminary Ideas

After the architect has learned about needs, he or she begins jotting down preliminary (beginning) ideas and plans. Some architects make rough paper sketches of several possible designs. Others use specialized architectural CAD software. Fig. 15-10. Some of these early ideas may be discarded almost immediately. Others may be saved and reviewed later. As more facts are gathered, certain of these ideas may begin to show greater promise. Frequently, ideas are combined from various sketches. The most promising ideas and sketches move into the next stage of design.

FIG. 15-9 Identifying the needs of the client is the beginning of the design process. This couple is looking over plans of other homes to see some features they might like to include in their home's design.

22. **Extension.** Have students make sketches of ideas for their houses to go along with their written descriptions of their ideal home.

FIG. 15-10 After determining client needs, the architect begins making preliminary plans. These plans may be developed by making several rough sketches on paper or by using special architectural CAD software.

Refining Ideas and Analyzing the Plan

[23] Architects review and revise preliminary design ideas. They consider:
- the special needs and desires of the client
- the money available
- the site on which the structure will be built

As they refine their ideas, they may prepare renderings or models. This depends on the type of project and the people making the decisions.

For individual residential projects, the architect will usually again meet with the clients to show them the preliminary ideas and together agree on which ideas seem to best meet the buyer's needs.

In commercial construction, one or more engineers may be asked to analyze the preliminary plans. As you learned in Chapter 14, engineers figure out how a structure will be built and what structural materials will be used. Sometimes, ideas must be eliminated because they simply cannot be built strongly enough with currently existing structural materials. Basically, engineers deal with matters related to the function and strength of structures. Questions they need to consider might include:

- What is the soil type; how much weight can it carry? Will special supports be needed?
- What are climate conditions? Are high winds or heavy snows common? Fig. 15-11. Are temperatures extremely hot, or do they vary greatly? [24]
- Will the structure serve its planned purpose as it is currently designed? For example, will a sports stadium efficiently accommodate large crowds?
- Could different materials or construction techniques be used to reduce costs and yet maintain quality?

Answers to these and other engineering questions help the architect refine the preliminary plans to make sure that they will work with available modern building materials.

23. **Background.** Architects refer to this as "design development."
24. **Reinforcement.** Make sure students understand the importance of climate when choosing structural materials and design. Direct students to the illustration of the snow-covered house and discuss the snow's effects on various roof styles. Have a class discussion about other climatic factors that could affect design and/or materials.

CHAPTER 15: Planning Construction

FIG. 15-11 Snow is very heavy. Houses in cold climates usually have peaked roofs so snow will slide off and not damage the structure.

FIG. 15-12 Renderings are refined drawings that help clients visualize the appearance of the planned structure. This is a 3D CAD rendering.

Soon a more nearly finished (though still preliminary) design is developed. Most architects complete this stage on computers with CAD software. Before CAD systems existed, making changes to designs required architects to completely redraw paper plans. This was time-consuming and therefore expensive. When plans are made, stored, and printed with computers, architects can make small or large changes to these preliminary designs relatively easily.

Renderings of the refined ideas help clients visualize what the structure will look like. Fig. 15-12. Sometimes models are made. However, this is usually done only when a presentation must be made before a group. Fig. 15-13.

Virtual reality technology now allows architects to further refine the design process. Powerful modern computers and software provide clients with the ability to "walk thorough" 3D cyberspace versions of planned buildings. Fig 15-14. The software can even simulate the location of furniture and the light provided by the sun at various times of the day and year. This allows architects and their clients to make changes that might better meet the clients' needs before the actual construction process begins. Changing a design at this stage is much less expensive than making changes after construction has begun.

Preparing Final Plans and Specifications

The architect's refined preliminary plans are reviewed by the client and the engineers. When plans are approved, final drawings and specifications are prepared.

25. **Background.** Architects refer to these final plans and specifications as "construction documents" or just "CDs."

FIG. 15-13 Models are useful in helping people understand complicated or unusual construction projects.

Working Drawings

The working drawings contain the information needed to construct a project. They are drawn to scale. In **scale drawings**, a small measurement is used to represent a large measurement. For example, one-fourth inch (¼") may represent one foot (1'). Measurements given on the drawings are for the actual size, however.

There are four types of architectural working drawings:

- The **site plan** shows where the structure will be located on the lot. Boundaries, roads, and utilities are included. Fig. 15-15.

FIG. 15-14 3D virtual reality software allows architects and their clients to visually experience the "feel" of a proposed structure. The client can take a simulated walk through a building before construction ever begins.

FIG. 15-15 The site plan. Notice that the scale used for the drawing is shown at the bottom.

- **26** • The **floor plan** shows the locations of rooms, walls, windows, doors, stairs, and other features. Fig. 15-16.
- **Elevations** are drawings that show the finished appearance of the outside of the structure, as viewed from the ground level. A separate elevation is made for each side of the structure. Fig. 15-17.
- **27** • **Detail drawings** are special drawings of any features that cannot be shown clearly on floor plans or elevations, or that require more information to be constructed. Fig. 15-18.

FIG. 15-16 Notice this floor plan is labeled "First Floor." A separate floor plan is made for each story of a building.

FIRST FLOOR PLAN
SCALE ¼" = 1'-0"

26. **Extension.** Identify symbols and other features on this drawing to help students learn to read drawings.

27. **Enrichment.** Mention that section drawings are frequently used to help show spatial relationships between rooms and spaces within buildings.

CHAPTER 15: Planning Construction

FIG. 15-17 This elevation shows the front view of a church. Other elevations must be drawn to show the sides and the rear of the church.

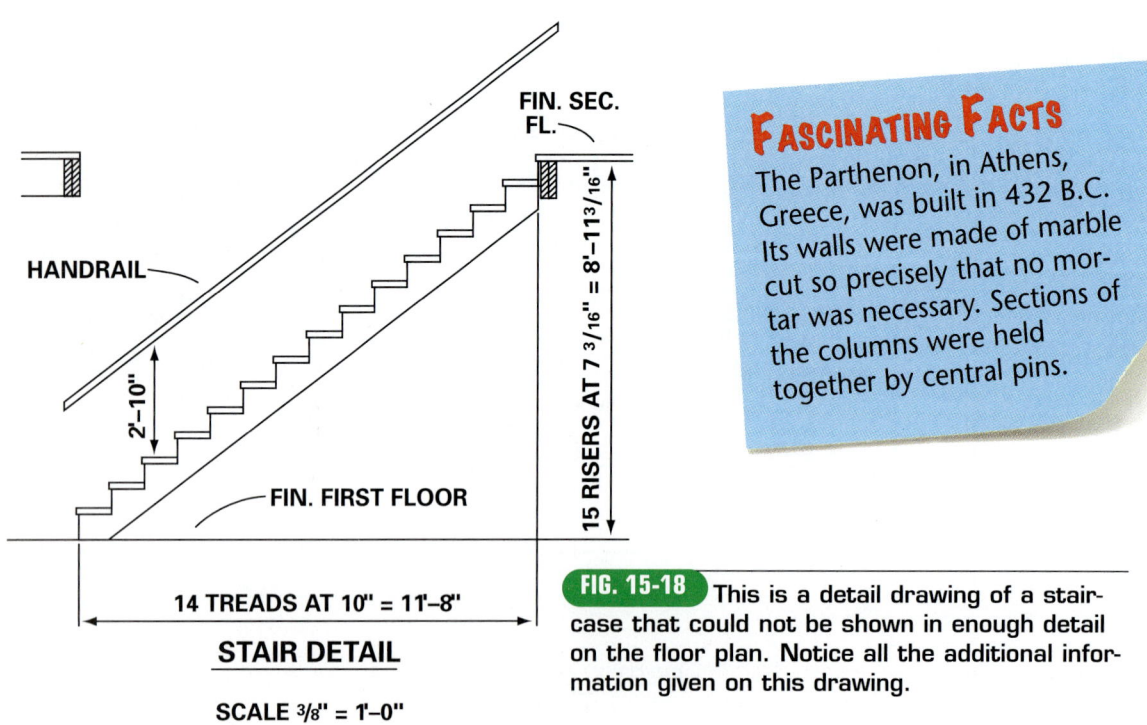

STAIR DETAIL
SCALE 3/8" = 1'-0"

Fascinating Facts

The Parthenon, in Athens, Greece, was built in 432 B.C. Its walls were made of marble cut so precisely that no mortar was necessary. Sections of the columns were held together by central pins.

FIG. 15-18 This is a detail drawing of a staircase that could not be shown in enough detail on the floor plan. Notice all the additional information given on this drawing.

344 SECTION 4: Construction Technology

Specifications

In addition to all the working drawings, a set of specifications must be prepared. In architecture, *specifications* are written details about what materials are to be used for a project, as well as the standards and government regulations that must be followed. They describe or list the size, number, type, and (if appropriate) model number and color of every item to be included in the finished building. Construction details and materials that could not be shown on the drawings are given in the specifications. Fig. 15-19.

Specifications are extremely important. They provide owners with an accurate description of the materials and services they are buying. Contracting firms use specifications when calculating costs and when building the structure. The specifications serve as a guide or set of instructions.

Division 4: Carpentry and Millwork

Sec. 1. Scope. This division includes the furnishing and installation of all carpentry, millwork, as indicated on the drawings and/or hereinafter specified.

Sec. 2. Materials

a. Rough lumber shall be Framing Lumber—No. 3 Grade Southern Pine.
b. Exterior millwork. Horizontal siding shall be 6" bevel redwood.
c. Sheathing shall be $5/8$" aluminum-foil-backed foam as made by Dow Chemical.
d. Attic insulation shall be 12" fiberglass batts installed in accordance with manufacturer's instructions.
e. Interior trim shall be of pine. Pattern selected by owner.
f. Interior doors shall be as indicted on door schedule.
g. Exterior door frames shall be $1 3/4$" thick, rabbeted with $1 1/8$" outside casings with white pine.
h. Interior door frames shall be $7/8$" thick pine.
i. Double-hung sash shall be Andersen or equal.
j. Kitchen cabinets shall be birch. Countertops Formica or equal.
k. Closets and wardrobes shall have $7/8$" by 12" shelving and one clothes pole running length of space.
l. Living room, hall, bedroom floors to be straight-line oak; select grade, finished natural; kitchen, vinyl composition tile over plywood underlayment; bathroom, ceramic tile set in rubber in squares, cemented to plywood subfloor; entrance hall, imitation slate.

FIG. 15-19 Sample specifications for work to be done on a project. Each part of the project is specified in detail.

---

28. **Enrichment.** Obtain a set of specifications. Develop a set of questions about the specified project and have students find the answers within the specifications.
29. **Critical Thinking.** Ask what could happen if specifications were not exact.

Career File—Cost Estimator

EDUCATION AND TRAINING

All estimators receive a lot of on-the-job training. Many start as construction workers. Employers prefer to hire estimators with considerable construction work experience, especially those with additional training in estimating or a college degree in civil engineering or building construction.

AVERAGE STARTING SALARY

Income varies widely by experience and education. A cost estimator with limited training can expect to earn from $16,000 to $24,000 a year. With a bachelor's degree, the salary can approach $30,000.

OUTLOOK

Employment is expected to grow at an average rate through the year 2005, according to the Bureau of Labor Statistics. Even if there is an overall decline in the construction industry, there will always be a need for cost estimators to accurately predict costs in all areas of the business.

What does a cost estimator do?

Cost estimators develop dollar estimates for proposed construction projects. They review sketches or blueprints, determine the amount of time, labor, materials, and machines required, and prepare total costs and recommendations. The estimates are used to submit a bid for the project.

Does most of the work involve data, people, or things?

Most tasks involve collecting and analyzing data. A computer is commonly used to perform the mathematical analyses that are often required.

What is a typical day like?

The activities of any particular day depend on the phase of the estimating process. To gather project requirements, the estimator may consult with engineers. To determine unit costs, the estimator may check price books and catalogs or make telephone calls to suppliers. To prepare total costs, the estimator uses his or her analytical skills and usually a computer.

What are the working conditions?

Most of the time a cost estimator works in an office. However, frequent visits are made to construction sites that are dirty and cluttered with debris.

Which skills and aptitudes are needed for this career?

Cost estimators like to work with numbers, and they should be familiar with computers. They should also have a thorough knowledge of the construction industry—materials, equipment, tools, and procedures.

What careers are related to this one?

Other workers who analyze mathematical data include appraisers, economists, financial analysts, operations research analysts, and cost accountants.

One cost estimator talks about her job:

Estimating is interesting because no two jobs are the same. It's always a learning process.

Chapter 15 REVIEW

Reviewing Main Ideas

- Construction may be initiated by the private sector or the public sector.
- A city planner may study a community and analyze its needs, then make recommendations to planning commissions.
- Zoning laws and building codes are drawn up to make sure that any structures built will be in the best interest of the community.
- The site is acquired through direct purchase or the power of eminent domain.
- Architects and engineers design and plan structures. Working drawings and specifications are prepared.

Understanding Concepts

1. What factors do community planners study in the course of their work?
2. What three zones are communities divided into?
3. Name at least five factors that should be considered when selecting a home's site.
4. What is shown on a site plan? What is shown on a floor plan?
5. What are specifications? Why are they so important to construction planning?

Thinking Critically

1. Suppose you are on the planning commission for your community. What would you recommend if a new shopping mall was proposed to be built near your school? Explain your reasoning.

2. Suppose you are on the planning commission for your community. What would you recommend if a clothing manufacturing company wanted to build a new facility that would employ 5,000 people? Explain your reasoning.
3. In what kinds of situations do you feel it is justifiable to exercise the power of eminent domain?
4. Discuss why you think laser survey tools are replacing levels, transits, and theodolites.
5. Suppose you are an architect. Make a list of questions that would need to be answered before you could begin to design a new high school.

Applying Concepts and Solving Problems

1. **Science.** Compare the load a flat roof can support to the load a peaked roof can support. Construct roof shapes from scrap wood. Hang a bucket from the center of each. Pour sand into the bucket until the structure falls.

2. **Technology.** Choose a room in your house that you'd like to remodel. Make a detailed sketch of the planned room, and write a brief description of the changes that would be made.

3. **Internet.** Do a search on the Internet to identify construction companies that use recycled materials in the construction of homes or businesses.

30. **Resource.** Answers to "Understanding Concepts" and "Thinking Critically" can be found in the Teacher's Manual at the front of this *Teacher's Annotated Edition* and in the *Teacher's Resource Binder*.

CHAPTER 16

Managing Construction

Objectives

After studying this chapter, you should be able to:
- describe how a contractor is selected.
- explain the responsibilities of a contractor.
- describe the three basic types of schedules.
- describe how construction is monitored.
- explain the importance of quality control.
- identify what inspectors look for and discuss the importance of inspections.

Terms

bid
contractor
monitor
Occupational Safety and Health Administration (OSHA)
purchasing agent
schedule
subcontractors

Technology Focus

Breaking Tradition: Women in Construction

Traditionally, construction has been a man's world, but women are breaking down barriers to employment in the construction industry. Women make up only a small percentage of the construction workforce, but their numbers are increasing. The National Association of Women in Construction (NAWIC) is preparing to assist women in gaining employment in this male-dominated field.

NAWIC Offers Opportunities for Women

NAWIC sponsors pre-apprenticeship programs, providing students with instruction in the skills necessary for working in the construction industry. These women study blueprint reading, math skills, and safety. They also receive hands-on training in four building trades. The NAWIC Education Foundation offers other courses, including a home-study course in construction management, which can lead to becoming a Certified Construction Associate.

Toward a Quality Workforce

Most women don't consider the construction industry as a career choice because they are not aware of its many opportunities. NAWIC has begun working with elementary schools by sponsoring "Block-Kids," a national block-building competition. Winners receive prizes of U.S. savings bonds ranging from $100 to $2,500 to use as a down payment toward their education.

The organization also sponsors Career Days to acquaint junior high and high school students with the career opportunities in construction. NAWIC's 6,000 members believe that if they can reach one out of every ten graduating high school students, then they can secure the female workers needed to maintain a quality workforce.

NAWIC also contacts women already employed in other areas of the workforce. The construction industry has traditionally provided higher salaries than other industries. NAWIC believes that many clerical workers would thrive on the challenges and rewards of the construction industry.

Take Action!
Visit NAWIC online at: http://www.nawic.org and discover upcoming events, membership information, and employment opportunities.

Resources
- Chapter 16 Lesson Plan in the *Teacher's Annotated Edition* and in the *Teacher's Resource Binder.*
- Chapter 16 Review in the *Teacher's Resource Binder.*

Who Manages Construction Projects?

Management procedures may vary according to the size and type of construction project. Large projects are usually overseen by an architect or engineer. Fig. 16-1. Smaller projects may use existing designs that have already been made and checked by architects and engineers. These projects are then overseen by a **contractor** who owns and operates a construction company. Contractors are responsible for the actual building of projects. They must follow the plans and specifications developed by the architect(s) or engineer(s). Fig. 16-2. Whoever manages the construction—the architect, the engineer, or the contractor—is responsible for seeing that the owner's wishes are carried out.

The contractor is chosen after planning is complete. The person or group planning the project announces or advertises that the job is available. Qualified contractors submit bids for the contract. A contract gives the chosen contractor the right to build the project. A **bid** is a price quote for how much a contractor will charge. The amount equals the total of the contractor's best estimate of what it will cost to build the project according to the owner's plans and specifications, plus the amount of profit the contractor hopes to make. Generally, the contractor who submits the lowest bid is awarded the contract. However, a contractor with a history of poor-quality construction may not be awarded the contract even if he or she submits the lowest bid. The contractor must control costs in order to stay at or below the bid price. At the same time, the contractor must use quality materials and make sure that high-quality work is done in order to develop (and keep) a good reputation. A good reputation will help obtain future projects.

Contractors may also hire subcontractors. **Subcontractors** specialize in certain types of construction work. Fig. 16-3. For example, the electrical systems in a building are usually installed by subcontractors.

FIG. 16-1 An architect or an engineer usually oversees a large construction project.

2. **Extension.** Bring to class a few ads from your local newspaper that are "invitations to bid."
3. **Extension.** Point out that possible increases in the costs of materials and labor must be factored into the bid.
4. **Critical Thinking.** Ask students to discuss why a contract should or should not be awarded to contractors with a history of poor quality.

1. **Enrichment.** Invite a local contractor to speak to the class about a contractor's work and about construction from a contractor's point of view.

FIG. 16-2 Contractors must be sure that the plans are carried out down to the finest details.

FIG. 16-3 Plumbing is usually installed by a subcontractor.

Scheduling Construction

6 All the workers, materials, and equipment needed to complete a construction project must come together at the right times. To accomplish this, the contractor must prepare schedules. A **schedule** is a plan of action that lists what must be done, in what order it must be done, when it must be done, and, often, who must do it. Fig. 16-4. Of course, schedules must have some degree of flexibility and allow time for unexpected events like bad weather or design changes.

The three main schedules that govern (control) a construction project are the work schedule, materials delivery schedule, and financial schedule.

5. **Extension.** Discuss the value of a good work reputation.
6. **Reinforcement.** Discuss the importance of scheduling in general, such as planning a schedule of classes.

CHAPTER 16: Managing Construction

CONSTRUCTION SCHEDULE

	JUNE	JULY	AUG.	SEPT.	OCT.	NOV.	DEC.	JAN.	FEB.	MAR.
EXCAVATION AND GRADING	——	——								
FOUNDATIONS - FORMWORK - WALLS	——	——	——	——						
REBAR	——	——	——							
CONCRETE	——	——	——							
SUSPENDED SLABS			——	—	—					
SLAB ON GRADE			——	—						
PLUMBING UNDERGROUND	——	——								
PLUMBING ABOVE GROUND				——	——	——	——	——	——	——
ELECTRICAL UNDERGROUND	——	——	——	——	——					
ELECTRICAL ABOVE GROUND		——	——	——	——	——	——	——	——	——
MECHANICAL - AIR CONDITIONING				——	——	——				
ROOFING					——	——				
LATH AND PLASTER				——	——	——	——			
MILLWORK - DOORS - WINDOWS							——	——	——	——
PAINTING								——	——	——
PAVING AND LANDSCAPING - PARKING									——	——
HARDWARE							——	——	——	——
FINAL INSPECTION - PICK UP COMP.										——

FIG. 16-4 Bars have been plotted across this schedule to indicate the planned time frame for each aspect of the construction process—from excavating and grading to final inspection and turning over the completed structure.

Work Schedule

Before a useful work schedule can be prepared, the contractor must first analyze the job. That is, he or she must know and understand every phase of the total project. A contractor must know what must be done and where, how it should be done, and when each phase must be finished. Also, the contractor must decide who should do each job. By studying both the drawings and the specifications, a contractor can determine:

- the number of workers needed
- the skills or crafts needed
- the equipment needed
- the time required for each process
- the order or sequence of jobs

Using this information, the contractor can then prepare a written work schedule. Computers are often used to help plan work schedules as well as materials delivery and financial schedules.

Fascinating Facts

The Egyptian pyramids at Giza are more than 4,000 years old. Of the seven wonders of the ancient world, only these pyramids have survived. It is believed that the largest took twenty years to build.

7. **Brainstorming.** Ask students to think of any other reasons construction might be delayed after it has already begun.

8. **Reinforcement.** Stress the importance of being well organized.

SECTION 4: Construction Technology

Materials Delivery Schedule

Suppose a contractor was hired to build a brick house. Then, on the day the bricks were to be laid, only the bricklayers showed up. No one delivered the bricks.

To avoid such problems, the contractor prepares a materials delivery schedule. This is a list of every kind of material needed to complete the building. It includes the quantity, style, color, and price of each material. It also shows where and when the materials must be delivered.

Financial Schedule

A contractor pays wages to workers on a regular basis. Materials and supplies must also be paid for. This means a regular and dependable source of money is needed.

How and when a contractor will be paid is worked out in advance with the owner or people who initiated the project. Generally, contractors are paid certain amounts as the work progresses. The amount is usually a percentage of the total price of the job. However, what if money that was expected does not arrive in time? Suppose no arrangements were made? Then the contractor could be in trouble.

To avoid these problems, the architect or engineer prepares a financial schedule. It lists amounts and dates for payment. The financial schedule is reviewed and approved by the owner and, possibly, the bank.

Monitoring Construction

Careful scheduling alone is not enough to assure a successful operation. In order to make sure that all the terms of the contract are properly met, the project must be carefully monitored by the contractor or those hired to do the monitoring. To **monitor** a construction project means to watch over and inspect it to ensure safety and quality.

Monitoring Materials

A **purchasing agent** (buyer) is responsible for obtaining the right materials at the right price. The materials must be of the proper kind and in the correct quantity. They must be reasonably priced. The purchasing agent works closely with the various materials suppliers.

FIG. 16-5 This purchasing agent uses inventory software to help keep track of materials.

MATHEMATICS CONNECTION
Using Inventory Records

Companies keep inventory records of their materials on hand. The records help the companies keep track of:

- how many of each item they have available
- when more items must be reordered
- how much money has been spent on the items

Taking a physical inventory means counting all the items on hand in each category. Because most companies now use computers to keep track of inventory, a physical count does not have to be taken as often. Each time an item is purchased or used, the information is input and the computer automatically adjusts the inventory record.

The number of items on hand at the beginning of the accounting period added to the number purchased during the period gives the total items available for use. At the end of the period, the number of items remaining is subtracted from the total to find the number used. Multiplying the cost of each item times the total number of items gives the cost of materials in that category.

Try It!

1. How many gallons of adhesive did Rashstone use during January?
2. How much did that adhesive cost altogether?
3. How many gallons did the company have on hand January 20?
4. How much was their inventory of adhesive worth at the end of the month?

Rashstone Construction Company Inventory Record

Item: Waterproof adhesive, one-gallon size   **Shelf:** 8
Supplier: Valley Adhesive Co.   **Maximum:** 25   **Minimum:** 5

DATE	PURCHASED			USED			BALANCE		
	Units	Unit Cost	Total Cost	Units	Unit Cost	Total Cost	Units	Unit Cost	Total Cost
Jan. 1							7	10.00	70.00
Jan. 15				5	10.00	50.00	2	10.00	20.00
Jan. 20	23	10.00	230.00				25	10.00	250.00
Jan. 31				3	10.00	30.00	22	10.00	220.00
Total	23		230.00	8		80.00			

SECTION 4: Construction Technology

FIG. 16-6 The contractor checks to make sure that all work is getting done on schedule. Here, a contractor is checking the progress of the electrical subcontractor.

The purchasing agent also prepares and monitors the materials delivery schedule. Fig. 16-5. He or she checks all materials as they are received. Materials must be:
- those that were ordered
- delivered to the correct site
- delivered on time
- in good condition
- of acceptable quality
- in the quantity ordered

All of these requirements must be met each time materials are purchased, delivered, and paid for.

Monitoring Job Progress

Keeping a construction project on schedule is important. This requires careful supervision and close monitoring of the work schedule. Suppose bad weather interrupts work. The contractor must find ways to make up for time lost. If needed materials are not on hand, the purchasing agent must notify the supplier at once. Careful monitoring of job progress is vital to the success of the project. Fig. 16-6.

Monitoring Quality

Quality is vital in all aspects of construction systems. Contractors monitor the quality of both the materials and the work to make sure that everything is being constructed according to the plans and meets all specifications.

Suppose the quality of the work does not meet the standards described in the original contract. Then the owner may not be required to pay for the work that was done. The unacceptable work would have to be repaired or replaced. Making extensive repairs is very costly and could put the entire project way behind schedule. The contractor may have to bear the cost of the repairs. Contractors are often legally responsible for quality. For example, if a bridge should later collapse, the contractor may be sued.

9. **Extension.** Try to obtain a scheduling program for students to try out on the computer.

CHAPTER 16: Managing Construction

Quality cannot be added or attached to a structure. It must be part of it from the beginning. Everyone involved in the project must be committed to quality. Most contractors encourage all their employees to take pride in their work and help monitor every stage of construction projects as part of quality control. Committed workers who are proud of the structures they build are essential to quality construction. This continuous monitoring by and commitment from workers has allowed some construction companies to become known for their dedication to quality.

Inspecting Construction

All work done must meet the requirements of the contract and the standards set by building codes. To ensure this, the structure is inspected carefully by someone who knows what the correct results should be and what conditions should be met.

Inspections are usually done on a regular basis during the building process. This way problems can be spotted early. Corrections can be made without losing much time and are less costly.

What Do Inspectors Look For?

It is usually not necessary to inspect all parts of a project. Trying to do so would be difficult, expensive, and time-consuming. Inspections are usually limited to the more critical parts of the project. For example, inspectors routinely check structural, electrical, and mechanical elements.

They check the quality and appropriateness of the materials being used. They make sure work is being done properly. In short, inspectors are concerned with the three main aspects of any construction project:

- materials
- methods
- quality

Quality must be present in the design of the structure, in the engineering, in the materials used, and in the way the work was done. That's what inspectors look for. They make certain that each detail of the construction process has met or exceeded (gone beyond) the standards required for quality construction.

Who Inspects?

The types of inspectors depend upon the nature of the project. Inspections may be conducted by local building inspectors or quality-control specialists. Sometimes

Fascinating Facts
The Plains Indians of North America once lived in tepees. Tepees consisted of poles that leaned together and were fastened at the top. The poles were wrapped in buffalo skins and fastened to the ground with rocks or pegs. Tepees were portable and wind-resistant.

10. **Critical Thinking.** Ask students what would happen if standards were not set and enforced.

insurance agents or bank representatives inspect projects. All projects are inspected by representatives of the government. Fig. 16-7.

All levels of government—city, state, and federal—have set safety regulations. The **Occupational Safety and Health Administration (OSHA)** has been established by the federal government. It is part of the U.S. Department of Labor. OSHA sets safety standards for the workplace. Its representatives visit sites to see that those standards are met.

FIG. 16-7 Inspectors make sure that safety standards are met.

Fascinating Facts
The Sears Tower is no longer the tallest building in the world. Petronas Towers, in Malaysia, is currently the tallest at 1,476 feet.

11. **Reinforcement.** Discuss the importance of safety, both in construction and in the school lab.

CHAPTER 16: Managing Construction

Career File—Construction Supervisor

EDUCATION AND TRAINING
A high school diploma is required, and a college degree in construction management or engineering is preferred. In some cases, a person with several years' experience as a construction worker and with some training in management can rise through the ranks to become a supervisor.

AVERAGE STARTING SALARY
Beginning construction supervisors earn about $18,500 annually.

OUTLOOK
According to the Bureau of Labor Statistics, employment of construction supervisors is expected to rise along with the number of new construction projects through the year 2005.

What does a construction supervisor do?
The primary responsibility of a construction supervisor is to ensure that the work on a construction project is done properly, efficiently, and on time. Supervisors create work schedules, monitor activities, ensure safety procedures are followed, and keep track of materials used. They may also train new workers. They are often called *foremen* or *forewomen*, *superintendents*, or *crew chiefs*.

Does most of the work involve data, people, or things?
Construction supervisors work mostly with people and machinery. They manage other workers and discuss problems and solutions with other managers. They are also responsible for ensuring that equipment and machinery are in good working order.

What is a typical day like?
The construction supervisor is the boss. He or she is often the first to arrive at the construction site in the morning, review schedules, and make any necessary adjustments to the work plan. Throughout the day, the supervisor inspects the work, discusses problems and solutions, and handles worker grievances.

What are the working conditions?
Most of the work is done at a construction site, which is usually dirty and cluttered with debris. A hard hat is almost always required.

Which skills and aptitudes are needed for this career?
Being well organized, having leadership qualities, and being able to deal with different situations and different types of people are important qualities of construction supervisors. They must be able to motivate employees, maintain high morale, and command respect.

What careers are related to this one?
Other workers with supervisory duties include sales managers, bank officers, hotel managers, and surveyors.

One construction supervisor talks about her job:
I like the variety of tasks and the challenge of learning something I didn't know before. It's a stressful job at times, but I enjoy it.

358 SECTION 4: Construction Technology

Chapter 16 REVIEW

Reviewing Main Ideas

- Architects or engineers oversee most large building construction sites.
- Contractors are responsible for the construction of the project. They may hire subcontractors to do specialized work.
- To make sure all the terms of the contract are met, careful scheduling and monitoring are necessary.
- Quality control systems are set up to help ensure quality.
- Various people, including government representatives, periodically inspect construction.

Understanding Concepts

[12]

1. Briefly describe the responsibilities of a contractor.
2. Explain how a contractor is chosen. How do contractors go about getting the contract for a project?
3. What information does a contractor need to consider when making a work schedule?
4. What are the six things a purchasing agent checks when materials are received?
5. What are the three critical parts of the project on which inspectors concentrate their checks?

Thinking Critically

1. What traits would a contractor need to do a good job?
2. What characteristics would an inspector need to do a good job?
3. A contractor must be able to make accurate estimates of job expenses. What are the risks if the estimates are way off?
4. Describe a schedule that you have had to adapt to. Did you have to make any adjustments of your time? Did a schedule help you make better use of your time?
5. What advantages might a contractor gain by encouraging his/her workers to participate in quality control?

Applying Concepts and Solving Problems

1. **Mathematics.** Matt has his own home-repair service. He charges $12.50 an hour. Matt is currently bidding on a job for which he estimates that the materials will amount to $475. He has figured that it will take him 6 hours to complete the job. What bid will Matt submit?

2. **Technology.** Concrete must be mixed, poured, finished, and cured properly. Make two 1½" x 1½" x 14" forms for concrete. Mix two batches of concrete. Mix one batch according to the directions for the correct proportion of each ingredient, and carefully pour it into one of the forms. Use more water than directed when mixing the second batch, and pour this mixture into the other form. Allow the concrete to cure for seven days. Conduct testing to compare the strength of the two samples.

12. **Resource.** Answers to "Understanding Concepts," "Thinking Critically," and "Applying Concepts and Solving Problems" can be found in the Teacher's Manual at the front of this *Teacher's Annotated Edition* and in the *Teacher's Resource Binder*.

CHAPTER 17

Constructing Homes and Other Buildings

Objectives

After studying this chapter, you should be able to:
- describe preparation of a construction site.
- explain how foundations and superstructures are constructed.
- describe how interiors are finished.
- discuss finish work and landscaping.

Terms

batter boards
footing
foundation
insulation
joists
landscaping
load-bearing wall structures
modular construction
prefabrication
rafters
roof trusses
sheathing
studs
superstructure

Technology Focus

Smart Houses—The Future Is Now

Imagine yourself living in a "smart" house. You enter the house and put on an electronic pin that contains information about you. The pin is programmed to tell the house how to meet your needs and respond to you.

It is dark inside, but a moving beam of light illuminates your way as you walk. As you enter an unoccupied room, the lights are off, but as you step through the door, the lights turn on.

Your house's entertainment system uses remote controls throughout, with nearly invisible screens. You select the specific movies, TV shows, or recordings you want to see or hear, and music of your choice accompanies you as you walk. People in other parts of the house, however, do not hear the music you hear unless they want to. In fact, they can listen to different music at the same time.

Then the phone rings, and you know it's for you. How do you know? The smart house automatically routes calls to the phone nearest the person being called.

Science Fiction or Science Fact?

If you think this house sounds like something out of science fiction, think again. It already exists in Washington state and belongs to Bill Gates, Microsoft founder. Other smart houses also exist, although they may have different high-tech features.

Smart houses use a centralized, electronic system to control security, energy management, entertainment, communications, and lighting. This makes the house energy-efficient, saving both energy and money. A smart house also conserves its occupants' energy by performing tasks such as turning lights on and off, adjusting the thermostat, or, in a matter of seconds, finding and turning on music, TV shows, or movies.

A *Business Week* article by Heather Millar described some of the joys of living in a smart house:

Instead of getting jarred out of bed by a blaring alarm clock, soft music from a compact-disc player nudges you into consciousness. Your curtains glide open automatically to let in the light. You crawl out of bed into a perfectly controlled climate. In the bathroom, the shower is on, set to just the right temperature. Downstairs, the coffeemaker has stirred to life.

Sounds pretty cool, doesn't it?

Now would you like to live in a smart house?

Take Action!

Troubleshoot the "electronic pin" system mentioned at the beginning of this article. Think of three ways the system might get crossed up in ordinary use. Write brief solutions.

Resources
- Chapter 17 Lesson Plan in the *Teacher's Annotated Edition* and in the *Teacher's Resource Binder*.
- Chapter 17 Review in the *Teacher's Resource Binder*.

FIG. 17-1 Using explosives is a fast way to demolish tall buildings. This series of photos shows the stages in a building's collapse.

Preparing the Construction Site

Construction of a structure begins with preparing the site. If the site has not already been surveyed to establish and mark the property lines, surveying will need to be done as the first step in site preparation. (You read about the surveying process in Chapter 15.) Next, the site is cleared of anything that might interfere with construction. Then the building's position on the site is laid out.

Clearing the Site

The site must be cleared of anything in the way of new construction. This might include trees, old structures, rocks, and/or excess soil.

Trees the owner wants to keep are marked. Then bulldozers can be used to remove unwanted trees. Owners and companies who are environmentally aware try to save as many trees as possible.

Demolition

Demolition processes are used to rid the site of old buildings. *Demolition* involves destroying a structure by tearing it down or blasting it down with explosives. A crane that swings a large, steel wrecking ball against the structure may be used to demolish large structures. A bulldozer may be used to demolish small structures. Tall buildings, smokestacks, and similar structures can be demolished quickly by blasting. Fig. 17-1.

The rubble (pieces of concrete, brick, and other building materials) that remains must then be cleared away. Fig. 17-2. Most of it is usually hauled away in dump trucks. Some of it may be bulldozed to low areas of the site and used as fill.

Not everything that is removed from a construction site is destroyed. For example, signs, trees, and even whole houses can sometimes be salvaged (saved) and relocated. Even when a building is to be

FIG. 17-2 Once an old structure is demolished, the rubble must be removed. In this photo, water is being applied to the rubble to help reduce the heavy dust.

attempt is being made to disturb fewer trees and other natural features on a building site.

1. **Review.** You may wish to review the Chapter 15 discussion of surveying with students.
2. **Reinforcement.** Emphasize that, with the trend toward protecting the environment, a greater

FIG. 17-3 The tripod holds a laser-emitting device. Electronic receivers attached to the grader blade receive a signal from the laser. This allows the laser to control the height of the grader blade and enables the grader to level the earth to an exact elevation.

Earthmoving

In *earthmoving,* excess earth and rock are cleared away and the remaining earth is leveled and smoothed. This is done with heavy and powerful equipment, such as bulldozers and graders. A laser beam can help earthmoving equipment operators level the ground to the right height, or elevation. Fig 17-3.

Occasionally, swampy areas or ponds must be drained or pumped dry. Then the area is filled in or covered with a deep layer of earth or sand. Also, small creeks or streams may need to be blocked or redirected.

Laying Out the Site

Laying out the site is the process of identifying and marking the exact location of the structure on the property. The site plan shows how far the building will be from the edges of the property. Using the site plan as a guide, workers take measurements from the stakes surveyors placed at each corner of the property to the corners and edges of the proposed building. They mark these new locations with stakes.

The boundaries of the proposed building are usually marked using batter boards. **Batter boards** are boards held horizontally by stakes driven into the ground. These boards are placed four to five feet outside the building's boundaries. String is used to connect batter boards at opposite sides of the building. The attached strings cross directly over the boundary stakes at the corners of the building. Fig. 17-4. The corner stakes can then be removed, and the batter boards are used to guide excavation.

demolished, there is often much that can be saved. Doors, windows, lighting fixtures, and cabinets are examples of items that are often salvaged before the building is demolished.

3. **Reteaching.** Have students identify objects in the lab or school that could be salvaged if the school were to be demolished.

4. **Review.** Remind students that zoning laws must be observed when locating a building on a site.

FIG. 17-4 Batter boards and string are used to lay out the building's location. They are also used to guide excavation.

The Foundation

Most structures have two major parts: the foundation and the superstructure. The **foundation** is the part of the structure that rests upon the earth and supports the superstructure. A foundation may be above or below ground level, depending upon the type of structure and the local climate. Foundations are usually made from concrete. The **superstructure** rests on the foundation. In buildings, the superstructure usually consists of everything above ground.

Excavating for the Foundation

Excavating, or digging, for the foundation is done by heavy equipment such as backhoes, front-end loaders, and trenchers. Fig. 17-5. The size, shape, and depth of the excavation depends upon the design of the building. For example, a ranch-style house may be designed to rest on a simple concrete slab. It will require a

FIG. 17-5 Heavy equipment like this backhoe is used to excavate for the foundation.

CHAPTER 17: Constructing Homes and Other Buildings

wide but shallow area below the surface of the ground. On the other hand, a two-story house with a full basement will require a wide but deep opening. The excavation for a skyscraper must extend straight down, deep into the ground.

The soil may be checked by an engineer to make sure it will be able to support the structure. If the soil is soft or very loose, it may have to be compacted (packed down to make it firm). Soil can also be made firm sometimes by adding certain chemicals to it.

Parts of the Foundation

The two important parts of the foundation are the footings and the walls.

The **footing** is the part of the structure below the foundation wall that distributes the structure's weight. Fig. 17-6. The footing is usually twice as wide as the foundation wall so it can distribute the weight over a wider area. (See inset in lower right-hand corner of Fig. 17-6.) Footings are made of reinforced concrete.

The foundation walls transmit the weight of the superstructure to the footing. Foundation walls may be made from concrete blocks or from poured concrete that has been reinforced with steel. In buildings with basements, the foundation walls become the basement walls. (Refer again to Fig. 17-6.)

Building the Superstructure

Once the foundation is finished, work can begin on the superstructure.

Types of Superstructures

In buildings, the superstructure is usually one of two types: framed structure or load-bearing wall structure.

FIG. 17-6 The main parts of the foundation are the footings and the foundation walls.

5. **Reteaching.** Use soil in a plastic freezer container to demonstrate the effects of compacting.
6. **Extension.** Explain that footings must be located below the frost line, which is the depth to which soil freezes in winter. (The depth of the frost line varies with the climate.) This is because the ground swells upwards as it freezes and then settles again as it thaws. This movement can disrupt footings.

FASCINATING FACTS

The tallest building in the *United States* is the Sears Tower in Chicago. Completed in 1973, it reaches 1,454 feet and has 110 stories.

Framed Structures

A framed building has a main "skeleton," or framework, that supports the weight of the building. The framework consists of various structural members and other supports fastened together. These give the building its particular shape. The members may be concrete, steel, or wood.

Most houses have a wood-frame structure. The frames of the walls, floors, and roof are made from the following framing members (Fig. 17-7):

- **Studs.** These are parallel, evenly spaced, vertical boards that form the frame of exterior and interior walls. Studs are nailed at the top and the bottom to horizontal boards called *plates*.
- **Joists.** These are parallel, evenly spaced, horizontal boards that form the frames that support floors and ceilings.

FIG. 17-7 This illustration shows most of the framing members of a wood-frame superstructure as well as the layers that make up the finished roof.

CHAPTER 17: Constructing Homes and Other Buildings

- Rafters or roof trusses. **Rafters** are sloping roof framing members cut from individual pieces of dimension lumber. They extend from the *ridge* (the horizontal beam along the roof's peak) downward over the side walls of the building. Fig. 17-7. **Roof trusses** are preassembled triangular frames that are used to frame the roof. The sloping sides of the truss's triangular frame serve as rafters; the base of each triangular frame forms a ceiling joist. Fig. 17-8.

In a wood frame, the members may be nailed, screwed, or bolted together.

A steel frame is used for large industrial or commercial structures, such as office buildings. Fig. 17-9. The steel for these frames is prepared in a fabricating shop. Then it is delivered to the site. Ironworkers assemble and erect the steel framing members on the site according to the building plan. The steel parts are bolted, riveted, or welded together to make a rigid frame.

FIG. 17-9 A steel frame is used for most large buildings.

FIG. 17-8 Using pre-built roof trusses saves time on the construction site, forming both the rafters and the ceiling joists of the structure's frame.

7. **Critical Thinking.** Ask students why they think that trusses are increasingly being used in place of rafters.

SECTION 4: Construction Technology

Load-Bearing Wall Structures

In **load-bearing wall structures**, the heavy walls support the weight of the structure. There is no frame. Bearing walls are usually made of concrete blocks, poured concrete, or precast concrete panels. Fig. 17-10. Bearing wall construction is best suited for low buildings of one or two stories.

Enclosing the Superstructure

The wall and roof frames of most frame buildings are first enclosed with layers of sheathing. **Sheathing** is a layer of material, such as plywood or insulating board, that is placed between the framing and the finished exterior. After the windows and exterior doors are installed, decorative finish materials such as wood paneling, vinyl or wood siding, stone, or brick are placed over the wall sheathing. The sheathing over the roof frame is first covered with roofing felt. Then roofing materials such as asphalt or wood shingles or tiles are applied. (See Fig. 17-7 on page 367.)

Installing Floors

A *subfloor*, consisting of sheets of plywood, is nailed or glued to the floor joists. The subfloor serves as a base for the finish flooring. It also provides a surface for workers to walk on when completing other parts of the building. An additional layer of plywood, called *underlayment,* is nailed or glued to the subfloor before the finish floor is applied. Fig. 17-11.

FIG. 17-10 This building has no frame. The outer walls provide its support.

8. **Reinforcement.** Point out that sheathing materials come in large panels for faster installation.

CHAPTER 17: Constructing Homes and Other Buildings

FIG. 17-11 Floors are installed in several layers.

Installing Utilities

The term *utilities* is used to refer to various service systems in a building. They include:

- electrical systems
- plumbing systems
- heating, ventilating, and air conditioning (HVAC) systems

Utilities are installed in two stages. First they are *roughed in*. Parts such as wires, pipes, and ductwork are placed within the walls, floors, and ceilings before the interior surfaces are enclosed. Ductwork for HVAC is usually installed first, then piping, and finally, wiring. Fig. 17-12.

After the interior walls are enclosed, the utilities are finished. Such things as light switches, plumbing fixtures, and temperature controls are installed.

FIG. 17-12 Utilities must be roughed in before interior walls can be finished.

9. **Extension.** You may wish to discuss utility systems in greater detail at this time.
10. **Reinforcement.** Explain that roughed-in utilities must be inspected by the appropriate building inspector before they are covered. If they are covered before inspections are made, the contractor may have to uncover them at his or her own expense.

SCIENCE CONNECTION
Water Down Under

Groundwater is water below the earth's surface. Groundwater comes from rainwater, as well as water from lakes and ponds, that soaks into and filters through the soil. It makes up about 20 percent of all fresh water in the United States. It is usually found in rural areas and is a source of water for wells and springs.

Groundwater is the result of the water cycle, a natural recycling process, in which water is used over and over again. The water cycle begins with evaporation. As the sun heats it, water from lakes, streams, and other sources turns into water vapor. The water vapor is then suspended in the air, sometimes as clouds, until the air cools. Cooling causes condensation, and the water vapor changes into rain. The rain collects in bodies of water, is taken up by plant life, or soaks into the ground. Then the process repeats itself.

Threats to Groundwater

Although it can be replenished, groundwater is not unlimited. The *water table* (the level of groundwater) drops when water is drawn up faster than nature can replace it. This occurs mostly in areas with large populations or little rainfall. Wells dependent upon groundwater may run dry. In coastal areas, the lowering of the water table can be dangerous. When the water table gets low enough, salt water can leak into the groundwater, causing health hazards.

Pollution can also threaten water beneath the ground. When industrial chemicals and radioactive wastes leak into the groundwater supply, they can contaminate it, causing sickness in plants, animals, and people.

Try It!

1. Contact your local water department. Find out where the water in your home or school comes from. Report your findings to the class.
2. Find out what other sources are used for water for homes and other buildings.

Answers to these questions can be found in the Lesson Plans in the Teacher's Manual at the beginning of this book.

CHAPTER 17: Constructing Homes and Other Buildings

FIG. 17-13 Various types of insulation can be used to help keep a building warmer in winter and cooler in summer.

Installing Insulation

Insulation is material that is applied to walls and ceilings to help keep heat from entering the building in the summer and leaving it in the winter. This helps make the house more energy-efficient. The ideal amount of insulation used in a building depends on local climate. For example, a house in Seattle, Washington, where the weather is moderate year-round, requires much less insulation than one in northern Minnesota, where winters are very cold.

Insulating materials are labeled according to their R-value. The higher the R-value, the better the insulating qualities. Insulation comes in various forms (Fig. 17-13):

- *Batts or blankets.* These are thick fiberglass sheets or rolls, with a paper or foil backing, designed to fit snugly between framing members.

FIG. 17-14 This worker is filling in the spaces between the drywall sheets.

- *Rigid panels.* These are large sheets of plastic foam or natural fibers.
- *Loose fill.* Loose fill is fibrous or granular material that is blown into place using a special hose.

Finishing the Interior

After the utilities are roughed in and the insulation is installed, the interior is ready to be finished.

Ceilings and Walls

Most walls and ceilings are first enclosed with drywall. Drywall is a general term used for plasterboard or wallboard. It is a heavy, rigid sheet material. Drywall is fastened directly to the wall studs and ceiling joists. It may be fastened with screws, nails or adhesives. Holes must be cut in the drywall for electrical outlets, light switches, and lighting fixtures.

A filler is placed into any dents as well as the spaces or seams between the drywall panels. Fig. 17-14. These areas are then taped. Another layer of filler is applied. When the filler is dry, the surfaces are sanded smooth. Now the walls and ceilings can be finished with wallpaper or paint. Paneling and tile are also popular wall finishes.

Floor Coverings

Finish flooring is usually installed over the underlayment after the walls and ceilings have been finished. This is done to avoid damage to the flooring. Wall-to-wall carpeting, sheet vinyl, and vinyl or asphalt tiles are commonly used for floor covering. Wood, ceramic tile, and flagstone are also popular. The various materials are installed in different ways. Special fasteners or adhesives may be required.

Trim and Other Finish Work

Trim is the woodwork, baseboards, and moldings used to cover edges and the joints where the ceilings, walls, and floors meet. Trim is also applied around doors and windows. Most trim used in homes is wood trim. Plastic and metal trims are also available. Fig. 17-15.

FIG. 17-15 Trim is used to cover joints where ceilings, walls, and floors meet. It gives the room a finished appearance.

11. **Enrichment.** Collect samples of various types of floor coverings and pass them around the class. Explain how they compare in cost, durability, quality, and ease of installation.

12. **Extension.** Show the class various types of moldings used for trim.

FIG. 17-16 Installing accessories, such as this light fixture, is the last step in finishing the inside of a structure.

Room doors and closet doors are put in place next. Then hardware is installed. Hardware includes items such as doorknobs and towel bars.

Installation of accessories is the last indoor finishing task. Kitchen and bathroom cabinets are major accessories. Plumbing and electrical fixtures are also major accessories. Fig. 17-16. Shelving, countertops, and other built-ins are also installed as part of finish work.

Prefabricated Structures

Components of structures and even whole structures are now being built in factories. The term used to describe these processes and products is **prefabrication.** The prefabricated parts or sections are shipped from the factory to the construction site, where they are assembled. Site preparation and the building of the foundation will already have taken place while the structure itself was being constructed at the factory.

Prefabrication methods include panelized construction, modular construction, and manufactured housing.

Panelized Construction

In *panelized construction,* the floors, walls, and roof all consist of prefabricated panels that have been made in factories. Fig 17-17. These panels are then shipped to the construction site, where they are assembled to produce the framed and sheathed shell of the structure. Prefabricated roof trusses and floor joists are usually used with the panels. Insulation, utilities, and siding may be either prebuilt or added at the site in the traditional manner.

Modular Construction

In **modular construction** entire units, or modules, of structures are built at factories and shipped to the site. The modules are assembled at the site to produce a finished structure. Fig. 17-18.

This construction method has been used to build motels. Identical modules—complete with plumbing, electricity, heating, and air conditioning—are delivered to

13. **Extension.** Discuss the ways in which modern manufacturing techniques aid in quality control in prefabricated structures.
14. **Critical Thinking.** Panelized construction is widely used in Japan and Europe but is not yet common in the U.S. Ask students if they think it will become common in the U.S. Discuss why or why not.
15. **Reteaching.** Explain to students the difference between panelized and modular methods. Show photos of each.

SECTION 4: Construction Technology

FIG. 17-17 These workers are assembling panelized wall sections.

FIG. 17-18 These prefabricated sections are being quickly assembled into a complete house.

the site and assembled. In a very short time, the motel is set up and can turn on its "Vacancy" sign.

Manufactured Housing

Many new single-family homes are entirely manufactured in factories and then shipped to the site. These structures are called manufactured housing. Usually, these homes are built in two or more sections to produce a finished structure.

16. **Reteaching.** Explain that the housing industry frequently refers to all prefabrication methods as "manufactured housing." For purposes of better understanding the various processes, the term *manufactured housing* is used here in its most specific definition.

Post-Construction

After construction is complete, several other tasks remain. The site must be finished and the project transferred to the client.

Finishing the Site

The site as well as the structure must be finished. The two major types of outdoor finishing that need to be done are paving and landscaping.

Paving

Several areas around homes and other buildings need to be paved. Driveways, walkways, parking lots, and patios are examples. These are usually made of concrete or asphalt. Their locations, sizes, and shapes are indicated on the working drawings. Workers mark or stake out these areas on the ground. Then the earth is carefully leveled and compacted. Where concrete is used, forms that will hold the concrete while it cures may be set up. Finally, the concrete and/or asphalt is installed.

Landscaping

Landscaping begins after all the debris from construction has been removed. Landscaping involves changing the natural features of a site to make it more attractive. It includes shaping and smoothing the earth with earthmoving equipment as well as planting trees, shrubs, grass, and other vegetation.

> **FASCINATING FACTS**
> Windsor Castle in Great Britain, where Queen Elizabeth lives, is the largest inhabited castle in the world. It covers 24 acres. Most of the castle was constructed in the 12th century.

Landscaping is done according to a landscaping plan. This plan shows how the finished site should look. Fig. 17-19.

Transferring the Project

The last step in the construction process is formally transferring the completed project to the owner. After construction is complete, the project is given a final inspection. This is done to make certain that the terms of the original contract and specifications have been fulfilled. The quality of the work must also be acceptable.

A number of important legal matters must be attended to. All outstanding bills for the construction of the project must be paid by the contractor. There must be no claims of money owed against the property.

After all requirements have been satisfied, the final payment is made to the contractor. Then the keys to the building are given to the new owner. The new owner now takes over full responsibility for the property.

---

17. **Mathematics.** Explain that concrete is sold in cubic yards. To figure the amount needed for a particular area the formula is l' x w' x th" ÷ 27 = cubic yards. Give students a problem to solve.

18. **Problem Solving.** Hand out a scale drawing of a house on a lot that includes some topographical features. Have students plan the landscaping. Distribute nursery catalogs and have them figure landscaping costs.

FIG. 17-19 Landscape plans are prepared to show how the finished site will look.

CHAPTER 17: Constructing Homes and Other Buildings

Career File—Carpenter

EDUCATION AND TRAINING
A high school diploma is desirable, including courses is carpentry, shop, mechanical drawing, and general mathematics. An apprenticeship program is considered the best way to learn carpentry. Most of these programs are sponsored by labor unions and last three to four years.

AVERAGE STARTING SALARY
Apprentice carpenters start earning about $13,000 per year.

OUTLOOK
According to the Bureau of Labor Statistics, employment of carpenters is expected to increase, although more slowly than average, through the year 2005. The increased use of prefabricated components, such as pre-hung doors, reduces the need for as many carpenters.

What does a carpenter do?
Carpenters are skilled workers who cut, fit, and assemble wood and other materials in the construction of buildings, highways, bridges, industrial plants, and other structures.

Does most of the work involve data, people, or things?
Carpenters work with wood and other construction materials (plastic, fiberglass, tile, and drywall, for example). They also work with tools to measure, cut, and join the materials. Their contact with other people is usually limited to construction supervisors and other construction workers.

What is a typical day like?
Each carpentry project is somewhat different, but most tasks involve the same basic steps. Working from blueprints or supervisor instructions, carpenters first measure and cut materials. They then join the pieces and check the accuracy of their work. Their work must often mesh with that of electricians and other specialists.

What are the working conditions?
Carpentry work can be strenuous. Prolonged standing, climbing, bending, and kneeling are often necessary. Carpenters risk injury from falls, from working with sharp or rough materials, and from the use of sharp tools and power equipment. Outdoor work is common.

Which skills and aptitudes are needed for this career?
Manual dexterity, eye-hand coordination, physical fitness, a good sense of balance, and the ability to solve arithmetic problems quickly and accurately are all important.

What careers are related to this one?
Other workers in skilled construction jobs include bricklayers, electricians, pipefitters, plumbers, and plasterers.

One carpenter talks about his job:
I like to take a set of blueprints and a stick of lumber and build something new. I feel proud of what I've built. It's like producing a work of art.

Chapter 17 REVIEW

Reviewing Main Ideas

- Preparing the site involves clearing away unwanted objects, leveling and smoothing, and laying out the site.
- The foundation, which supports the superstructure, consists of footings and walls.
- The superstructure may have a wood or a steel frame or may be a load-bearing wall structure.
- Once the structure is framed, it is enclosed, the subfloor is laid, the utilities are roughed in, and insulation is installed. Then the interior is finished.
- Components of structures and even entire structures may be prefabricated in factories, then shipped to construction sites to be assembled.
- Finishing the site involves paving and landscaping. Then the project is transferred to the owner.

Understanding Concepts

1. Describe the process of laying out the site.
2. Describe the two main parts of the foundation. Tell the function of each.
3. Compare a frame structure to a load-bearing wall structure. Name and describe the framing members.
4. What are the three types of utilities discussed in the text?
5. Briefly describe what is done during landscaping.

Thinking Critically

1. What do you think might happen in a construction project if errors were made when the site was surveyed?
2. What might be some disadvantages of a wood frame for a house?
3. Sometimes people who are having a home built want to do some of the work themselves. If you were having a home built, what jobs would you do?
4. Would you buy a prefabricated house? Discuss why or why not?
5. What are some of the important things a landscape architect needs to consider when developing a landscape plan?

Applying Concepts and Solving Problems

1. **Science.** Obtain literature or a video on a large piece of earthmoving equipment. Identify at least six simple machines used to create the equipment.

2. **Mathematics.** Fiberglass insulation has an R-value of 3.33 per inch. One family's attic insulation should have a value of R-26. The fiberglass insulation in their attic now is 3½" thick. What is its current R-value? (Round to the nearest whole number.) About how much insulation must be added to reach R-26? (Round to the nearest ½".)

3. **Technology.** Design and construct a scale model of a wall section using only recycled materials. Consider safety and cost factors.

19. **Resource.** Answers to "Understanding Concepts," "Thinking Critically," and "Applying Concepts and Solving Problems" can be found in the Teacher's Manual at the front of this *Teacher's Annotated Edition* and in the *Teacher's Resource Binder*.

CHAPTER 18

Other Construction Projects

Objectives

After studying this chapter, you should be able to:
- describe how highways, dams, canals, tunnels, pipelines, bridges, and air-supported structures are built.
- identify and describe the seven types of bridges.
- discuss construction in space.

Terms

abutments
air-supported structure
arch bridge
beam bridge
cable-stayed bridge
cantilever bridge
cofferdams
irrigation canals
lock
movable bridge
reservoir
suspension bridge
truss bridge

Technology Focus

A Giant among Dams

What is 726 feet tall, 1,244 feet wide, and 660 feet thick at the bottom but only 45 feet thick at the top? Don't know? Here's another hint: It contains over four million cubic yards of concrete—enough to pave a 16-foot-wide highway from New York City to San Francisco, California, a distance of 2,930 miles. Give up? It's Hoover Dam, one of the tallest dams in the world and one of the largest concrete structures ever built.

Hoover Dam stands in the Black Canyon of the Colorado River, between the states of Arizona and Nevada. This giant structure was originally named Boulder Dam for nearby Boulder City, Nevada. However, it was renamed in honor of Herbert Hoover, who was president of the United States when the dam was under construction.

Water, Electricity, and Lots of Fun!

Dams are structures built across rivers to control or block the flow of water. The water that collects behind the dam creates a *reservoir* or lake. The reservoir created by the Hoover Dam is Lake Mead. This lake is 115 miles long and averages 200 feet in depth. It contains more than 10 trillion (10,000,000,000,000) gallons of water, making it one of the largest man-made lakes in the world.

Water from Lake Mead is used to irrigate more than one million acres of farmland in Arizona, Nevada, and California. Thanks to this irrigation, these areas produce hundreds of millions of dollars worth of crops each year. This reservoir created by Hoover Dam also provides many millions of gallons of drinking water for a number of cities in southern California. The water travels through a 240-mile-long aqueduct, itself a large-scale construction project.

Hoover Dam was not only built to provide water, but also to create electricity through hydroelectric energy. Hydroelectric energy is produced by the force of water falling from a dam onto a *turbine* (wheel with fins) that drives an electric generator. The power plant at the base of this dam has a capacity of over 1¼ million kilowatts. Much of the electricity produced is carried by power lines to Los Angeles, 250 miles away.

In addition to providing water and electrical energy, Hoover Dam has also created a recreational bonanza. Along its 550-mile shoreline, are six major recreational centers and a host of smaller resorts.

Take Action!

Do research to learn about any impacts—positive and negative—this giant construction project may have had on the environment.

Resources
- Chapter 18 Lesson Plan in the *Teacher's Annotated Edition* and in the *Teacher's Resource Binder.*
- Chapter 18 Review in the *Teacher's Resource Binder.*

Highways

Construction involves many projects besides buildings. Highway construction includes the building of any type of highway, street, or road. Before actual construction begins, the route is determined by environmental, financial, and land use issues as well as by the *topography* (the land's surface features, such as hills, streams, large rocks, and soil type.)

Field survey information is used to assist in the determination of the proposed route. After the acceptable route has been determined and the project has been awarded to the construction contractor, a survey team stakes the center line and edges of the planned roadway. Fig 18-1.

The route must first be cleared of any trees, vegetation, and large rocks. Then the roadway route is graded to meet the design specifications that have been determined by the project engineers. The design is controlled by the physical features of the land and by the speed limits proposed for the route. Earthmoving equipment is used to remove excess soil from areas that are too high and place it in areas that are too low. Fig. 18-2. These fill areas must be compacted (firmly packed) to keep the soil from settling in the future.

Two major types of roadway surfacing are used for the finished roadway: flexible and rigid. Fig. 18-3. Flexible roadbeds use materials like asphaltic concrete (commonly known as *asphalt*) to create the smooth finished surface, which can flex, or give slightly, to absorb heavy loads. Usually, a thick bed of gravel or similar granular material is used as a subbase. It is placed on the subgrade before the asphalt pavement is installed. Depending upon the subsoil properties, this subbase may be up to two feet thick. This gravel-type subbase transfers and spreads the loads on the highway into the soil below.

Portland cement concrete (PCC) is an example of a rigid roadway material. PCC (commonly known as concrete) pavement

FIG. 18-1 Survey teams mark the exact location of a planned access road.

1. **Extension.** Tell students that nearly 5,000 years ago, the Egyptians built roads to transport materials needed to construct the pyramids. The Romans built an extensive highway system to support military activities. Some of the original Roman roads have been made into modern European highways.
2. **Enrichment.** Ask a local road construction contractor to talk to students about choosing roadway surfaces.

FIG. 18-2 This bulldozer is preparing the ground so a new highway can be constructed through these hills. Building a road requires extensive earthmoving.

FIG. 18-3 Cross-sections of a flexible and a rigid roadway.

FLEXIBLE ROADWAY
- PAVEMENT (ASPHALT LAYER)
- SUBBASE (SAND OR GRAVEL)
- SUBGRADE (PACKED SOIL)

RIGID ROADWAY
- PAVEMENT (CONCRETE)
- STEEL BARS
- SUBBASE (SAND OR GRAVEL)
- SUBGRADE (PACKED SOIL)

3. **Reteaching.** Explain the difference between concrete and cement—many people incorrectly use these terms interchangeably.

CHAPTER 18: Other Construction Projects

may be either reinforced or nonreinforced. Reinforced concrete pavement is produced by placing steel reinforcing bars on metal supports so that the steel is in the middle of the poured PCC slab. Nonreinforced PCC pavement sometimes has steel bars across joints to keep those joints from moving upward or downward. Loads applied to rigid pavement are spread through the slab and transferred to a large area under the slab.

Once the finished surface has been installed, lights, pavement markings, signs, and possibly traffic control signals are added to complete the road.

Bridges

Bridges are structures built to allow people and vehicles to pass over something else. When we think of bridges, we most often think of those built over water. However, bridges may also extend over valleys, highways, or railroad tracks.

Fascinating Facts

The Humber Estuary Bridge, located near Hull, England, and opened in 1981, currently has the longest span (distance between supports) in the world. It is 4,625 feet long and took 7 million construction hours to build.

Substructure of Bridges

The substructure of a bridge consists of abutments, piers, piles (sometimes), and road deck. Fig. 18-4.

Abutments are the supports at the ends of a bridge. They not only support the bridge, but also the earth at the ends of the bridge. Abutments are usually made of reinforced concrete.

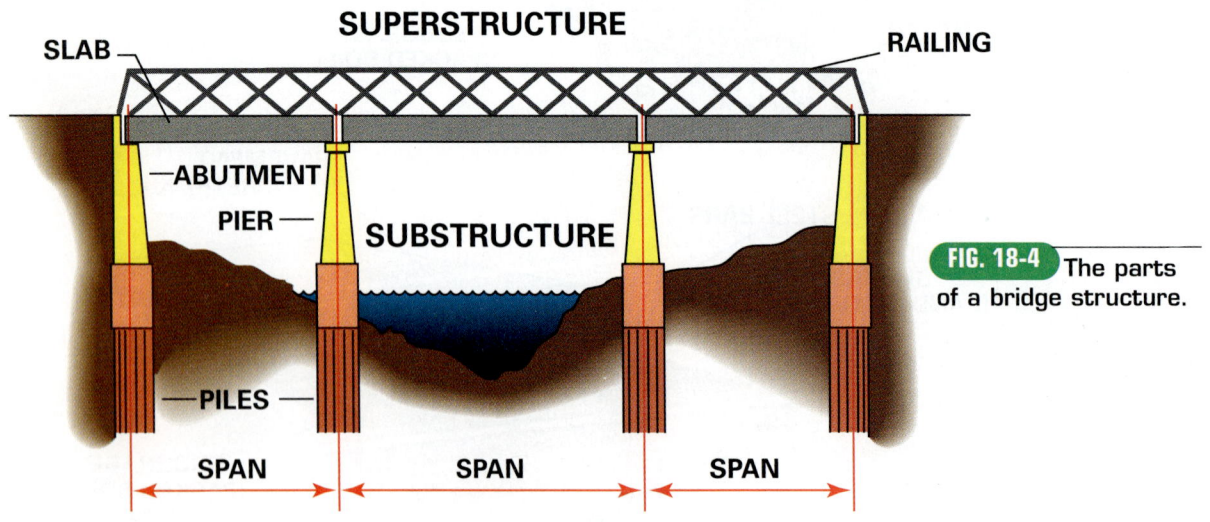

FIG. 18-4 The parts of a bridge structure.

SECTION 4: Construction Technology

Piers are the vertical structural supports placed between abutments in longer bridges. They are positioned to keep the longer bridge deck from sagging. The distance between each pier or between a pier and an abutment is called a *span*. (Note that the word *span* is also used to mean "extend across," such as "the bridge spans the Spoon River.")

Supports (piers and abutments) for a bridge must rest on a solid surface. When the earth under the bridge is not solid, piles are used. *Piles* are wood, metal, or concrete members that are driven down into the earth to a solid base. Piers and abutments are placed on top of these piles.

Types of Bridge Superstructures

The type of bridge constructed depends on how long the bridge must be and on the weight it must support. There are seven common types of bridges. Fig. 18-5.

- **Beam bridge.** Piers support beams that support spans of reinforced concrete slabs. The beams may be reinforced concrete, rolled steel girders or fabricated steel. This is the most frequently used type of bridge because it is normally the least expensive.
- **Arch bridge.** The load of the bridge is transferred along the arch (curved portion) to the abutments or piers at the end of the arch. Single or multiple arches may be used, depending on the distance the bridge must span (extend across).
- **Truss bridge.** As you learned in Chapter 17, a truss is a triangular framework. Trusses may be used above or below the roadway to support the bridge. Trusses are also used with other bridge types, such as suspension bridges, to give them additional support.
- **Cantilever bridge** (CAN-tul-EE-vur). Beams called cantilevers extend from each end of the bridge. They are connected by a section called a suspended span. To remember this kind of bridge, think of two diving boards at opposite sides of a pool being connected by placing a board across the top of them.
- **Suspension bridge.** These have two tall towers that support main cables that run the entire length of the bridge. The cables are secured by heavy concrete anchorages at each end. Suspender cables dropped from the main cables are attached to the roadway. These bridges are used to span long distances. Probably the most famous suspension bridge is the Golden Gate Bridge in San Francisco.
- **Cable-stayed bridge.** These are similar to suspension bridges except the cables are connected directly to the roadway. To date, most cable-stayed bridges have been built outside the United States. This is because some of our engineers are concerned about the strength and durability of these bridges.
- **Movable bridges.** These bridges are designed so that a portion of the roadway can be moved to allow large water vessels to pass underneath. Bascule (BASS-kyool) bridges open by tilting upward. Lift bridges have a section of roadway that moves up between towers. Swing bridges have a section that moves sideways.

---

4. **Reinforcement.** Ask students to bring in pictures of bridges for a bulletin board display. Classify the bridges in various ways: type of construction, purpose, materials, etc.
5. **Extension.** Have students research the 1989 San Francisco earthquake and its effect on the bridges.
6. **Critical Thinking.** Ask students why castles needed a movable bridge.

CHAPTER 18: Other Construction Projects

FIG.18-5 Common types of bridges.

Dams

A dam is a structure that is placed across a river to control or block the flow of water. The water that collects behind the dam creates a reservoir. A **reservoir** is a lake in which water is stored for use. That is one of the main reasons for building a dam—to provide a dependable water supply for nearby communities. As a bonus, the lake can also be used for recreation. Fig. 18-6. The other main reason for constructing a dam is to collect water to power the water turbines in a hydroelectric power station at the base of the dam.

Dams may be made of earth, concrete, steel, masonry, or wood. Usually a combination of materials is used.

The three main parts of a dam are its embankment, outlet works, and spillway. The *embankment* blocks the flow of water. *Outlet works* are used to control the flow of water through or around the dam. When water is needed downstream, gates of the outlet works are opened to allow water to flow through. A power plant may be part of the outlet works. The *spillway* acts as a safety valve that allows excess water to bypass the dam when the reservoir becomes too full due to flooding. If water could not bypass the dam during flooding, the dam would break.

Constructing a large dam is a complicated process. Temporary watertight walls, called **cofferdams**, must be built to keep the construction site dry. As work progresses and construction in the area protected by the cofferdam is completed, another cofferdam is built farther out in the river, and the first cofferdam is removed. The dam is built in carefully planned stages. Fig. 18-7.

Fascinating Facts

Scheduled to be completed in 2014, the Gorges Dam will cross the Yangtze River in China. The world's largest hydroelectric dam, it will be possible to see it with a telescope from the moon.

FIG. 18-6 A dam's reservoir provides a dependable water supply for communities. The reservoir can also be used for recreation.

7. **Reinforcement.** Refer to the appropriate section (A, B, C, etc.) of the illustration as you discuss the stages of building a dam.

CHAPTER 18: Other Construction Projects

FIG. 18-7 Stages in building a dam. A. Building abutments. B. Building the first cofferdams. C. Building the first part of the outlet works and spillway. D. Building another cofferdam and channeling water through the outlet works. E. Completing the outlet works and spillway. F. Completed dam.

Canals

Canals are artificial waterways that are built for irrigation or navigation. **Irrigation canals** carry water from a place where water is plentiful to another place where water is needed. These are constructed to supply water to land that otherwise could not be used to grow crops. Fig. 18-8.

Navigation canals connect two bodies of water. Navigation canals may also be constructed when a river has a portion that either bends too much or is too shallow to navigate. The canals allow ships to bypass those parts of the river.

Construction of a canal requires a great deal of earthmoving. Once the excavation is complete, clay, concrete, or asphalt is usually used to line the canal. This prevents leaks and washing away of the soil.

Navigation canals may also require the construction of locks if the two waterways being connected are at different elevations. A **lock** is an enclosed part of the canal that is equipped with a gate. Fig 18-9. The level of water within the lock can be changed in order to raise or lower ships from one water level to another. For example, the Welland Canal, which connects Lake Erie and Lake Ontario, contains eight locks along its twenty-eight mile route to help ships change levels safely.

Tunnels

A tunnel is an underground passageway. Tunnels are built to allow people, vehicles, or materials to pass through or under an obstruction. For example, tunnels may be built under busy city streets for subways. They may be built under rivers or through mountains for railroads or highways. A tunnel may also be built to carry water around a dam. There are three common types of tunnels: earth, immersed, and rock.

FIG. 18-8 These irrigation canals help make more land available for farming.

8. **Extension.** Show students where Panama is on a map. Point out how the Panama Canal cut the distance a ship had to travel from one coast of the U.S. to the other.
9. **Extension.** Tell students Amsterdam, Holland, and Venice, Italy, are cities known for their transportation canals. Have students find these cities on a map or globe.
10. **Reteaching.** Make a model or obtain diagrams to demonstrate how locks work.

CHAPTER 18: Other Construction Projects

FIG. 18-9 Locks allow ships to be raised and lowered. This allows navigation of canals and rivers that have changes in water levels.

FIG. 18-10 Giant boring machines use disk cutters to cut tunnels through soft rock. A conveyor system and elevator transport the rock and earth that is cut away up to the surface.

Earth tunnels are constructed in soil or sand. Because sand and soil can be unstable, these tunnels are hazardous to build. As earth tunnels are dug, concrete sections can be installed to prevent collapse.

For immersed tunnels, pre-manufactured sections are floated to the tunnel site. Here, they are sunk into trenches that have been scooped out at the bottom of the waterway. The sections are then connected to form the tunnel.

In rock tunnels, material is removed by blasting or by using giant boring machines. Conveyors and small boxcars on rails carry the cut-away rock out the ends of the tunnel. Fig. 18-10. This was the method used to cut the three parallel tunnels that make up the Chunnel, which runs under the English Channel to connect England and France. Fig. 18-11. The largest tunnel ever constructed, the Chunnel is 31 miles long. Twenty-three miles of that distance is under water—at some points, as deep as 450 feet under the sea bed. Two of the three interconnected tubes that make up the Chunnel carry passengers and vehicles; the smaller central tunnel is used for maintenance. Fig. 18-12.

FIG. 18-11 This photo shows one of the two huge boring machines used to construct the Chunnel.

11. **Extension.** Discuss the relationship of the techniques used in mining to those used in tunneling.
12. **Background.** The idea of a cross-channel tunnel was first proposed in 1753. In the 1870s, shafts were begun but were abandoned.
13. **Extension.** Have students do research to learn how the Chunnel has affected the economy in both France and England and what impacts the Chunnel may be having on the lives of French and English citizens.

FIG. 18-12 The 31-mile-long Chunnel (the English Channel tunnel) allows trains carrying passengers and vehicles to travel quickly and comfortably between England and France.

SERVICE TUNNEL

TRAIN TUNNELS

Pipelines

Pipelines are an efficient way to transport products such as crude oil, refined petroleum, and natural gas. Most pipelines are buried underground. There are above-ground pumping stations along the pipelines that are used to maintain the pressure needed to keep the product moving.

Once the route for a pipeline has been surveyed and marked, backhoes and trenchers dig the trenches that will hold the pipe. Sections of pipe one to four feet in diameter are manufactured in factories.

These are transported to the site and laid next to the trench. Then the sections of pipe are welded together. Fig. 18-13. Cranes then lift the pipe and place it in the trench. Fig. 18-14.

Before being covered with soil, the joined sections must be inspected for leaks. This is usually done with X-ray or ultrasound equipment. After any necessary repairs are made, the trenches holding the pipeline are covered. The earth must be packed down all around the pipe. Otherwise, the ground may later "settle," causing the position of the pipes to shift. This could cause damage to the welded joints, resulting in leaks. A tamper may be used to pack the earth as the trench is gradually filled. A roller or the wheels of a heavy tractor may be used to finish tamping and leveling the surface.

14. **Extension.** Explain to students that the Trans-Alaska Pipeline is buried in some sections, but due to permafrost conditions in other areas, it must be above ground and supported by bridges. To control the temperature and keep the crude oil flowing, the pipeline had to be insulated, and heating and cooling units had to be installed.

FIG. 18-13 Pipefitters weld sections of pipe together.

FIG. 18-14 Cranes with special pipe-holding devices lift the pipe into the trench.

CHAPTER 18: Other Construction Projects

MATHEMATICS CONNECTION
Costs of Workplace Illnesses

When calculating the cost of doing business, construction companies, like other employers, must include the expense of workplace illnesses. Not only do companies help pay for employee insurance, but they also bear the cost in lost productivity when employees are absent. Calculations usually involve working with percentages and dollar amounts.

Following are several categories of work-related illnesses:

Skin diseases. Skin diseases caused by chemicals and other irritants account for 15-20% of work-related illnesses. In a recent year there were 66,000 cases nationwide. Cost in lost workdays and productivity is $1 billion per year.

Occupational asthma. Workplace substances, such as certain types of dust, can cause airway diseases. Over 9 million such cases are associated with asthma, a breathing disorder. In a recent year, 92,000 people died of asthma. The cost to American companies is $400 million annually.

Injuries from Accidents. On an average day in the U.S., 17,000 workers are injured in accidents. The cost to U.S. companies is $121 billion per year.

Try It!

1. Using a separate sheet of paper, create a chart like the one shown here. Drawing from the above information, fill in the third column.
2. Assume that you work for BCC Construction, which has 3,400 employees. Lost workdays and drops in productivity cost BCC $72,688 last year. Using the percentages given, make the necessary calculations to complete predictions for this year in the last two columns on the chart.

Type of Illness	Percentage of U.S. Workers Affected Annually	Annual U.S. Cost in Lost Workdays and Productivity	Estimated Number of BCC Workers Affected	Estimated Cost to BCC in Lost Workdays and Productivity
Skin diseases	Less than 1%			
Asthma	6%			
Accidental injuries	4%			

Answers to these questions can be found in the Lesson Plans in the Teacher's Manual at the beginning of this book.

Air-Supported Structures

An **air-supported structure** is just what the name implies—there is no frame or load-bearing wall to support the roof. Instead, air pressure supplied by a fan supports the walls and the roof. Fig. 18-15. An air-supported structure is made up of four elements:

- a structural membrane
- a means of supporting the membrane
- a means of anchoring the membrane to the ground
- a way in and out of the building

The structural membrane is usually made of a strong, synthetic (human-made), impermeable (can't be penetrated by gases or fluids) material. This material is supported by introducing an air supply that will create an uplift of the membrane, distributing tension evenly through the structure. An electrically driven fan usually blows air into the structure for this purpose. All parts of the membrane must be kept under tension to withstand external air pressure. To accomplish this, the fan keeps the air pressure within the structure at a level high enough to counteract the air pressure outside the structure.

Because the air within the structure creates an uplift, the structure must be suitably anchored to keep it from pulling away from the ground. The membrane is

FIG. 18-15 Air-supported structures can be used to provide shelter for sporting events and exhibitions.

15. **Reteaching.** A parachute (if available) or a large sheet of plastic can be used to illustrate uplift. Have the students distribute themselves around the chute or plastic, raise it into the air, and draw it down quickly. They should feel the uplift created by the upward pressure of the air.

FIG. 18-16 Air-supported technology can also be used to help lower construction costs. The Minneapolis Metrodome's roof is air-supported.

usually attached to a rigid concrete base (buried underground) to keep it firmly anchored at ground level.

Getting in and out of an air-supported structure is different from getting in and out of conventional buildings. If a traditional door were used in an air-supported structure, air would leak out of the building every time someone opened it. Also, if the door opened inward, people would have to push hard to overcome the internal pressure. If it opened outward, the internal pressure would make the door pop open with extra force. One method of solving this problem is to use an air lock. To create an air lock, two sets of doors are used. People enter through an outer set, which is then closed before opening the inner set. The air lock this creates provides an opportunity for pressure to be equalized when people are going in and out. In large air-supported structures like the Minneapolis Metrodome, revolving doors can be used. Fig 18-16.

The main advantage of air-supported structures is that they can be set up and taken down easily. Their portability is one of their greatest assets. For this reason, they are often used for exhibition buildings. These exhibition buildings can be easily transported, set up, and taken down to transport again when an organization tours from one location to another.

Construction in Space

Building a structure in space is not like building one on Earth, where thousands of years of history and experience have made design and construction problems relatively simple to solve. There are many factors in space construction that make any space structure a difficult project. Space structures must provide all of the environmental and life-support systems needed by humans surrounded by the hostile environment of space. Huge panels of solar cells are needed to produce power for the space structures. To perform any construction tasks in space, complex space suits and specialized tools and machines must be used. Fig 18-17. Even a tiny leak in the structure or in the suit could be fatal. In addition, all tools and materials must be blasted into orbit with expensive rockets or shuttles.

Space Stations

In spite of these daunting challenges, teams of astronauts and scientists have been living in small space structures for many years. These space structures, called space stations, were prefabricated on Earth as modules or whole structures. They were launched aboard powerful rockets and "parked" in earth's orbit. So far, only two countries have built space stations: the United States and Russia. The two most famous stations are the American *Skylab* and the Russian *Mir*.

Skylab

The *Skylab* was a small prefabricated module that was launched into orbit by a huge Saturn V rocket in 1973. Fig 18-18. Astronauts arrived aboard Apollo-style vehicles, which docked in orbit with *Skylab*. The longest stay by any astronaut crew was 84 days. Astronauts' brief stays on *Skylab* allowed scientists to study how the human body reacted to the weightless conditions in space. After six years in orbit, *Skylab* fell back to Earth and was destroyed.

FIG. 18-17 Working in space requires extremely specialized tools and machines.

16. **Critical Thinking.** Ask students to discuss as many reasons as possible that make construction in space so difficult. Develop a list on the chalkboard.
17. **Extension.** Discuss the specialized tools and machines needed for construction in space, including robotic arms, manned maneuvering units, and hand tools that have been specially modified to be used with the heavy gloves astronauts must wear
18. **Critical Thinking.** Ask students to discuss whether (and why) they would have any desire to live in space.
19. **Reteaching.** Make sure students realize that *Skylab* was not constructed in space—essentially, it was a large space vehicle parked in orbit.

CHAPTER 18: Other Construction Projects

FIG. 18-18 NASA's *Skylab* was a prefabricated module that allowed astronauts and scientists to live in space.

Mir

Russia's *Mir* space station was assembled in orbit from several elements or modules. Fig 18-19. The current *Mir* is actually a complex of several elements that have been attached to special docking ports on the original main module. The main module was launched on February 20, 1986. It has been continuously occupied since 1992 by Russian cosmonauts and scientists who arrive aboard *Soyuz* space vehicles. American astronauts and scientists have also been transported to *Mir* by Space Shuttle missions. These Americans stayed on the *Mir* for only a few weeks or months at a time. The *Mir* has allowed scientists to experiment even further with how well humans can survive and function in space. In 1997 many repairs were made on *Mir* so that it could remain in operation.

FIG. 18-19 The Russian *Mir* space station was constructed in space from several modules.

20. **Extension.** Discuss the many problems *Mir* has encountered as it has aged, such as the 1997 onboard fire and collision with a supply vehicle.

SECTION 4: Construction Technology

The International Space Station

The United States, Canada, Japan, the European Space Agency, and Russia are constructing a new, much larger space station—the International Space Station. Fig. 18-20. The National Aeronautics and Space Administration (NASA) has led the development and production of this project.

Scheduled to be completed in 2002, the Space Station is to be 356 feet across and 290 feet long. It will weigh about 950,000 pounds. At any one time, as many as seven people will be able to live on it. The Space Station is designed to be a laboratory for research into advanced industrial materials, communications technology, medical research, and other new technologies.

The space station has been redesigned several times since it was first conceived. The International Space Station is being constructed in several stages. It will be launched in stages as well. Elements or modules will be carried into orbit by American shuttlecrafts and Russian rockets. Astronauts and cosmonauts will use the knowledge and skills gained from *Skylab* and *Mir* to assemble the new station in space. The first element scheduled for launch is called the Functional Cargo Block. It is a 20-ton, 43-foot-long module that contains propulsion, command, and control systems and living space.

FIG. 18-20 The International Space Station will be the largest structure ever constructed in space.

21. **Reteaching.** Help students understand that the International Space Station will be the first such structure on which actual construction will take place in space.

22. **Extension.** Discuss that funding cuts have forced NASA to redesign the International Space Station several times.

CHAPTER 18: Other Construction Projects

Career File—Civil Engineering Technician

EDUCATION AND TRAINING
Most employers prefer to hire people with training from a vocational or technical institute or the Armed Forces. Persons with college courses in science, engineering, and mathematics may qualify for a beginning position, but some on-the-job training is usually required.

AVERAGE STARTING SALARY
Beginning wages for civil engineering technicians can vary from $15,000 to $18,500 per year, depending on education, experience, and area of the country.

OUTLOOK
According to the Bureau of Labor Statistics, job opportunities for civil engineering technicians are expected to grow more slowly than other occupations through the year 2005. As the population increases, there will always be a need for civil engineering technicians to help repair and develop more efficient water, transportation, and pollution-control systems.

What does a civil engineering technician do?
Civil engineering technicians help civil engineers plan and build structures for public use—roads, airports, tunnels, bridges, sewage systems, and buildings.

Does most of the work involve data, people, or things?
Civil engineering technicians often work as part of a team. They may also work with CAD software, measuring devices, and construction materials.

What is a typical day like?
Technicians usually work under the leadership of a civil engineer. They may perform surveys or special studies for the engineer. Some inspect water and wastewater treatment systems to ensure pollution control requirements are met. Others estimate construction costs and specify materials to be used.

What are the working conditions?
Many civil engineering technicians work at least part of the time at a construction site, which may offer temporary office space. The construction site may pose hazards which require the use of hard hats or other protective equipment.

Which skills and aptitudes are needed for this career?
Creativity and an active imagination are desirable for technicians involved in design work. Good communication skills and the ability to work well with others are important since technicians often work as part of a team.

What careers are related to this one?
Other occupations that require similar training and use similar scientific principles include drafters, surveyors, and science technicians.

One civil engineering technician talks about her job:
I love working in the field with the team. I would never want a desk job. Although some days when it's bitter cold, I think maybe desk work isn't so bad!

Career File

400 SECTION 4: Construction Technology

Chapter 18 REVIEW

Reviewing Main Ideas

- Surveying, earthmoving, and paving are used to build roads and highways.
- Types of bridges include beam, arch, truss, cantilever, suspension, cable-stayed, and movable.
- Dams provide a dependable water supply, help control flooding, and provide hydroelectricity.
- Canals are built for navigation or irrigation and may require locks.
- The three main types of tunnels are earth, immersed, and rock.
- Most pipelines are buried underground, with only the pumping stations above ground.
- Air-supported structures consist of a structural membrane supported by air pressure.
- Space structures require special transportation and building methods.

Understanding Concepts

1. Describe the two types of roadbed used in highway construction.
2. Name and describe the seven types of bridges.
3. What is a reservoir? Name two uses for a reservoir.
4. What is a cofferdam? When are they used?
5. Describe the Chunnel and how it was constructed.

Thinking Critically

1. What impacts might occur if a major highway that goes directly through the center of a town were to be replaced with a new highway that would bypass the town?
2. What needs to be considered when deciding on the type of bridge to build?
3. What factors would you consider if you had to decide whether or not to build a dam in a specific location?
4. Discuss some negative and positive impacts of various types of construction (highways, dams, etc.) on farming.
5. Do you feel that it is worthwhile for humans to live in the dangerous environment of space? Discuss your reasons.

Applying Concepts and Solving Problems

1. **Mathematics.** Nick wants to install a driveway that will measure 13½' x 24'. If he intends to pour the concrete 4" thick, how many cubic yards of concrete will he need? (Hint: there are 27 cubic feet in a cubic yard.)
2. **Science.** Plants will grow differently in space than they do on earth. Use a moistened piece of paper towel inside a glass to sprout a bean seed. Note the orientation of leaves and roots. Now turn the glass over. What happens in the next few days? Why?

24. **Resource.** Answers to "Understanding Concepts," "Thinking Critically," and "Applying Concepts and Solving Problems" can be found in the Teacher's Manual at the front of this *Teacher's Annotated Edition* and in the *Teacher's Resource Binder.*

1 DIRECTED ACTIVITY

Learning about Construction in Your Community

Context

As we go about our daily lives, we often take only passing notice of construction in progress—unless we happen to get in the middle of a traffic snarl caused by road construction! One day, a structure just seems to have appeared out of nowhere. Other times, especially in the case of large construction projects or an eagerly awaited store, we watch more closely, noting the progress each time we pass by. It's fun watching the project take shape.

Goal

In this activity, you will look more closely at the various types of construction in your community. You will photograph different construction sites at various stages of construction. Then you will do research to learn how many people and what types of services it took to make this construction project possible. You will open your eyes and observe how construction is changing the world in your own "backyard," your community.

> **Equipment and Materials**
> - 35 mm camera to take color slides or photos
> - sufficient film to take one or more slides or photos of each construction project
> - a map of your community
> - a telephone book

Procedure

Note: If you live in a small community where little or no construction is underway or has been recently completed, you may need to adopt a community for this activity. If you live in a large community with a great deal of construction underway, it may be necessary for the class to be divided into teams, with each team responsible for a different section of the community.

1. Obtain the needed equipment and materials.

2. Locate examples of the different categories of construction—residential, industrial, commercial, and public works—in your community. (The construction may be completely new, or it may be an addition or a ren-

ovation.) Mark these sites on your map, using a code of separate symbols or colors for each of the four categories of construction.

3. Visit and photographically record each site in at least one stage of construction. If possible, prepare a series of photographs of one or more construction sites showing various stages of construction. (*Note:* In some situations, it is best to have the permission of the owner before taking the photographs.)

4. Obtain the names of the companies or contractors doing the construction, and contact them to find out how many people are employed on these projects.

5. If possible, talk to the general contractor on at least one project. Find out what kind of services had to be employed to help complete the project. It is not important to have the names of the organizations providing this help unless the general contractor volunteers the information.

6. Organize the information you have obtained during this study. Create a display comparing the four categories of construction, based on your observations.

Evaluation

1. Were you able to find each of the four categories of construction in your community? If not, what did you do?

2. Which of the four categories was the easiest to find? Which was the hardest?

3. Which of the four categories employed the most workers?

Useful Skills across the Curriculum

1. Language Arts: Write a news article for your school paper about one of the construction sites you visited. Mention at least two impacts that you think the new structure will have on the community.

2. Mathematics. Find the mileage scale on your map. What scale is used? Pick two sites on the map and determine the distance in miles between them.

2 DIRECTED ACTIVITY

Estimating and Bidding a Construction Project

Context

In order to know how much a construction job will cost, an estimate is prepared for all the materials and labor required. The contractor can then make a bid on the project. The bid takes into account the costs of materials and labor as well as a certain percentage for profit. The contractor who makes the lowest bid usually gets the job, so accuracy in the estimate is important.

An estimate is based on information found in the architectural drawings and specifications. The information on these drawings is used to determine not only the quantity of materials (paint, in this case) that will be needed but also the time—in terms of labor—that will be needed. The cost of painting a given area will vary depending upon its size, since the larger the area, the more time and materials that will be needed.

Goal

For this activity, you will estimate the cost of materials and labor needed to paint several rooms in a house. You will then submit a bid for the project.

Equipment and Materials
- pencil
- notebook paper
- calculator
- paint supply catalogs or price lists

Procedure

1. Carefully study the floor plan and tables for this activity. When estimating painting materials, the job should be figured room-by-room.

2. First, look at Table A, Approximate Paint Requirements. (Note that the amounts indicated on the table are for *one coat* of paint. Frequently, two coats will be needed.) The first column in the table lists several room perimeters. (The *perimeter* is the distance around the room.) For example, the floor plan indicates that the den is 10' x 12'. To find the den's perimeter, you would add the length of each wall (10 + 10 + 12 + 12) for a total distance of 44 feet. In the first column of Table A, find the perimeter that is nearest to that distance.

SECTION 4: Construction Technology

FLOOR PLAN
CEILING HEIGHTS ARE 8'

TABLE A. APPROXIMATE PAINT REQUIREMENTS*		
Room Perimeter	**For Walls with 8' Ceiling Heights**	**For Ceiling**
30 ft.	⅝ gallon	1 pint
35 ft.	¾ gallon	1 quart
40 ft.	⅞ gallon	1 quart
45 ft.	⅞ gallon	3 pints
50 ft.	1 gallon	3 pints
55 ft.	1⅛ gallons	2 quarts
60 ft.	1¼ gallons	2 quarts
70 ft.	1⅜ gallons	3 quarts
80 ft.	1½ gallons	1 gallon
Baseboard trim	⅙ the amount required for the room	
Window frames (average size)	¼ pint each	
36"-wide doors	½ pint each	

*For one coat

SECTION 4: Construction Technology

3. The next column in Table A indicates the amount of paint needed for *one coat* of paint for all the walls of a room of a given perimeter and an 8' ceiling height (distance from floor to ceiling). The amount of paint needed for the walls of a room with a 44' perimeter and 8' high ceilings is ⅞ gallon.

Note that the bottom section of the table indicates the amount of paint needed for the baseboard trim as well as for each window frame and each door. These areas are usually painted in a different color or are stained. When estimating the amount of paint needed for the *walls*, you should, therefore, reduce the amount of wall paint needed by ¼ pint for each window in the room and by ½ pint for each door. There are two doors in the den and one window, so the amount of paint needed for the walls can be reduced by 1¼ pints (¼ + ½ + ½). This amount can be subtracted from the ⅞ gallon of wall paint needed. (When purchasing paint for only one room, this amount is often too small to matter. However, if you are purchasing paint for an entire house, it can make a difference.)

4. If the trim, door(s), and window frame(s) are to be painted a contrast color (rather than stained), you will need to include paint for these areas when determining your *total* paint needs.

5. The final column in Table A indicates the amount needed for one coat of paint for the ceiling of a room of a given perimeter. The den has a perimeter of 44 feet, so you would need approximately 3 pints of paint for one coat on the den's ceiling.

TABLE B. LABOR REQUIRED FOR INTERIOR PAINTING*	
Baseboard Trim	2.5 ft. per minute
Walls and Ceiling (with roller)	5 sq. ft. per minute
Window Frame (average size)	3/4 hour per coat
Door	1/2 hour per coat

*Does not include time for preparing surfaces or setting up equipment.

6. Table B shows the labor requirements for interior painting. Note that labor for the trim is indicated in *linear* feet while the labor for the walls and ceilings is indicated in *square* feet. In the case of the den, you already determined (in Step 2) that the linear distance around the room

is 44 feet. The table indicates that a painter could paint the baseboard trim at a rate of 2.5 feet per minute, so the den's trim could be painted in about 17½ minutes per coat.

7. To find the wall area, you multiply the length of each wall by its height (ceiling height). To determine the *total* wall area of a room, you need to calculate the area for each wall in the room and then add these four areas. In the den, two walls are 10' long and 8' high. The area of each of these walls is 80 square feet (8 x 10). The other two walls are 12' long and 8' high, so the area of each of these walls is 96 square feet. Add the square footage of each wall to find the total wall area. For the den, the total wall area is 352 square feet (80 + 80 + 96 + 96).

For ceilings, the area is found by measuring the length times the width. For the den in our example, the ceiling's width is 10 feet and its length is 12 feet. The ceiling area is 120 square feet (10 x 12). Combine this area with the total wall area. The combined area of the walls and ceiling in the den is 472 square feet. Table B indicates that a painter could apply one coat of paint at a rate of 5 square feet a minute, so it would take about 1 hour and 34 minutes to apply one coat of paint to the den's walls and ceilings.

8. Next, determine the labor required for painting the window frame(s) and door(s) of the room. Refer again to Table B. The table indicates that it takes ¾ of a hour to paint each window frame and ½ hour to paint each door. The den has one window and two doors, so it would take 1¾ hours to paint them. Add this to the 1 hour and 34 minutes for painting the walls and ceiling and the 17½ minutes for painting the trim. The total labor for painting the den is about 3 hours and 37 minutes per coat.

9. Table C shows the rooms in the house (see the floor plan) that need painting and the paint colors chosen. Study this table carefully. Which totals can be combined when determining the amount of paint needed for the entire job and which cannot?

10. Using separate sheets of paper and paint supply catalogs (or information provided by your teacher), calculate the *total* cost of painting each room indicated in Table C—the cost of the paint (for walls, trim, doors, and window frames) plus the labor costs (assume that the painter earns $18.50 per hour). *Your costs should reflect that each area will need two coats of paint.* Then develop and complete a bid estimate form similar to the one in Table D. (Please create your own table on a separate sheet of paper and do not write in this book.) Be sure to add 10 percent to the totals to allow for profit; this is shown as "Extended Cost" (actual cost plus profit) in Table D.

TABLE C. JOB REQUIREMENTS

Room	Wall Color	Trim Color	Ceiling Color	Door Color	Window Frame Color
Family room	yellow	white	white	white	white
Living room	blue	white	white	white	white
Dining room	green	white	white	white	white
Bedroom A	blue	white	white	white	white
Bedroom B	pink	white	white	white	white

TABLE D. CONSTRUCTION BID ESTIMATE FORM

Paint Color	Quantity	Unit Cost	Total Cost	Extended Cost

Total: $ _____

Room	Hours of Labor	Unit Cost	Total Cost	Extended Cost

Total: $ _____

Total Bid for the Job: $ _____

11. At the end of the activity, all the bids should be collected and analyzed. Which bids were lowest? Were they accurate? Did they include all the necessary materials and labor?

Evaluation

1. Did you have any problems converting inches to feet when determining the wall and ceiling areas for the bedrooms?

2. Assume you are going to paint your own room. What are three things you would do before buying the paint?

Useful Skills across the Curriculum

1. Mathematics. Suppose you are a professional painter who works for $18.50 a hour. You and one *un*skilled helper can paint a house in five eight-hour days. With a *semi*-skilled helper, you could do the job in three days. Unskilled help earns $5.50 a hour. Semi-skilled help earns $12 a hour. Calculate the total labor costs with both unskilled and semi-skilled help. Which is most economical and would produce the best bid?

2. Science. Do some research to learn about the chemistry of paint. What ingredients are used and what qualities do they add to the mixture?

---

1. **Answers.** With semi-skilled help $732; with unskilled help $960; using semi-skilled help is more economical.

SECTION 4: Construction Technology

DESIGN AND PROBLEM-SOLVING ACTIVITY

Designing and Building Modular Hotel Rooms

Context

Not all buildings are constructed entirely on site. Some items, such as trusses and wall panels, may be manufactured elsewhere and then shipped to the construction site. Modular construction involves manufacturing a module, or box-like structure, having several rooms. Modules are shipped to the construction site and joined together to create such structures as hotels and houses. Modular sections are commonly no more than 12 feet wide so they can be transported by semi-trailer.

> **Health and Safety Notes**
> Before doing this activity, make sure you understand how to use the tools and materials safely. Have your teacher demonstrate their proper use. Be sure knives and scissors are sharp. Follow all safety rules.

Goal

For this activity, you and your teammates will design and build two connecting modular room units for a hotel.

1. State the Problem

Design and build modular room units for a hotel.

Specifications and Limits

- Construct models of two identical modular hotel rooms. The actual rooms would measure 12 ft. x 16 ft. The ceilings would be 7 ft. high. The models should be built to a scale of ½" = 1'0".
- The scale models must be stackable, so wires, pipes, and fixtures must be aligned.
- Each unit must include a sleeping area, bathroom, closet/entry area, and sitting area. The wall opposite the entry must have a window. All windows must be on exterior walls.
- The scale models must include mockups of bathroom and electrical fixtures.
- Hallways should not be a part of the module.
- You must turn in a floor plan and a wall section drawing showing major parts, such as joists, sills, and headers.

SECTION 4: Construction Technology

2. Collect Information

With your teammates, read through this entire activity. You may want to copy the specifications outlined above. This will help you remember them later.

Has any member of the team stayed in a hotel or motel? Discuss the experience and the room's features.

3. Develop Alternative Solutions

Talk with your teammates about possible room designs. Each of you should make sketches of two or three that might work. Determine how to build your model. The list of equipment and materials at the right may be of help.

Equipment and Materials
- straws
- styrofoam
- modeling clay
- foamboard
- utility knives
- scissors

4. Select the Best Solution

As a team, select the design you like best. Check it against the specifications. Is anything missing?

5. Implement the Solution

Create a floor plan and a wall section drawing.
Build your scale models.

6. Evaluate the Solution

As you worked, did you perceive some advantages to using modular construction? Explain. What might be some disadvantages?

Was any part of the model construction more difficult than others? If so, why? What would you do differently if you were to design another modular room?

Useful Skills across the Curriculum

1. Mathematics. Calculate the square footage of the life-size module. Why would knowing square footage be important during actual construction?

2. Science. What forces of nature act upon structures? How must designers compensate?

Credit: Brigitte G. Valesey

SECTION 4: Construction Technology

4 DESIGN AND PROBLEM-SOLVING ACTIVITY

Investigating and Designing Highway Construction

Health and Safety Notes
Before doing this activity, make sure you understand how to use the tools and materials safely. Have your teacher demonstrate their proper use. Be sure knives and scissors are sharp. Follow all safety rules.

Context
Highway construction usually occurs in response to local or regional transportation needs. Roads may become more heavily traveled, accidents and fatalities may increase, and traffic tie-ups may be a daily commuter problem. New roads and changes in existing roads are the solutions to these problems, and they have many impacts on the community. The changes may make the roads safer. Wider shoulders and guide rails may be added. Travel from local neighborhoods connected to main roads may be easier and more efficient. However, there are also disadvantages. Buildings may have to be torn down, utility poles set back, and private property adjoining the roads reduced. Increased traffic noise and parking problems may also result. Each community must consider many issues when designing a highway.

Goal
For this activity, you will select a roadway in your community that needs improvement, make a map of it, re-design it, and create a new map and model showing your changes.

1. State the Problem
Study a local roadway needing improvement, make a map of it, re-design it, and create a new map and model showing your changes.

Specifications and Limits
- You must create two maps, one of the current roadway and one showing your improvements.
- The maps must use the appropriate civil engineering symbols and show:
  -traffic patterns
  -road design features (such as drainage areas and high-occupancy-vehicle [HOV] lanes)
  -environmentally sensitive areas next to or within five miles of the roadway

- The maps can be drawn on paper with technical drawing tools or on computer using CAD software.
- The model can be a traditional, physical model or one made on the computer.

2. Collect Information

Consider roadway sites in your community that are scheduled for improvement, have already been improved, or have become hazardous. Obtain a map of the site. Your local transportation department, department of public works, or local library should have this information. Your local parks and planning office or environmental groups can provide information about environmentally sensitive areas along the roadways.

As you gather information, create a map of the current roadway showing key features and environmentally sensitive areas with such things as shading or color overlays. Use this map as your guide to what needs to be remedied.

Study local roadways that are similar but seem to have fewer problems. What contributes to their success?

3. Develop Alternative Solutions

Consider several solutions to the problems. Would such things as more traffic lanes, shoulders, guard rails, landscape features, or overpasses help? What impacts would result from each solution? For example, if a roadway is widened, will it come too close to adjacent buildings? Will it mean the destruction of many mature trees?

4. Select the Best Solution

Select the solution that seems to solve the most problems with the fewest negative impacts.

5. Implement the Solution

Create the second map showing your improvements. Check to make sure you have followed all the specifications.

Build your model. Check to be sure the model contains everything shown on the map.

6. Evaluate the Solution

If you were to design another roadway, what would you do differently? What did you learn about maps from this activity?

Useful Skills across the Curriculum

1. Mathematics. Visit your roadway site during rush hour in the morning or evening and count the number of cars that pass during a fifteen-minute period. From this number, estimate how many cars use the roadway on an average weekday.

2. Language Arts. Write a letter to the editor of your local newspaper describing the problems occurring at the site you have studied and outlining your solutions.

Credit: Brigitte Valesey

DESIGN AND PROBLEM-SOLVING ACTIVITY 5

Designing and Building Model Space Structures

Context

A *satellite* is an object that orbits a larger object. For example, the moon is a satellite of earth. Today, most satellites that people manufacture and send into orbit are communication tools. They transmit information. Some detect weather patterns. Others relay telecommunication messages and transmit television signals from around the world. Plans are now being made for another type of satellite. It will be a space station in which people will live and work—an actual space structure. Figure A shows one possible design.

Health and Safety Notes: Before doing this activity, make sure you understand how to use the tools and materials safely. Have your teacher demonstrate their proper use. Follow all safety rules.

Goal

In this activity, you will research the needs that must be met to make a space station livable and functional. You will also design and build a three-dimensional scale model. Along the way, you will develop your skills in research, design, and development.

SECTION 4: Construction Technology 415

1. State the Problem

Design and build a three-dimensional scale model of a space station.

Specifications and Limits

- The work areas on the station should include a research laboratory, a manufacturing facility, and an area for satellite repair.
- The living areas should include provisions for eating, sleeping, personal hygiene, and exercise.
- Life support areas should include a greenhouse and a power plant.
- Remember, this will be a low-gravity environment. People will be able to stand, work, or sleep on the "ceiling" as easily as on the "floor."

2. Collect Information

Research books, magazines, scientific journals, the NASA web site, and other sources for information about the design and possible uses of space stations.

3. Develop Alternative Solutions

Prepare three or four thumbnail sketches of possible exterior designs. Keep the designs fairly simple. Remember, you'll need to construct a model of the final design. Consider what equipment and materials you might use.

4. Select the Best Solution

Select one design. Make a rendering of the exterior of your space station. Develop and draw a floor plan for each level. Two ideas are shown here, but try to create your own design. Use a computer drawing program if one is available.

5. Implement the Solution

Plan how to build your model. Determine the best way to show both the inside and the outside of your space station. For example, you may want to construct your model so that the top lifts off, giving you a "bird's eye view" of the inside.

Construct your model. Prepare a presentation of your ideas, using your rendering, drawings, and model to show your ideas as you explain them.

Equipment and Materials

When selecting equipment and materials, consider their cost, availability, and suitability to the design.

- paper
- posterboard
- drafting instruments or CAD system
- scissors or utility knife
- colored pencils
- glue
- tape
- mylar film
- aluminum foil
- paints
- string
- contact cement

SECTION 4: Construction Technology

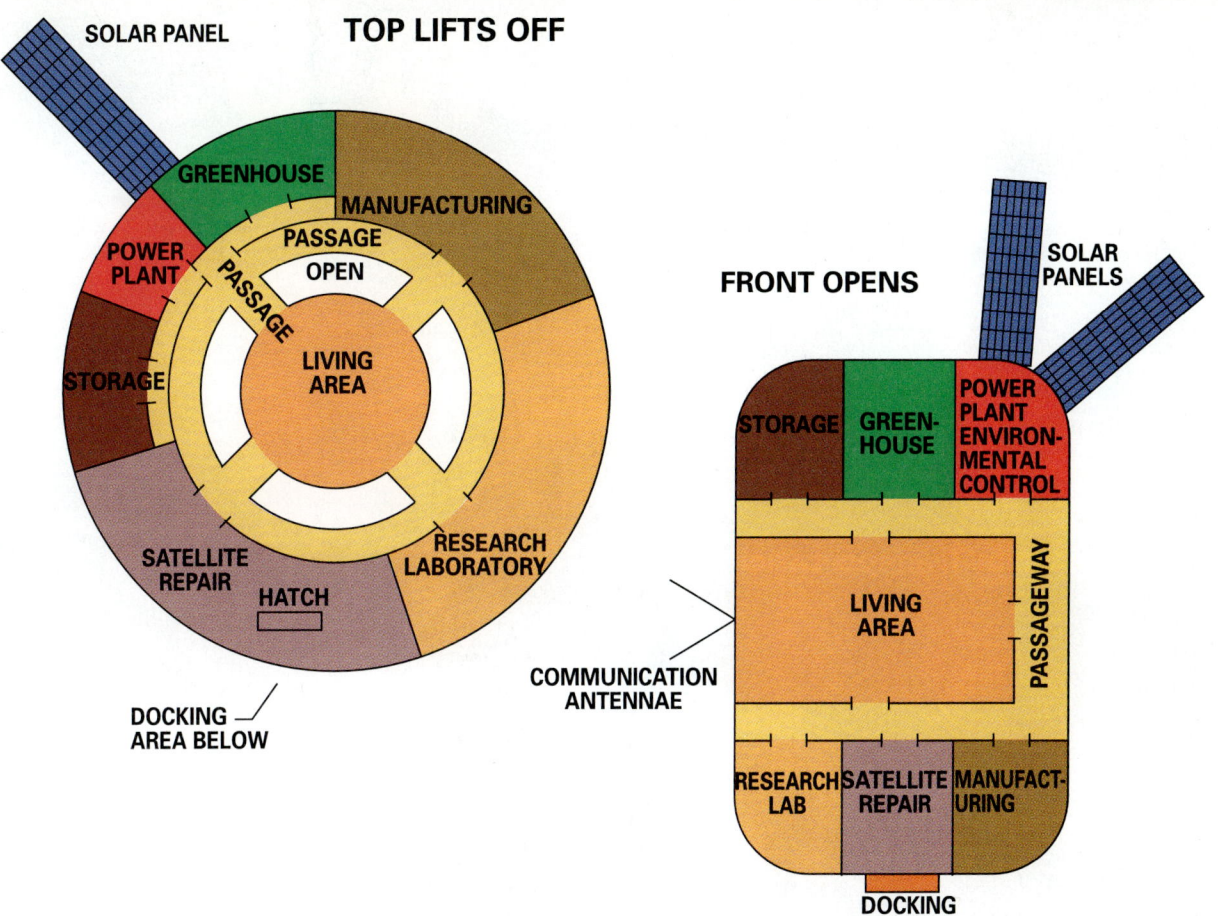

6. Evaluate the Solution

Review the research you did on space stations. How have the possible designs of space stations changed since they were first planned?

How livable and functional is your space station? Look at your drawings and finished model. Are there any parts that you have forgotten to include?

If you could design and build another space station, what would you do differently?

Useful Skills across the Curriculum

1. Science. Research what materials are preferred for building space structures. What are their properties?

2. Social Studies. What problems might arise among people living in close quarters in a space structure. How could such problems be avoided?

SECTION 5

CHAPTERS

19 Transportation Systems

20 Types and Modes of Transportation

21 Intermodal Transportation

22 Power in Transportation

TECHNOLOGY TIME LINE

3000 B.C.	550 B.C.	1500s	1662	1769	1869	1885
Wind is used to power boats.	Anaximander of Miletus makes the first map of the known world	Incas create 12,000 miles of roadway in South America.	First public transportation system is established in Paris, France.	First steam-powered car is developed in France.	Transcontinental railroad completed in U.S.	J.K. Stanley develops the basic model for the modern bicycle in England.

418 SECTION 5: Transportation Technology

Transportation Technology

*T*ransportation involves moving people or goods from one place to another. A large variety of vehicles and methods are used. In this section, you'll learn the advantages and disadvantages of different kinds of transportation. You'll find out about the engines and motors that make transportation possible.

1903
Wright brothers fly a gasoline-powered plane.

1947
Capt. Charles Yeager breaks the sound barrier in a jet plane.

1954
U.S.S. Nautilus becomes the first nuclear-powered submarine.

1957
Soviet Union's *Sputnik I* becomes the first artificial satellite to orbit Earth.

1989
Voyager II heads for interstellar space.

1997
NASA Pathfinder lands on Mars.

SECTION 5: Transportation Technology

CHAPTER 19

Transportation Systems

Objectives

After studying this chapter, you should be able to:
- define transportation.
- describe how transportation is used to provide services.
- explain how transportation affects the economic value of products.
- name the parts of transportation systems.
- identify some of the positive and negative impacts of transportation.

Terms

break bulk cargo
bulk cargo
cargo
freight
gridlock
on-site transportation
routes
time and place utility
transportation
vehicle
ways

Technology Focus
Transportation Lifelines

Transportation systems serve many uses. They bring groceries to your local stores. They take you to vacation spots. They can even save your life. In a medical emergency, every moment counts. In minutes, today's transportation systems can bring an accident victim to the hospital. In a few hours, they can bring life-saving medicine from one country to another.

Taking Patients to the Hospital

An ambulance transports sick or injured people to hospitals or other places where they can get medical care. The first ambulances were horse-drawn wagons, and they were used in wartime to carry wounded or sick soldiers. The Spaniards used ambulances as early as 1487. However, a French military surgeon named Dominique-Jean Larrey made ambulance service part of standard wartime medical practice. Larrey, who lived from 1766 to 1842, was the principal surgeon of the French Army under Napoleon Bonaparte. The first nonmilitary hospital to use an ambulance was Bellevue, in New York City. It began ambulance service in 1869.

Early ambulances simply brought people to the hospital as fast as they could. Many people died because they did not receive help soon enough. Today's ambulances carry advanced equipment and are staffed by emergency medical technicians (EMTs). The EMTs have been trained to provide emergency care at the scene and on the way to the hospital.

What if a regular ambulance can't get to the emergency scene? Air ambulances began during World War II, when helicopters were used to rescue wounded soldiers. Today, both helicopters and airplanes are used to transport sick or injured people.

Bringing the Hospital to the Patients

In 1928, the world's first "flying doctor" service was started in Australia. Today such services exist in many parts of the world. They transport doctors and other medical caregivers to remote areas, such as the Arctic, that have few or no hospitals.

Médecins Sans Frontières (Doctors Without Borders) provides medical care to victims of war, epidemics, or natural disaster. It was started in 1971 by a group of French doctors. It is now an international organization that sends about 2,000 volunteer doctors and nurses to more than 70 countries each year. Within 24 hours, MSF can send airplanes with medical equipment and personnel to the site of a crisis.

Medical transport has come a long way since the days of the horse-drawn ambulance, but the purpose is still the same: using transportation technology to save lives.

Take Action!
Visit the web site of Doctors Without Borders at http://www.dwb.org/index/htm. What are some of their current missions?

Resources
- Chapter 19 Lesson Plan in the *Teacher's Annotated Edition* and in the *Teacher's Resource Binder*.
- Chapter 19 Review in the *Teacher's Resource Binder*.

What Is Transportation?

To *transport* means to carry from one place to another. **Transportation** is the movement of people, animals, or things from one place to another using vehicles. A **vehicle** is any means or device used to transport people, animals, or things. Some examples of vehicles include buses, trucks, airplanes, ships, railroad cars, and bicycles. Fig. 19-1.

The distance traveled to transport someone or something need not be long. **On-site transportation** always happens within a building or a group of buildings. Moving from one story of a building to another in an elevator is an example of on-site transportation. Loading or unloading freight using an industrial truck is another example. Fig. 19-2. You are performing on-site transportation tasks when you carry your schoolbooks from your locker to the classroom.

People see and use the services of transportation each day. For example, how do you and other students get to school? Look around you. Everything in your classroom has been transported. Transportation affects almost all parts of our lives.

FIG. 19-1 More and more trucks are being used to carry goods on interstate highways.

1. **Discussion.** Develop a lively discussion that draws each student into sharing his/her experiences with transportation.

FIG. 19-2 A variety of on-site transportation vehicles are available for efficient movement of people and cargo.

Using Transportation

All vehicles need a place to operate and facilities to support their operation. Transportation is used to move people or cargo, and it can increase the value of what is moved.

Ways and Routes

Vehicles are usually operated on ways and routes. **Ways** are the actual spaces set aside especially for use by the transportation system. The way may be a specific strip of land or measured altitude above the earth. Pipeline right-of-ways, highways, and railways are examples of ways. **Routes** are particular courses traveled. Shipping lanes and air routes are examples of routes.

The air, water, highway, and railroad transportation industries are operated on ways and routes that are owned or controlled by the local, state, or federal government. Fig. 19-3. The railroad and pipeline industries own and maintain their own ways.

Support Facilities

Transportation support facilities include a variety of terminals and warehouses, rail yards, landing strips, ports and docks, repair garages, and hangars. Your nearby gas station is a support facility for a transportation system.

In every method of transportation, the costs of operating the support facilities are greater than the costs of buying and operating the vehicles. Most of the people employed in the transportation industry work at support facilities. Fig. 19-4.

2. **Reinforcement.** Discuss familiar ways and routes as well as one or two that students are not likely to think of.
3. **Brainstorming.** It takes far more people to keep an airplane flying than to actually fly the plane. Have students name as many "behind the scenes" people as they can.

CHAPTER 19: Transportation Systems

FIG. 19-3 The interstate highway system was built and is maintained by federal and state governments working together.

FIG. 19-4 The airport, such as this one in Denver, is the main support facility for air transportation.

Passengers and Cargo

Everything that is transported in vehicles is classified as either passengers or cargo. Passengers are people who are moved from one place to another. Everything else transported, from automobiles to zoo animals, is **cargo**. Cargo is also called **freight**. There are two basic types of cargo: bulk cargo and break bulk cargo.

Bulk cargo is *loose* cargo. It may be a solid material like sand or a liquid such as oil. It may even be a gas. Bulk cargo is not packaged, and it is usually never mixed with another cargo in a transportation vehicle. Fig. 19-5.

Break bulk cargo consists of single units or cartons of freight. Fig. 19-6.

FIG. 19-6 Break bulk cargo must be packed with care to make efficient use of space and to prevent damage. These sweaters must be protected from dirt and rips or tears.

FIG. 19-5 Bulk cargo, such as coal, is loaded and unloaded quickly.

Books, bicycles, and meat are packaged and shipped as break bulk cargo. Almost all the items in the stores where you shop were delivered as break bulk cargo.

It generally costs more to transport materials as break bulk than as bulk. Packaging costs are higher, and more handling is required. Cost is not the only factor that must be considered, however, when deciding how freight should be transported. For example, a school could charge less for milk if each student filled his or her own glass from a milk tank truck (bulk). However, imagine the problems that would cause! It is more convenient to provide milk in small cartons (break bulk).

4. **Reteaching.** Have students give examples of times when they have been passengers.
5. **Reteaching.** Have students give examples of bulk cargo they have seen being transported.

CHAPTER 19: Transportation Systems

TRANSPORTING A PRODUCT FROM THE FACTORY TO THE CUSTOMER

AT THE FACTORY, SNAP 'N CRACKLE CEREAL IS LOADED ON TRUCKS.

THE TRUCKS TAKE THE CEREAL TO THE WHOLESALER, WHERE IT IS STORED IN WAREHOUSES. WHOLESALERS BUY LARGE QUANTITIES OF ITEMS AT A DISCOUNT AND RESELL THEM TO RETAIL STORES.

WHEN EAT-RITE FOODSTORES PLACES AN ORDER, THE WHOLESALER LOADS THE CEREAL ON ANOTHER TRUCK.

THE TRUCK DELIVERS THE CEREAL TO EAT-RITE'S WAREHOUSES.

AS INDIVIDUAL EAT-RITE STORES PLACE ORDERS FOR CEREAL, SNAP 'N CRACKLE TAKES YET ANOTHER TRUCK RIDE.

A CUSTOMER BUYS THE CEREAL AND TAKES IT ON ITS FINAL JOURNEY TO HER HOME.

FIG. 19-7 What needs do the different locations shown here fill?

SECTION 5: Transportation Technology

Economic Value of Using Transportation

Transportation affects the economic value of services and products. ("Economic" refers to money.) The services of a person or the worth of an object may increase in value by being transported. Two things are required to make the value increase. First, transportation must satisfy a *need* for a person or item to be moved to another location. Fig. 19-7. Second, the transportation must happen at the right *time*. For example, oranges cannot be grown in Canada. Orange growers in southern states transport their fruit to Canada. However, there would be no profit if it took too long to transport the oranges and many of them rotted while in shipment. Another example is seasonal items such as Christmas decorations. If they are delivered December 30, they are of little value.

Change in value caused by transportation is called **time and place utility**. For the most value, items should be delivered to the place they are needed at the time they are needed. Fig. 19-8.

Inputs

Transportation is a system. Fig. 19-9. In our world, there are few systems larger than the transportation system. Inputs are things that are "put into" a system. All systems of technology need seven basic types of inputs to operate: people, information, materials, tools and machines, energy, capital, and time.

People

In all systems of technology, people are the most important input. Without people to supply the knowledge and skills, transportation could not take place.

FIG. 19-8 Items should arrive where they are needed, when they are needed. Fish that is old would not be popular!

6. **Critical Thinking.** See if students can identify examples in your local area to illustrate time and place utility. Stress that these factors must be considered carefully when making transportation decisions.

7. **Reteaching.** Have students make up a mnemonic device to help them remember the seven inputs like: *People In My Tech Ed Class Today.*

CHAPTER 19: Transportation Systems **427**

FIG. 19-9 A transportation system is made up of these parts.

The people involved in transportation are either providers or users. Sometimes the same people can both provide and use the services of transportation. You are a user of transportation. How many times have you used the benefits of transportation today? If you've ever delivered newspapers or other items, you have provided transportation.

Do you know someone who works in the transportation industry? Truck drivers, airline mechanics, dock workers, and train engineers all work in transportation. Other examples include the travel agents who sell airline tickets, the railroad detectives who investigate freight theft, and the people who plan bus routes. You may have a job in the transportation industry right now. Do you work at a car wash? In a service station? For a company that delivers freight? There are many career opportunities in the transportation industry. Fig. 19-10.

8. **Reinforcement.** Have students name transportation-related jobs held by friends or family members. Stress the many jobs in transportation.

FIG. 19-10 The transportation industry is one of the largest employers in the world, requiring many workers like this mechanic.

Information

Information is necessary for every type of transportation system. People who control the systems need information to gain knowledge and skills. Workers need information about how to safely operate tools and vehicles or to load cargo. Managers need information to decide where to send freight. Without the information provided by road signs and maps, imagine trying to travel to or around an unfamiliar city. Fig. 19-11.

FIG. 19-11 Clearly presented information, like these superhighway signs, helps keep traffic flowing smoothly.

9. **Discussion.** Ask students where they get their information.

CHAPTER 19: Transportation Systems **429**

MATHEMATICS CONNECTION
Calculating Distances on a Map

Transportation companies must often calculate distances from one location to another. An ordinary road map is one means of figuring these distances. Road maps usually come with at least one of three common devices for the purpose.

The most frequently used device is a map scale. You use the scale by holding a pencil along its length and measuring off a certain distance, such as 10 miles, by placing your thumb at that point on the pencil. You then hold the pencil over the map, measuring off 10-mile increments until you've covered the distance. Your total of 10-mile increments determines the total mileage.

Most road maps also print notations along each route indicating mileage between points. On the map shown here, find the towns of Natrona and Powder River. Below the route between them is written the number 9. This indicates the miles from one to the other. Some maps use two sets of numbers. One set indicates mileage between cities. A second set shows mileage between the small towns in between.

Some maps also include a third device, a mileage estimating chart. These charts usually give distances only between large cities. If you wanted to learn the distance between Atlanta and Cleveland on this chart, you would place one finger on the Atlanta column and another finger on the Cleveland column. Then you would run your fingers along each column until they met. That number would give the distance—on this chart, 774 miles.

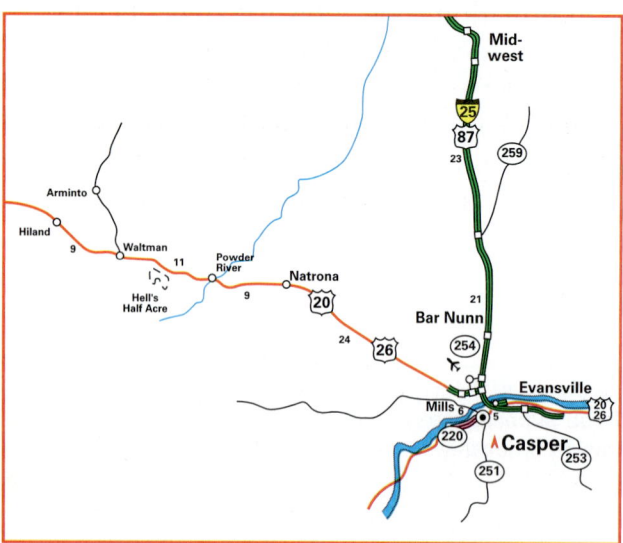

Try It!

1. Using the map, calculate the distance between Waltman and Mills.
2. Using the chart, find the distance between Dallas and Albuquerque.

Answers to these questions can be found in the Lesson Plans in the Teacher's Manual at the beginning of this book.

Materials

You have already learned that any material transported by the system is called cargo or freight. Materials are also needed as inputs for transportation systems. Water is used to cool engines, clean vehicles, and supply drinking water for people. Sand and salt are used to make icy roads safe for driving.

Tools and Machines

A transportation system needs tools and machines of all types. Most of the vehicles used in transportation are actually very complex machines. Without tools to maintain and repair these machines, no transportation system could function very long. Fig. 19-12.

Energy

Transportation systems consume large amounts of energy in many forms. Gasoline, diesel fuel, propane, and kerosene are the most common energy forms used to power vehicles. All of these fuels are made from crude oil, which is a limited natural resource. Electricity is used to power some vehicles, and it also supplies the energy for traffic lights, street lights, and computers.

Capital

Capital is another name for money or investments. A company's capital consists of working capital and fixed capital. Fig. 19-13. Working capital is the money that companies use to buy things and pay their employees. It includes more than cash. Other examples of working capital are shares of stock, money certificates, and property titles. All of these items can be exchanged on an open market.

Fixed capital includes items such as vehicles, facilities, and equipment owned by a company. The buses owned by your school are fixed capital. The computer that records your attendance and grades is another example of the school's fixed capital.

FIG. 19-12 This mechanic needs tools ranging from hand tools to computers to work on modern automobiles.

Time

Time used in transportation systems represents how long it takes to get from one place to another. It may be measured as the seconds it

CHAPTER 19: Transportation Systems

FIG. 19-13 Working capital is money. Fixed capital includes vehicles like this ship.

takes to go from one floor to another in an elevator or as the years it took the *Voyager* spacecraft to travel to another planet.

Saving time often means saving money for an industry. Transportation systems that save time usually replace slower methods. Just as the truck replaced the horse drawn wagon, something faster may someday replace the truck.

Processes

Processes are the things that are done to actually cause the movement of passengers and cargo. In the system of transportation, there are two basic groups of processes: management processes and production processes.

Management Processes

Management processes include the activities the company does to keep the movement of passengers and cargo organized. Computers are important tools in the management processes. The three main management processes are:
- planning
- organizing
- controlling

When *planning* transportation, people decide what must be done and how and when it should be done. They also consider costs.

Organizing processes include deciding who will do the tasks. Deciding which team of workers will do which loading operations is an example of an organizing process.

10. **Extension.** Explain that at walking speed (3 miles per hour) it would take over 42 days to go the 3,000 miles from New York City to San Francisco. A passenger jet takes about 6½ hours. The Space Shuttle, at its fastest speed (17,000 mph), could do it in about 9 minutes.
11. **Critical Thinking.** Have students suggest other examples.

432 SECTION 5: Transportation Technology

During the *controlling* process, people perform a variety of tasks. For example, they may keep records of cars on freight trains, follow directional signals, or record how much money employees earn.

Production Processes

Production processes are the most visible parts of a transportation system. The flight of an airplane and the movement of trucks and cars on the highways are examples of transportation production processes.

Production processes are divided into three types of activities:
- preparing to move
- moving (operation)
- completing the move

Preparing to move involves all the activities accomplished just prior to actually moving. Airline companies load the passengers' baggage into the plane. Shippers lock and seal cargo containers. The school bus driver may scrape the snow from the windshield before leaving the parking lot.

The actual operation (*moving* of the vehicles) is the most exciting process of a transportation system. Two kinds of activities are involved: those needed for vehicle operation and those providing en-route services. Fig. 19-14.

Vehicle operation includes a wide range of activities. One example is opening and closing valves on a pipeline. Another is driving an 18-wheel truck. Even slowing down or speeding up the elevators inside the Statue of Liberty is vehicle operation. How many different transportation vehicles have you ever operated or ridden in? What activities were involved in the vehicle's operation?

Providing en-route services is part of the moving process. These services include such things as serving food on an airline flight or changing a tire at a truck stop. Delivering mail to a ship while it is steaming across the ocean is also an en-route service.

The third process of providing transportation is *completing the move*. Cargo is unloaded. Passengers leave a plane. Vehicles are shut down.

FIG. 19-14 When vehicles are actually moving passengers and cargo, they contribute to the process part of transportation.

12. **Discussion.** Ask students for examples of what they do as they prepare to move or transport something.
13. **Discussion.** Discuss students' experiences with operating vehicles.
14. **Enrichment.** Many students work in service stations or have jobs related to transportation services. Ask for comments about work tasks.

FIG. 19-15 Passengers leave, but workers must stay and prepare the plane for the next trip.

Outputs

The outputs of a system are the results achieved because the system completed the planned processes. In transportation, the basic output is that passengers and cargo are moved. Let's break this down further.

When the move is efficiently completed, changes occur. These are outputs. Perhaps the value of a cargo (time and place utility) increases. Maybe people make money for completing their part in the process. Maybe business people sell and buy merchandise. Perhaps it's just that family and friends get together.

Among the outputs of a system are the effects, or impacts, that system has on people and the environment.

Impacts of Transportation

Transportation has both negative and positive impacts. Let's first consider the positive.

Transportation systems are seldom idle, though. The production process may start all over again immediately. As soon as passengers leave an aircraft, workers begin to prepare for the next flight. Fig. 19-15.

FIG. 19-16 Remotely operated vehicles (ROVs), carrying lights, TV cameras, and other equipment, allow us to explore other worlds from the safety of earth. This picture shows Sojourner, NASA's explorer vehicle, as it roamed about Mars.

15. **Review.** Review the time and place utility concept.

Positive Impacts—The Rewards

Improvements in the technology of transportation have had a tremendous impact on our society. Think about the history of the United States. Developments in transportation supported the development of the country. Canals, railroads, and highways enabled people to explore and settle the far regions of the country. Today, we are still exploring. We are traveling down into the depths of the oceans and out into the far universe. Fig. 19-16.

Many benefits of transportation affect our daily life. Producers expand their markets by transporting their products, which generally causes sales to increase. As sales increase, the demand for the product or one like it also increases. Increased demand also encourages others to go into business. This is *business growth*.

New businesses are also created. The invention of the automobile, for example, brought about the manufacture of tires and traffic lights.

As people and cargo are moved from one location to another, services and products become available to more people, allowing people to choose among products. Each provider or producer tries to offer the best price. This is *competitive pricing*.

Because of transportation, the countries of the world are able to trade products. American producers have exchanged countless shiploads of timber for Japanese electronic equipment, such as VCRs. This could not happen without the services provided by ships, trains, and trucks.

Improvements in vehicles, ways, and facilities have enabled people to travel longer distances in less time. Business people meet more customers. Families can travel to different locations and see new things. Students from the South can snow ski in New York. People from one country can meet people from other countries. All of this and more happen because of an efficient transportation system.

Modern transportation has also improved and expanded public services. Many cities have public transportation, such as subways. In emergency situations, specialized vehicles bring help quickly. Fig. 19-17.

FIG. 19-17 Modern transportation vehicles help us cope with medical emergencies.

16. **Reinforcement.** Discuss the economic principles of supply and demand.
17. **Discussion.** Ask students what products they own that were transported from another country.

FASCINATING FACTS

When Orville Wright made history by piloting the first manned flight in history, he was lying on his stomach. There was no seat in the airplane.

Negative Impacts—The Challenges

Just as transportation can increase the value of an item, it could also cause the value to decrease. If too much of the same product were available at a particular place, the value would go down.

Americans have come to rely upon the automobile as our most popular transportation system. A negative impact of this is that most major cities experience large traffic jams each day. Sometimes a traffic jam becomes a gridlock. A **gridlock** means that no vehicles can move in any direction.

The costs of transportation can be high. People must pay high taxes for the construction and maintenance of public ways and transportation facilities. Other costs cannot be measured as easily as dollars. Homes and land must sometimes be given up to make way for new roads and airports. Thousands of lives are lost each year due to transportation accidents.

Transportation also creates pollution. Discarded or abandoned vehicles cause land and visual pollution. Vehicle exhaust smoke pollutes the air. However, new technology is being developed to help control this. Fig. 19-18. People must do their part in protecting the environment from these types of pollution. New businesses have been started that deal with the recycling of tires and other auto parts.

As you can see, transportation has both positive and negative impacts. Most people feel there are more positive than negative impacts. Understanding how transportation systems work will help us make decisions about how to improve transportation and reduce the negative impacts.

FIG. 19-18 Locomotives used to produce a great deal of smoke. Today, thanks to advances in technology, modern locomotives contribute much less to air pollution.

18. **Critical Thinking.** Compare the positive and negative effects of transportation. List the examples on the chalkboard. Encourage students to try to think of examples not given in the text.

SECTION 5: Transportation Technology

FASCINATING FACTS

Have you ever been seasick? Travel sickness is not caused by food. It is a response to the messages transmitted to the brain by your sense of touch and your sense of balance.

world by satellite. Fig. 19-19. If the package was late, a part of the shipping process may need to be changed.

Feedback also comes from customers. Imagine that two airlines fly between the same two cities and one has a much lower ticket price. Most people would probably fly on the cheaper airline. Low sales would tell the more expensive airline to lower prices or go out of business. The number of tickets sold provides a feedback loop.

Feedback

All systems are made to get a desired output. We need some way to see if the results of the system are what we wanted. The feedback loop is the part of the system that checks the output. If the output is what was desired, the system worked correctly. If the output is not what was desired, the inputs or the processes of the system must be changed.

In transportation systems, feedback can take many forms. If a package was to be shipped to a specific place on a specific day, a feedback loop would check if it arrived on time. The feedback might be as simple as a phone call to the destination or as complex as a computer-controlled system that tracks the package around the

FIG. 19-19 This device can help tell the status of a package at all times.

19. **Reinforcement.** Try to provide as many examples as possible of feedback loops. Ask students if a thermostat on a furnace is part of a feedback loop. Have students come up with examples of their own.

CHAPTER 19: Transportation Systems **437**

Career File—Truck Driver

EDUCATION AND TRAINING

A high school diploma is preferred. In addition, courses in automotive mechanics may help drivers make minor roadside repairs. Drivers of large trucks and drivers who transport hazardous materials need a commercial driver's license. Completing a tractor-trailer driver training program may also be required.

AVERAGE STARTING SALARY

Truck driver jobs vary greatly in terms of earnings. Long-distance drivers, for example, are generally paid by the mile, and the rate per mile can vary widely from employer to employer. The size and type of truck also affects a truck driver's salary. In general, a beginning truck driver might earn about $16,000 per year.

OUTLOOK

The availability of truck driver jobs through the year 2005 is expected to grow about as fast as the average for all jobs, according to the Bureau of Labor Statistics. However, competition for these jobs will be high.

What does a truck driver do?

Truck drivers transport food, materials, and manufactured products from one place to another—usually between manufacturing plants, warehouses, distribution centers, and retail stores. Some truck drivers deliver goods, such as water, frozen foods, or express delivery parcels, directly to consumers.

Does most of the work involve data, people, or things?

Most long-distance truck drivers have little contact with people. They spend most of the day behind the wheel, alone or with a helper. However, drivers with shorter, local routes and several delivery stops meet people at each stop.

What is a typical day like?

Before leaving on a trip, drivers check the truck for fuel and oil, inspect the brakes and lights, and make sure that safety equipment is on board. They make sure cargo is loaded properly so it will not shift during the trip. Some drivers may have to load or unload cargo themselves from the truck. After reaching their destination or returning to home base, truck drivers complete a report about the trip, the condition of the truck, and details about any accidents.

What are the working conditions?

Truck drivers spend most of their working life on the road. Driving for many hours at a stretch can be boring, lonely, and tiring. On long runs, drivers may be away from home for a week or more.

Which skills and aptitudes are needed for this career?

Employers look for responsible, self-motivated individuals, since drivers work with little supervision. A good driving record is necessary. Many employers prefer drivers who are at least 25 years old.

What careers are related to this one?

Other driving occupations include ambulance driver, bus driver, chauffeur, and taxi driver.

One truck driver talks about his job:

I enjoy getting paid for traveling. I've seen most of the United States. My truck is my home on wheels.

Chapter 19 REVIEW

Reviewing Main Ideas

- Transportation is the movement of people, animals, and things from one place to another using vehicles.
- Ways, routes, and support facilities are needed in order for vehicles to operate.
- People who are transported are passengers. Any other transported item is either bulk or break bulk cargo.
- A change in an item's value caused by transportation is called time and place utility.
- A transportation system consists of input, processes, output, and feedback.

Understanding Concepts

1. What is transportation?
2. What is the difference between a way and a route?
3. What is the difference between bulk and break bulk cargo? Give an example of each.
4. Explain how time and place utility can cause the value of an item to increase.
5. List three positive impacts and three negative impacts of transportation. Try to include one or more not given in the book.

Thinking Critically

1. Highways are owned and maintained by federal, state, and local governments. The money for these roads comes mainly from taxes. Is it fair to someone who does not own an automobile to pay these taxes? Why or why not?

2. Is the freedom autos give us worth the pollution that they produce? Explain.
3. Is space exploration worth the billions of dollars it costs? Why or why not?
4. What can or should you do now if you are considering a career in the transportation industry?
5. Most transportation accidents are caused by people. What is the responsibility of vehicle operators and people who maintain and repair vehicles?

Applying Concepts and Solving Problems

1. **Social Studies.** Use a road map to chart a trip from Springfield, Illinois, to Denver, Colorado. Write down all the roads and highways you will take. Estimate how many miles your trip will be.

2. **Science.** Tires inflated with air provide a smoother, safer ride because air is elastic. Fill a small plastic bag with air and hold it closed. Place a heavy object such as a book on it. Apply pressure to the book. What happens?

3. **Mathematics.** To calculate gas mileage, divide the miles traveled since the last fill-up by the number of gallons needed to fill the tank. Ernie's odometer read 39,406.3. He filled his gas tank with 14.2 gallons. The next time he filled up, his mileage was 39,860.7. How many miles did he get to the gallon?

20. **Resource.** Answers to "Understanding Concepts," "Thinking Critically," and "Applying Concepts and Solving Problems" can be found in the Teacher's Manual at the front of this *Teacher's Annotated Edition* and in the *Teacher's Resource Binder*.

CHAPTER 20

Types and Modes of Transportation

Objectives

After studying this chapter, you should be able to:
- name the five modes of transportation.
- explain the reasons for selecting one mode over another.
- name different types of vehicles used to move people and cargo.
- tell which facilities are needed for each mode.

Terms

AMTRAK
barge
booster rockets
classification yards
Global Positioning System
payload
rolling stock
sea-lanes
slurry
space shuttles
ton mile
tractor-semi-trailer rig
unit train

Technology Focus
A High-Tech Nightmare

When Denver International Airport (DIA), the first new airport built in the United States in 20 years, finally opened in 1995, it was 16 months late and almost $3 billion over budget. The costs and delays were caused primarily by the airport's computerized, state-of-the-art automatic baggage-handling system.

Despite warnings from consultants against the system, Denver city officials decided to use it. They believed it would speed up takeoffs and landings, thus saving the airlines hundreds of millions of dollars in the long run and set the standard for airports of the future.

In theory, the $300-million system would handle up to 30,000 pieces of luggage daily. A coded destination tag would be attached to each bag. The tag would be scanned and the bag placed in a computer-controlled car. The car would travel along its 22 miles of tracks to the waiting plane. Everything would happen much faster than in a conventional system.

What Went Wrong

The automated system malfunctioned over and over again during testing and delayed the airport's opening four times. When DIA finally opened, the baggage system created a nightmare for passengers, the airlines, and airport management.

One problem was the scanner's inability to read tags covered by a bag's straps. As a result, for a while, only about 70% of the baggage arrived at its correct destination. Another recurring problem involved bags falling out of their cars and onto the track. Another car would come along at 17 miles per hour and hit the bag, doing serious damage. One computer company executive's $7,500 computer, which had been stored in a protective case, was returned to him in a plastic bag! Handlers had to resort to a conventional system that had been installed as a backup.

Perhaps the lesson to be learned from the experience of Denver International Airport is that a sophisticated new technology requires sophisticated de-bugging. The more complex the system, the more that can go wrong.

Take Action!
Suppose you are a consultant to the Denver airport. How would you solve the problem of straps covering the luggage tags?

Resources
- Chapter 20 Lesson Plan in the *Teacher's Annotated Edition* and in the *Teacher's Resource Binder*.
- Chapter 20 Review in the *Teacher's Resource Binder*.

Air Transportation

Air transportation is the fastest mode of transport in operation today. Airplanes can travel across the United States from coast to coast in five hours or less. To do this, the engines on the planes use large amounts of fuel. Fuel costs, plus the high price of building the aircraft, make air travel for short trips very expensive. However, long trips by air, usually over 500 miles, have become a good value for passenger and some cargo transportation. They save time, which, in turn, helps save money.

Types of Air Transportation

Air transportation is divided into three basic types:
- commercial aviation
- general aviation
- military aviation

Companies involved in *commercial aviation* provide air transportation for a profit. Most commercial aviation planes are large. Some planes now being designed will carry up to 800 passengers. The more people traveling in a plane, the higher the profits. Fig. 20-1.

General aviation includes all privately owned airplanes. These may be used for personal or business reasons. General aviation planes are usually smaller than commercial vehicles. Personal and business flights do not usually involve large groups of people.

Military air transportation includes many kinds of vehicles, large and small. Helicopters, fighter planes, bombers, and surveillance aircraft are some examples of military aircraft.

FIG. 20-1 These operating statistics are for some commonly used aircraft for a recent year.

AIRCRAFT OPERATING STATISTICS

Aircraft	Number of Seats	Speed	Flight Length	Fuel (Gallons per hour)	Operating Cost Per Hour
B747-400	403	534	4,375	3,361	$7,098
L-1011-100/200	293	495	1,519	2,362	4,072
DC-10-10	282	488	1,387	2,189	4,056
A300-600	266	474	1,206	1,686	3,917
B767-300	224	489	2,087	1,485	3,384
A320-100/200	149	441	944	757	1,868
B727-200	148	430	689	1,242	2,263
MD-80	142	419	662	889	1,842
B737-300	131	415	633	739	1,768
DC-9-50	122	387	453	890	1,640
F-100	99	361	388	690	1,445

1. **Extension.** Ask interested students to research some of the newest types of military aircraft and report on their findings.

SECTION 5: Transportation Technology

Aircraft

Vehicles that fly are called aircraft. Airplanes, helicopters, airships, and rocket vehicles are among the different examples of aircraft that transport various types of cargo and passengers.

Airplanes are the most common aircraft, and come in a variety of sizes. They are flying vehicles that are heavier than air. If they are heavier than air, how can they fly? Air helps them. Let's see how it's done.

The plane's engines move the plane down a strip of pavement called a runway. Air flows over and under the wings. The shape of a wing increases the speed of the airflow on the upper surface. This reduces the pressure. At the same time, the pressure on the lower surface increases. The resulting upward movement is *lift*, and the plane flies. Fig. 20-2.

FIG. 20-3 The U. S. Navy's V-22 *Osprey* uses tilt-rotor technology that allows the aircraft to take off like a helicopter and fly like an airplane.

FIG. 20-2 The shape of an airplane wing actually helps the vehicle fly. It causes the air traveling over the wing to go faster than the air traveling under the wing. This creates lift, and the plane rises.

LIFT
↑
LOW PRESSURE

HIGH PRESSURE
→
FORWARD MOVEMENT OF PLANE

Some aircraft use the lifting shape of a wing differently. The blades of a helicopter are shaped like an airplane wing, but they spin to provide lift. Helicopters are an example of V/STOL (vertical/short takeoff and landings) aircraft. Some new designs give airplanes the ability to rotate their wings or engines so that they can take off like a helicopter and then fly like an airplane. Other airplanes use special wings that can provide great lifting power. They can take off and land on very short runways. Fig. 20-3.

Another type of aircraft flies because it is filled with a gas that "floats" on air. You have probably seen a helium-filled balloon. If you let go of the string, the balloon floated away because helium gas is lighter than air. Hot air is also a lighter-

2. **Extension.** Aerodynamic design is discussed in Chapter 22.
3. **Resource.** Refer to the activity at the end of this section entitled "Investigating the Aerodynamics of Gliders and Airplanes."

CHAPTER 20: Types and Modes of Transportation **443**

than-air gas. A hot-air-filled balloon will rise just like one filled with helium.

Lighter-than-air vehicles can be either rigid or nonrigid. Rigid vehicles, called dirigibles, have a strong internal skeleton. They were used frequently from 1900 until the 1930s to transport people and cargo. Fig. 20-4. Nonrigid, lighter-than-air vehicles are usually either hot-air balloons or blimps. These craft have no rigid structure. The strength of the material that contains the gas or hot air is what forms their shape. The largest civilian airship, the *Spirit of Akron*, holds 247,800 cubic feet of helium.

The blimp is usually self-powered by engines and propellers like an airplane. Blimps can transport people or cargo efficiently, but somewhat slowly. Today, blimps are most often used to carry television cameras and for advertising. Fewer than 20 civilian airships are currently registered in the U.S.

FIG. 20-4 Filled with highly flammable hydrogen, the *Hindenburg*, during its trip to America in the 1930s, burned and crashed. From then on, blimps and dirigibles have been filled with helium, which does not explode or burn.

Airways

The area above the earth is termed *airspace*. For safety, aircraft moving through airspace follow specific routes, or airways. There are different routes for commercial, general, and military aircraft. The Federal Aviation Administration (FAA) is a government agency that controls all air traffic above the United States.

FIG. 20-5 Different types of ground vehicles are used to service an airplane at an air terminal.

4. **Extension.** Stress the role that the FAA continues to play in keeping busy airways organized. Planes fly west at even-numbered altitudes; planes fly east at odd-numbered altitudes.

Facilities and Support Services

Transportation facilities are buildings or areas in which activities take place to keep the transportation system working. In air transportation, many ground vehicles as well as facilities are needed. Ground vehicles provide a variety of services. Fig. 20-5. Some bring supplies. Some clean the craft inside and out. Others move passengers and their baggage. Cargo and the planes themselves are moved by still other ground vehicles.

The airport is the main ground facility in air transportation. Aircraft move on taxiways and runways. Terminals, control towers, automobile parking lots, hotels, fire stations, and airplane hangars make up different parts of an airport. Fuel storage areas are usually located nearby. However, they must be far enough away for safety.

Rail Transportation

Some advantages of rail transportation can be seen as you look at a train. Trains are strong and tough! They have steel wheels that move on steel rails. These are strong enough to efficiently transport large, heavy loads. Fig. 20-6.

Trains offer other advantages, too. A loaded railroad car is a "rolling warehouse." It protects the shipper's cargo during transport. In addition, a single train engine hauling several cars of freight uses a lot less fuel than the several trucks it would take to haul the same amount of freight. Also, trains do not get into traffic jams in crowded cities, which can mean a big savings of time.

FIG. 20-6 Rail transportation can carry heavy loads because very little friction is produced when the hard wheels roll on a smooth surface.

5. **Enrichment.** Arrange a field trip to a nearby airport.

Passenger Service

Rail passenger service is a good example of how government and private industry work together to provide transportation. Today, all long-distance rail passenger service in the United States is provided by **AMTRAK**. The AMerican TRavel trAcK system is owned by the federal government. The trains are operated by the railroad lines that own the track the trains are using.

In the United States, most trains carry cargo rather than passengers. Most other industrial countries make much more use of railroads for passenger transportation than we do. Some European countries and Japan have high-speed passenger trains that can go more than 200 miles per hour. These are discussed in Chapter 22.

High-speed rail systems that will travel over 300 miles per hour are currently being developed and operated. Some of these systems are now being planned to link various American cities with high-speed passenger service. The cost of traveling on these modern trains will be competitive with air transportation and handier for the traveler because train stations are located within cities whereas airports are generally located outside of cities.

Regional and city rapid transit trains provide commuter service in some large cities. *Commuter service* is regular back-and-forth passenger service. For example, in New York City, many students travel to and from school on commuter trains. Fig. 20-7.

Freight Service

The railroad industry offers more freight (cargo) service than passenger service. Trains serve over 50,000 towns and cities in North America. There are two basic types of freight trains: unit trains and regular freight trains. Fig. 20-8.

FIG. 20-7 Subway trains provide fast commuter service below the busy city streets.

6. **Enrichment.** Discuss AMTRAK. Gather sample schedules and fare and route information. Help students understand the benefits and costs.

7. **Reinforcement.** Discuss types of commuter services with which students have had experience.

SECTION 5: Transportation Technology

FIG. 20-8 A unit train (left) is owned by a company and carries only that company's products. A regular train (right) may be made up of many different types of cars going to many different locations.

A **unit train** provides efficient transportation. It carries the same type of freight in the same type of car to the same place time after time. Some unit trains are not owned by the railroad lines. The owners may pay the lines to pull their cars.

Regular freight trains offer a variety of short- and long-haul services. Cars are arranged on the train according to the type of cargo and final destination. The cars may be switched to several different trains before they are finally unloaded.

Rolling Stock—the Vehicles

Rail transportation vehicles are called **rolling stock**. There are three different groups of rolling stock: engines, railroad cars, and maintenance vehicles.

Engines today are much safer and more comfortable than engines of the past. However, these modern locomotives still serve the same purpose—they pull the train. Most locomotives use diesel engines to turn electrical generators. The electricity produced then powers traction motors that turn the wheels. These combinations of engines and motors are called *diesel-electric locomotives*. [8]

Railroad cars can transport many different forms of cargo. The most common car is the boxcar, a fully-enclosed car that can carry about anything that can be packed into it. Other cars such as liquefied gas cars are more specialized.

Maintenance vehicles are used to keep rails clean and safe. Track inspection cars, brush-cutting machines, and hoist cars are examples of maintenance vehicles.

In the United States, all tracks are built to the same gauge (distance between rails). This allows engines and cars from one line to interchange and roll on the tracks of any other line. [9]

8. **Extension.** Stress respect for moving trains and railroad ways and crossings. A moving train cannot stop quickly. NEVER try to "beat" a train to a crossing and never walk within the railroad way.

9. **Research.** Have students research to see if other countries around the world build all their tracks to the same gauge.

Facilities

Classification yards, shops, and terminals are facilities in rail transportation. They are connected by railroad tracks.

Classification yards are also called *switch yards*. This is where trains are "taken apart and put together." A train pulls into the yard and the cars are disconnected and sorted according to their destinations. The reclassified cars are rolled onto tracks with other cars going in the same direction. *Switch crews* and *dispatchers* are responsible for arranging the cars according to their final destinations. Fig. 20-9. Modern classification yards use computers to efficiently sort and classify cars. When the trains are made up, *road crews* connect multiple engines to the train. Multiple engines are used to produce more power to pull large/long trains.

> **FASCINATING FACTS**
>
> In 1902, Mary Anderson felt trolleys were unsafe because it was difficult to see out of the windshield when it was raining or snowing. To solve the problem, she invented windshield wipers.

Each railroad line operates many repair and maintenance shops. Some are portable. These can be moved to where they are needed. Others are permanent. They are located at a particular place and cannot be moved.

In repair facilities, worn places on wheels are ground until the wheel is round again. Broken parts are replaced or rebuilt. Tons of grease and oil are applied to moving parts.

Terminals serve as meeting places for train crews. Business offices and communication facilities are housed in terminals. Two types of terminals are used, freight terminals and passenger terminals. *Freight terminals* are loading docks and storage places for cargo. At *passenger terminals*, people buy tickets, check their baggage, and wait for trains.

FIG. 20-9 A dispatcher keeps each train on the right track and makes sure the movement of freight is organized and continuous.

SECTION 5: Transportation Technology

Water Transportation

Water transportation was the first major mode of moving people and cargo. Using the natural lakes and rivers, explorers were able to travel into areas where pack mules could not go.

Water transportation costs less than most other forms of transport. Cost is figured by the ton mile. A **ton mile** means moving one ton a distance of one mile. To take advantage of this low-cost transportation, a shipper must expect slower travel. Also, water transportation is limited to areas where there are navigable waterways. A waterway that is navigable is wide enough and deep enough to allow ships or boats to pass through.

Types of Water Transportation

There are two main types of water transportation companies: cargo lines and passenger lines.

Like rail transportation, most water transportation business involves moving cargo. Cargo movement that takes place on water within the United States is called *domestic inland shipping*. Foreign shipping is either *importing* (bringing in) goods from another country or *exporting* (sending out) goods to another country.

Passenger lines include luxury or cruise ship lines, paddlewheel riverboats that offer short recreational trips, and public transportation. Fig. 20-10. Lake and river ferries are good examples of public water transportation.

FIG. 20-10 The twin hull design of this passenger ship places the engines in areas below the waterline. This results in less vibration and quieter areas for passengers.

10. **Critical Thinking.** Ask why water was the first major mode of transportation.
11. **Extension.** Using a map of the U.S., show that most navigable waterways are in the northeastern and central states. Western states have very few.

CHAPTER 20: Types and Modes of Transportation

FIG. 20-11 A towboat (bottom) has a flat bow, or front end. It *pushes* barges. A tugboat (right) has a pointed bow. It *pulls* barges.

Vehicles

Water vehicles are also called *vessels*. There are several different types of water transportation vehicles. The "workhorse" of domestic water transport does not even have an engine. It is a large floating box called a **barge** or a lighter. (A barge is *lighter* in weight than a ship.) Barges are tied securely together to form a *tow*, which towboats and tugboats then push and pull through many of the major rivers, canals, and lakes in North America. Fig. 20-11. Barges can even be loaded onto ships and transported across oceans. Barges are used to ship bulk cargo such as petroleum products, coal, and grain.

Towboats and tugboats are relatively small vessels. They operate like railroad locomotives. These boats move barge tows and larger ships.

Other types of ships include containerships, tankers and supertankers, general cargo ships, and bulk carriers.

12. **Critical Thinking.** Ask students why crews on river vessels usually work 6 or 7 months and are off 5 or 6 months.

13. **Extension.** Discuss what happened to the Exxon *Valdez* supertanker.

Cruise ships are like floating hotels. These passenger ships contain restaurants, recreation facilities, staterooms (bedrooms), water treatment plants, storage warehouses, and communication stations.

Some communities located on rivers or large lakes have large paddlewheel boats. Passengers are taken on short trips, usually lasting only a few hours. The trips may offer scenic tours, dinner, dancing, and/or other recreational activities. Ferries are used to transport people, and sometimes their cars, across rivers, lakes, and bays.

Waterways and Sea-Lanes

Water transportation follows either inland waterways or sea-lanes. *Inland waterways* are navigable bodies of water such as rivers and lakes. *Channels* (deep paths) must be kept deep enough for large vessels, and these channels must be marked with buoys, lighthouses, and other markers. In the United States, the U.S. Army Corps of Engineers is responsible for channel maintenance. Locks and dams, needed to keep rivers navigable, are also built and maintained by the Corps of Engineers.

Sea-lanes are shipping routes across oceans. The countries of the world agree in general on certain boundaries and locations of sea-lane routes.

Facilities

Major water transportation cargo facilities are ports, docks, and terminals. These places provide areas for loading and unloading equipment, needed services and supplies, cargo warehouses, and ship company offices. Passenger ships require the same facilities. Terminals built for passengers are more comfortable than those built for cargo.

14. **Reinforcement.** Stress the safety of following standard routes in all modes of transportation.
15. **Critical Thinking.** Ask students why this is true. (The main reasons are that we have the freedom to travel and the money to pay for it, but there are many others.)

FASCINATING FACTS
When facing the bow of a ship, port is left and starboard is right. In port, the left side of a ship faces the port.

Highway Transportation

The most common mode of transportation is highway transportation. Cargo and people are transported over millions of miles of highways, streets, and roads. The major advantage of highway transportation is the independence given the operator. As long as there is a road, a person can travel almost anywhere at any time, day or night. A major disadvantage is that highways all across the United States are becoming overcrowded leading to traffic accidents, traffic jams, and air pollution.

Personal Transportation

People commonly use highway transportation for their own personal travel. This is especially true in the United States. More people here use the highways than people in any other country in the world.

Commercial Transportation

Commercial use of highways is very big business. This use is divided into two classes: "inter" and "intra." *Inter* means between. *Intra* means within. Commercial users of highways often travel *inter-city* or

CHAPTER 20: Types and Modes of Transportation

inter-state. This means they travel between cities or between states. They may also travel *intra-city* or *intra-state*—that is, within a city or within a state.

Commercial transport can involve passengers or cargo. Bus and taxi companies typically offer commercial passenger service. However, the most common commercial use of the highways is for freight, or cargo, service.

There are two types of freight operations: motor freight carriers and owner-operators.

Many companies own vehicles and hire people as drivers. As you travel the highway, you see many trucks with the same company name. These companies are *motor freight carriers*. Some carriers offer regularly scheduled pickup and delivery.

FIG. 20-12 These trucks are all freight-hauling vehicles.

MATHEMATICS CONNECTION

Horsepower

As you know, the work an engine can do is measured in horsepower (hp). The term was first used by the Scottish engineer James Watt in describing the power of his steam engines. He compared his engines to the number of horses they could replace for pumping water out of coal mines.

All power is the rate of doing work. The basic unit of power in the customary system of measurement is foot-pounds per second. A foot-pound is the amount of work required to move one pound a distance of one foot. One horsepower equals 550 foot-pounds of work per *second*.

Horsepower can be figured using this formula:

$$hp = \frac{pounds \times feet \div seconds}{550}$$

For example, if an engine lifts a 1,100-pound object to a height of 2 feet in 1 second, its horsepower can be calculated by finding the number of foot-pounds and dividing by 550:

1100 x 2 ÷ 1 = 2200 (foot-pounds per second)

2200 ÷ 550 = 4 (horsepower)

Try It!

If an engine lifts a 1,650-pound object to a height of 3 feet in 1 second, at what rate of horsepower is it working?

In recent years, a new type of commercial trucking has become popular. The owner of a truck also operates it. He or she is an *owner-operator*. An owner-operator usually owns only one truck. This vehicle may be both office and home for the owner. Owner-operators hire out their vehicles and their driving services.

Freight-Hauling Vehicles

A variety of vehicles are used in the trucking business. The cargo and the length of the trip determine the type of vehicle used. Fig. 20-12. Three common types are: single unit, tractor-semi-trailer rigs, and trailer units. *Single unit* trucks are

Answers to these questions can be found in the Lesson Plans in the Teacher's Manual at the beginning of this book.

CHAPTER 20: Types and Modes of Transportation

one-piece vehicles. They are also called *straight* trucks. The engine is in front, and the drive wheels are in the rear. The bed is permanently mounted over the drive wheels. Single units are generally used for carrying one type of cargo and for local hauling.

A **tractor-semi-trailer rig** is a combination of a tractor and a semi-trailer. The *tractor* is the base unit that pulls the *semi-trailer*, which contains the cargo. They are called *semi* because the trailers have no front wheels. The rear wheels of the tractor support the front end of the trailer through a "fifth wheel" hook-up arrangement. Any semi-trailer can be hooked up to any tractor. Tractor-semi-trailer rigs are sometimes called "18-wheelers." (The tractor has a total of 10 wheels, and the trailer has two axles with 4 wheels on each axle.) These trucks can haul many different loads, depending upon the type of semi-trailer.

Sometimes *trailer units* are connected to the back of a semi-trailer to allow a single tractor to pull more freight. These units have both front and back wheels. Some states allow as many as two trailer units behind a tractor. The total length could be over 140 feet long. Imagine the size of the engine that the tractor would need to efficiently pull that much!

A new diesel truck engine is usually controlled by an on-board computer. The computer also monitors the transmission, the cab environment, and the direction or location of the vehicle. In some ways, these computers control the operation of the truck more than the drivers do. For example, the computer can be set to allow a parked truck to idle for a specific amount of time, then automatically turn off the truck after that time has passed. This controls and records the frequency of the driver's rest time.

FIG. 20-13 At this toll booth, a fee is collected for use of the roadway. Trucks and other heavy vehicles pay more than cars.

16. **Enrichment.** Invite a local trucker to bring a "rig" to school and explain its various parts and operations. Have the trucker tell about the business of trucking.

17. **Critical Thinking.** Ask students how trucks driven in the mountains need to be different from trucks driven in the plains states.

A technician can preset the computer to prevent the engine from being operated over a certain number of revolutions per minute for an extended length of time. This would limit the cruising speed of the truck without preventing the driver from increasing the power in an emergency situation. Computers can also be used to record all the speeds, downtime, and other driver-initiated uses of the vehicle. This information can be reviewed immediately by the driver in the cab or, through a satellite link, by shop mechanics, dispatchers, and customers. For example, a terminal manager can check the refrigeration unit on a truck carrying meat products while that truck is on the highway. Owners can see how the truck is operated, mechanics can tell if the truck is running properly, and customers can determine where their shipments are.

Facilities

Commercial use of highway transportation continues to increase. The need for physical facilities is also increasing. Services must be provided before, during, and after trips. Cargo must be loaded, unloaded, and stored. Terminals and truck stops offer these services for both company and independent operators.

Good road systems are necessary for a highway system to be effective. In the past, little planning was given to road construction. Roads were built to connect cities and towns as they were needed. Today, in the United States, an interstate superhighway system, designed for safe travel up to 70 or more miles per hour, is used by the millions of automobiles, motorcycles, and trucks. Highway designers and engineers must carefully plan how to try to keep an ever-increasing number of vehicles moving smoothly and safely.

Highways are built and maintained by local and state governments using tax money collected on gasoline and diesel fuel. Some states help pay for their highway systems with toll roads, where drivers must pay a certain fee, or toll, to use the road. Fig. 20-13. The heavier the vehicle, the more the fee per mile. The fee is collected at toll booths, which may be automated or run by a person.

18

Pipeline Transportation

Pipeline transportation offers four special characteristics:
- It is the only mode in which the cargo moves while the vehicle stands still.
- Most pipelines are buried in the ground, so they are unseen and quiet. Also, they cause no traffic congestion or accidents.
- Pipelines are laid out in straight lines across the country. This decreases the transportation time.
- Theft from pipelines is difficult. Also, damage or contamination of cargo is rare.

Using Pipeline Transportation

Certain types of cargo, such as oil products, natural gas, coal, wood chips, grain, and gravel, can be transported very economically through pipelines. However, pipeline transportation is not as flexible as other modes. Fig. 20-14. Service depends upon how close a customer is to a line.

18. **Reinforcement.** If there are any toll roads in your area, discuss these.

Cargo is shipped in batches. Solid materials are moved in a slurry. A **slurry** is a rough solution made by mixing liquid with solids that have been ground into small particles. The batch is put into the system. As it goes in, the amount is measured and recorded. The material is then pumped through the lines.

Pipeline transportation is one-way. If products also need to flow in the opposite direction, two lines are installed.

Pipelines

There are over one million miles of pipelines crossing the United States and Canada. Fig. 20-15. These lines range in size from two inches in diameter to fifteen feet in diameter. Most of these lines carry natural gas or petroleum products.

Pipelines are made of either steel or plastic. There are three types of lines, gathering, transmission, and distribution.

Gathering pipelines are used to collect the cargo from the suppliers. These lines meet at central holding tanks and pumping stations.

Transmission pipelines are the main long-distance lines that transport the cargo. The batches end up at terminals where cargo can be stored temporarily.

Distribution pipelines deliver the cargo from the terminals to the customers. Small distribution lines probably bring water and natural gas right into the building where you live.

Facilities

The facilities that service the pipelines are above ground. These include:

- pumping stations
- control stations
- measuring stations
- exchange stations

Pumping stations use pumps to move the cargo down a pipeline. Natural gas may be compressed up to 2000 pounds per square inch. It does not travel more than 15 miles per hour. Liquids such as crude oil travel 2 to 5 miles per hour. To keep

FIG. 20-14 Pipelines are an economical way to transport certain materials.

19. **Extension.** Pipelines must run full. Students usually ask what keeps products apart in a pipeline. Actually they don't stay completely separated. A small portion "interfaces" or mixes. When products arrive at their destination, the mixed portion is separated.

MAJOR INTERSTATE PRODUCTS PIPELINES

• REFINERY LOCATIONS

FIG. 20-15 Many miles of underground pipelines transport products and materials throughout the country.

cargo moving, pumping stations pump the batches every 30 to 150 miles along the pipeline. Fig. 20-16.

Control, measuring, and *exchange stations* are also located along pipelines. These facilities make sure the customers safely receive the correct size batches of their cargo.

Pipelines can become clogged. To prevent this, a scraper tool called a "pig" is regularly pushed through the pipeline. Pigs may be round like a ball or tubular shaped. Fig. 20-17. Some pigs have brushes on the outside. Others carry instruments that take readings on conditions inside the pipeline.

FIG. 20-16 Here, an operator checks the pumping station, but most operations are remotely controlled by computers many miles away.

20. **Extension.** If you are right above a pipeline when a pig is in use, you will probably be able to hear it. Pipeline workers say that dogs sometimes bark at and chase the sound.

CHAPTER 20: Types and Modes of Transportation

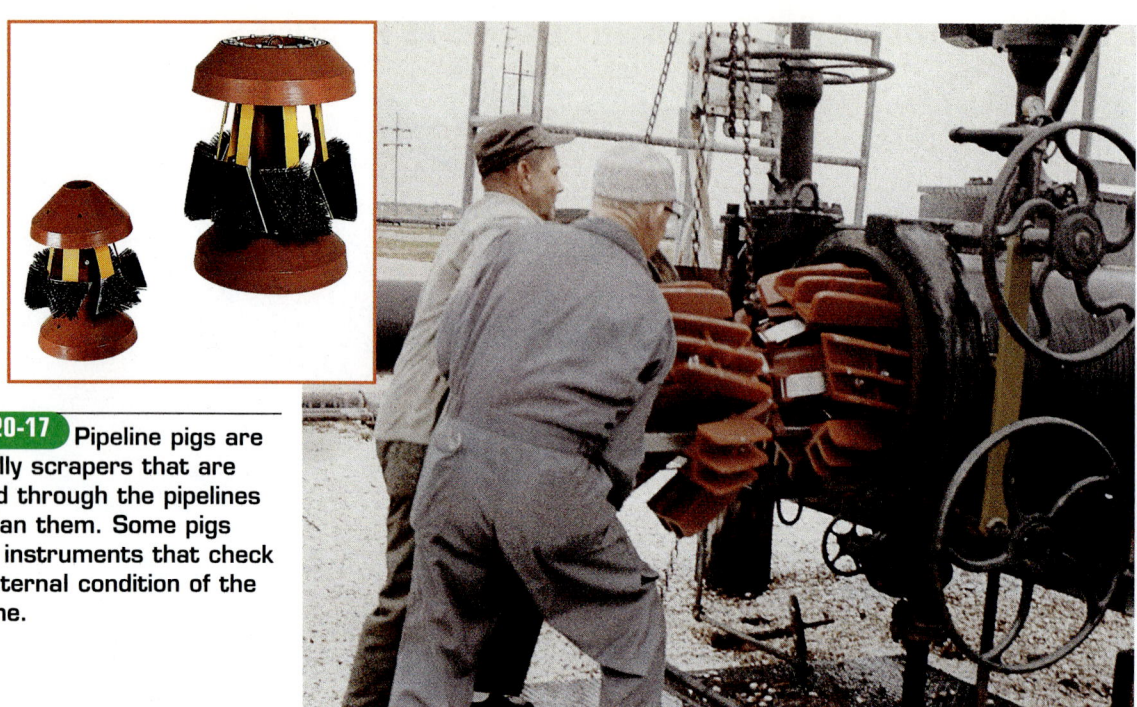

FIG. 20-17 Pipeline pigs are actually scrapers that are moved through the pipelines to clean them. Some pigs carry instruments that check the internal condition of the pipeline.

Space Transportation

In July 1958, the National Aeronautics and Space Administration (NASA) was founded to plan and operate the U.S. space program. In 1969, Neil Armstrong and Edwin Aldrin landed on the moon in an Apollo spacecraft.

However, space transportation is still very new. The future of human space travel is still hard to predict.

Manned Spacecraft

There are two main types of space transportation systems—manned and unmanned. Manned systems carry humans. Unmanned systems do not have human crews or passengers. Manned space flights are still much less common than unmanned flights. Manned spacecraft are of two basic designs: nose-cone-mounted spacecraft and space shuttles. They must provide everything necessary to keep their passengers alive in the vacuum of space.

Nose-cone-mounted spacecraft were the first type used to transport humans. These spacecraft consist of sealed containers put onto the nose of a booster rocket and launched into space. **Booster rockets** are rockets used to push a payload. A **payload** is anything transported. When the rocket reaches space, the container separates from the booster rocket. Fig. 20-18. These nose-cone-type spacecraft have no wings and have no ability to fly through the atmosphere on their own.

Space shuttles can be called the first true spaceships because they take off, maneuver in space, and fly back through

458 SECTION 5: Transportation Technology

FIG. 20-18 Nose-cone-mounted spacecraft like this *Apollo* vehicle sit on top of a booster rocket. After the booster rocket has pushed the vehicle high enough, they separate. The spacecraft continues into orbit, and the booster rocket burns up while falling back to earth.

the atmosphere for landing. Shuttles have their own powerful engines, but they still need the help of a booster rocket to get into space. Fig. 20-19. Each shuttle can be used over and over again. They are versatile and powerful enough to be called "space trucks." Most shuttle flight missions include launching satellites.

Unmanned Spacecraft

Unmanned systems use nose-cone-type vehicles. These nose cones are usually just hollow containers. Once in orbit, the containers open and the payload is

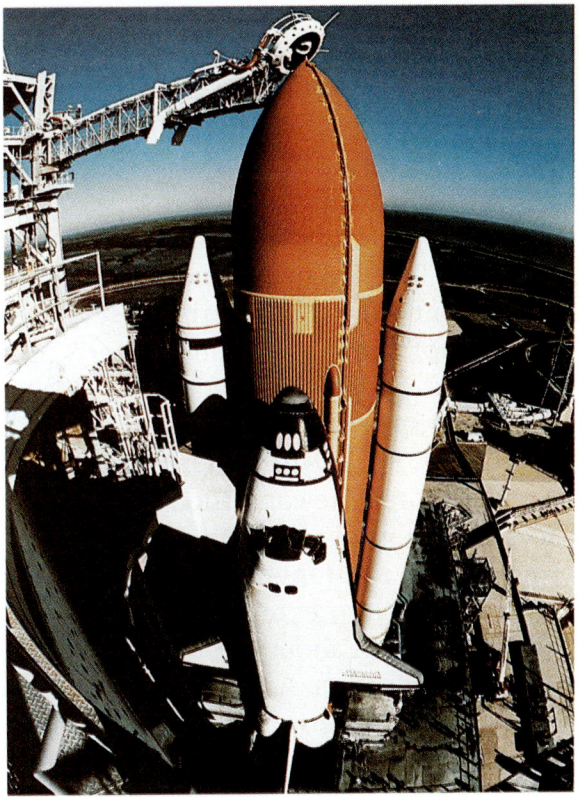

FIG. 20-19 The shuttle *Endeavour* contains fuel in the large red middle tank to power its main engines during launch. Reusable booster rockets are mounted on both sides of the fuel tank.

FIG. 20-20 This was the capsule used by the *Pathfinder* rocket. Inside, it carried the little Sojourner vehicle and other important equipment.

maneuvered into position with tiny rocket engines. These vehicles and their booster rockets can be used only once.

The most common cargo on unmanned space transportation vehicles is satellites. There are now hundreds of satellites in orbit around the earth. Most of them are for communication. In 1997, NASA succeeded in sending a second unmanned spacecraft, *Pathfinder*, to the planet Mars. Fig. 20-20. Inside the craft was a small, remote-controlled land vehicle called Sojourner. (See Fig. 22-28 on page 507.) In constant communication with earth, Sojourner reported back much new information about the "red" planet.

Space transportation is by far the most expensive transportation system. It costs thousands and sometimes millions of dollars to put just one pound of payload into space. The future of space transportation will really be determined by how much money governments and corporations are willing to spend to explore space. Many people feel that things like mining and manufacturing in space will make space transportation worth the cost. Others feel that space transportation will always remain too expensive.

Computerized Navigation

As the world population grows, so do the numbers of vehicles on the roads, on waterways, and in the air. Traffic congestion is increasing faster than the funds to build new ways to handle it. For this and other reasons, researchers have been experimenting with computerized navigation. With computerized navigation systems, travelers can find their location by means of a computer. They can also learn where congested areas are and how to avoid them.

The Global Positioning System

The **Global Positioning System (GPS)** was first developed for ships and planes. It is a system of 24 satellites orbiting the earth at an altitude of about 11,000 miles. The satellites are spaced evenly so that a minimum of four are above the horizon everywhere on earth, 24 hours a day. Signals from a vehicle allow the satellites to track it. Then, using mathematical calculations, GPS computers determine the vehicle's location.

Using GPS, travel is safer. Under crowded conditions, planes can fly closer together with less chance of accidents. Travel is also more cost-effective. Pilots do not stray from the most direct route, which saves fuel. Travel can be faster as well. Ambulance and delivery drivers waste no time searching for the right street.

Today the benefits of GPS can be used for almost every mode of travel, including cars and buses. Even some hikers and backpackers, equipped with special devices, have used it to find their way around in the wilderness. People with handicaps can also be given the proper guidance. Fig. 20-21.

Smart Highways/Smart Cars

Traffic engineers in large cities around the country are developing Intelligent Vehicle Highway Systems (IVHS). These systems allow a car or truck to "communicate" with the highway. Some systems include beacons spaced along the road that send signals to vehicles giving traffic information. Others make use of GPS technology.

In Chicago, 1,800 wire detectors are embedded in 118 miles of expressway pavement. When a large piece of metal, such as a car, passes over them, the detectors transmit a signal to computers. The more cars, the more signals are sent. Using this information, the computers determine how much congestion there is. Based on the level of congestion, they then estimate travel times from one location to another. The computers send the information to local radio stations, which broadcast it to listening motorists.

Some cars are now being built that include dashboard computer screens on which city maps can be displayed. The car's exact position can then be seen on the screen. Some computers can even suggest the best route from one spot to another.

FIG. 20-21 This man who is blind can find his way around using the global positioning system, which relies on satellites orbiting in space. The system tracks his location and, with the aid of a voice synthesizer in his backpack computer, can tell him how to get to his destination.

Career File—Billing Clerk

EDUCATION AND TRAINING
A high school diploma is required. Employers prefer graduates who have experience working in an office.

AVERAGE STARTING SALARY
Billing clerks generally start earning around $16,500 per year, but salaries vary depending on the size of the company and geographic location.

What does a billing clerk do?
Billing clerks prepare bills and invoices for services rendered. They keep records of orders, sales, charges, and payments. They use this data to calculate the amount due from each customer and prepare the printed invoices.

Does most of the work involve data, people, or things?
Billing clerks work with a lot of numerical data. Preparing bills and invoices requires keeping track of charges and payments together for each customer. Most billing clerks use a computer to retrieve this information and prepare the bills.

What is a typical day like?
Billing clerks first retrieve the information they need—charges and credits—from a computer or from paper files. They do this for each invoice, which usually contains a detailed list of charges and credits. They may have to consult a rate book for per-unit costs to compute some of the charges, such as trucking rates for machine parts. Then they print the invoice. The entire operation may be done with paper, pencil, and typewriter. More frequently, most or all of these activities are done on a computer.

OUTLOOK
Although the number of billing clerk positions will stay about the same through 2005, many jobs will be available, according to the Bureau of Labor Statistics. Most positions will open up because people leave the company, are promoted, or transfer to a different job. The increased use of computers, which makes people more productive, will keep the number of jobs from growing.

What are the working conditions?
Billing clerks work in an office, usually alongside other clerical workers. They may have to sit for long periods of time using a computer, which may cause eye and muscle strain.

Which skills and aptitudes are needed for this career?
Billing clerks should be careful, orderly, and detail-oriented to avoid making mistakes. They should be good at working with numbers and be familiar with computers.

What careers are related to this one?
Similar occupations that maintain financial data in computer and paper files include bookkeeping, accounting, auditing, and working with payrolls.

One billing clerk talks about her job:
I've always enjoyed mathematics. This job is a good place to start when building a career.

Chapter 20 REVIEW

 ## Reviewing Main Ideas

- Transportation is made up of air, rail, water, highway, and pipeline systems.
- The three basic types of air transportation systems are: commercial, general, and military aviation.
- Rail transportation is used to efficiently transport large, heavy loads.
- Water transportation is usually slow, but it is very inexpensive per ton mile.
- In the U.S., highway transportation is the most common mode.
- Pipeline transportation is primarily used to transport liquids and gases. The three types of lines are: gathering, transmission, and distribution.

 ## Understanding Concepts

1. Name the transportation mode that most closely matches the description:
   a. Most economical mode, considering the weight moved per mile.
   b. The operator has more independence than other modes of transportation.
   c. The cargo moves while the vehicle stands still.
2. List at least one advantage and one disadvantage of each of the five types of transport.
3. How would you ship one ton of lumber from Seattle to Dallas? Explain your choice.
4. Name at least two types of vehicles for each mode of transportation.
5. Describe several examples of facilities that are common to all five modes of transportation.
21. **Resource.** Answers to "Understanding Concepts" and "Thinking Critically" can be found in the Teacher's Manual at the front of this *Teacher's Annotated Edition* and in the *Teacher's Resource Binder*.

 ## Thinking Critically

1. Explain in your own words how an airplane wing works.
2. Why do you think other countries use passenger railroad transportation more than the U.S. does?
3. What type of transportation system would you use to ship (a) coal, (b) oil, and (c) urgently needed medicines?
4. Describe what type of commuter service would be best for your area.
5. What efforts can be made to maintain good transportation service and still protect the environment?

 ## Applying Concepts and Solving Problems

1. **Social Studies.** Research how the railroad industry developed in the U.S. Report your findings. What trends do you see for the future use of railroads?
2. **Science.** An object's buoyancy is equal to the weight of the water displaced by the object. If buoyancy is greater than the weight of the object, the object floats. If it is less, the object sinks. Try immersing a variety of objects in water. In each case, tell which is greater: the weight of the object or the upward force of the water (buoyancy).
3. **Mathematics.** Visit this Internet web site to learn more about the mathematics used in the Global Positioning System: www.trimble.com/gps/

CHAPTER 20: Types and Modes of Transportation

CHAPTER 21

Intermodal Transportation

Objectives

After studying this chapter, you should be able to:
- define intermodal transportation.
- describe how intermodal semi-trailer systems combine rail and highway transportation modes.
- define containerization and explain its advantages.
- describe how liquid and solid bulk cargo are transported intermodally.
- describe how people can be transported on intermodal systems.

Terms

containerization
container on flatcar (COFC)
containerships
conveyor
escalator
intermodal transportation
piggyback service
trailer on flatcar (TOFC)

Technology Focus

RR Stands for RoadRailer

A RoadRailer is a combination vehicle, half boxcar and half semi-trailer, with steel wheels to ride the rails and rubber tires to roll on the highways. It is not as strong and heavy as a railroad boxcar. However, it is stronger and heavier than a highway semi-trailer.

Changing from highway mode to rail mode takes only about four minutes. A semi-tractor pulls the RoadRailer to the rail terminal and into position over the tracks. A set of flanged, steel wheels called a couplermate is positioned under the front and back of the RoadRailer. Then the rubber wheels are lifted up, and it's ready to be added to a train like any other freight car. When the RoadRailer reaches its rail destination, the steel wheels are removed, and a semi-tractor pulls the RoadRailer onto the street.

The Combination Tradeoff

Like any combination vehicle, the RoadRailer pays for its versatility by sacrificing top performance in other ways. A RoadRailer can't compete with standard tractor-trailers in long hauls over the highway because of its extra weight. Nor can a RoadRailer trailer last as long as conventional boxcars, which are much heavier and more durable. However, in certain circumstances, a combination vehicle may be exactly right for the job.

A RoadRailer has three major advantages. First, it is faster to unload than either a boxcar or a "piggyback" flatcar (semi-trailer on a flatcar). Second, a RoadRailer gets better mileage than a conventional railcar because one engine can pull a full RoadRailer train. Third, it is cheaper. RoadRailers cost less per mile to move, there is less chance of theft, and less damage to goods. The right application for a RoadRailer is one in which the best choices are rail and trucking, *and* rapid offloading is important.

RoadRailers Deliver Mail

Rapid offloading is a premium for the U.S. Postal Service, which needs city-to-city mail delivered to regional sorting centers. Highway trucking offers "to your door" delivery, but rail claims an advantage in reliability, being less susceptible to weather delays and traffic congestion. The Postal Service is giving RoadRailer a try on three regular routes: Philadelphia to Jacksonville, Chicago to Philadelphia, and Chicago to St. Louis.

Take Action!

Make a sketch of a combination vehicle (real or imagined) and include notes about any special features. Write a brief description of circumstances in which your combination vehicle would be practical.

Resources
- Chapter 21 Lesson Plan in the *Teacher's Annotated Edition* and in the *Teacher's Resource Binder*.
- Chapter 21 Review in the *Teacher's Resource Binder*.

CHAPTER 21: Intermodal Transportation

What Is Intermodal Transportation?

In the past, transportation companies were very protective of their businesses. Every transportation company competed with every other transportation company. A truck company not only competed with other truck companies, but also with railroads, ship companies, and airlines. Today there is still competition, but other things are quite different.

The service offered by each of the different modes is becoming more specialized. Companies cooperate and often own other types of companies. Railroads also own and operate trucks, ships, and barges. Airlines own trucklines and automobile rental agencies. Ship companies own shares of railroads and trucking companies.

The process of combining transportation modes is called **intermodal transportation.** Fig. 21-1. A combination of two or more of the five major modes of transportation (air, rail, water, highway, and sometimes pipeline) are used together to efficiently transport passengers and cargo. There is increased efficiency because much less time and labor is spent in loading and unloading cargo. This results in reduced shipping time and costs.

Fascinating Facts

At the end of their empire in the 4th century, the Romans had built more than 50,000 miles of roads through Europe and the Middle East. The roads were so good that they were not replaced until the 19th century—1400 years later.

FIG. 21-1 During shipping, cargo is often transferred from one mode of transportation to another. Here, goods from the trains are being loaded on the ship.

SECTION 5: Transportation Technology

Intermodal Cargo Transportation

Let's take a look at intermodal cargo transportation and see how it works. We'll discuss several types, including trailer on flatcar, containerized shipping, moving liquids, and moving coal and gravel.

Trailer on Flatcar

Trailer on flatcar (TOFC) is an intermodal method of transporting cargo involving highway and rail modes. Truck trailers full of cargo are carried on railroad flatcars.

TOFC is interesting to see. A trucking company uses tractors to move semi-trailers to the loading dock of the company wanting to ship the cargo. At the loading

FIG. 21-3 Piggyback service is an efficient and economical intermodal operation.

FIG. 21-2 A giant TOFC loader is used to load semi-trailers onto the flatcars.

dock, the semi-trailers are carefully loaded and sealed by the shipper.

The trucking company moves the semi-trailers to a railroad yard. Here, they are loaded onto specially-built flatcars. Fig. 21-2. The flatcars are connected to trains and are moved by the railroad to another yard close to the final destination. There, the semi-trailers are unloaded and connected to tractors. These units travel on the highways to the customers' loading docks. The semi-trailers are unsealed and off-loaded. ("Off-loaded" is another term for unloaded.)

Perhaps you know the common name for this popular intermodal method of transporting cargo. TOFC is often referred to as **piggyback service**. Fig. 21-3. Use of TOFC service has increased greatly since it was first introduced in 1950. Entire unit trains are used by TOFCs. All the cars carry semi-trailers.

Containerized Shipping

Containerization (or containerized shipping) is an efficient method of handling cargo. Cargo is loaded into large boxes or other containers before it is transported. Containerization is used by all modes but pipeline.

The most popular container used by the transportation industry is basically a large metal box. The standard size box is eight feet high, eight feet wide, and forty feet long. The frame of the box is strong. The corners are reinforced. There are holes in the corners to provide a means of grasping the box during loading, securing, and unloading operations.

Containerization has created new forms of intermodal transportation. The same standard size container is used in several different ways by the highway, rail, and water transportation modes. Fig. 21-4. A smaller container is used in air transportation.

Containers are used in the following ways:

- Containers can be made into semi-trailers. This is done by placing them on a

FIG. 21-4 A standard container can be carried on (top) a highway truck frame, (right) a containership, or (left) a special rail flatcar.

1. **Reinforcement.** Reinforce the ideas that the use of containers is an efficient method of handling cargo.

2. **Review.** You may wish to review the concept of standard size and standard stock, which students were introduced to in Chapter 9.

SECTION 5: Transportation Technology

FIG. 21-5 Containers are loaded into the hatch of a plane. Different airplanes require different sizes and shapes of containers.

frame that has sixteen ordinary wheels and a fifth wheel. Each "semi-trailer" can then be driven to its destination or lifted onto railroad flatcars and moved piggyback style.

- Containers can also be loaded directly upon railroad flatcars. When they are fastened directly to the flatcars, the method of transportation is called **COFC** for **container on flatcar**.
- These same standard containers are also loaded onto barges and ships. Many oceangoing ships, called **containerships,** are specially designed and built to carry containers.
- The airlines use containers that are shaped to fit into different models of aircraft. These are smaller and lighter than the standard containers. They also fit into standard highway vehicles. Some of these air-mode containers have round ends and some have square ends.

Like the larger containers, air-mode containers can be loaded and sealed by the shipper. They are transported by truck to and from airports. Loading containers into aircraft and then unloading them requires special equipment. The containers are moved into the plane through the cargo hatch. Fig. 21-5.

Advantages of Containerization

There are many advantages to containerized shipping. The most important advantage is less handling of cargo. Fig. 21-6. To illustrate this, let's compare the typical transport of a product, first without a container and then in a container.

3. **Review.** Review the discussion of fifth wheels in Chapter 20.
4. **Extension.** TOFC loaders can also handle COFC operations.
5. **Reinforcement.** Give other examples that show advantages of packing and moving efficiently from one place to another.

CHAPTER 21: Intermodal Transportation

FIG. 21-6 Containerized shipping permits less piece-by-piece handling of cargo. These cartons are wrapped together in a single unit.

Suppose a factory in Peoria, Illinois, produces special automobile transmissions (speed-changing gears). The factory has just received a large order from a company in Germany. How will the order be shipped?

Method 1. The products will travel as break bulk.

Workers crate each transmission separately. The crates are loaded, one at a time, into a semi-trailer or trailer unit. A trucking company moves the trailers to a rail yard. TOFC service is not available. The transmissions have to be unloaded from the truck and reloaded into a boxcar. The boxcar is placed in a train and moved to the seaport. The transmissions are unloaded from the boxcar and again placed into a truck.

The truck delivers the transmissions to the overseas dock. There, the transmissions are unloaded from the truck and placed in a warehouse. When enough cargo for a full shipload has been gathered in the warehouse, the ship comes. The transmissions are loaded into the holds (storage areas) of the ship.

When the ship arrives at a port in Europe, the transmissions are again unloaded and packed, one at a time, into a railroad car. When they arrive in Germany, they are loaded into a truck and delivered to the automobile company. The automobile company unloads the truck.

Using the break-bulk method of cargo transportation to send the transmissions is not efficient. Too much loading and unloading is required.

Method 2. The transmissions will be shipped in a standard container.

An empty container fastened to a highway wheel frame is delivered to the factory. The transmissions are placed in plastic bags and then packed into the container. Shipping crates are not needed. Special racks within the container hold the transmissions in place. After the container is completely loaded, the workers lock and seal it.

6. **Review.** Review the meaning of break bulk discussed in Chapter 19.

SECTION 5: Transportation Technology

A trucking company picks up the container and moves it to a railroad yard. There, the entire container is placed on a flatcar and moved to the overseas dock. The container is stacked on the dock with other standard containers. When the ship arrives, the container is hoisted onto the ship and transported across the ocean.

In Europe, the locked and sealed container is unloaded. It is placed on a railroad car or on a wheel frame to be hauled by a tractor-semi-trailer rig. When the container arrives at the German automobile factory, the customer breaks the seal and unloads the transmissions.

Do you see the reduced amount of handling? Using a container saves a great deal of labor and money. Also, there is less chance for theft. The chances for damage to cargo from dropping, bumping, or exposure to bad weather are also decreased.

The Seagirt Marine Terminal in Baltimore, Maryland, provides a good example of the benefits gained when computers, containers, and intermodal transportation are combined. Fig. 21-7. At Seagirt, the world's largest cranes are used to carry cargo from dock to ship and from ship to dock. The cranes stand 20 stories tall, and their legs can span a six-lane highway. They can handle up to 55 containers an hour—more than twice as many as the next largest cranes and twice as fast. At the truck terminal, computer software helps keep track of thousands of containers. Clerks can check truckers in and out in half the time it used to take.

Moving Liquids

It is not practical to build pipelines to every city and town. By using rail or highway modes, pipeline companies are able to expand their range of service to almost anywhere. Some tank truck cargo companies actually call themselves "rolling (or highway) pipelines."

FIG. 21-7 The boom on this huge Sumitomo crane is longer than a football field and can lift 60 tons.

7. **Extension.** Containerization also requires less paperwork.

8. **Review.** Compare two methods of shipping.

CHAPTER 21: Intermodal Transportation **471**

FIG. 21-8 Refined oil products are being loaded into highway tank trucks for transport to customers.

Oil is often shipped in this way. First the crude or refined oil products are shipped by underground pipelines. When the product reaches a terminal, it may be temporarily stored in large tanks or pumped directly into railroad tank cars or tank semi-trailers. Fig. 21-8. When the cargo is delivered to the final destination, pumps are again used to unload the oil.

Oil is not the only product that is moved by this type of intermodal transportation. Large newspapers and printing companies purchase ink in great quantities. Their most efficient delivery system uses pumps and tanks to transfer the ink from a tank truck to a storage tank.

Milk tank trucks transport milk from the farm to the dairy. The milk is never exposed to air. Milking machines collect milk from the cows. The milk travels through pipes to holding tanks. From there, it is piped into tank trucks. At the dairy, it is pumped through hoses and pipelines.

Moving Coal and Gravel

Another form of intermodal transportation combines on-site conveyors, trailers, and railroad cars. In the northwestern United States, coal and gravel are mined for customers in other parts of the country. As these materials are gathered, they are loaded onto very long conveyors. A **conveyor** is a continuous chain or belt that moves materials over a fixed path. In this case, the material is moved across rugged land. Some of these conveyors are

9. **Background.** Explain that the milk is chilled in the holding tanks.

SECTION 5: Transportation Technology

over a mile in length. Finally, the conveyor dumps its load into trucks or railroad cars. Fig. 21-9.

In rail transportation, hopper cars are used in unit trains. Sometimes the trains don't even stop moving while they are being loaded. The engine slowly moves the cars under the conveyor. The conveyor loads the cars so fast that the train can keep moving.

The hopper car unit trains travel to the customer's location where they are unloaded. There are two main methods of unloading cars. In one method, the cars are moved over elevated sections of track that have open spaces between the rails and the ties. Hopper doors on the bottom of each car are opened. The load is dropped onto a conveyor. The conveyor carries the material to a storage area and dumps it.

FASCINATING FACTS
The United States has more train track than any other country. The train tracks in the U.S. could go around the world eight times.

Another method of unloading the railroad cars involves rolling each car individually onto a section of mechanical track. Special clamps lock the car in place. The track and car rotate together until the car is upside down. Fig. 21-10. The load dumps out all at once. The mechanical track continues rotating until the car is upright again. Then the empty car is moved ahead and replaced by the next full car. The procedure is repeated until all cars are unloaded. The train must stop and start as each car is uncoupled, unloaded, and moved on. However, this is still a very fast method of unloading cargo.

FIG. 21-9 This conveyor is carrying coal from the mine to the train, where it dumps its load into the waiting cars.

CHAPTER 21: Intermodal Transportation

FIG. 21-10 An unloader dumps the entire railcar load at one time.

Other Intermodal Cargo Transportation

Other forms of intermodal cargo transportation include the vacuum-operated transfer of flour or grain among ships, railcars, and trucks. This system is much like cleaning with a vacuum cleaner.

Military equipment comes in many odd shapes and sizes. The U.S. military must have an even broader system of intermodal transportation. Fig. 21-11. A container of supplies may be shipped by highway, rail, water, and air. Then finally it is dropped by parachute from an airplane into a camp that is hard to reach.

Engineers and designers continue to think of new methods of handling materials. Use of intermodal transportation will increase and become even more efficient.

FIG. 21-11 Military vehicles must be able to accommodate cargo of all sizes and shapes.

MATHEMATICS CONNECTION

How Far, How Fast, How Expensive?

The decision as to which type of cargo carrier to use depends upon three main factors: cost, distance covered, and speed. All of these factors are influenced by the type of cargo to be shipped. For example, a load of steel beams would be shipped differently than a human organ transplant. The beams would probably go by ship, depending on their destination. The human organ would be shipped by air freight. Today, many shippers use freight consolidators, companies that combine more than one mode in order to achieve the best time and price.

The average speed (miles per hour) for each type of transport can be calculated using this formula:

$$\text{speed} = \text{distance} \div \text{time}$$

Try It!

Suppose you have to ship two tons of watermelons from New Orleans to New York. Using the chart, determine the best choice of cargo carrier. Calculate its speed using the formula. Why did you choose that carrier?

Carrier	Cost Per Ton	Miles Traveled	Travel Time to NY
Ocean freight	$90	1,800	7 days
Railroad	$210	1,350	3 1/2 days
Truck	$324	1,000	34 hours
Air cargo	$2500	975	2 3/4 hours

Intermodal Passenger Transportation

People also use intermodal transportation when they travel. Let's see how a person named Bill traveled from his home in New York City to the Superdome in New Orleans, Louisiana.

First, Bill walked to a bus stop. From there, the bus (highway mode) carried him to a city transit train station. Bill boarded the train (rail mode) and was transported across town to a station near the airport. He took a taxi (highway mode) to the airport. Inside the airport, a moving walkway (on-site) transported him to his departure gate. Fig. 21-12.

10. **Reteaching.** Relate Bill's trip to students' own travel experiences.

CHAPTER 21: Intermodal Transportation

FIG. 21-12 A moving walkway is a conveyor device used for transporting people. It is also called a moving sidewalk.

When it was time to board the plane, Bill walked down a covered telescoping ramp called a *jet way*. Fig. 21-13. After a few hours on the plane (air mode), Bill's first stop was at the Dallas-Fort Worth airport in Texas. Bill was scheduled to change planes there. He traveled from one gate to another by a computer-controlled passenger vehicle that travels on special tracks (rail mode). Fig. 21-14. At the next gate, he boarded another plane for the final part of his trip to New Orleans.

When he arrived in New Orleans, Bill walked out of the plane through another jet way. He rode another on-site method of transportation, a moving stairway called an **escalator**, to the next level. From the escalator, he walked to a bus stop. The bus delivered him to the entrance ramp of the Superdome.

Bill's trip used many different modes of transportation. He will rely on all of them again when he returns home.

FIG. 21-13 Passengers walk through a covered, telescoping ramp called a jet way when boarding or leaving an airplane.

SECTION 5: Transportation Technology

Fascinating Facts

Highways are not named randomly. Even-numbered interstate routes run east-west. Odd-numbered routes run north-south. Routes that have three digits and start with an even number are routes through or around a city. A route with three digits that starts with an odd number is a spur into a city.

Improved Airport Service

Passenger and cargo intermodal transportation are much alike. The major difference is that the transfer of cargo between modes is the task of the transportation company. Passengers must usually search out the connecting links between modes themselves. That may be changing. More and more cities are creating direct links between airports and city centers, tourist areas, and other high-traffic locations. As a result, passengers will have to do less of the work necessary to arrange transportation. For example,

- The Port Authority of New York and New Jersey is planning to link all three New York-area airports using people-mover connections and existing commuter trains and subways.
- A high-speed train has been built to carry passengers between Orlando International Airport and an intermodal transportation center. The center is almost 14 miles away, but the train covers the distance in six minutes.
- An automated electric people-mover will carry up to 5,000 passengers an hour between Los Angeles International Airport and remote ticketing and check-in facilities. The first of the 12 facilities will be 25 miles away.

FIG. 21-14 This computer-controlled automatic passenger vehicle travels on special tracks.

CHAPTER 21: Intermodal Transportation

Career File—Dispatcher

EDUCATION AND TRAINING
A high school diploma with courses in business is preferred. Trainees usually develop the skills they need on the job. Such informal training lasts from several days to several months, depending on the complexity of the job.

AVERAGE STARTING SALARY
The salary for beginning dispatchers varies by industry. About $13,000 per year is usual.

OUTLOOK
Employment of dispatchers is expected to grow about as fast as average through the year 2005, according to the Bureau of Labor Statistics. As the population increases, so does the need to coordinate the movement of a larger amount of goods and services.

What does a dispatcher do?
Dispatchers receive requests for service and take action to provide the service. Duties vary, depending on the needs of the employer. For example, truck dispatchers schedule and coordinate the movement of trucks and drivers to ensure that they arrive on schedule. Ambulance dispatchers handle calls from people reporting medical emergencies. Taxicab dispatchers relay requests for cabs to individual drivers.

Does most of the work involve data, people, or things?
Dispatchers work with all three. They have a lot of contact with people, mostly over the telephone. They maintain service schedules, usually on a computer.

What is a typical day like?
Incoming telephone calls bring requests for service—a truck to pick up and deliver goods, for example. The dispatcher, who maintains a master schedule, matches a suitable truck and driver to the request, enters it on the schedule, and notifies the driver in time for the pickup to be made. Dispatchers of emergency services and taxicabs must notify the service providers without delay.

What are the working conditions?
Dispatchers typically work in an office. They spend a lot of time on the telephone or two-way radio. They can experience back discomfort and eyestrain as a result of sitting for long periods in front of a computer. The work of dispatchers can be very hectic when a large number of calls come in at the same time.

Which skills and aptitudes are needed for this career?
Good reading and writing skills and the ability to work under pressure are important. They must be able to deal with disruptions in schedules caused by bad weather, road construction, or accidents. Familiarity with computers is desirable.

What careers are related to this one?
Other occupations that keep track of various operations include stock clerks, inventory clerks, and shipping and receiving clerks.

One dispatcehr talks about his job:
It takes an organized person to do this job. A lot of drivers depend on me to give them the right directions.

Chapter 21 REVIEW

Reviewing Main Ideas

- Intermodal transportation is the process of using more than one transportation mode to move people or cargo.
- Trailer on flatcar (TOFC) companies often combine highway semi-trailers and rail flatcars to ship cargo.
- Containerization is the use of special containers for shipping cargo efficiently.
- Liquids are frequently shipped by intermodal systems of pipelines, tank trucks, and railroad tank cars.
- Solid bulk cargos are often moved by conveyors, truck trailers, and rail cars.

Understanding Concepts

1. How does intermodal transportation work? Give two main advantages.
2. Define and briefly describe TOFC.
3. Describe containerization. What are its advantages?
4. What is a "rolling pipeline"?
5. Give examples of four modes of transportation that a passenger might use when traveling across country.

Thinking Critically

1. Why would industrial companies be interested in buying different types of transportation systems?
2. If transport companies did not cooperate, would intermodal transportation be practical? Explain.
3. Do you think it would be practical to develop airplanes large enough to carry standard size containers? Explain.
4. Early American settlers traveling across the United States often used intermodal transportation. What modes were used?
5. What disadvantages might there be to intermodal transportation systems?

Applying Concepts and Solving Problems

1. **Language Arts.** Interview a delivery person. Where does he or she pick up cargo? How did it arrive at that point? How many deliveries are made in one day? Present your findings to the class in an informative speech.

2. **Science.** When transferring loads of sand or gravel at first, particles slide through any openings. After a time, however, no more particles fall because the material has developed a stable slope. The angle at which this happens is known as the *angle of repose.* Using small samples of sand, sugar, or flour, find the angle of repose for each. Pour the material until you form a pile with the steepest possible sides. Stick an index card through the pile, and trace its slope on the card. The angle formed between the sloping line and the bottom of the card is the angle of repose.

11. **Resource.** Answers to "Understanding Concepts" and "Thinking Critically" can be found in the Teacher's Manual at the front of this *Teacher's Annotated Edition* and in the *Teacher's Resource Binder.*

CHAPTER 21: Intermodal Transportation

CHAPTER 22

Power in Transportation

Objectives

After studying this chapter, you should be able to:
- describe the two main types of engines and tell how they work.
- describe electric and fluid-powered motors.
- describe several alternative sources of power.
- explain how design affects vehicle performance.

Terms

aerodynamics
diesel engine
engine
external-combustion engine
fuel cells
horsepower
hybrid vehicles
hydraulics
internal-combustion engine
maglev systems
motor
photovoltaic cells

Technology Focus

World's Biggest Trucks

Monster trucks draw crowds to rodeos and conventions, but the biggest trucks in the world drive along roads most people never see. They stand as high as a house and carry loads the size of a garage. Although their top speed is only around 40 miles per hour, their payloads may be as heavy as 320 tons. They are called "haul" trucks and are used in the mining industry to carry rock and ore.

A Special Drive System

The power needed to move trucks this size is enormous. For 120-ton or larger loads, horsepower requirements may be as high as 2,000 to 2,500 hp. They require a transmission with super-massive components made of strong specialty steels. For this reason, truck makers may install the same type of diesel-electric system that powers locomotives.

In a diesel-electric system, the engine runs at a constant speed and is used to generate a steady flow of electricity. The electric current goes to the wheel motors, which turn the truck's wheels. The operator controls the flow of current to the wheel motors by means of floor pedals in the cab.

A mechanical system with a transmission offers greater drive train efficiency, because the gears provide mechanical advantage. However, a diesel-electric system has no transmission to service or wear out. The wheel motors turn faster or slower according to the amount of electric current they receive. The largest trucks also use alternating current (AC) rather than direct current (DC) to gain an extra two to three percent efficiency.

Is Bigger Really Better?

Payload capacities of haul trucks have increased from 200 to 300 tons in the last ten years. Is bigger really better? Mining companies must decide which truck—a smaller one or a larger one—will move the dirt and rock for the lowest cost. Many factors are considered: purchase price, taxes, financing and insurance costs, fuel needs, operator salary, maintenance and repair requirements, and conditions at the work site. When these factors are studied, a few big trucks prove to be a better investment than a lot of smaller trucks.

Take Action!

Visit an auto dealership that sells trucks. Ask what the load capacity is on their largest model. Calculate how many of those trucks would be needed to haul 300 tons of ore.

Resources
- Chapter 22 Lesson Plan in the *Teacher's Annotated Edition* and in the *Teacher's Resource Binder*.
- Chapter 22 Review in the *Teacher's Resource Binder*.

Prime Movers

Prime movers supply the power (the use of energy to create movement) in transportation. A prime mover is a basic engine or motor.

The terms engine and motor are often used to mean the same thing. This is not wrong, but it is possible to be more exact. In this book, we will use the following definitions:
- An **engine** is a machine that produces its own energy from fuel.
- A **motor** is a machine that uses energy supplied by another source. Fig. 22-1.

Many different types of engines and motors are used in transportation. Most engines burn fuel to create heat. (This burning is called *combustion*.) The primary fuels used are fossil fuels—gasoline, diesel fuel, oil, natural gas, propane, and coal. The heat creates pressure. The engine uses the pressure to create mechanical force or motion. There are two basic types of engines: internal-combustion engines and external-combustion engines.

In this chapter, we will examine the most common engines and motors. Alternative power and the effect of vehicle design on power will also be discussed.

Internal-Combustion Engines

Internal-combustion engines are the most common engines used in transportation. They power most of the cars, trucks, buses, and motorcycles on the road today. In an **internal-combustion engine**, fuel is burned within the engine itself. (Internal means within or inside.)

FIG. 22-1 An engine (left) creates its own energy from fuel (usually gasoline). An electric motor (below) gets energy from an outside source, such as an electric outlet or batteries.

1. **Reinforcement.** Emphasize the difference between an engine and a motor.
2. **Extension.** Discuss the six forms of energy and how they are related. Note that each form can be converted into another form. Here, the chemical energy of fuel is changed to heat energy, which in turn creates mechanical energy.

FIG. 22-2 Three types of motion can be produced by internal-combustion engines. Reciprocating motion is a back-and-forth or up-and-down motion. Rotary motion is circular. Linear motion is motion in a straight line.

Types of Motion

An internal-combustion engine can produce three different kinds of motion. These are reciprocating motion, rotary motion, and linear motion. Fig. 22-2.

FIG. 22-3 In a reciprocating engine, the piston moves up and down, turning the crankshaft in a circular motion.

Reciprocating Motion

Most transportation vehicles are powered by reciprocating internal-combustion engines. These include most highway and rail vehicles, propeller-driven aircraft, and smaller water vehicles.

A reciprocating engine uses pistons, connecting rods, and a crankshaft. During operation, the *reciprocating motion* (up-and-down or back-and-forth motion) of a piston is changed into the circular or rotating motion of a crankshaft. Fig. 22-3.

Not all of the energy used by the piston becomes crankshaft power. Some power is lost while the engine is operating. This is due to friction. The parts rub together and create resistance to motion between them.

The speed of the engine is determined by the size of the parts. The larger the piston (weight and diameter), connecting rods, and crankshaft, the slower the speed

CHAPTER 22: Power in Transportation **483**

of movement. Very large reciprocating internal-combustion engines used in locomotives or on ships may turn at only 100 to 400 RPM (Revolutions of the crankshaft Per Minute). Very small reciprocating internal-combustion engines, like those used in model airplanes, may turn at well over 20,000 RPM.

Reciprocating internal-combustion engines have several major advantages.

- The RPM of the engine can be quickly and easily controlled. This makes them well-suited for highway stop-and-go vehicle movement.
- The cost of manufacturing a reliable engine is low. Also, the cost of rebuilding or repairing is low compared to the cost of repair for other types of internal-combustion engines.

There are also disadvantages to this type of engine:

- Only a limited amount of energy can be economically produced by a reciprocating internal-combustion engine.
- The amount of horsepower that can be provided by each piston is limited. As you know, *horsepower* is a unit of measurement of power. High-horsepower engines usually have many pistons. Fig. 22-4. They have more moving parts than smaller engines and they weigh more. These factors make large engines more expensive to produce.
- The faster the RPM of the engine, the shorter its usable life span.
- Reciprocating-motion internal-combustion engines use quick-burning fuels. These fuels are more expensive than the slow-burning fuels used in some other engines.

Rotary Motion

Rotary-motion engines produce more horsepower than the same size reciprocating engines. This is partly because their working parts have less mass (weight). They also produce less friction.

Rotary motion is a circular motion. Rotary-motion engines turn because exploding fuels form expanding gases that push against the blades of a turbine or

FIG. 22-4 A modern automotive piston engine is quite powerful. Power is a measurement of how quickly work can be done.

3. **Background.** James Watt, who developed the first workable steam engine, established this unit to show a steam engine could do more work than a horse. He calculated a horse could pull 550 pounds one foot per second.
4. **Extension.** Do an experiment to show how friction creates heat. Relate this back to the discussion about transforming energy into various forms.

FIG. 22-5 These drawings show the operation of a rotary engine.

rotor. A *turbine* is a wheel with evenly spaced blades or fins attached to it much like a fan. A *rotor* is the triangle-shaped part of a rotary engine that revolves in a specially shaped combustion chamber. Fig. 22-5. The turbine or rotor is usually mounted on the same straight shaft as the device that is being powered.

Rotary-motion engines work best where the same RPM is held for extended periods of time. However, some automobile manufacturers are using small internal-combustion rotary-motion engines for stop-and-go road use. A few trains use gas turbines. The turbines operate the electric generators that power the trains. Gas turbines used on ships turn the propeller shaft that moves a ship through the water.

The following are some advantages of rotary-motion internal-combustion engines:

- There are fewer moving parts to maintain or cause friction than in a reciprocating engine.

5. **Reteaching.** Use a simple pinwheel toy to demonstrate the turbine concept.

CHAPTER 22: Power in Transportation **485**

- Horsepower can be increased by enlarging the rotor or turbine. No additional parts are required.
- A wide variety of fuels can be burned to produce the expanding gases.
- A constant RPM can be maintained for very long periods of time without damage to the engine parts.

Some disadvantages of rotary-motion engines are the following:

- Quick changes in RPM cannot be made efficiently.
- The cost of manufacturing and rebuilding rotary engines is greater than for reciprocating engines.
- Because of the high RPM attained, the rotor or turbine must be kept well lubricated and in perfect balance.

Linear Motion

Linear-motion engines are generally referred to as *jet* or *rocket engines*. They are the most powerful type of internal combustion engine. *Linear motion* is motion in a straight line. Fig. 22-6.

In a jet engine, power is created by the internal combustion of the fuel and expanding gases. Fig. 22-7. The more gas that is expanded, the faster the vehicle will go. The expanding gas is aimed behind the vehicle. The force of the gases escaping rearward makes the vehicle move in the opposite direction—forward. This resulting forward force is called *thrust*. Linear-motion internal-combustion engines are used on large aircraft, rockets, and spacecraft.

Rocket engines work much like jet engines. All engines need to mix oxygen with fuel for combustion since nothing can burn without oxygen. The main difference between jet and rocket engines is that jet engines take in air (which is part oxygen) from the atmosphere. Since rocket engines are usually designed for use in spacecraft outside earth's atmosphere, a rocket engine must carry all of its own oxygen and fuel supplies. The fuel supply may be either solid fuel or liquids that are carried as two separate chemicals. When the chemicals are combined, combustion takes place and hot gases expand through a nozzle to produce thrust. A liquid fuel rocket engine may be turned on and off. Once a solid fuel rocket is started, combus-

FIG. 22-6 A jet engine produces linear motion. Powerful jet engines can propel aircraft at very high speeds.

6. **Reinforcement.** Refer to rocket launches shown on TV as you explain linear motion.
7. **Reteaching.** To demonstrate that nothing can burn without oxygen, place a candle in a jar and light it. Then put a lid or piece of aluminum foil over the top of the jar to seal out the air.
8. **Extension.** Discuss how chemical reactions can release oxygen from solids.

FIG. 22-7 In the operation of a simple jet engine, the burning fuel provides thrust.

tion will continue until all the chemicals are used.

There are some advantages to using the linear-motion internal-combustion engine:
- Very high horsepower or thrust can be developed.
- Several different types of solid or liquid fuels can be burned to make the expanding gases.
- There are no moving parts needed to transfer the power to the vehicle.

Some disadvantages of the linear-motion internal-combustion engine include the following:
- Manufacturing and rebuilding costs are high, and some rocket engines are built to be used only once.
- No directional change of thrust or motion is possible. The force is generated in one direction only.

Engine Systems

To understand how internal-combustion engines work, let's examine one example—a single-cylinder, four-cycle, small engine. Other types of internal-combustion engines are more complex but operate in basically the same way.

Most engines have six systems:
- mechanical system
- lubrication system
- fuel system
- ignition system
- starting system
- cooling system

The *mechanical system* contains the piston, connecting rod, and crankshaft. Fig. 22-8. The reciprocating motion of the piston is changed into the rotary motion of the crankshaft.

The function of the *lubrication system* is to decrease friction. Lubricating (oiling) parts makes them slippery. Lubricating is done in basically three ways. On some small engines, a small oil pump pushes oil through channels to the moving parts. Other engines rely on simple splash-and-dipper devices to oil the parts. Certain engines are lubricated by oil that is mixed into the fuel and supplied by the fuel system.

9. **Extension.** Coordinate this section with a lab experience centered around small gas engines.

10. **Extension.** Perform an experiment in class in which students slide an object over greased and ungreased surfaces.

FIG. 22-8 This drawing shows the basic parts of an engine.

FIG. 22-9 These two types of systems provide fuel to an engine.

The *fuel system* supplies fuel to the cylinder. It also mixes the fuel with air before it enters the cylinder for combustion, and it carries away exhaust. Fig. 22-9. Parts of a fuel system include:

- an air cleaner to prevent dirt from entering the engine
- a fuel tank
- a *carburetor* or fuel mixer to mix the air and fuel accurately to ensure efficient engine operation
- *intake valves*—mechanical devices that open and close to control when the air-fuel mixture flows into the cylinder
- *exhaust valves* and pipes to allow spent gases to escape from the engine
- a muffler to quiet the engine noise

The *ignition system's* job is to ignite the air-fuel mixture at the correct moment. The ignition system includes:

- *spark plugs* to supply a spark to begin combustion
- a *coil* to increase the spark plug's spark
- *breaker points* and *condenser* or other electronic devices that switch the electricity flow
- wires for the electricity to flow along
- a *generator* or *magneto*—devices that produce electricity from motion of the engine

If the mechanical, lubrication, fuel, and ignition systems are all working correctly, the engine can be started.

The *starting system* may be as simple as a rope on a pulley connected to the *flywheel* (the part in a small engine that is set into motion by the action of pulling a starting rope). It may also be as complex as a starting motor that works from a storage battery. Most small engines have recoil (rewind) rope starters. Fig. 22-10. The starting system must stop operating once the engine begins running.

While the engine is running, heat from combustion and friction must be kept under control. This is the job of the *cooling system*. Small engines are usually air-cooled. An air-cooling system includes a fan (usually part of the flywheel) and metal shrouds (covers). The shrouds contain and direct the cooling air movement. Fig. 22-11.

Larger engines are usually water-cooled. This type of system is more complex. It contains a water pump, radiator, hoses, and a water jacket. The water jacket is usually built into the crankcase and cylinder head.

FIG. 22-10 With a rope starter, the operator turns the crankshaft by pulling the rope. This starts the engine, which then keeps the crankshaft turning.

11. **Reinforcement.** Explain why a mixture of air and fuel is needed.
12. **Extension.** If possible, bring a fuel pump to class.

CHAPTER 22: Power in Transportation

SCIENCE CONNECTION
Friction

Friction is the force that resists the motion of one object along the surface of another. Friction is very important in everyday life. Without friction, you would not be able to walk. Your shoes would slip on the sidewalk. Friction also allows automobile tires to grip the road and the wheels of a locomotive to travel along the rails.

Friction also has disadvantages. To overcome it requires extra power. It is also an undesirable source of heat that may cause damage. For example, in a motor, the heat created when the moving parts rub against one another may cause those parts to wear. This is why oil and other lubricants are used. The lubricants fill in the rough places on part surfaces. The parts move more easily and produce less friction and less heat.

The three major types of friction are sliding friction, rolling friction, and fluid friction. *Sliding friction* is produced when two surfaces slide across each other, such as when a box is slid across a floor. The friction between an automobile tire and a street is *rolling friction*. Rolling friction is about 1/100 as great as that due to sliding friction. *Fluid friction* is the friction between moving fluids or a fluid and a solid.

Try It!

Experiment with rubbing objects having different surfaces together to see how much heat they produce. Make a list of those that produce little heat and those that produce much heat. What conclusions about surfaces can you draw from this?

FIG. 22-11 An air cooling system keeps heat from combustion and friction under control.

SECTION 5: Transportation Technology

Engine Operation

Internal-combustion engines can be either two-stroke cycle or four-stroke-cycle designs. Almost all automobiles and large motorcycles use four-stroke engines.

Four-Stroke-Cycle Engines

A four-stroke engine completes four *strokes* of the piston for one combustion *cycle*. Fig. 22-12.

Stroke 1: The air-fuel mixture is drawn into the cylinder (intake stroke).

Stroke 2: The mixture is compressed (compression stroke).

Stroke 3: The spark plug fires and the burned gases expand and force the piston down (power stroke).

Stroke 4: The piston comes back up, pushing the spent gases out of the cylinder (exhaust stroke).

These four strokes together make two revolutions. While the engine is operating, the cycles repeat over and over again.

Two-Stroke-Cycle Engines

Two-stroke-cycle engines complete the four operations in only two piston strokes, or cycles. Small boats and snow-

FIG. 22-12 There are four motions in a four-stroke cycle engine: intake, compression, power, and exhaust.

13. **Reinforcement.** Have students follow the steps in Fig. 22-12 as you explain them.

CHAPTER 22: Power in Transportation   **491**

FIG. 22-13 There are two motions in a two-stroke cycle engine: compression and power.

mobiles use two-stroke-cycle engines. Fig. 22-13. During the first stroke, the piston goes up. It allows the air-fuel mixture to enter the cylinder and compresses it at the same time. The spark plug fires when the piston is at the top. This pushes the piston down for the power stroke. When the piston is at the bottom of the stroke, the exhaust is let out of the cylinder.

This design allows two-stroke engines to have fewer moving parts. The shorter cycle and fewer parts allow the two-stroke engine to run at higher RPMs than four-stroke engines. A disadvantage of the two-stroke engine is that it usually produces more pollution than a four-stroke engine.

14. **Reinforcement.** Have students look at Fig. 22-13 as you explain how the four operations are completed in two strokes.

Diesel Engines

A special type of two- or four-stroke-cycle engine, called a **diesel engine**, does not need a spark plug. The special low-grade diesel fuel is ignited by heated, compressed air. Fig. 22-14. Because of the high

FIG. 22-14 In a diesel engine, the compression cycle forces the air-fuel mixture together. The added compression creates heat that causes the air-fuel mixture to burn.

Why are diesel engines in demand? There are several reasons. Diesel fuel requires less refining and costs less. It will not explode easily and needs no ignition system. It burns completely which means more miles to the gallon, and diesel engines do not require exhaust pollution control devices. The higher cost of manufacturing and need for larger lubricating and filtering systems does make a diesel engine cost more to produce and install. However, a well-maintained diesel engine will go over 300,000 miles before needing an overhaul. With regular overhauls, many diesel truck engines have logged over 3 million miles.

compression necessary, diesel, or "heat," engines are usually heavier than gasoline engines.

Although the diesel engine was first designed and patented in Germany in 1892, the early version was not efficient. It took another 45 years of development before a profitable diesel engine could be produced and marketed. Today, most highway trucks, about half of all local delivery trucks, and nearly 10 percent of all personal light trucks are powered by diesel engines. Diesel engines also propel most commercial ships, while burning many different types and grades of fuel. As improvements in manufacturing, design, and electronics are found, a smaller diesel engine may eventually power more family automobiles.

External-Combustion Engines

An external-combustion engine was one of the original manufactured devices that provided power for transportation. In **external-combustion engines**, the fuel is burned *outside* the engine. External-combustion engines can provide reciprocating and rotary motion, but not linear motion.

Steam-Driven Engines

Almost all of the external-combustion engines currently used in transportation are steam-driven. Water is heated inside a closed container called a boiler. The water becomes steam under high pressure. This steam is used to drive pistons or turn turbine blades. Fig. 22-15. The size of the steam boiler must be appropriate for the size of the vehicle. Using a heavy boiler in a compact automobile, for example, would not be efficient.

CHAPTER 22: Power in Transportation **493**

Steam moves from the boiler into cylinder. It pushes against the piston, causing the piston to move to the right.

The slide valve cuts off the steam supply to cylinder A. The steam enters cylinder B and pushes against the other side of the piston. The piston moves to the left, the slide valve opens, and the cycle is repeated.

FIG. 22-15 These drawings show the operation of an external combustion steam engine.

The steam engines used in vehicles operate as a closed system. After steam has provided power, it is allowed to cool. The steam condenses into water. The water is reheated and again becomes steam. This process keeps repeating. In a closed system, there is little noise or water contamination.

The heat needed to boil the water can come from many sources. Early vehicles burned coal and wood. Most modern vehicles burn petroleum products. Some modern military vessels use nuclear fuel to boil the water. Because very little nuclear fuel is needed, these vessels do not need to refuel often.

Advantages of an external-combustion steam-operated engine include the following:

- Almost any type of fuel can be used to produce the steam.
- The boiler and the engine can be placed in different physical arrangements.
- Many different designs of steam-driven engines are available to serve specific needs.

There are several disadvantages to external-combustion steam-operated engines:

- Boilers require a great deal of space. Their use is limited to large vehicles.
- Even when handled carefully, steam is dangerous.
- The steam and water cause metal to break down. Continuous care and preventive maintenance are necessary.

Modern Uses for Steam Power

Steam-powered railroad locomotives were once very important to rail transportation. Their "chug-chug" sound resulted when steam was released from the reciprocating engine. They are still used in many parts of the world. However, few remain in use in the United States.

Some modern locomotives use steam in a rotary-engine. Steam is produced and piped into a chamber where it is directed at the fins of a turbine rotor. The pressure of the steam turns the rotor in a circular direction. The more steam, the faster the rotor turns. Other designs use gas turbines to turn the rotors.

The rotor is connected to an electrical generator. The rotating generator produces the electricity needed to power electric motors. These electric motors then turn the axles and wheels of the locomotive.

Reciprocal and rotary engines are both used to power ships. Ships have room for large boilers and various methods of creating heat. Some still use coal, but most use oil. The steam then turns a turbine, which turns the propeller.

Motors

Electric- and fluid-powered motors are used to move vehicles. Most on-site transportation vehicles, such as subway trains, are powered by motors.

> **FASCINATING FACTS**
> The steam engine was first used in England. People thought they were dangerous. For safety, a man carrying a red flag walked in front of the vehicle, warning anyone in the area.

15. **Background.** Note that pure water is used in order to keep corrosion and residue buildup at a minimum.
16. **Enrichment.** Bring to class some books about old steam railroads.

Electric Motors

For many years, electric motors have powered urban mass transportation commuter vehicles. The oldest operating electric street car line runs daily on tracks in New Orleans. Electric power is supplied to each car's motors by an overhead electrified cable. Fig. 22-16 (right). Other lines may use a "hot" third rail on the ground.

Electrically powered passenger and freight trains are relatively common in the northeastern section of the United States and all across Europe. In addition, several large cities have electrically-powered mass transportation systems. Washington, D.C., Atlanta, Georgia, and Orlando, Florida, are examples. Fig. 22-16(left).

At the present time, electric cars depend on batteries for power. Batteries currently available are heavy, need frequent recharging, and lose power in cold conditions. New types of batteries are being developed to overcome these drawbacks.

As batteries are improved, more and more electrically-powered vehicles will be operated on streets and highways. Small cars that operate efficiently on batteries for short distances are available now.

There are many advantages to electrically-powered vehicles. The most important ones are:

- Except for pollution caused by electrical generating plants, they do not pollute the environment.
- Operating costs are comparatively low.
- They require less maintenance and repair work than other power systems because they have fewer moving parts.
- Little or no noise is produced by electrically-powered equipment.

FIG. 22-16 Compare the electric streetcar and electric train. How do they differ?

17. **Enrichment.** If possible, have an electric car brought to the school. Do a side-by-side comparison of a gasoline engine and the electric motor.

- Electric motors will generally run in either direction. No transmissions (gears) are needed.

Vehicles that operate on electricity have one big disadvantage. Supplying electricity to each vehicle is often difficult and inconvenient. This problem is especially serious for vehicles that are operated independently, such as cars and trucks. However, a train on a track does not change its route, and electricity can be supplied more easily. Some researchers are working on an electric car that obtains power from a flywheel.

Pipeline Motors

The power to move gases, liquids, or slurry through pipelines is produced by pumps turned by electric motors. The pumps may be of several different types, but they all have the same job—"pushing" the pipeline contents along. The pumps add pressure to the pipeline in almost the same way that a bicycle pump adds pressure to a tire.

Because there is friction between the contents and the pipeline walls, pressure must be added to the pipeline regularly. Pumping stations along the pipeline continuously add pressure to replace what is lost to friction.

Fluid-Powered Motors

Fluid-powered motors can develop considerable horsepower, but not much speed. They are used in heavy equipment and other slow-moving vehicles that require a lot of power.

Fluid-powered motors are controlled by increasing or decreasing the pressure of the fluid. Using the reactions of fluid under pressure to develop motion is called **hydraulics.** A fluid pump produces the pressurized fluid. It can be powered by almost any rotating power source. Most fluid-powered vehicles use internal-combustion engines to power the pumps.

Like electric motors, most fluid-powered motors will also reverse. They run efficiently in either direction.

Motor and Diesel Engine Combinations

Most modern locomotives use *diesel-electric power*. These locomotives work just like the rotary-engine-powered models except that a diesel engine is used to turn an electric generator. The generator produces the electricity needed to power the electric motors that turn the wheels. Diesel-electric engines are very efficient and long-lasting. They need much less space than steam engines, so the locomotives can be smaller. Diesel-electric engines also accelerate faster than steam engines. Fig. 22-17.

Diesel-hydraulic locomotives use diesel engines to drive a torque converter instead of a generator. (*Torque* is any force that produces or tends to produce rotation.) The torque converter, which includes a pump and a turbine, uses fluids under hydraulic pressure to transmit and regulate power received from the diesel engine. The pump forces oil against the blades of the turbine. This action causes the turbine to rotate and to drive a system of gears and shafts that moves the wheels. Diesel-hydraulic locomotives are not commonly used in the United States, but they are widely used in other countries, such as Germany.

See Fig. 22-18 for a table showing different modes of transport and the types of power commonly used.

18. **Extension.** One of the prime uses of the pipeline industry is to pump fluids.

CHAPTER 22: Power in Transportation **497**

FIG. 22-17 Even though alternate forms of power are being found for locomotives, diesel-electric engines still power most modern locomotives today.

Mode of Transportation	Type of Power Used
Trucks, buses	Gasoline engine Diesel engine
Cars	Gasoline engine Diesel engine Electric motor
Propeller-driven planes	Reciprocating, internal combustion engine
Jet planes, spacecraft, missiles	Linear-motion internal combustion engine
Locomotives	Steam engine Electric motor Gas-turbine engine Diesel-electric combination Diesel-hydraulic combination
Ships	Steam engine Gas turbine engine
Pipelines	Electric motor

FIG. 22-18 More than one type of power may be used for certain types of vehicles.

Alternative Power

One of the biggest problems facing transportation is finding power sources that are in plentiful supply and do not harm the environment.

Wind Power

Wind-powered sailing ships traveled the seas long before engines were developed. Today, modern technology is finding new ways to use wind to power ocean-going ships. Turbosails, like those shown in Fig. 22-19, produce almost four times as much thrust as the best traditional sails.

Wind energy is also being turned into electrical energy. Some small sailing yachts have windmills and electrical generators installed on their masts. Even when winds are light, the electricity produced by these windmill units is enough to charge the batteries of the boat. The same designs, on a larger scale, are used to provide additional electricity on larger ships. Power is limited only by battery capacity.

Solar Power

The sun is a potential source of natural energy for transportation. **Photovoltaic cells** are devices that can change the energy from the sun's light (solar energy) into electricity. These have already been used to power experimental vehicles. Some vehicles use the power from the cells to directly power an engine. More often, the electricity from the photovoltaic cells does not go directly to a motor. Instead, it is used to charge high-capacity batteries. Fig. 22-20. The photovoltaic cells charge the battery whenever the sun shines on them. This allows the vehicle to be used even when the sun is not shining. The charged batteries also produce a steady flow of electricity to the motor. This makes it much easier to maintain constant speeds and gives the vehicle a longer range. Fig. 22-21.

FIG. 22-19 This ship, the *Alcyone*, uses computer-controlled turbosails.

19. **Background.** Explorer Jacques Cousteau and his son, Jean-Michel, have sailed this ship. They also helped design it. Students may see it in TV specials.
20. **Enrichment.** A small inexpensive cell can be used to demonstrate the development of electrical power.

FIG. 22-20 The Solar Baby, a solar-powered minicar, is used in Malaysia. Each wheel has a separate motor. Its top speed is 45 mph.

As with wind energy, solar-produced electricity is limited by battery capacities. Most batteries that are powerful enough to store enough electricity to power vehicles are either very heavy or very expensive. Several types of new battery designs that may soon solve this problem are now under development.

Alternative Fuels

The use of alternative (non-fossil) fuels in internal combustion engines is very possible in the near future. However, new fuels must be safe to use, readily available, and economical.

FIG. 22-21 The GM Sunraycer uses its photovoltaic cells to charge special batteries. The batteries then provide electrical power to the motor.

SECTION 5: Transportation Technology

Methanol

In the United States, corn is converted into methanol. *Methanol* is a chemical that can be used as fuel in internal-combustion engines. Usually it is mixed with gasoline to make a mixture called *gasohol*. Gasohol can be used in automobiles just like straight gasoline, and it produces less pollution. Because the methanol in gasohol is derived from corn, gasohol production also provides a new market that helps keep farms profitable. In some countries, other crops are used to produce methanol. The government of Brazil is attempting to produce enough methanol from sugar cane to fuel the country's automobiles.

Methane Gas

Methane gas is a waste product given off when plant or animal waste decays. It is similar to natural gas and propane. Propane is already used to power many vehicles. Experiments have shown that methane may also work well. Some large garbage dumps already collect methane for fuel purposes.

Recyled Fuels

Some previously used substances can be recycled as fuels. Steam-powered engines on boats and trains run well on waste motor oil. Diesel-quality fuel can be refined from used cooking oil. Other fuels may be obtained from old tires or plastic bottles.

Maglev Propulsion

Maglev systems are rail systems that operate on the scientific principle that like poles of a magnet repel each other. The word maglev is short for *mag*netically levi-tated. (Levitate means to rise or float in the air.)

Magnets in the maglev guideway (rail) repel magnets of like polarity on the bottom of the maglev vehicle. Fig. 22-22. This causes the train to levitate above the guideway creating a nearly frictionless riding surface. Magnets are also involved in the vehicle's propulsion. Changing the polarity of the magnets on the train and the guideway at the proper moments speeds up or slows down the train.

Because there is so little friction and the vehicles are designed to reduce air resistance, the trains can easily reach speeds of 300 mph. They glide quietly, smoothly, and swiftly along the guideway using relatively little energy.

A maglev system is very different from our present "steel-wheel-on-steel-rail" system. A maglev is already in use in Germany. Fig. 22-23. In the United States, maglev trains are being considered for the routes between Las Vegas and Los Angeles and between Baltimore and Washington, D.C. However, the expense of building the guideway has made development slow.

NASA is studying a rocket launcher, called MagLifter, that would use magnetic ing electromagnets could make maglev train systems much more economical.

> **FASCINATING FACTS**
> In the early 1800s, it was believed that going faster than 12 mph on a train could cause mental disorders or even suffocate people. People feared that such high speeds would suck all the air out of passengers' lungs.

---

21. **Extension.** You may want to teach basic principles of magnetism.
22. **Extension.** Note that maglev systems would require all new construction. They are also very expensive to construct.
23. **Background.** The development of superconduct-

CHAPTER 22: Power in Transportation **501**

FIG. 22-22 One set of magnets levitates the train. Another set propels it along the track.

propulsion as a substitute for first-stage rockets. A high-speed maglev test track is also being built by the Air Force. The track is expected to move test payloads at a speed of 6,840 mph.

Fuel Cells

Fuel cells are batteries that generate power by means of the interaction between two chemicals, such as hydrogen and oxygen. Fuel cells don't need to be

FIG. 22-23 Maglevs are safer than ordinary trains. Hooked onto the track, they cannot derail.

Hybrid Power

Flexible-fuel vehicles, also called **hybrid vehicles**, are being designed to run on more than one kind of fuel, such as both gasoline and compressed natural gas.

Toyota's new hybrid car, Fig. 22-24, is powered by both gasoline and electricity. At low speeds the car relies on the electric motor, which gets its power from batteries. At higher speeds, the gasoline engine takes over, recharging the batteries at the same time. The result is fuel efficiency increased to 66 miles per gallon. U.S. automakers are also working on hybrid designs.

plugged into an electric socket for recharging. They are also non-polluting.

Fuel cells have been used by NASA in space, and some city buses are operated with fuel cells. Daimler-Benz, a German automaker, has produced Necar II, a minivan powered by a fuel cell.

FIG. 22-24 Toyota's hybrid car is powered by a gasoline engine and an electric motor. At the top are the engine and transmission; below is the inverter that switches power from gasoline to electricity.

CHAPTER 22: Power in Transportation

The Effect of Vehicle Design on Power

Design can make vehicles more efficient, more economical, and safer. Today, most design considerations involve aerodynamics, size and weight, materials, and electronics.

Aerodynamics

Aerodynamics (AIR-oh-die-NAM-iks) is a science that deals with the interaction of air and moving objects. Air tends to resist or slow movement. Vehicles meet this resistance as they move through the air. The faster they travel, the greater the resistance. A strong, power-robbing force is created called *aerodynamic drag*. Design engineers look for ways to reduce this drag. Doing so eases the workload of the engine and thus improves fuel efficiency. Much progress has been made.

Shape is the critical factor. Engineers describe an aerodynamically efficient shape as "slippery." They try to design vehicles that will slip smoothly through the air. Air should glide smoothly over, under, and around a vehicle. Even a mirror or door handle that sticks out can disturb the air and increase drag. Fig. 22-25.

Car aerodynamics can be further improved by making the underside of the car smooth, reducing turbulence. Body surfaces can be grooved or dimpled to speed airflow. Outside mirrors can be replaced by TV camera "eyes."

Aircraft must also be aerodynamic. The basic wing (cross section) shape is not likely to be changed. However, length, positioning, angle, and other factors may be changed in relation to the total design. A major goal is to improve lift and further reduce drag.

FIG. 22-25 Aerodynamic design helps reduce drag on this car, making it more fuel efficient. It is General Motors' prototype, the Ultralite.

24. **Reinforcement.** Ask students if they have ever noticed bicycle enthusiasts bending down low over their handlebars. This is done to reduce drag. The airfoil mounted on top of the cab in front of the semi-trailer on an 18-wheeler is also to reduce drag.

Size and Materials

American cars weigh about 1.5 tons. This immense weight requires a large engine to push it along. Lightweight vehicles are generally more fuel-efficient than heavy vehicles. One way to achieve weight reduction is to produce smaller vehicles. Today, most personal transportation vehicles are smaller and slower than vehicles built in the 1950s. Even trucking companies often use smaller vehicles called "hotshots." These cost less and are cheaper to operate.

Design engineers are also looking for ways to reduce weight without reducing size. Lightweight materials such as aluminum or plastics are used instead of heavier materials such as steel.

Plastics

Many, new, smaller aircraft are made primarily of special fiber-reinforced plastics. They are light, but strong. Parts of larger planes are also made of these materials.

Plastics have long been used in car and truck interiors. It is expected that more and more plastics will also be used to form exterior parts. It is estimated that up to 70 percent of new cars will have plastic body panels by the year 2000. Advanced car designs are using "ultralight" technology, which includes components and bodies made from carbon fibers, composites, and advanced engineering plastics.

Ceramics

Ceramics, such as earthenware and porcelain, are made from nonmetallic minerals that have been fired at high temperatures. Ceramics are used in spacecraft because they can withstand extreme temperatures and because they insulate well.

FIG. 22-26 This engine contains mostly ceramic parts.

Some car engines include ceramic parts. Fig. 22-26.

Engineers hope to be able to use ceramics in trucks, because the diesel engines used in trucks produce great amounts of heat. As a result, the front of a truck must be built large and flat like a "wall" to accommodate cooling airflow. However, this shape increases aerodynamic drag and reduces fuel efficiency. In order to give trucks a more aerodynamic shape, the engine must be redesigned. If ceramics could replace metal parts or protect them from the heat, air-cooling would not be necessary. The design of trucks could be changed.

Before ceramics can be widely used, two major problems must be overcome. Ceramic materials are so hard that special diamond-coated tools must be used to cut them. They are also brittle. A part with even a small blemish may break under stress.

25. **Extension.** Watch for developments in the technology of ceramics. Some people in industry expect these developments to have long-term significance.

Composites

A composite is a new material made by combining two or more materials. Each component retains its own properties, but the new material has more desirable qualities.

Composites are being used more and more today to make parts for different types of transportation. For example, fiberglass reinforced with plastic is now being used for boat hulls and automobile bodies. Kevlar™, which is a composite that is very difficult to cut, is being used to make reinforcing belts for tires. Carbon/graphite composites are now being used to make lightweight bicycle frames and to produce several new types of aircraft.

Electronics

Computers, microprocessors, and other electronic devices are revolutionizing vehicle operations. Electronic controls can reduce the amount of air pollutants in vehicle exhaust. The controls precisely regulate the air-fuel mixtures in the engine. Not only does this reduce waste, it also improves engine performance. Electronics are also used in many vehicle subsystems, such as brakes, power steering, transmission, and heating and air conditioning.

AMTRAK has introduced a new high-speed train that can take curves at 112 mph, reducing travel times. Microprocessors in each car feed information from sensors to a hydraulic system that tilts the cars. The tilting keeps passengers from falling out of their seats when the train rounds a curve. Fig. 22-27.

Vehicles used in space flight and exploration rely completely on electronic signals and controls. The little rover vehicle used by NASA during the 1997 tour of Mars is an excellent example. Fig. 22-28.

FIG. 22-27 The centrifugal force that would act on passengers when the train rounds a curve is counteracted by the car's ability to tilt.

26. **Discussion.** Discuss the advantages of anti-lock brakes.
27. **Extension.** A spacecraft has to travel faster than 17,000 miles per hour in order to escape earth's gravity. This speed is called earth's "escape velocity." For more information on this subject, see the handout "Escape from Planet Earth" in the *Teacher's Resource Binder*.

FIG. 22-28 Sojourner, about the size and shape of a skateboard, was packed with equipment that gathered information about the surface of Mars. It was powered by solar energy.

Fiber Optics

The optical fibers used in vehicles are usually thin strands of plastic. They carry signals used to control such items as power windows and door locks.

Signals from several controls can be combined, and a pair of optical fibers can replace many wires. This saves space and weight. It also greatly simplifies the wiring system.

Eventually, fiber optics might be used to link electronic operations in engines. However, the fibers currently available cannot take the heat produced by today's engines. As engines are redesigned, their use may increase.

CHAPTER 22: Power in Transportation

Career File—Aircraft Mechanic

EDUCATION AND TRAINING

Most mechanics are certified by the FAA and learn their skills in a trade school certified by the FAA. Schooling generally lasts from two to two-and-a-half years. For mechanics to retain their certification, they must take 16 hours of additional training every two years.

AVERAGE STARTING SALARY

Earnings of aircraft mechanics vary with the size of the airline. The average annual salary for beginning mechanics ranges from $18,000 at smaller turbo-prop airlines to $28,000 at major airlines.

OUTLOOK

Aircraft mechanic jobs are expected to increase about as fast as average through the year 2005, according to the Bureau of Labor Statistics. Competition for jobs with major airlines will be high because they offer high wages and travel benefits.

What does an aircraft mechanic do?
Aircraft mechanics keep airplanes and helicopters in peak operating condition. They perform scheduled maintenance, make repairs, and complete inspections required by the Federal Aviation Administration.

Does most of the work involve data, people, or things?
A mechanic works mostly with things—the various parts of an aircraft and the tools required to service and repair the aircraft. Most mechanics fill out inspection and repair reports. On large planes, mechanics work with other mechanics.

What is a typical day like?
When an aircraft is scheduled for maintenance, the mechanics inspect and make any necessary repairs to engines, landing gear, instruments, brakes, valves, pumps, air conditioning systems, and other parts of the aircraft. A mechanic often relies on the pilot's description of a problem to find and fix faulty equipment.

What are the working conditions?
Mechanics usually work indoors in hangars, but they may also work outdoors—sometimes in unpleasant weather—when repairs must be made quickly. Noise and vibration are common when testing engines, so eye and ear protection is needed. Mechanics have a great responsibility to maintain safety standards, which can be stressful.

Which skills and aptitudes are needed for this career?
A high degree of mechanical aptitude is required. Aircraft mechanics should be able to solve complex mechanical problems under time pressure. Flexibility is important because they often stand, lie, or kneel in awkward positions. Occasionally they work on scaffolds or ladders. Frequently, mechanics must lift or pull objects weighing as much as 70 pounds.

What careers are related to this one?
Other occupations that involve similar mechanical and electrical work include electricians, elevator repairers, and telephone maintenance mechanics.

One aircraft mechanic talks about her job:
I can't imagine doing anything but fixing and flying these incredible machines. My dream is to build and fly my own airplane.

Chapter 22 REVIEW

Reviewing Main Ideas

- Power is using energy to create movement.
- The two basic types of engines are internal- and external-combustion engines.
- The three types of motion produced by internal-combustion engines are reciprocating, rotary, and linear.
- Internal-combustion engines operate in either two- or four-stroke cycles.
- Steam engines are the most common external-combustion engines.
- Motors used in transportation systems are either electric- or fluid-powered.
- Alternative power sources include wind and solar power; alternative fuels, such as hydrogen gas, methanol, methane gas, and recycled fuels; maglev propulsion; fuel cells; and hybrid power.
- Design engineers look for ways to improve vehicles through changes in aerodynamics, size and weight, materials, and electronics.

Understanding Concepts

1. What is the major difference between internal-combustion and external-combustion engines?
2. Explain how a reciprocating engine differs from a rotary engine and a linear-motion engine.
3. Name some alternative sources of power for transportation.
4. How do size and materials affect the aerodynamics of a vehicle?
5. What is the principle upon which maglev propulsion operates?

28. **Resource.** Answers to "Understanding Concepts," "Thinking Critically," and "Applying Concepts and Solving Problems" can be found in the Teacher's Manual at the front of this *Teacher's Annotated Edition* and in the *Teacher's Resource Binder*.

Thinking Critically

1. Would it be practical to develop large, electric-powered trucks? Explain.
2. Discuss the effects of friction on engines and on pipeline transportation.
3. If you were going to design a "perfect" transportation vehicle, what type of engine would you put in it? Why?
4. Select different examples of vehicles and determine the most efficient and economical method of power for each.
5. Which alternative source of power seems to you to be the most likely source used in the future? Explain your answer.

Applying Concepts and Solving Problems

1. **Science.** Raise a book off the table with a balloon. Place the book on the balloon and blow up the balloon. What stroke of a four-stroke-cycle engine is represented here? Which part of the engine is represented by the book? Which event is represented by your blowing air into the balloon?
2. **Mathematics.** Current flow in an electrical circuit is calculated by using Ohm's Law: *Amperage = voltage ÷ resistance*

   If you apply 24 volts to a circuit that contains 24,000 ohms of resistance, what will the amperage be?
3. **Technology.** Some people want to limit transportation in large cities to bicycles. People in many cities in Asia already depend upon bicycles. Redesign a bike so that it would make a practical vehicle for working people in the U.S.

CHAPTER 22: Power in Transportation

1 DESIGN AND PROBLEM-SOLVING ACTIVITY

Designing a Transportation System for the 21st Century

Context

Increased air and land traffic, poor air quality, road and bridge maintenance needs, and space for parking are a few of the transportation-related problems facing communities today. In the future, public and private transportation will have to be different. Solutions, such as car pooling, new vehicle pathways, smart highways, alternative vehicles, high-speed rail, and monorail systems, are all being studied. Low-cost, nonpolluting alternatives for commuting short distances, such as bicycles, may one day be common.

Health and Safety Notes: Before doing this activity, make sure you understand how to use the tools and materials safely. Have your teacher demonstrate their proper use. Follow all safety rules.

Goal

For this activity, you and your teammates will research and develop an alternative transportation system for your community for the year 2015. You will then demonstrate your system in *one* of these ways:
- build a model of the system showing topography, landmarks, and new system features
- create a multimedia presentation showcasing your solutions
- create a computer-animated scenario
- build a prototype of an alternative vehicle and demonstrate its use

1. State the Problem

Research and develop an alternative transportation system for your community for the year 2015 and demonstrate it in one of four ways.

Specifications and Limits

- The system must include vehicles, vehicle pathways, guidance systems, and control systems. The guidance system can be the global positioning system. Controls can be such things as signs, signals, video monitoring, and remote sensing.

1. **Cooperative Learning.** Each student on a team can be assigned a particular subsystem. Subsystems can then be combined.

SECTION 5: Transportation Technology

- At least five alternative forms of transportation must be included, such as gas-, electric-, or solar-powered vehicles; hovercraft; maglev or high-speed trains; people movers; and bicycles. All five forms must be nonpolluting and use energy efficiently.
- The system must allow a user to travel anywhere within the community using one of the five alternative forms of transportation.

2. Collect Information

With your teammates, read through this entire activity. You may want to copy the specifications outlined above. This will help you remember them later.

Interview a transportation or public works official, a member of your local planning board, or a civil engineer with expertise in alternative transportation design. Find out what is involved in making changes to the present system, what the present concerns are, and what future plans may be.

Obtain maps of your community showing transportation systems and major routes, like the one shown on page 512. Maps or engineering drawings may be available from your local or regional transportation department, community planning board, bus or rail companies, or your local library. Some maps may be available on CD-ROM. If necessary, enlarge the map by means of photocopying or computer scanning and printing.

Determine the key elements in the present system. Include inputs (people, fuel, vehicles, roads, waterways, parking, signs and signals), processes (methods of moving people and goods), and outputs (passenger and freight mobility; impacts on people, environment, economy, and community).

3. Develop Alternative Solutions

Discuss the positive and negative features of the present system, then brainstorm alternatives. Consider factors such as cost, feasibility, storage, availability of energy resources, and public acceptance. Consider at least five alternatives. Use a decision-making tool, such as the decision matrix shown on page 513, to help you consider all the solutions. Consider your needs for equipment and materials.

Equipment and Materials
- sketching paper and pencils
- computer system
- CAD, animation, or multimedia software, such as Astound® or Power Point®
- materials and tools for building a prototype or model

4. Select the Best Solution

As a team, select the best solution or combination of solutions. If you are combining more than one solution, discuss how that should be done. Check your solution against the specifications. Is anything missing?

5. Implement the Solution

Create your demonstration model, multimedia presentation, animated scenario, *or* prototype.

6. Evaluate the Solution

Compare your solution to those of your classmates. How effective are the solutions presented? What are the positive and negative impacts?

What did you learn about materials from this activity?

How would you persuade members of your community to use your alternative transportation system?

DECISION MATRIX

Rate each vehicle HIGH, MEDIUM, or LOW	Maglev	High-speed Rail	Electric Car	Hovercraft
Energy needs				
Cost				
Environmental impact				
Social impact				

Useful Skills Across the Curriculum

1. Science. Identify the prime movers in each of the forms of transportation used in your system.

2. Mathematics. Research the typical cost for a gallon of gas in your community. Based on an average of 20 miles to the gallon, how much would it cost a car to travel back and forth each day over one of the major routes in your community for a period of one year?

Credit: Brigitte Valesey

② DIRECTED ACTIVITY

Designing and Building a Solar-Powered Vehicle

Context

Most of today's vehicles use petroleum products for fuel. As you know, experiments are being done to find alternative power sources that are in plentiful supply and do not harm the environment, such as solar energy. Solar energy must be converted to electricity and transmitted to an electric motor, which generates mechanical energy. The mechanical energy then causes the wheels to rotate and move the vehicle.

Goal

In this activity, you will build a vehicle that uses solar energy as its power source. You'll also need indicators, which provide feedback on fuel levels, speed, etc.

Equipment and Materials

- 2"-thick Styrofoam™ polystyrene
- razor knife (pointed craft-type)
- coping saw
- fine sandpaper
- solar cells (2 to 10)
- 1/4" nut driver
- 22-gage solid wire (blue and red)
- wire strippers (set to 22 gage)
- voltmeter
- toggle switch
- rotary selector switch
- 1"-diameter dowel rod
- Stanley center square
- 1/8"-diameter bare welding rod (for axle)
- plastic straws (for axle housing)
- 1/4"-long slices of 1"-diameter dowel rod with a 1/8" hole in the center
- drill press with 1/8"-bit
- gear motor
- Liquid Nails® construction adhesive
- 1/4"-thick paneling scraps (for mounting motor and front axle assembly)
- #4 x 1/2" combination-head sheet metal screws (for attaching things to paneling)
- #1 Phillips screwdriver
- latex or acrylic paint
- 1" foam brush

SECTION 5: Transportation Technology

Procedure

1. This is a team activity. You and your teammates should read this entire activity before performing any of it.

Developing the Propulsion System

2. Carefully read the assembly instruction card that comes with the gear motor. Assemble your motor so that it produces the greatest mechanical advantage.

> **Health and Safety Notes**
> Use all tools and equipment carefully and safely. Also, make sure that you know how to wire properly. Ask your teacher for help if you are in doubt about any procedure.

3. Carefully study Fig. A to make sure that you know how to connect the solar cells in series, parallel, and series-parallel formations.

Fig. A

4. Using the wire stripper, cut an 8" red and an 8" blue piece of 22-gage wire for each of your solar cells. Then strip ½" of the insulation from each end of each wire. Bend one end of each wire into a loop.

5. Using the 1/4" nut driver, gently (using only your index finger and thumb on the handle to tighten the nuts) attach the loop of one red wire to each plus (positive) terminal of each solar cell. Similarly, attach the loop of one blue wire to each minus (negative) terminal of each solar cell.

6. Wire your solar cells in series, parallel, and series-parallel to determine which configuration of cells provides the best voltage for your motor. Just twist the wire ends together to make the connections. Use a voltmeter to make your measurements.

7. Using this information, choose an electrical configuration for your solar cells and gear motor. Make a note of your decision for later.

Developing the Structure

8. The solar cells need to be on the top, front, back, or sides of the vehicle so that *all* solar cells in any one series will have maximum illumination. A shaded or blocked solar cell will act like a 5,000 ohm resistor, effectively switching the power off.

9. Sketch what you would like your vehicle to look like. Keep in mind the angle of the sun at the time of day you intend to run your vehicle. A solar cell works best when the sun's rays are perpendicular (at right angles) to the face of the cell. Now, draw the placement of the solar cells on your sketch and dimension your drawing. Lightweight, stable, aerodynamic vehicles have been found to work best.

10. Have your teacher approve your drawing and dimensions.

11. Obtain the necessary tools and supplies.

Fig. B

CONTROLS

SECTION 5: Transportation Technology

Fig. C

12. Build the vehicle body from 2"-thick Styrofoam. Use construction adhesive to attach anything to the vehicle.

Developing Control and Suspension

13. Suggested methods for controlling your vehicle's on/off state, speed, and direction are shown in Fig. B.

14. Suggested methods for making and mounting your vehicle's front wheels are shown in Fig. C. Methods for mounting the wheels on the gear motor and then mounting that assembly on the structure are shown in Fig. D.

Assembling Your Vehicle

15. Use the Phillips screwdriver and screws to mount the gear motor to a small piece of paneling.

16. Use the adhesive to mount the motor assembly to the structure. For a more streamlined design that will have less wind resistance, you may want to recess your motor assembly.

17. Use the adhesive to mount the steering system to the structure. To aid in streamlining, you may want to recess your steering assembly.

18. Assemble the solar cells in the configuration you chose earlier (Step 7).

19. Mount the solar cells on the structure using staples made from scraps of 22-gage electrical wire.

20. Wire and mount the toggle switch and rotary switch if you have decided to use them.

The Support Team

21. You are the support system for your vehicle. You must troubleshoot problems, obtain any extra parts that are needed, and repair your vehicle when it breaks down.

Fig. D

22. Use the foam brush and paint to decorate your vehicle. Make sure you don't get paint in any of the working parts.

23. Test your vehicle. If necessary, make appropriate changes in wiring, gearing, switching, solar arrangement, etc.

Evaluation

1. How does the angle of the sun's rays, with respect to the surface of the solar cells, affect energy output?

2. Do you think that an aerodynamic shape is an important consideration when designing a vehicle? Why?

3. Using the universal systems model, analyze your vehicle as a whole and its power system. If you could design another solar-powered vehicle, what would you do differently?

Useful Skills across the Curriculum

1. Science. Research the differences between series and parallel circuits. Report your findings to the class.

2. Mathematics. If a circuit carries 16 amps and its resistance is 2, what is its voltage? [1]

Credit: Daniel B. Stout

1. **Answer.** 2. 30 V

DIRECTED ACTIVITY 3

Investigating the Aerodynamics of Gliders and Airplanes

Context

Many inventors in the late 1800s and early 1900s studied the theory of light. Some, like Otto Lilienthal of Germany, Clement Ader of France, Sir Hiram Maxim of England, and Octave Chanute and S.P. Langley in the United States, actually built and flew non-powered gliders based on their observations and discoveries. Probably the most famous aircraft inventors are Orville and Wilbur Wright of the United States.

Health and Safety Notes Before doing this activity, make sure you understand how to use the tools and materials safely. Have your teacher demonstrate their proper use. Follow all safety rules.

All of these inventors believed that air would support the large wings of their gliders and hold the gliders aloft. After many experiments, they discovered that gliders and powered aircraft can remain aloft because of the set of scientific principles called aerodynamics. Air in motion interacts with a moving object. This interaction produces or results in a force. All aircraft in flight are constantly being pulled toward Earth by gravity. Other aerodynamic forces acting on an aircraft are lift, thrust, and drag. Gliders and powered aircraft are designed to use these forces to control flight. These forces are considerations in aerodynamic design. Fig. A.

Fig. A

RUDDER
ELEVATOR
AILERON
CENTER OF GRAVITY

VERTICAL AXIS YAW (TURN)

LATERAL AXIS PITCH (CLIMB OR DIVE)

LONGITUDINAL AXIS ROLL (BANK)

SECTION 5: Transportation Technology

Goal

This activity is designed to help you examine the theory of flight. You will build two different types of simple, model gliders and test-fly them.

Equipment and Materials
- 8½" x 11" typing paper or construction paper
- paper clips
- balsa wood (C grade) ³⁄₁₆" x 3" x 14", ³⁄₁₆" x ½" x 14 ½", ¹⁄₁₆" x 3" x 8"
- wood glue
- modeler's glue or cement
- scales (12")
- triangles
- modeling knives
- pencils
- sandpaper
- 1" x 1" x 4" blocks of wood
- straight pins
- tape measures
- stopwatch

Procedure

1. Make *two* paper gliders. Carefully fold 8½" x 11" paper into the proper shape. Fig. B.

2. One of the gliders is to be made for speed. The purpose of this glider is to show how sleek, narrow shapes "flow" through the air with little drag. The wings are shaped so there is enough lift to push on them, but not enough drag to slow the glider during its flight.

3. The second paper glider is to be made for maximum time aloft. The shape of the wings must be altered to catch more lift and use the drag over the wings more efficiently. Modify the wings of the second glider by folding them into a dihedral, as shown in Fig. C. Place a paper clip on the bottom center of the glider as shown.

Fig. B

SECTION 5: Transportation Technology

WEIGHT (PAPER CLIP)

BEND WING UPWARD SLIGHTLY TO FORM DIHEDRAL ANGLE

Fig. C

4. Test-fly your gliders in a large area such as outside or in a gymnasium. Time all flights with a stopwatch. Record the time aloft of each glider. Measure the distance each glider traveled in the air.

Wood Glider

5. Next, make a model glider from balsa wood. Balsa wood is stronger than paper and weighs more. However, it is still light enough for use in the design of scale model gliders. Begin construction by referring to the working drawing shown in Fig. D.

6. Begin with the wings. These are made from one piece of balsa wood 3/16" x 3" x 14". Cut the balsa wood to the shape and size shown.

7. Use sandpaper to shape the leading edge of the wings into an airfoil shape. An airfoil enables the wings to develop a lower air pressure against the top surface of the wing, providing lift needed to keep the glider aloft.

8. To increase stability, the wings must be formed into a dihedral angle. This is done by carefully cutting the 14"-long wing in half, sanding a 7 1/2° angle on both sides of the wing along the cut line, and gluing the wing back together. Sand from the bottom edge of the wing to the top edge. Assemble the dihedral airfoil wing according to the assembly detail drawing.

9. Use the 3/16" x 1/2" x 14 1/2" piece of balsa wood to construct the body or fuselage of your glider. Refer to the drawing of the fuselage in Fig. D for dimensions. (*Note:* Only the side view is shown.) Lightly sand the edges of the fuselage along its length. Remove the square edges only!

10. Make the tail section (horizontal stabilizer) from a 1/16" x 3 1/2" x 5 1/2" piece of balsa wood. Cut it to the dimensions shown in the drawing. Lightly sand the edges of the tail section.

SECTION 5: Transportation Technology

Fig. D

11. Make a rudder as shown. The grain of the wood must run vertically through the rudder. Lightly sand the leading and trailing edges. Do not sand the bottom. Leave it square for proper assembly.

12. Carefully assemble the glider as shown in Fig. E. Make sure the wings and horizontal stabilizer are attached to the fuselage at a 90° angle. The leading edge of the wings should be 2¾" from the nose of the body. The rear stabilizer and rudder should be attached flush (even) with the end of the fuselage.

13. Test-fly your glider and record the time aloft and distance flown.

Evaluation

1. How much faster did the first paper glider go than the second paper glider? Did altering the shape of the wings change the aerodynamics of your second paper glider? Did the air currents or wind have any effect on its performance? Try moving the paper clip on your second paper glider. How does this affect its flight?

2. How well did your wood glider fly? Did it appear to be balanced? What changes would make it fly better?

Useful Skills Across the Curriculum

1. Mathematics. Suppose you have a piece of balsa wood that measures 5½" x 8¼". How many horizontal stabilizers (shown in Fig. V-3D) could you cut from it?

2. Science. Look up Bernoulli's Principle in an encyclopedia. Write a paragraph describing how Bernoulli's Principle affected your gliders.

Fig. E

Credit: Developed by Richard Seymour for the Center for Implementing Technology Education, Ball State University, Muncie, Indiana 47306

1. **Answer.** I. 5 horizontal stabilizers

SECTION 6

CHAPTERS

23 Bio-Related Systems

24 Applications of Bio-Related Technology

TECHNOLOGY TIME LINE

11,000 B.C.	6500 B.C.	3000 B.C.	1500	1796	1857	1892	1928
Humans first use sickles to gather grain.	Plants are cultivated in the Near East.	The first plow is used by the Romans.	Native Americans teach settlers to raise corn.	First vaccine against smallpox developed in England.	Pasteur of France discovers bacteria.	Russian Dmitry Ivanovsky discovers viruses.	Penicillin discovered by Scotland's Alexander Fleming.

524 SECTION 6: Bio-Related Technology

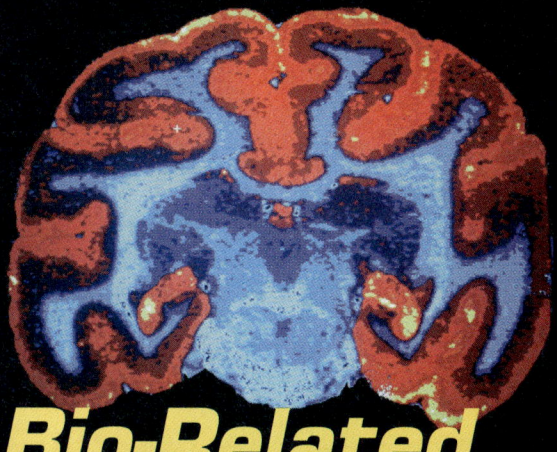

Bio-Related Technology

Although bio-related technologies have been around for thousands of years, they have attracted more attention recently. One reason is that people have become more aware of the earth as a vast network of connections. We now realize that what we do affects the entire planet as a *biosphere*. In this section you will learn about how diverse bio-related technologies are and how they affect you and the world around you.

1941	1960s	1970's	1978	1982	1990	1997
Human factors engineering grows in importance.	Researchers try cross-breeding and hybridization.	AIDS virus migrates from Africa to rest of world.	Genentech in the U.S. clones human insulin in the lab.	Barney Clark receives the first artifical heart.	Over 500 corporations spend $5 billion per year on bio-related technology.	Cloning of adult mammals becomes a reality.

SECTION 6: Bio-Related Technology

CHAPTER 23

Bio-Related Systems

Objectives

After studying this chapter, you should be able to:
- define bio-related technology.
- explain how bio-related technology developed.
- discuss the inputs, processes, outputs, and feedback of several bio-related technologies.
- discuss the impacts of the newer bio-related technologies.
- discuss bioethics.

Terms

adaptation
bioethics
bio-related technologies
cell
cloning
conversion
DNA
genes
genome
harvesting
propagation

Technology Focus

Fighting Pests with Technology

Three technologies of the mid-twentieth century—the atom bomb, penicillin, and DDT—encouraged us to believe in miracle cures. Half a century later, we've come to recognize that technology can't make problems just go away. Superweapons spur demand for even more terrible weapons. Consider, for example, pesticides. Universal pesticides have seeped through the environment and are poisoning the poisoners.

The U.S. Department of Agriculture (USDA) has a new strategy for using technology against pests. In Integrated Pest Management (IPM), the problem that arises is treated rather than all problems that *might* arise. IPM is a knowledge-based approach: know the enemies and their mating habits.

How IPM Works

Integrated Pest Management is based on a better understanding of natural systems. Rather than automatically spraying powerful pesticides on fields, the fields are checked regularly to identify which pests, if any, pose a threat this year. Monitoring is a key component of IPM.

Timing is also important. For example, the best time to prevent the spread of insects is during mating or egg-laying season, before there is a new generation and a bigger problem.

Cooperation is important, too, because it doesn't do any good to use insecticide in one farmer's field if the insects continue to lurk in the neighbor's field. Both fields must be treated. Through this type of cooperation, cotton farmers in North and South Carolina have overcome the problems of the legendary boll weevil.

New Approaches to Old Problems

Sometimes biological control, using natural predators and parasites against pests, works as well or better than chemicals. For example, sparrow hawks like to eat grasshoppers, and so does a friendly fungus called *Beauveria bassiana*.

Another form of biological control is to interfere with insect reproduction. Sterile males may be released into the environment, or trees may be sprayed with sex hormones so that confused males don't know where to find a mate.

Other methods depend upon planting practices. Researchers are testing the idea of alternating groves of apples with groves of peaches to slow the spread of apple moths. Another experimental idea is to plant ground cover along with crops to provide hiding places for pest-eating predators.

Take Action!

Visit an agriculture-related agency, such as the County Extension Service, to find out how IPM strategies work in your area.

Resources:
- Chapter 23 Lesson Plan in the Teacher's Annotated Edition and in the Teacher's Resource Binder.
- Chapter 23 Review in the *Teacher's Resource Binder*.

What Is Bio-Related Technology?

In the 1960s, the most exciting scientific field was physics. People wanted to learn about the origins of the universe. They wanted to develop technologies that would get us to the moon and the other planets. Today, the emphasis has shifted from outer space to inner space. All the excitement is in the fields of biology and bio-related technologies, where a revolution is taking place in how we think about living organisms.

As you learned in your science classes, biology is the study of living things. **Bio-related technology**, then, includes all technologies with a strong relationship to living organisms, such as agriculture, health care, and waste management. Fig. 23-1. When the chef at the pizza parlor uses yeast organisms to make a pizza crust heaped with toppings, the chef is using a bio-related technology.

FIG. 23-1 Bio-related technologies have a strong link to living things.

1. **Extension.** Explain that "biotechnology" is the term used by the scientific community for the work they are doing with living organisms. The term denotes a narrower area of interest.

SCIENCE CONNECTION
Emerging Viruses—A Deadly Mystery

Some of the most mysterious and dangerous microorganisms are the viruses. So small that billions can fit in a drop of blood, viruses can usually be seen only with an electron microscope. Most microorganisms, like bacteria, can live independently. A virus, however, depends on a host organism. Outside its host, the virus appears lifeless and cannot reproduce. The host provides the life processes the virus lacks. Even though they have no nerves and no brain and none of the five senses, each species of virus can somehow detect the special host for which it is designed. Once inside its host, it does what viruses are made to do—it duplicates itself to create additional viruses. Sometimes this process damages or kills the host.

Dangerously Talented Organisms

Although much is being learned about them, viruses remain mysterious. Of an estimated 30 million types, only 4,000 have been classified. Of these, about 150 are known to cause disease in humans. However, new *emerging* viruses that have not been connected with human disease before are being discovered all the time. Viruses have an incredible ability to *mutate*, or change. They can also hop from species to species and, in their travels, help other microorganisms mutate. For example, a virus living in a harmless bacteria that is penicillin-resistant can "jump" to another bacteria that is deadly. By simply exchanging genetic material, the virus can create a new bacteria that is both deadly and resistant to penicillin. Viruses are also capable of altering the human immune system, turning it against itself.

Emerging disease-causing viruses pose one of the greatest dangers in today's world. This is because few viral infections can be treated successfully with drugs, and vaccines are not very effective either. Our inability to conquer the AIDS virus is an example.

Try It!

Read one of these books about emerging viruses and report on what you learn to the class:

Virus X by Frank Ryan, M.D.
The Hot Zone by Richard Preston
The Coming Plague by Laurie Garrett

The bio-related technology attracting the most attention today is health care. As researchers learn more about how living cells work, they hope to find newer, more effective cures for human ills. (A **cell** is the smallest structural unit of an organism that can function independently.)

The Development of Bio-Related Technologies

Although many recent breakthroughs have brought bio-related technology to the public's attention, it is not new. Bio-related technologies have been around for as long as humans have been selectively breeding animals and plants, using yeast to make bread, and using bacteria to make cheese. Agriculture itself has been practiced since about 8000 B.C., and, as early as 1557, a book titled *Points of Husbandry* was printed about breeding animals. However, the difference between what was known then and what is known now is very great. This growing body of knowledge includes the efforts of many scientists. This chapter will discuss the work of a few key people.

Louis Pasteur

In 1857 Louis Pasteur, a French chemist, discovered that tiny organisms, which we now call bacteria, were responsible for turning milk sour. He also found that heat-treating the milk destroyed many of the organisms, and the milk stayed fresher longer. The process was named after him, and today milk is still *pasteurized* before it is sold. Fig. 23-2.

Pasteur went on to identify many other microorganisms, and, as a result, helped doctors begin to control such diseases as anthrax and rabies. He developed vaccines made from weakened forms of the organisms, which, when injected into healthy people or animals, made them immune to the illness.

Joseph Lister

Joseph Lister, an English surgeon, learned about Pasteur's work and took it a step farther. During Lister's time, surgery was performed by doctors who wore their

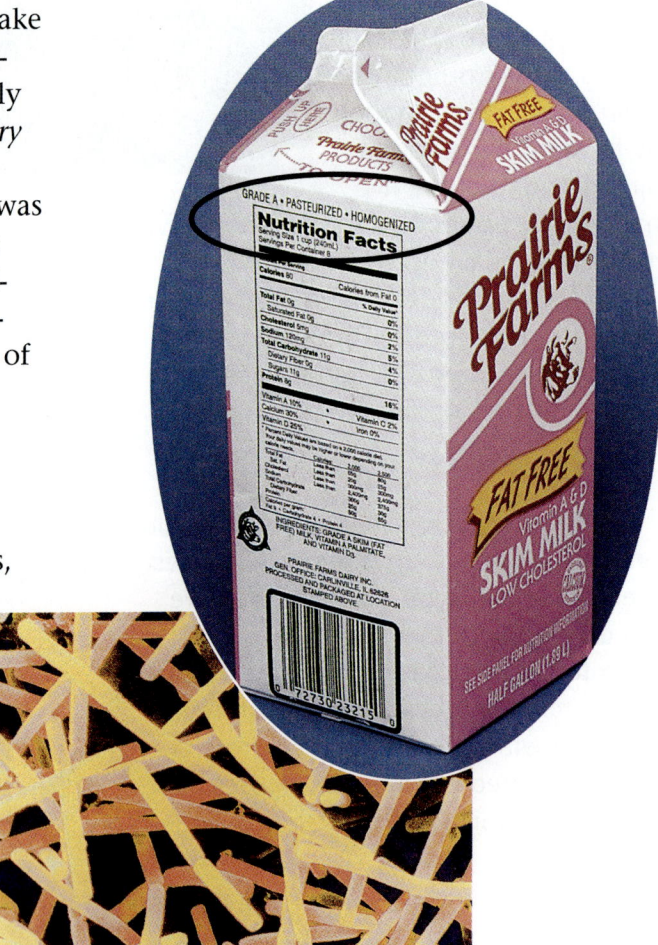

FIG. 23-2 Pasteurization destroys the bacteria that turn milk sour.

2. **Reteaching.** Point out to students that selective breeding is similar to redesigning a product so that it better meets the needs of the consumer.
3. **Extension.** Ask students to look up "micro" in a dictionary for the meaning of "micro-organism."
4. **Extension.** Ask students to look up anthrax and rabies in an encyclopedia.

FIG. 23-3 Because early doctors did not suspect that microbes caused disease, they did not use antiseptic methods.

everyday clothes and did not wash their hands or even their instruments. Many patients died, not from the operation but from infection. Fig. 23-3.

Lister reasoned that if microorganisms were causing the infections, perhaps prevention was possible. In 1865 he began experimenting with chemicals that destroyed the microorganisms before they entered wounds. His experiments were successful, and in 1867 he published a paper "On the Antiseptic Principle in the Practice of Surgery." Lister went on to develop many disinfectants and the technique of sterilization, which destroys germs with boiling water.

5. **Extension.** Explain to students that it was commonly believed at this time that these infections were caused by gases that presumably settled around hospitals.

Gregor Mendel

Another pioneer in bio-related technology was Gregor Mendel, an Austrian monk, who began experiments in heredity in 1857. By selectively breeding different varieties of garden plants, Mendel was able to develop plants having predetermined characteristics.

Mendel theorized that cells from the parent plants contained factors or elements that were passed down to their offspring. These carriers of heredity were what we today call **genes**. They contain all the genetic (hereditary) information necessary to reproduce an organism. Fig. 23-4.

CHAPTER 23: Bio-Related Systems

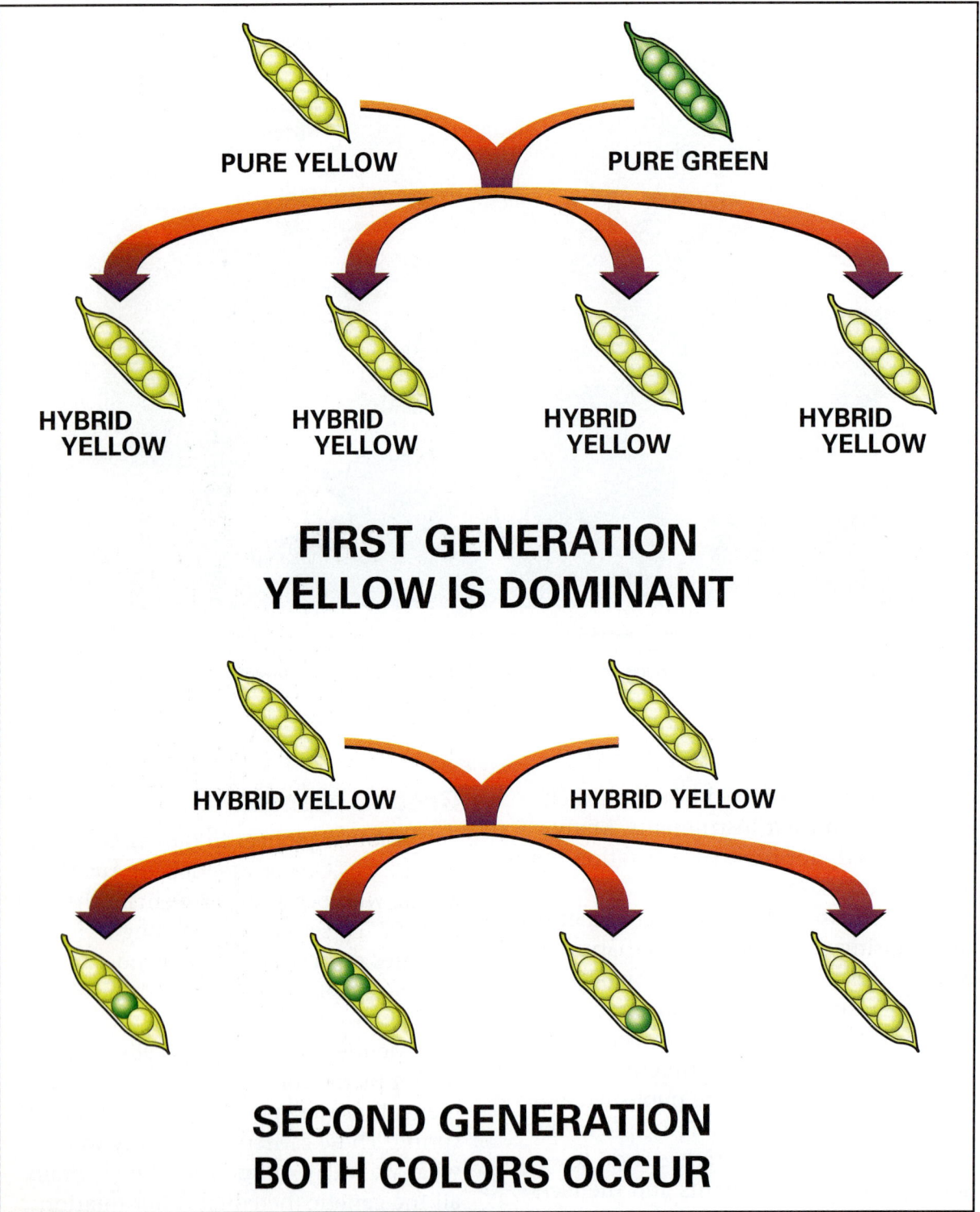

FIG. 23-4 Gregor Mendel's experiments with peas helped him learn how characteristics were passed from generation to generation.

FASCINATING FACTS
There are an estimated 450,000 species of plants on earth. Of these, 325,000 reproduce by spores instead of seeds. Ferns, moss, and seaweed are among the plants that reproduce through spores.

Watson, Crick, and Wilkins

The modern successors to Gregor Mendel are James Watson (American), Francis Crick (English), and Maurice Wilkins (English), three scientists who won the Nobel Prize in 1962 for working out the structure of DNA. **DNA** (deoxyribonucleic acid) is a long chain of four simple molecules within a gene that actually carries the genetic information. Fig. 23-5. The order, or sequence, in which the molecules are linked determines what the information is and what type of organism (dog or cat, for example) will result. All the DNA in an organism is called that organism's **genome**.

Genetics is an explosive field today. Gene researchers in at least 17 countries are attempting to decipher the information contained in the genes of a wide variety of living things, from microbes to human beings. In the United States, the attempt to discover all the 60,000 to 80,000 human genes is called the Human Genome Project. One of the project's goals is to determine the complete sequence of the three billion DNA subunits. When the sequence is finished, it will require 200,000 pages of small type to print it. Because genetic information is so complex, the project will take a number of years to complete.

After the information is obtained, scientists hope that they will be able to alter the genetic code of an organism. In some cases, their purpose would be to cure disease by "repairing" a person's genetic code. Cystic fibrosis is one disease caused by a missing or damaged gene. Another

FIG. 23-5 Genes are segments of DNA found in the chromosomes of every cell.

(Labels: CHROMOSOME, INDIVIDUAL PROTEINS, GENE, NUCLEUS, CHROMOSOME, CELL)

6. **Enrichment.** Point out to students that the DNA in one human genome is about one meter long; as of this writing, researchers have sequenced about 2.5 centimeters of it.
7. **Internet.** Access http://hugo.gdb.org/ for information about countries doing gene research.
8. **Extension.** Have students create their family trees, going back as far as they remember. Use this to point out how genetic characteristics are passed on.

CHAPTER 23: Bio-Related Systems

possible use would be to alter the genes of certain deadly bacteria so as to render them harmless. The important point to remember is that early scientists were busy *identifying* organisms and genetic processes. Today we have the ability to *change* those organisms.

How Does Bio-Related Technology Affect Your Life?

Although it is easy to think of bio-related technology as something that goes on in laboratories, it is an important part of our everyday lives. Fig. 23-6.

FIG. 23-6 Bio-related technologies touch almost every aspect of our lives.

534   SECTION 6: Bio-Related Technology

When you get up in the morning, one of the first things you probably do is get a drink of water. That water has been treated with chemicals and other substances to make it safe for you to drink. If you pour some of the water down the drain, it goes to a waste management center where it is recycled, another bio-related technology.

The clothes you put on may also be bio-products. Do you wear cotton? It comes from a plant. Do you wear wool? It comes from sheep. Both cotton and wool fibers are processed and turned into yarns. The yarns are woven into clothing.

The food you eat for breakfast is the result of other bio-related technologies. If you live in a house made from wood, your house is a bio-product. If the car or bus that took you to school uses gasohol, the gasohol came from a plant, such as corn, which is a product of agriculture, a bio-related technology. The list goes on and on.

We take many of the products and processes of bio-related technology for granted, but they all have an effect on us and our environment. As the ability to tinker with nature increases, it is important to remember that changing one thing can change something else unexpectedly. It is easier to understand these consequences if you think of bio-related technology as a system. Like the other systems you have learned about in this course, it has inputs, processes, outputs, and feedback.

9. **Extension.** Have students research how water is treated in their community.

Inputs

Bio-related technology systems need the same seven resources required by other technology systems—people, information, materials, tools and machines, energy, capital, and time.

For example, suppose a dairy company wants to develop and market a new flavor of cheese. Some of the inputs required might include the following:

- People: Food researchers are needed to develop the flavor in the lab; production workers are needed to make the cheese mixture, monitor it, and package it; salespeople are needed to sell it to grocery stores.
- Information: Researchers must test different flavors with consumers to learn which they'll buy; production workers need instructions for making the mixtures; sales people need to know prices and shipping dates.
- Materials: Milk and flavorings are needed to make the mixture; cheese-producing bacteria are needed to turn the mixture into cheese.
- Tools and machines: Measuring devices are used to measure ingredients; large vats are used for mixing; heating devices are used to warm the mixture and encourage the bacteria to grow.
- Energy: Production workers provide human energy; electrical energy is used to heat the plant and run the machinery.
- Capital: Investment capital provides money to buy materials and equipment; working capital is used to pay workers.
- Time: Time is required to make the cheese mixtures; bacteria require time to grow; the cheese must "age" to develop flavor.

CHAPTER 23: Bio-Related Systems

Processes

Seven basic processes are used in bio-related technology. They include the techniques necessary to propagate, grow, maintain, harvest, adapt, treat, or convert living organisms.

Propagation

Propagation is the technique of reproducing a living organism. It would include breeding animals or planting crops. Fig. 23-7.

For example, hybrid plants are those with special characteristics developed by breeding two different varieties. This breeding technique is a type of propagation.

Cloning is also a propagation method. A cell from a plant or animal is used to create a duplicate.

Growth

For a living organism, growth is the period after conception and before maturity. Scientists can alter the growth period of a plant or animal by speeding it up or slowing it down. Penicillin and other antibiotics have been used in meat-producing animals because they promote growth.

Maintenance

Maintenance has to do with providing food and water for an organism, as well as making sure conditions in its environment support its growth. A typical greenhouse is an example of a special environment created for plants to support their growth in a cold climate. Some of the fish sold in supermarkets come from fish "farms," where water environments are maintained. Fig. 23-8. Other animals are raised in special housing that enables workers to control what they eat and how they live.

Maintaining environments for humans is also a bio-related technology. Heating or air-conditioning a home is an example.

FIG. 23-7 These young steers have been bred to produce more meat.

10. **Extension.** Explain that hybrids can be developed only within a species. Crosses between species produce few, if any, offspring.

11. **Enrichment.** Ask students who have owned aquariums to describe some of the tasks involved in maintenance.

Harvesting

Harvesting is the gathering of organisms and preparing them for use. For example, farmers harvest bean crops by going into the fields with giant machines that "pick" the beans from the plants and funnel them into a giant hopper. Later the beans are separated from any stalks and other waste products, dried, and sent to processing plants.

Animal products may be harvested as well. Eggs are gathered regularly from chickens. Fig. 23-9. Each spring, sheep are sheared of their wool.

FIG. 23-8 Maintaining water-based environments is called aquaculture. These shrimp have been raised at a fish "farm."

FIG. 23-9 Gathering eggs is a type of harvesting. The eggs shown here fall onto a conveyor (right) and a worker sorts them (left).

CHAPTER 23: Bio-Related Systems **537**

Adaptation

Adaptation is an organism's ability to change. The change is usually made to react to something in its environment. Most organisms adapt naturally. A bear, for example, grows a heavier coat of fur for the winter.

Organisms can also be changed artificially. Plants, for example, can be genetically altered to resist certain pests or diseases. Humans who live in cold climates can put on sweaters and coats.

Treatment

Organisms can be treated to improve their characteristics, remedy problems with their growth, or cure them of disease. Have you ever taken medicine for an illness? If so, you were being treated by your doctor.

Animals are also given medicines for illnesses. Plants, too, can be treated in this way. For example, they may be sprayed with *pesticides* (poisons that destroy insects), or soil conditions may be improved to stop the growth of molds.

Conversion

Conversion is change. In bio-related technology, it means that the form of the organism is altered in some way to prepare it for use. A steer, for example, may be converted into steaks and hamburger. Kernels of corn may be washed, flattened, combined with other ingredients, and baked to turn them into breakfast cereal.

Outputs

As you know, outputs are the result of a system. If microorganisms have been added to waste material to break it down and obtain a usable gas, then the usable gas is the desired output. If you have taken a medicine to cure a disease, then your restored health is the desired output.

Not all outputs are positive. Some of them may even be unexpected. For example, years ago a pesticide called DDT was used to kill harmful insects. No one realized that it was gradually entering the food chain and that it would be dangerous to fish and birds. Fig. 23-10. In 1997, several diet drugs, which had proven very successful in helping people lose weight, were found to cause heart problems in some of the people who used them. The treatment created a worse problem than the problem it was designed to fix.

Feedback

Feedback occurs when information is learned about the outputs of a system. Sometimes, as in the case of DDT, the information is not acquired until much later. For this reason, care must be taken in studying the results of a system before it is put into use.

Some of the excitement people feel about bio-related technology centers

> **Fascinating Facts**
> About 400 million years ago, the first plants appeared. They had no leaves, flowers, or seeds. They resembled a type of fern native to tropical and subtropical regions.

12. **Enrichment.** Explain to students how a plant mold was found to be very useful in the development of penicillin.
13. **Enrichment.** Ask a volunteer to research and report on Fen/Phen and Redux, the diet drugs in question.

around unexpected outputs. They want to be sure that those who work with these technologies follow a set of rules so that living beings are respected and any potential misuse is discouraged.

Bioethics

Ethics are a set of rules or standards that guide the conduct of a person or group of people. For example, when we say a certain doctor is ethical, we mean that she follows the standards set by the medical profession in doing her work. **Bioethics** are those rules or standards that apply to people and companies who engage in bio-related science and technology. Fig. 23-11.

FIG. 23-10 The use of DDT was responsible for the near-extinction of the bald eagle.

FIG. 23-11 Many people are concerned about the potential abuses of bio-related technologies.

14. **Enrichment.** Ask students to name other professions that follow a code of ethics.

CHAPTER 23: Bio-Related Systems

Because the bio-related technologies are changing so rapidly, the setting of ethical standards has not caught up with them. Many important issues must be evaluated so that rules of behavior can be established. These issues fall into three important categories—medical knowledge, commercial use of living beings, and environmental interactions.

Medical Knowledge

When scientists are able to map human genomes, they will be able to tell if a certain person is *at risk* for a particular disease, such as breast cancer. A gene that increases this risk has already been identified. On the surface, this seems like worthwhile information to have. However, the situation is not that simple.

Being at risk for a disease does not mean you will automatically get it. It only means that you have a better chance of getting it than many other people. Should a woman who possesses this gene get a *mastectomy* (breast removal surgery) rather than take the chance she will get cancer? What if she possesses other, as yet unidentified genes that will cancel out the risk? What if it turns out that genetics play only a small part in getting cancer?

These issues become even more complex when other factors, such as medical insurance, are considered. Insurance companies argue that they need genetic information in order to refuse insurance to people who will cost them a lot of money. Do they have a right to this information? Do they have a right to deny coverage to a person who possesses the breast cancer gene? What other abuses might result?

15. **Critical Thinking.** Have students respond to these questions for a class discussion.
16. **Extension.** Invite someone from the health insurance industry to discuss the risks insurance companies assume when they insure unhealthy people.
17. **Enrichment.** Point out that creating special breeds

540 SECTION 6: Bio-Related Technology

Commercial Use of Living Beings

Several years ago, the gene for making plastic was slipped into a type of mustard plant. Fig. 23-12. The gene turned the plant into a natural plastics factory. At least one plastics manufacturer is developing the idea commercially. The same can now be done with animals. Cows, for example, can be given the genes for producing certain drugs in their milk, turning the cows into drug factories.

For many years, mature plants have been cloned with the use of their genetic material. Then in 1997, a sheep was cloned. Agricultural researchers hope that cloning will enable farmers to do such things as make copies of a prize dairy cow to create a herd of champion milkers.

We already use plants and animals for food, so many people feel that altering their biology can be no worse. Are they right? What if other, unexpected changes take place later on as a result?

FIG. 23-12 By inserting a special gene into this plant (Arabidopsis thaliana), it can be made to manufacture plastic granules.

can backfire. Currently, most farmers raise only a few varieties of crops because of their hardiness and yield. When disease strikes, it can wipe out enough plants to cause famines.

18. **Update.** As of this writing, other researchers have questioned the cloning results.

Environmental Interactions

A bio-related technology of special interest to researchers is the use of microorganisms to remove pollutants from the environment. Already, certain biotech companies are specializing in the creation of microbes designed for different types of toxic waste. The microbes eat the waste, rendering it harmless. One such microbe has been released into the environment to clean up oil spills. Fig. 23-13.

Although no one would argue with the need to clean up pollutants, the introduction into the environment of a new species of microbe makes some people uneasy. Scientists claim that the microbes have been thoroughly tested, yet it is impossible to test for every possible interaction in the laboratory. Also, organisms do not always behave in nature as they do in laboratories. Will microbes that appear harmless today mutate tomorrow and create worse problems than those they were designed to remedy?

These and other questions must be carefully considered as we enter the bio-related technology era.

FIG. 23-13 Genetically altered bacteria have already been released into the environment to clean up oil spills like this one.

19. **Critical Thinking.** Have students respond to this question for a class discussion.

CHAPTER 23: Bio-Related Systems   541

Career File—Clinical Laboratory Technician

EDUCATION AND TRAINING
An associate's degree or a certificate from a hospital, vocational, or technical school is usually required. A few technicians learn their skills on the job.

AVERAGE STARTING SALARY
The first year's salary for a clinical laboratory technician is about $15,000.

OUTLOOK
The number of jobs for clinical laboratory technicians is expected to grow about as fast as the average through the year 2005, according to the Bureau of Labor Statistics. New, more powerful diagnostic tests will encourage more testing and sustain the demand for technicians.

What does a clinical laboratory technician do?
Clinical laboratory technicians, also called medical technicians, assist in performing most of the laboratory tests that are used to detect, diagnose, and treat disease. They examine body fluids, look for bacteria or abnormal cells, match blood for transfusions, and perform other tests.

Does most of the work involve data, people, or things?
Technicians work mostly with things—specimens (the blood or cells being examined) and laboratory equipment such as microscopes.

What is a typical day like?
A technician usually works under the supervision of a medical technologist. Technicians prepare specimens and operate machines that perform tests. They may try to determine the presence of bacteria, fungi, parasites, or other microorganisms. They may analyze samples for chemical content or determine blood cholesterol levels.

What are the working conditions?
Technicians work in clean, well-lighted laboratories. They may spend a great deal of time on their feet. Infectious specimens are a possible hazard, so proper safety precautions must be followed.

Which skills and aptitudes are needed for this career?
Technicians must be able to analyze information and work well under pressure. Close attention to detail is important because small differences in a test can affect patient care. Computer skills are also important.

What careers are related to this one?
Similar analysis and testing are done by crime laboratory analysts, food testers, veterinary laboratory technicians, and chemists.

One clinical laboratory technician talks about her job:
I enjoy putting to use what I learned in science in school. My work is important to doctors in their diagnosis of patients.

542 SECTION 6: Bio-Related Technology

Chapter 23 REVIEW

Reviewing Main Ideas

- Bio-related technology includes all technologies with a strong relationship to living organisms.
- Bio-related technologies have existed since humans first bred animals and plants and used living organisms to make other foods.
- Important events in the history of bio-related technology include the discovery of what causes disease, how antiseptic practices can control or prevent disease, how hereditary characteristics are passed on to new generations, and how genetic information is carried in DNA.
- Bio-related technology is a system having inputs, processes, outputs, and feedback.
- Bioethics are a set of standards of conduct that apply to people who work in bio-related technology.

Understanding Concepts

1. Define bio-related technology.
2. What is Louis Pasteur known for?
3. What is a genome?
4. Name at least three processes of bio-related technology.
5. What is propagation?

Thinking Critically

1. How has the way in which we think about living organisms changed since Watson, Crick, and Wilkins worked out the structure of DNA?

20. **Resource.** Answers to the Chapter Review questions can be found in the Teacher's Manual at the front of the *Teacher's Annotated Edition* and in the *Teacher's Resource Binder*.

2. Identify at least three ways in which bio-related technology affects your life.
3. How might a fish farm use the process of propagation? Of harvesting?
4. Most people follow a code of ethics. Name at least one rule in your own personal code.
5. Suppose you were a member of a panel to determine the standards for working with the DNA of sharks. What are some factors you would want to consider?

Applying Concepts and Solving Problems

1. **Internet.** Contact the web site http://www.amrivers.org/amrivers/ to see a listing of endangered rivers in the U.S. Research and write a short paper on one of the rivers and suggest a solution to the problem.

2. **Science.** Suppose you were responsible for recommending ways in which a biosphere could be created on the moon. (A biosphere is an area that contains all things necessary to support life—air, water, food, shelter, etc.) Research what has already been done on this topic. Then write your recommendations.

3. **Mathematics.** It has been theorized that there is a virus for every species of life on earth, or about 30 million viruses. So far only about 150 viruses have been identified as harmful to humans. What percentage of all viruses is this?

CHAPTER 24

Applications of Bio-Related Technology

Objectives

After studying this chapter, you should be able to:
- describe technologies used to control plant and animal production.
- discuss health care technologies.
- describe human factors engineering.
- describe methods for disposing of solid, sewage, and hazardous wastes.

Terms

biomaterials
controlled environment agriculture (CEA)
gene therapy
genetic engineering
human factors engineering
hybrid
immunization
integrated pest management (IPM)
recombinant DNA
vaccine
waste management

Technology Focus

Medical Orphans—Help at Last

Imagine finding out that you have a rare disease—a disease so uncommon that drug companies don't even look for a cure because it would not repay their investment of time and money. Believe it or not, about 20 million Americans, or one in every 12, suffer from such rare or "orphan" diseases.

What Is an Orphan Disease?

An orphan disease is one that affects fewer than 200,000 people in the United States. Currently, more than 5,000 rare disorders are classified as orphan diseases. Some of them may be familiar to you, while others may not.

Orphan diseases can strike anyone. Baseball great Lou Gehrig, who never missed a single game in 14 seasons, died from an orphan disease called amyotrophic lateral sclerosis (ALS). Today the disease is known as Lou Gehrig's disease. Woody Guthrie, a singer-songwriter best known for his song "This Land Is Your Land" suffered and eventually died from Huntington's disease, another orphan disease.

The National Organization for Rare Disorders (NORD)

Founded in 1983, the National Organization for Rare Disorders (NORD) functions as a source of information, referrals, assistance, and support for anyone suffering from or interested in orphan diseases. NORD, in its own words, is "a unique federation of more than 140 not-for-profit voluntary health organizations serving people with rare disorders and disabilities."

NORD provides assistance and leadership to affected people and their families, support groups, and health care professionals. It is "committed to the identification, treatment, and cure of rare disorders through programs of education, advocacy, research, and service." Those suffering from orphan diseases can find hope, help, and a home in NORD.

Take Action!
Ask for help at the library to find out the current costs and requirements for developing a new drug that meets FDA standards.

Resources
- Chapter 24 Lesson Plan in the Teacher's Annotated Edition and in the Teacher's Resource Binder.
- Chapter 24 Review in the *Teacher's Resource Binder*.

Agriculture

Agriculture is the practice of producing crops and raising livestock. Of course you know that agriculture provides food. You may not have thought about other agricultural products, such as cotton, linen, wool, and leather. Many industrial products begin on the farm. For example, many paints and varnishes include ingredients that come from soybeans.

Since the beginnings of agriculture, people have worked to develop improved varieties of plants and animals. Today's corn, for example, is very different from the wild corn that first grew in America. By selecting individual plants with desirable traits and planting the seeds, the Native Americans developed varieties of corn that were larger and produced more kernels. During the twentieth century, scientists in the United States developed corn hybrids. A **hybrid** is the offspring of two plants (or animals) of different varieties, breeds, or species. The hybrid has characteristics of both its "parents." Fig. 24-1.

Hybrids are one example of *selective breeding*, or artificial selection. In nature, plants and animals reproduce randomly. The offspring may or may not have traits that make them useful to humans. With selective breeding, human beings control the process. Selective breeding has produced plants and animals that produce higher yields, are more nutritious, and are resistant to disease and insect pests.

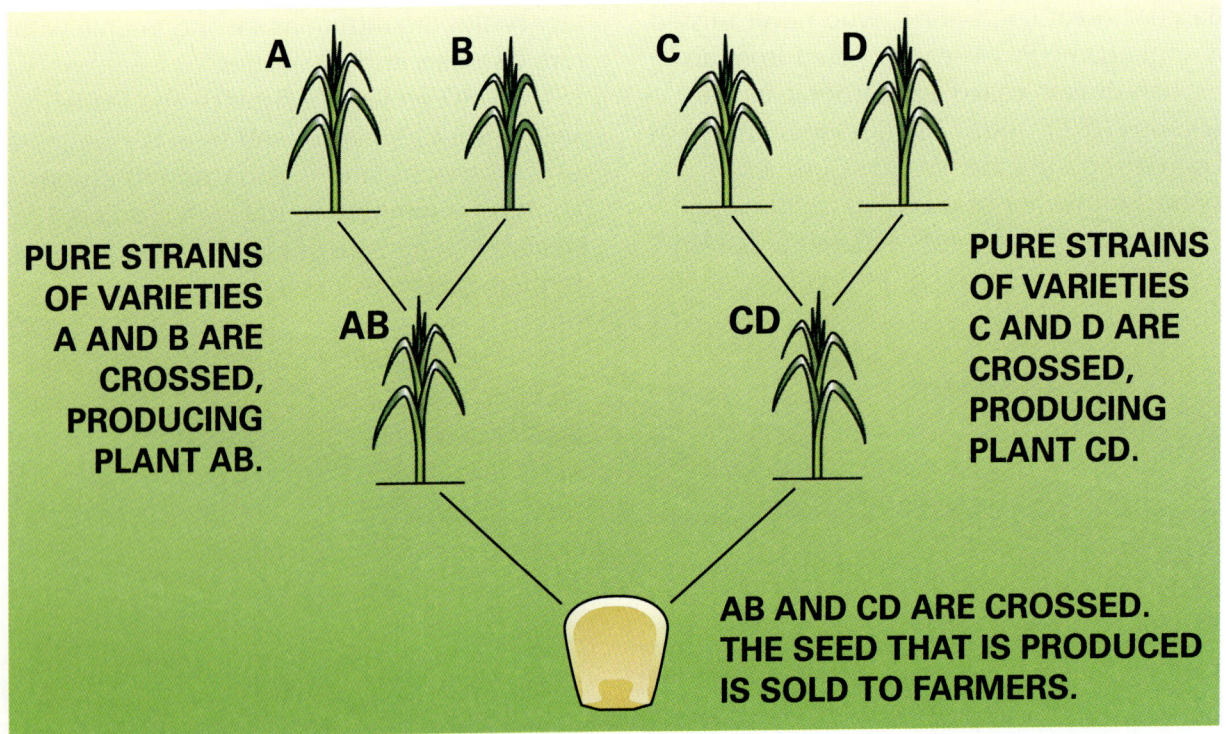

FIG. 24-1 Hybrid seed corn is produced by double crossing. The hybrid seed produces a higher yield than any of the earlier seeds.

1. **Extension.** Research the Green Revolution of the 1960s.

2. **Extension.** Today's hogs are much leaner. Today's broiler chickens are ready for market in 30% less time. Crop yields per acre are much higher.

A number of other technologies help humans to control plant and animal production. Among these are genetic engineering, cloning, controlled environment agriculture, and integrated pest management.

Genetic Engineering

Selective breeding works with an organism's natural traits. By contrast, **genetic engineering** gives organisms traits they never had naturally. One way to do this is by altering (changing) the organism's normal genes. To produce the Flavr Savr™ tomato, scientists at Calgene, Inc., altered the gene that causes softening. As a regular tomato ripens on the vine, it develops flavor. It also becomes soft and easily damaged. That's why tomatoes are picked and shipped before they are ripe. The Flavr Savr™ tomato can stay on the vine longer, develop more flavor, and still stay firm on its way to market. Fig. 24-2.

Genes can also be moved from one organism's DNA into another. The process is called gene splicing. In *gene splicing*, enzymes are used to remove the gene from one strand of DNA and glue it into another strand. Gene splicing was first done around 1978.

FIG. 24-2 The Flavr Savr™ Tomato was genetically engineered to stay firm while it ripens.

The technique of gene splicing makes recombinant DNA possible. **Recombinant DNA** means the joining (recombining) of two pieces of DNA from two different species. Fig. 24-3. The result is a *transgenic* plant or animal.

FIG. 24-3 In gene splicing, DNA molecules are cut at certain points. The ends are combined with the cut ends of other DNA molecules to make recombinant DNA.

DNA FROM ONE ORGANISM IS CUT.......

AND COMBINED WITH DNA FROM ANOTHER ORGANISM

3. **Extension.** The Flavr Savr™ tomato has not been a commercial success, primarily because its flavor is not much better than that of an ordinary hothouse tomato.

CHAPTER 24: Applications of Bio-Related Technology

> **FASCINATING FACTS**
> Humans have 22 pairs of identical chromosomes plus one pair of sex chromosomes (one from each parent). Each chromosome has 50,000 to 100,000 genes, and each gene has a precise location on its chromosome.

Calgene has spliced a gene from the California Bay tree into canola, an herb of the mustard family. The transgenic canola produces laurate, an ingredient used to make soaps, detergents, and shampoos.

Benefits and Risks

Genetically engineered plants and animals hold great promise. Plants that resist insects and disease will reduce the need for toxic pesticides. New foods, such as rice with a high protein content, will help feed a hungry world. Altered tobacco plants could be used to produce medicine instead of cigarettes. Mosquitoes could be bred with new genes that make them unable to spread disease.

However, there are also risks. As discussed in Chapter 23, there is concern that genetically engineered organisms can become a threat to our environment or our health. A food plant that can tolerate herbicides may be a good thing, but not if the food plant passes along its genes to a weed cousin. A food plant that is made too resistant to its natural enemies (insects and disease), might become a weed itself, spreading over large areas. As more and more genetically altered plants and animals enter our environment, new pests or new diseases might be accidentally created.

Cloning

A *clone* is an individual that is genetically identical to another individual. Cloning (making clones) is not new. It's been common in agriculture for centuries. All the MacIntosh apple trees in the world are clones derived from a single plant. In recent decades, however, the technology of cloning has advanced. It is now possible to clone many kinds of organisms, including some mammals.

One method of cloning plants is to take a single cell and place it in a medium that contains the right nutrients. This technique is called *tissue culture*. The plant that grows from the cell is genetically identical to the plant from which that cell came. One plant can provide enough cells for many clones. Fig. 24-4.

The technology for cloning embryos of animals such as pigs, cows, and sheep has been known since the early 1980s. In the late 1990s, scientists for the first time produced clones from *adult* mammals. However, animal cloning is not yet widely practiced. It's too costly and risky and raises serious ethical and economic questions.

Controlled Environment Agriculture

In **controlled environment agriculture (CEA),** a plant's or animal's surroundings are carefully monitored and adjusted. The plant or animal receives the right amount of humidity, temperature, light, and nutrients for the best growth. Two kinds of CEA are hydroponics and aquaculture.

4. **Brainstorming.** Since laurate is found naturally in palm kernel and coconut oils, why might U.S. companies be interested in canola that produces laurate?
5. **Extension.** There is also concern that widespread planting of such crops will hasten the development of resistant insects.
6. **Extension.** Tissue culture can also be used to produce vaccines from cultures of viruses.

FIG. 24-4 In plant tissue culture, one piece of a plant is used to grow a new plant. If the piece is a single cell, then the new plant will be a clone.

CELL IS REMOVED FROM PLANT AND PLACED IN GROWING MEDIUM

Hydroponics

Have you ever visited a greenhouse where flowers or vegetables were being grown in pots of soil? That's one type of CEA. Hydroponics is another. *Hydro-ponics* is the cultivation of plants in nutrient solutions without soil. The term comes from the Greek words for water *(hydro)* and labor *(ponos).*

Although hydroponics has been practiced for centuries, large-scale use began about seventy years ago. During the 1930s, Dr. William F. Gericke of the University of California experimented with hydroponics. He placed a wire frame over shallow tanks full of liquid nutrients. The roots of the plants on the frame descended through the mesh to feed on the solution below. With this technique, Dr. Gericke grew tomato plants that were more than twenty-five feet high.

There are several hydroponic systems. The system Dr. Gericke used was *water culture.* The plants' roots were submerged in water all the time. Another system is *aeroponics.* In this system, the plant hangs in the air, and its roots are sprayed with nutrient solution. In *aggregate culture,* the plant grows in a container filled with small pieces of mineral rock or with sand and gravel. Nutrient solution is added to keep the roots moist.

A hydroponic system can be set up much like a factory. The plants can be grown indoors, where temperature, humidity, and light can be controlled. The

7. **Enrichment.** Ask students what hydroponically grown flowers or produce they have seen in stores.
8. **Extension.** Hydroponics systems can be set up outdoors as well as in greenhouses.

CHAPTER 24: Applications of Bio-Related Technology

nutrient solution can be dispensed automatically. Computers can help manage the system, reducing the amount of human labor required. Crop yields are about the same as for soil-grown crops. Fig. 24-5.

As with other factories, the start-up costs are high. Also, energy is needed to run the system. In areas where the soil and climate are good, there are not many large-scale hydroponic systems.

Aquaculture

Aquaculture is the rearing of fish, shellfish, or plants in water under controlled conditions. The aquaculture farm may be a pond or concrete pool. It may be an area along the ocean's coast, with barriers set up to keep the fish in and the predators out. Some shellfish are raised in cages set in the water. Fig. 24-6.

You've probably seen farm-raised catfish in the grocery store. Other fish commonly raised by aquaculture are catfish, carp, salmon, and trout. Many aquaculture farms specialize in shellfish such as oysters, shrimp, and crayfish. Various types of red algae are grown commercially. The algae are used to make carrageenan, a white powder used as a binder and thickener in many products, from hot dogs to toothpaste.

Fascinating Facts

Scientists have experimented with using wastewater from a trout farm to grow strawberries. First, solid matter is removed from the wastewater. It is then piped into a greenhouse where strawberries are grown hydroponically. The wastewater provides phosphorus, an important nutrient, for the plants. The solids that were removed from the water are mixed with straw. The mix is used instead of rockwool or other aggregates for the hydroponic plants.

FIG. 24-5 These tomatoes are being grown on a large scale in a commercial hydroponic operation.

9. **Extension.** Aquaculture has been practiced since at least 500 B.C., but large-scale operations are fairly new. Since the 1960s, for example, catfish farming in the United States has quintupled.

550 SECTION 6: Bio-Related Technology

FIG. 24-6 This commercial aquaculture operation raises fish. Check out this website: http://www.aquafarm.com

Producing food for people is the goal of most aquaculture. However, many government agencies raise fish in order to stock lakes and rivers for sport fishing. There are also aquaculture farms that raise aquarium fish or fish that are used as bait.

Integrated Pest Management

Pests, such as insects, microorganisms, and weeds, destroy billions of dollars' worth of crops every year. Fig. 24-7. Chemical pesticides can be effective, but the pests can eventually build up a resistance. In addition, the chemicals may be harmful to other animals and to humans. One way to reduce use of pesticides is with integrated pest management.

Integrated pest management (IPM) combines various techniques to control pests. The overall goal is not to eliminate all pests but to minimize the damage.

- Mechanical controls. For example, reflective aluminum strips can be placed like a mulch in vegetable fields to reduce aphid attacks. This technique has been used to protect cucumbers, squash, and watermelons.
- Cultural controls. These have to do with the way a crop is grown. Crop rotation is one example. Clearing away fallen fruit from an orchard is another. Such techniques reduce the pests' supply of food or shelter.
- Biological controls. Bringing in natural predators that attack the pests is an example. Fig. 24-8. Earlier in this chapter, you read about genetic engineering to make plants that are resistant to pests. Another technique is to release sterile insects into the wild insect population. The sterile insects mate, but they can't produce offspring.

10. **Extension.** Salmon raised commercially contains fewer valuable nutrients than wild salmon. Discuss the implications of this for humans.
11. **Reteaching.** Pesticides include herbicides, fungicides, and insecticides.
12. **Extension.** Microbial controls (viruses, bacteria, fungi, etc.) are also used. Many of these attack only certain insects and are harmless to other species.

Insects

Many kinds of insects attack crops. One common insect is the Colorado potato beetle. It eats the leaves of potato plants, tomato plants, and eggplants. The beetle is found in North America, Europe, and Asia.

Diseases

Plant diseases may be caused by viruses, bacteria, or fungi. A very common disease is corn smut, a fungal disease that attacks the stalks and ears of corn plants.

Weeds

Weeds become pests when they invade a field where a crop is being grown. They take nutrients from the soil, and they cause problems at harvest time. One common weed is lamb's quarters.

FIG. 24-7 These common insects, diseases, and weeds are pests that invade crops.

- Chemical controls. When necessary, chemical pesticides are used. The strategy is to use the least amount necessary to do the job.

IPM requires constant monitoring of crops to catch problems early. Monitoring can include such things as inspecting crops for signs of pests, setting insect traps, and checking for weather conditions that are favorable to pests. When pests are found, treatment must be prompt and specific.

FIG. 24-8 Ladybug beetles eat aphids and other soft-bodied insects. They are useful in fruit orchards.

13. **Resource.** Refer students to the Technology Focus article in Chapter 23 for more information on IPM.

SECTION 6: Bio-Related Technology

Health Care

Bio-related technologies are used in health care to prevent, diagnose, or treat disease and disability.

Prevention of Disease

Have you noticed the words "Vitamin A & D" on milk cartons? The vitamins have been added to help make sure people will get enough. A shortage can cause eye, skin, and bone disorders. Adding vitamins to food is one example of disease prevention.

Another way to prevent disease is by immunization. **Immunization** is the process of making the body able to resist disease by causing the production of antibodies. *Antibodies* are proteins that attack foreign substances in the body, such as disease-producing microbes.

Immunization can happen naturally. People who have had measles are now immune to the disease. Their blood carries antibodies that attack and destroy measles viruses. It's possible, though, to become immune to measles without first getting the disease. That's because there is a vaccine that can make people immune.

A **vaccine** is made from killed or weakened microbes or from purified toxins the microbes produced. The vaccine does not cause disease, but it does stimulate the body's immune response. Antibodies are produced. If the person is later exposed to the disease, the antibodies recognize and attack the microbes.

DNA Vaccines

Scientists are now working on DNA vaccines. This type of vaccine uses a gene from the virus or bacteria rather than the whole microbe. It is therefore safer and more effective. When the vaccine is injected into a human being, the immune system will produce a response that protects the person from infection. Fig. 24-9.

Diagnosis of Disease

To *diagnose* means to identify by signs and symptoms. For many years, doctors had to rely on their own senses to diagnose disease. In 1819, the French physician R.T.H. Laënnec used a perforated wooden cylinder to listen to sounds from a patient's chest. That cylinder was the first stethoscope, and it made diagnosis easier. Today, there is a wide variety of tools for diagnosing disease. This section will discuss a few of them.

Laboratory Tests

Most laboratory tests examine body fluids. For example, a complete blood count (CBC) measures more than a dozen characteristics of blood. Among other things, it calculates the number of white or red blood cells in a cubic milliliter, the different types of white blood cells, and the volume of red cells.

Laboratory tests are valuable, but they are not perfect. All laboratory tests will sometimes yield a false-positive or false-negative result. A *false-positive* result means the test indicated the presence of disease but the patient does not really have that disease. A *false-negative* result means the person has the disease but the test did not show it. Additional tests, plus the knowledge and experience of the physician, are needed to make the correct diagnosis.

14. **Extension.** Ask students to name other foods that have vitamins or other additives to help prevent disease.

15. **Extension.** The smallpox vaccine was so successful that the disease has been declared extinct. Vaccination is no longer required.

CHAPTER 24: Applications of Bio-Related Technology

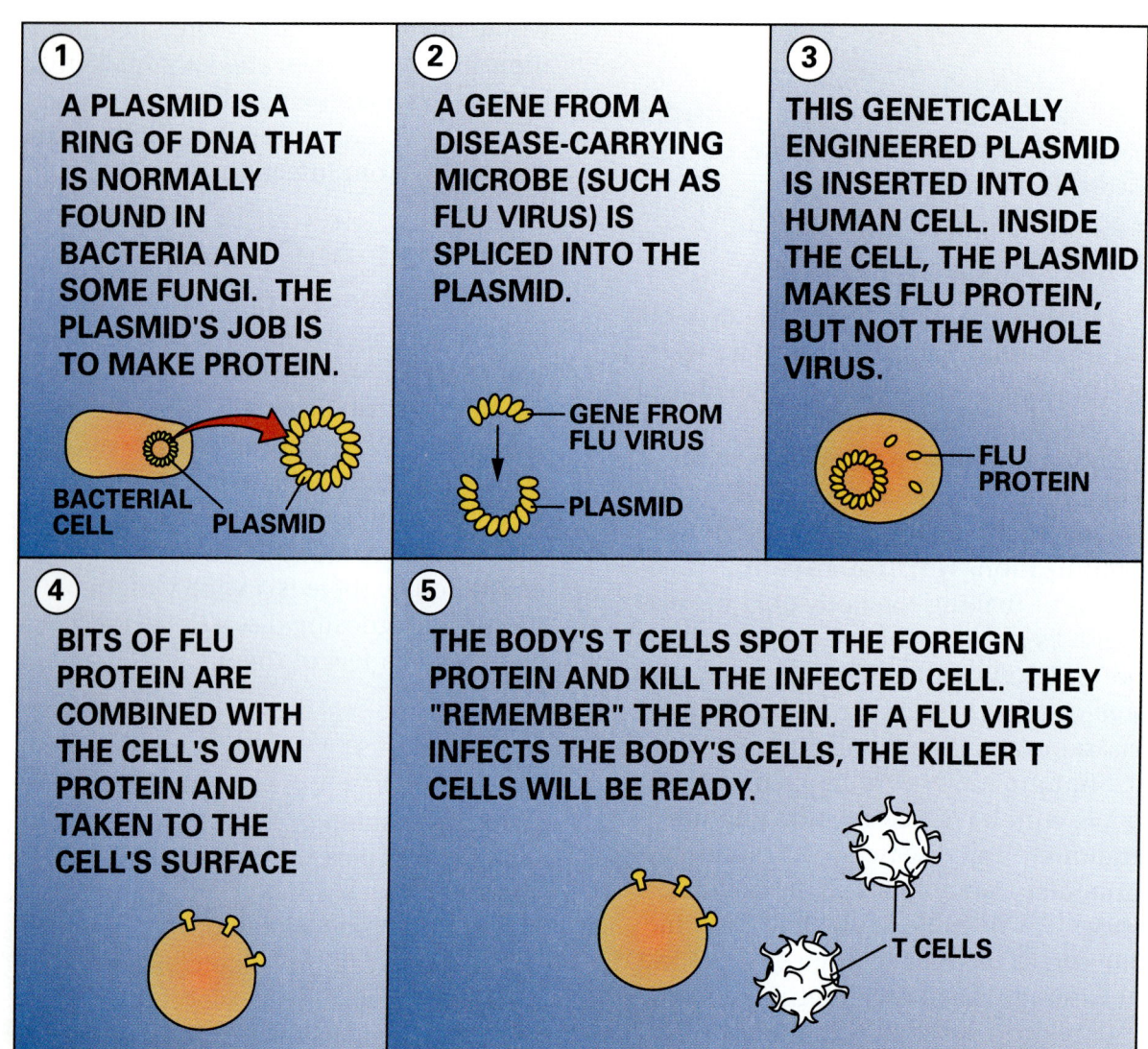

FIG. 24-9 This diagram shows how a DNA vaccine would work. Scientists hope to develop DNA vaccines against the flu, AIDS, malaria, and even some cancers.

Instrumental Screening

Many instruments can be used to examine the body for disease. One example is the gastroscope. This is a flexible lighted shaft used to examine the inside of the stomach. The patient is given a local anesthetic, and the tube is passed through the mouth and down into the stomach. The gastroscope can be used to check the stomach for ulcers or other disease. Fig. 24-10.

Some instruments measure electrical activity. An electrocardiograph measures electrical activity of the heart. It is used to detect abnormal heart actions. The instru-

16. **Extension.** Many of these scopes use optical fibers to carry the light waves.

17. **Resource.** Electromagnetic radiation is discussed in Chapter 4.

ment traces a graph to show the electrical activity. The graph is called an electrocardiogram (ECG or EKG). Fig. 24-11.

Medical Imaging

Medical imaging allows us to "see" inside the body. There are various imaging methods. Which one to use depends on what the health care professional is looking for. What all the imaging methods have in common is that they send energy waves into the patient's body. The waves' reactions to what they encounter produces the images. Fig. 24-12.

Surgical Examination

Some diagnostic techniques require surgery. One type of surgical examination is the biopsy. In a *biopsy,* a small amount of tissue is removed from the body and examined under a microscope.

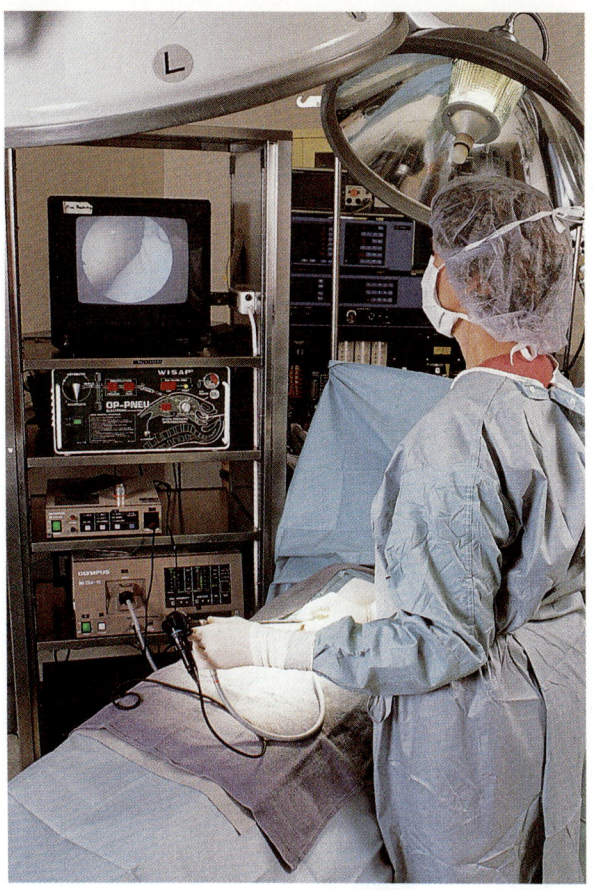

FIG. 24-10 A gastroscope is used to look inside a patient's stomach.

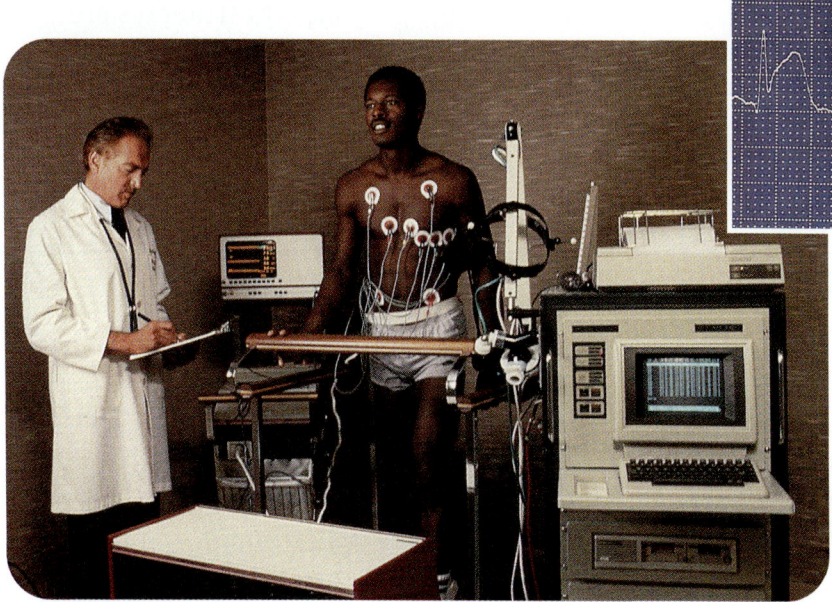

FIG. 24-11 As it beats, the heart produces a small amount of electric current. An electrocardiograph records this electrical activity.

18. **Extension.** Special substances called contrast mediums can be injected or swallowed to make soft internal structures show up. For example, a fluid can be injected to make arteries and veins visible on the X-ray photograph.

CHAPTER 24: Applications of Bio-Related Technology

Radiography, or X-ray Photography

X rays are short wavelengths of electromagnetic radiation. In radiography, the patient is placed between the X rays and a sheet of film. The X rays pass through some parts of the body and expose the film. Other areas, such as bones, block the X rays, and the film behind them is not exposed. The resulting picture looks like a negative, with areas that block X rays showing up light.

Computerized Tomography, or CT Scan

Tomography is a technique for obtaining X-ray images of deep internal structures. Computerized tomography (CT) uses a computer rather than film to generate the images.

In CT, the patient's body is scanned by an X-ray tube that circles the patient. The rays "slice" through the patient. Detectors collect data about those X rays. A computer interprets the data, and the image is graphically displayed on the computer screen. CT is also called computerized axial tomography, or CAT.

Positron Emission Tomography, or PET Scan

For a PET scan, a chemical containing a short-lived radioactive substance is injected into the body. As the substance decays, it emits gamma rays. The rays are sensed by detectors on opposite sides of the patient. The data from the detectors is analyzed by a computer, and an image is produced. A common use of PET scans is to diagnose disorders of the brain.

FIG. 24-12 Various imaging methods are used in health care.

If a simple biopsy cannot provide enough information for a diagnosis, exploratory surgery may be required. The surgery is done to examine and/or remove abnormal tissue.

Genetic Testing

Genetic testing is usually done for one of two reasons. The people being tested may want to know whether they have a condition that threatens their health, or

Sonography, or Ultrasound

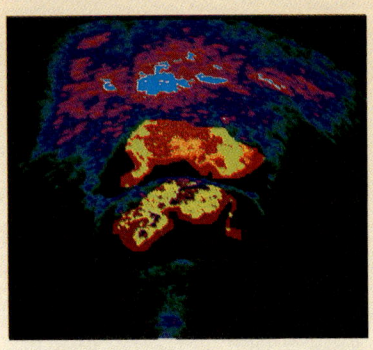

Sonography sends out sound waves and interprets the echoes that return. The sound waves are of a very high frequency, too high for humans to hear.

Different parts of the body reflect the sound waves in different ways. Fluids, for example, echo differently than solids. Echoes from things that are farther away take longer to return. A computer translates data about the strength and position of the echoes into an image called a sonogram.

Sonography is useful for cases in which X rays or chemicals would be damaging. For example, sonography may be used to determine whether a fetus is developing normally.

Magnetic Resonance Imaging, or MRI

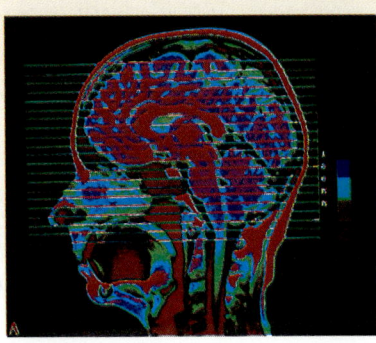

MRI is useful for showing soft tissue. (*Tissue* is a group of cells that form a structural part; for example, skin, bone, kidney.)

For an MRI exam, the patient is placed inside a tube-shaped magnet. The magnetic field generated by the tube causes the hydrogen atoms in the patient to line up in parallel. Radio signals then cause some of the atoms to tip over. When the signal stops, the atoms go back to their earlier position. As they do so, they emit a signal. The MRI machine reads the signals and translates them into images.

FIG. 24-12 Continued

---

they may want to know whether the health of future children is in doubt. For example, a woman with a family history of breast cancer may want to know whether she carries a gene that puts her at risk for developing this cancer. A couple with a family history of cystic fibrosis may want to know whether they are carriers who might produce children with this disorder.

Various procedures are used to do genetic testing. One procedure is *chromosomal analysis*. Human cells are grown in the laboratory. The cells are stained and sorted, and the chromosomes are then counted and displayed. Abnormalities in the chromosomes can predict the risk of inherited disorders such as Down syndrome. Fig. 24-13.

19. **Reteaching.** Point out that because MRI affects hydrogen atoms in water, areas of the body that have a lot of water show up as bright images. Areas with little water (such as bone) and areas where there's air (such as sinus cavities) are dark.

CHAPTER 24: Applications of Bio-Related Technology

MATHEMATICS CONNECTION
Using Scientific Notation to Write Large Numbers

One of the most useful tools in the field of science, including the medical field, is scientific notation. Of these two numbers—4,389,000,000 and 43,890,000,000—which is larger? It's hard to tell at first glance, isn't it? Large numbers like these are hard to read and understand. They are also hard to write accurately. If you were to add or multiply them, you might easily make a mistake. For scientists and others who work with them every day, another way of writing very large numbers is used. This method is called scientific notation.

The same two numbers are written below using scientific notation. See how compact and easy to read they are:

$$4.389 \times 10^9$$
$$4.389 \times 10^{10}$$

The exponent of the 10 tells you immediately the rough size or "order of magnitude" of each number. It's easy to tell which is larger, because 10 is larger than 9.

When written using scientific notation, a number has two parts. To the left of the decimal point is written only one (nonzero) digit. To the right of the decimal point is written the power of 10. In scientific notation, numbers are written as the product of these two parts.

For example, forests cover about 2,700,000,000 acres of the earth's surface. To write this number using scientific notation, first move the decimal point (which would ordinarily follow the last zero on the right) 9 places to the *left* so it follows the first nonzero digit, which is 2. Because you moved the point 9 places to the left, you would write " $\times 10^9$" after the decimal, so the value of the number is shown:

$$2,700,000,000 = 2.7 \times 10^9$$

What if you want to change 2.7×10^9 back to an ordinary number? The factor 10^9 means you must move the decimal point 9 places to the right to get 2,700,000,000.

Scientific notation is convenient for very small numbers, too. For example, the length of a bacterium is 0.000 003 5 meters. To write this length using scientific notation, place the decimal point to the right of the 3, because 3 is the first nonzero digit. Because the decimal point moves 6 places to the *right*, the exponent on the 10 is -6.

$$0.000\ 003\ 5 = 3.5 \times 10^{-6}$$

Try It!

Write the following measurements using scientific notation.
1. The length of a soybean field that measures 382 meters.
2. The length of a microbe that measures 0.000 275 centimeter.

Answers to these questions can be found in the Lesson Plans in the Teacher's Manual at the beginning of this book.

SECTION 6: Bio-Related Technology

FIG. 24-13 There are 46 chromosomes in a human cell. They are arranged in 23 pairs.

As discussed in Chapter 23, genetic testing has both risks and benefits. If people know they carry certain genes for disease, they can make lifestyle adjustments and lower their risk. However, the information may also be turned against them unless laws and ethical standards are established for its use.

Treatment of Disease

One goal of treatment is to cure disease. However, the disease may be unknown or incurable. In such cases, the goal of treatment is to relieve symptoms such as pain. This section will discuss some of the technologies used in the treatment of disease.

Recombinant DNA

As mentioned earlier, recombinant DNA means the joining of two pieces of DNA from two different species. In agriculture, recombinant DNA is used to give plants and animals new traits. In health care, recombinant DNA is used to make drugs.

One such drug is alpha interferon, which is used in the treatment of hepatitis B and C and some cancers. Alpha interferon is a protein that improves the body's ability to fight tumors and viruses.

Interferon is produced naturally in the body, but in very small amounts. Those amounts may not be enough to fight off serious diseases such as hepatitis. Recombinant DNA makes it possible to produce large amounts of interferon quickly. The human gene for interferon is spliced into the *E. coli* bacterium. The bacterium then produces interferon. Large numbers of the bacteria are grown, and the interferon is harvested from them.

Monoclonal Antibodies

As stated earlier, antibodies are proteins that attack foreign substances. They do this by binding to the substance. Monoclonal antibodies bind to one specific type of foreign substance. (*Mono* means "one.") The antibodies are clones of each other; they are all exactly alike.

In the 1970s, scientists developed a way to produce large amounts of monoclonal antibodies. The fact that these antibodies are very specific makes them useful tools in the diagnosis and treatment of disease. For example, certain monoclonal antibodies bind to cancer cells. Radioactive atoms can be added to these antibodies and given in tiny amounts to a patient. Because the antibodies bind to cancer cells, the radiation reveals the location of the cancer. Monoclonal antibodies are also being used experimentally to deliver drugs that kill cancer cells.

20. **Reteaching.** Point out that an incurable disease is not necessarily fatal, such as the common cold.
21. **Extension.** Insulin and some vaccines are also produced by recombinant DNA technology.
22. **Critical Thinking.** Ask students why monoclonal antibodies might be a better way to deliver cancer drugs. (Because the drugs would target the cancer cells and leave healthy cells alone.)

CHAPTER 24: Applications of Bio-Related Technology

Gene Therapy

Gene therapy involves the transfer of a normal gene into an individual who was born with a defective or absent gene. Cystic fibrosis is a hereditary disease caused by a defective or absent gene. It mainly affects the lungs, making breathing difficult. Experiments have been done in which genetically altered viruses are used to deliver healthy genes to the lungs of people with cystic fibrosis. Thus far, the treatment has been effective only for a few months. One day, there may be a permanent cure.

Tissue Engineering

Today, people whose kidneys fail may receive a new kidney from a donor. Imagine being able to grow a new kidney in the laboratory. With tissue engineering, this may soon be possible.

Tissue engineering is the science of growing living tissue to replace or repair damaged human tissue. Though still experimental, tissue engineering holds great promise as a way to replace or repair skin, bone, blood vessels, and organs.

In the laboratory, thin sheets of polymer (plastic) or other material are "seeded" with cells. The cells grow and form tissue, which is then transplanted into a patient. Skin patches have been grown this way and used on patients with severe burns. The skin patches cover the open wounds while the patient's own skin regrows. Fig. 24-14.

Polymers can also be used as "scaffolding," providing a place for cells to attach themselves and grow. Cartilage and bone have been grown this way in the laboratory. After being implanted in the body, the polymer eventually dissolves, leaving only the cartilage or bone.

Surgical Treatments

New surgical tools and techniques have been developed. They provide more effective treatment, with less pain and faster recovery, than traditional surgery. A few of these new methods are described below.

- Laser surgery. Because lasers can create intense heat, they can be used as a surgical tool. For example, surgeons use lasers to "weld" a detached retina in place or to stop the bleeding of an ulcer. Fig. 24-15.

FIG. 24-14 Tissue engineering has already produced skin patches, like these. Someday, entire organs may be grown in a laboratory.

23. **Resource.** The chemistry of plastics is discussed in Chapter 14.
24. **Resource.** Lasers are discussed in Chapter 15.

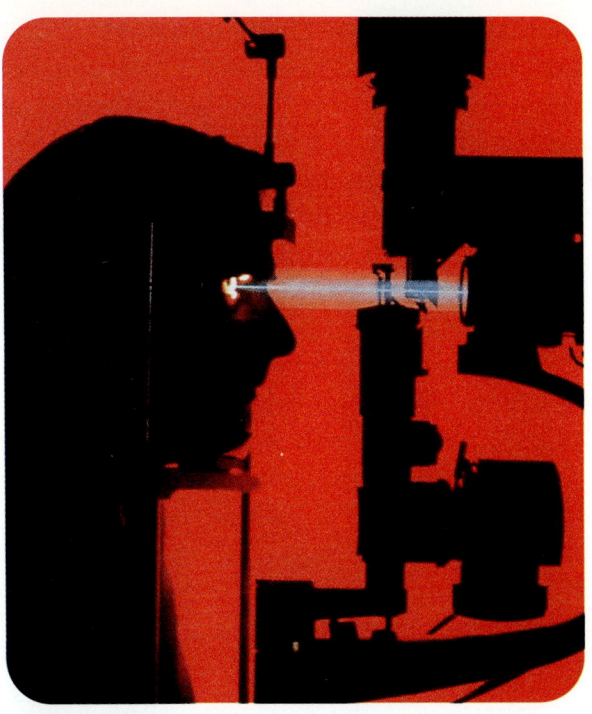

FIG. 24-15 This laser is being used to correct vision defects.

Artificial feet and legs were made as early as 500 B.C. Today there are more than 200 artificial parts, or *prostheses,* for the human body. Some of them use electric signals to communicate with natural muscles or nerves. This gives patients more and better movement abilities.

For example, a patient who loses a hand may be fitted with a *myoelectric* hand. This prosthesis looks very similar to a human hand. It is made of plastic and has a flesh-tone glove fitted over it. The myoelectric hand contains electronic sensors that detect signals from the nerve endings in the remaining portion of the arm. The sensors relay these signals to activate the wrist, hand, and fingers of the prosthesis. Fig. 24-16. Newer prostheses

- Cryosurgery. In cryosurgery, extreme cold is used to destroy tissue. For example, liquid nitrogen, whose temperature is about -195 degrees Celsius, is applied to warts and to some skin cancers.
- Ultrasound. You've read how ultrasound is used in medical imaging. Sound waves of even higher frequency can be focused on tissue. The sound waves cause the tissue to vibrate and heat, eventually destroying it. Kidney stones can be destroyed with bursts of focused ultrasound.

Prosthetics

Technology solves problems and extends human capabilities. This is very evident in the science of *prosthetics,* the artificial replacement of missing, diseased, or injured body parts.

FIG. 24-16 A myoelectric hand looks and operates like a real human hand. Electronic sensors detect signals from the nerve endings in what remains of the person's arm and relay them to activate the wrist, hand, and fingers.

25. **Extension.** Point out that some prostheses replace diseased tissue; for example, heart valves and hip, knee, or shoulder joints.

include sensors that relay to the wearer a sense of touch and the ability to distinguish hot from cold.

Prostheses can also help restore the function of a disabled body part. Cochlear implants, for example, improve hearing by translating sounds into electrical signals. The signals stimulate nerve fibers inside the ear. The nerves transfer the signals to the brain.

Other types of prosthetic implants are being developed. One kind delivers electric impulses to the brain to treat the shaking caused by Parkinson's disease. Another helps paralyzed people use their hands.

Biomaterials

Biomaterials are materials that are used for prostheses and that come in direct contact with living tissue. Most biomaterials are synthetic polymers, although metals and ceramics may also be used. Do you wear soft contact lenses? The lenses are made of a water-absorbing biomaterial called hydrogel. Other biomaterials are used to make artificial heart valves, grafts for blood vessels, joint replacements, and many other prostheses. Fig. 24-17.

One major concern with biomaterials is whether they are compatible with the body. Early heart valves, for example, included silicone rubber. The rubber absorbed fat from the blood and swelled, making the valve stop working. Biomaterials must also be strong enough to withstand constant use and last many years.

Fascinating Facts

A new surgical implant helps some quadriplegics use their hands. The device, called the Freehand System, allows paralyzed people to open and close their hands enough to pour coffee, hold a pen, or feed themselves.

FIG. 24-17 This artificial heart valve can be used to replace the valve in a living heart that is diseased.

FIG. 24-18 Some medicine bottles are designed to be easy to open. Who might benefit from such a design?

Human Factors Engineering

Human factors engineering is the design of equipment and environments to promote human health, safety, and well-being. It is sometimes called *ergonomics*.

Do you use a wrist rest with your computer mouse? That's an example of human factors engineering. The wrist rest helps prevent stress and strain on your wrist. It makes the interaction between human (you) and machine (the computer) easier and more comfortable.

What other measures might make working at the computer more comfortable? How about taking frequent breaks to relieve eye and muscle strain? What about making sure the room is not too hot or too cold? These measures all relate to human factors engineering. This bio-related technology is about the design of machines, work methods, and environments.

Machine Design

If a machine or tool is hard to use, the work takes longer. It may even be dangerous. Human factors engineering seeks to design machines and machine systems that fit the way humans move and think. Consider the way controls are placed on the dashboard of a car. Drivers can see and reach them easily. If they couldn't, it would be much harder to drive the car safely.

26. **Enrichment.** Have students name other examples of ergonomic design.

People come in many different sizes and shapes. They don't all have the same abilities. Some products therefore are made in different designs for different needs. Left-handed scissors are one example. Easy-open medicine bottles are another. Fig. 24-18.

Work Methods

Sometimes it's not the machines or equipment that cause problems but the way we use them. For example, people who work with desktop computers for long hours may experience eye strain, shoulder pain, and other discomforts. Using well-designed desks and chairs can help, but it's also important to adjust work methods. It's a good idea to take frequent short breaks to stretch muscles and rest the eyes.

27. **Extension.** Have students research *universal design*, which takes into account people of all ages and abilities.

CHAPTER 24: Applications of Bio-Related Technology

FIG. 24-19 The countertops in this kitchen are lower so that a seated person can reach them. Note the space under the sink, which allows room for a person's legs.

The Built Environment

Human factors engineering designs environments that improve comfort and safety. Suppose someone must use a wheelchair to get around. Homes can be designed with wider doorways so that wheelchairs can pass through easily. Kitchens can be designed with lower countertops so that a person seated in a wheelchair can reach them easily. Fig. 24-19.

Today, computers help keep our environment safe and comfortable. In *smart buildings,* sensors and computerized controls operate the lighting, heating, and security systems. These control systems improve efficiency and reduce the costs of operating the building.

28. **Resource.** Refer to the Technology Focus feature on smart houses in Chapter 17.

Waste Management

Waste management refers to all the operations involved in the collection, storage, and treatment of waste. Waste can be classified as solid, sewage, or hazardous. Each type requires different methods of handling.

Solid Waste

In the United States, we throw out more than one billion pounds of solid waste each day. Most of that is paper, glass, and other rubbish. Only about 10 percent is food waste.

29. **Extension.** This figure includes waste from all sources (commercial, industrial, residential, and institutional).

Most communities provide regular pick-up of solid waste from homes and businesses. In many communities, people are required to sort the waste before pick-up. For example, they may have to keep yard waste in a different container than household garbage.

Treatment and storage of solid waste varies.

- Incineration is the burning of waste. One advantage of this treatment is that the heat produced can be used to generate steam and electricity. However, the gases and ash produced by burning can be hazardous. Keeping the gases and ash from polluting the environment can be expensive.
- Composting may be done for yard waste and/or food waste. The waste is either stored outdoors or in large containers. Aerobic bacteria (bacteria that use oxygen) "digest" the waste, turning it into a crumbly product that can be sold as a soil conditioner or mulch. The chief disadvantage of this treatment is that there may not be much of a market for the compost.
- Sanitary landfill. Many communities bury solid waste in *sanitary landfills*. The landfill must have a liner at the bottom to keep the waste from polluting surface water or groundwater. Each day, a thin layer of waste is added to the landfill. Heavy equipment compacts it, and the layer is then covered with soil. The next day, another layer may be added over that one. When no more waste will fit in the landfill, it must be capped with a waterproof cover. Fig. 24-20.

Older landfills have caused pollution problems. *Leachate* is a very toxic liquid produced when water trickles through the decaying waste. It can pollute

FIG. 24-20 Modern landfills have several layers to keep trash and leachate away from groundwater.

30. **Resource.** At the end of this section is an activity in which students create and market their own compost.

CHAPTER 24: Applications of Bio-Related Technology

streams or underground water supplies. Landfills also produce methane gas, which is toxic and explosive. However, it is possible to collect the gas and use it as fuel.

Fascinating Facts

Organized garbage collection is fairly new. In England, a law was passed in 1714 that every city must have an official scavenger to collect waste and dump it outside the city.

- Recycling of solid waste can greatly reduce the amount that must be burned or sent to a landfill. Recyclable waste (paper, glass, metals, and plastics) can be sorted at a material recycling facility, or MRF (a "murf"). Once sorted, it can be sent to facilities that use it to make new products. Paper, for example, can be used to make new paper. What had been waste becomes a resource.

Sewage

Sewage includes the liquid and solid wastes that are carried by sewers. It includes waste from homes and business-

FIG. 24-21 In this aerial view of a sewage treatment plant, you can see the buildings and lagoons.

31. **Extension.** Investigate recycling facilities in your community or research the ways waste can be used to make new products.

es, factories, and storms. Sewage is mostly water, and the water can be returned to the environment. First, though, it must be treated to remove solids and toxic chemicals and to destroy harmful microbes. Fig. 24-21.

The sewage usually goes through several stages of treatment. The purified wastewater is then released into a lake or river. The solids left behind are called *sludge*. The sludge is treated to make it safe to handle and to remove odors. It may be sent to a landfill or incinerated. Some sludge is spread on agricultural fields, but it should not be used where food crops are grown. The sludge may contain toxic chemicals from industry.

Hazardous Waste

Hazardous waste is any waste that can harm human health or the environment. It may be chemical, biological, or nuclear waste. Hazardous chemical waste is usually a byproduct of industrial processes. Hazardous biological waste can include such things as used bandages or needles that carry harmful germs. Nuclear waste is produced mainly by nuclear power plants. The radiation it emits can cause illness or genetic mutations.

Special measures must be taken when storing, treating, or disposing of hazardous waste. The waste, which might be liquid, solid, or gaseous, must be stored in such a way that it is not released into the environment. For example, some liquid wastes are temporarily stored in ponds built for that purpose. The bottom of the pond has a liner to keep the liquid from seeping into the groundwater.

Some hazardous wastes can be treated to make them harmless. One treatment method, called *landfarming,* is used for petroleum wastes. The waste is brought to a plot of land, where it is mixed into the surface soil. Microbes that can "eat" the waste are added, along with nutrients. Sometimes the microbes used are genetically engineered bacteria. When this treatment is used on land that has been previously contaminated, it is called *bioremediation.*

Wastes that cannot be treated are either incinerated or buried. In both cases, special precautions must be taken. For example, waste that is solid or that has been placed in special containers may be sent to a *secure landfill*. This is a special type of landfill with double liners and a system for collecting and treating leachate.

---

32. **Extension.** An example is polychlorinated biphenyl (PCB). It was widely used in the 1930s and 40s in lubricants, paints, coatings, and electrical equipment. Improper disposal released PCBs into the environment. Although production of PCBs has been restricted since the mid-1970s, this long-lasting, toxic compound is still in our environment.

33. **Extension.** Microbes are not the only living things used to treat wastes. There is a class of plants called hyperaccumulators, which absorb heavy metals such as zinc. These plants can be grown on soil contaminated with such metals, thus cleaning the soil.

Career File—Forester

EDUCATION AND TRAINING
A bachelor's degree in forestry is the minimum requirement. Summer jobs that provide experience in forestry or conservation work are encouraged during a student's college years.

AVERAGE STARTING SALARY
Most people entering forestry with a bachelor's degree earn from $18,500 to $23,500 per year. Starting salaries are considerably higher for those with a master's degree or doctorate.

OUTLOOK
Forester jobs are expected to grow about as fast as the average through the year 2005, according to the Bureau of Labor Statistics. Demand will be ensured by continued concern for the environment.

What does a forester do?
Foresters manage and help protect lands that contain forests. Some foresters working in private industry manage the cutting and removal of timber. Other foresters supervise the planting and growing of new trees or manage public forests and parks.

Does most of the work involve data, people, or things?
Although some of their work is done alone, foresters deal regularly with landowners, loggers, government officials, and the public in general. To estimate future timber volumes, they use special tools to measure the height, diameter, and growth of trees. Computers are used for storing and analyzing information.

What is a typical day like?
A forester's work day depends on what he or she has been hired to do. For example, foresters who supervise planting new trees interact with workers and equipment used to clear the land, plant the trees, and monitor tree growth. Foresters who oversee national and state parks spend more time with the public, assisting visitors and enforcing park regulations.

What are the working conditions?
The work can be physically demanding. Many foresters work outdoors in all kinds of weather. They also may work long hours fighting fires.

Which skills and aptitudes are needed for this career?
Foresters must enjoy working outdoors, be physically hardy, and be willing to move where the jobs are. They must also work well with people.

What careers are related to this one?
Other workers who manage, develop, and protect natural resources include agricultural scientists and engineers, biological scientists, soil conservationists, and wildlife managers.

One forester talks about her job.
I like working outdoors. I feel like I'm contributing to the preservation of our natural environment.

Chapter 24 REVIEW

Reviewing Main Ideas

- The technologies used to control plant and animal production include selective breeding, genetic engineering, cloning, and controlled environment agriculture.
- Hydroponics involves the cultivation of plants without soil.
- Aquaculture is the rearing of fish, shellfish, or plants in a controlled water environment.
- Mechanical, cultural, biological, and chemical controls can all be used in integrated pest management to minimize pest damage rather than eliminate the pests entirely.
- Radiography, computerized tomography, positron emission tomography, sonography, and magnetic resonance imaging send different types of energy waves through a patient's body in order to examine the body for disease.
- Human factors engineering is the design of machines, work methods, and built environments to promote human health, safety, and well-being.
- Solid waste can be disposed of by means of incineration, composting, sanitary landfills, and recycling.
- Sewage waste is usually treated to remove toxic chemicals and to destroy harmful microbes.
- Hazardous waste is stored in secure places, treated, or buried.

Understanding Concepts

1. What is gene splicing and what is it used for?

34. **Resource.** Answers to "Understanding Concepts," "Thinking Critically," and "Applying Concepts and Solving Problems" can be found in the Teacher's Manual at the front of this *Teacher's Annotated Edition* and in the *Teacher's Resource Binder*.

2. What is aeroponics?
3. How does a vaccine work?
4. What type of energy does sonography use?
5. What makes a landfill "sanitary"?

Thinking Critically

1. How do you think a computer could be used to monitor a hydroponic system?
2. Why do you think crop rotation is considered a "cultural" control?
3. Why does integrated pest management choose *not* to eliminate all pests?
4. Humans have often considered themselves at the "top" of the food chain; in other words, they eat other organisms, but nothing eats them. Do you think this is true? Explain your answer.
5. Would you want to be genetically tested for a serious disease that has no cure? Explain your answer.

Applying Concepts and Solving Problems

1. **Science.** A mule is an example of a hybrid animal—a cross between a horse and a donkey. Do some research to learn what characteristics are obtained and how a mule differs from its parents.

2. **Mathematics.** Medical laboratory tests are done using the metric system of measurement. If an accident victim has lost 1.5 pints of blood, how many liters would be needed to replace it? (Pints can be converted to liters by multiplying by 0.47.)

1 DIRECTED ACTIVITY

Creating and Marketing Composted Soil

Context

Compost is nutrient-rich soil created from decaying vegetable matter and other ingredients. With the aid of moisture and warmth, bacteria in the vegetable matter gradually break it down into a dark, crumbly material. Good-quality compost can be made from such vegetable waste as peelings or spoiled produce. Grass trimmings, dead leaves, and ash from wood fires can also be added.

When the cycle of decay is completed, compost can be mixed into a garden or used as soil for houseplants. Sold in bags, it can provide a source of income for your technology class. It can also help waste management efforts at your school.

Health and Safety Notes: When working with decaying matter, wear gardener's gloves and be sure to wash your hands afterward. Do not use animal waste, such as meat scraps, in your compost pile.

Goal

For this activity you will build a compost bin, fill it, maintain it, and sell the resulting product.

Procedure

This is a class project. Teams can be responsible for the different duties involved in maintaining the compost pile.

1. Locate a site for your compost pile at a distance from school buildings so that odors and pests will not cause problems.
2. Clear a spot in the dirt with a circumference of 36", 48", or 60", depending on the space available. The size of the site should be determined by the amount of refuse it will contain. (If

Equipment and Materials

- 8 steel posts, about 48" high
- wire mesh or picket fencing
- plastic tie straps
- clear plastic sheeting
- grass clippings or decaying straw
- vegetable matter, such as peelings, cores, and waste from the cafeteria, and dead leaves
- additives, such as ash from a wood fire and purchased lime
- sprinkling can
- spade
- large plastic zip-lock bags

570 SECTION 6: Bio-Related Technology

you live in a cold climate and want to keep the pile active through the winter, you can dig a pit and cover the pile with several inches of soil.)

3. Set the eight steel posts equidistant from one another in a circle around the cleared area.

4. Wrap the wire mesh or picket fencing around the posts to create the sides of the compost bin. Attach the fencing to the posts with the plastic tie straps. Allow for the fence to be opened easily for maintaining the bin.

5. Line the sides of the bin with clear plastic sheeting.

6. Spread 2 to 4 inches of grass clippings or decaying straw on the dirt floor to make a base layer.

7. Add leaves and other vegetable matter to the compost pile each day. If possible, chop the material into smaller pieces so it will decompose more quickly. Mix in a scoopful of ordinary soil, which will provide the necessary bacteria. The more varied the mixture, the better the results. Do not use animal waste, such as meat scraps. It is not safe to handle and will attract animals.

8. For every 10 inches of vegetable matter, add another 2 to 4 inches of grass clippings or straw. This will add texture and block unpleasant smells.

9. Create a shallow depression in the center of the pile to allow rainwater to collect and penetrate. In dry weather, for every 10 inches of vegetable matter, sprinkle the compost with 2 gallons of water. The material should be moist but not soaking wet.

10. The center of the pile will become very warm (up to 150°) as decomposition takes place. Use the spade to loosen the pile occasionally to allow air to enter.

11. After 3 or 4 months, soil should begin to form at the bottom of the compost pile. Turn the pile with the spade so that the new soil is on top.

12. Bag the soil in large ziplock bags. Weigh one of the bags to determine how many pounds of soil it contains.

13. Create labels for the bags, identifying what they contain and the weight, and attach them.

14. Advertise and sell the composted soil. (To determine its price, you might want to compare it to soil sold in garden shops.)

Evaluation

1. How much time passed before soil began to form at the bottom of the compost pile? How do you think the action might be speeded up?

2. Look in the soil for earthworms. What purpose do they serve in creating soil?

3. What makes creating soil a bio-related technology?

Useful Skills across the Curriculum

1. Language Arts. Create a poster or radio commercial advertising the benefits of composted soil. Use the poster or commercial in selling your soil product.

2. Mathematics. Suppose the fencing you want to buy for the compost pile is sold by the centimeter. To convert inches to centimeters, you multiply by 2.54. If you need 16 feet of fencing material, how many centimeters will you buy?

Credit: Robert A. Daiber

1. **Answer.** 487.68 Centimeters

DIRECTED ACTIVITY 2

Designing and Making New Body Joints

Context

In recent years, human joint replacement has become very common. You may know of someone who has received an artificial knee or hip implant. These implants relieve the pain and suffering caused by injured or worn out joints. Over 50,000 hip implants alone are done each year. After surgery, patients can lead more active lives.

Implants are made of special metal alloys and plastics. The metal is secured to the bone tissue, and the plastic serves as a spacer that fills out the joint the way normal cartilage does. In order for the artificial joint to work almost as well as the natural joint, parts must be machined within critical tolerances. This means the patient must be accurately measured so all the parts fit correctly.

Goal

For this activity you will analyze a human joint, design a replacement for it, and create a model.

Equipment and Materials

- drawing materials
- tape measure
- outside calipers
- materials for constructing the joint, such as dowel rods, newspaper for papier mache, cardboard, thin plastic sheets or milk-jug plastic, and glue
- materials for holding body parts together, such as rubber bands
- equipment for mixing materials, such as mixing cups and sticks for stirring
- cutting tools, such as scissors, utility knives, and hand saws
- sandpaper

SECTION 6: Bio-Related Technology   573

Procedure

1. If possible, interview someone who has had a joint replaced. Ask about the process. Take notes.

2. Select a joint in the human body for which you will design a replacement part. Knee, hip, elbow, wrist, and ankle are the most commonly replaced.

3. Research the anatomy of the joint you've selected in a reference book, such as a biology text or encyclopedia, or on the Internet. If the biology lab in your school has a model of the human skeleton, ask to study it.

4. Make a sketch of the joint you've chosen. Label the bones, tendons, and cartilage correctly.

5. On your own body, measure the approximate size of this joint using a tape measure and outside calipers.

6. Develop a full-scale mockup of the joint, using such materials as dowel rods and papier mache. If you use papier mache, sand the material after it dries.

7. Assemble the joint, using rubber bands to hold it together.

8. Try to make the joint work by moving the parts back and forth.

9. Cut a piece of thin plastic and shape it to fit where cartilage would normally be placed.

10. Label the different parts of the joint.

Evaluation

1. Write a short paragraph analyzing the joint's movement. Is it more limited than a natural joint? If so, in what ways?

2. Into which branch of bio-related technology does this activity fit?

3. If you were to repeat this activity, what would you do differently?

---

1. **Enrichment.** Invite someone who has had a joint replacement to talk to the class.

2. **Resource.** Students can search for the Pacific Northwest National Laboratory website on the Internet for information on artificial joints.

Useful Skills across the Curriculum

1. Mathematics. Suppose you work for a company that manufactures body parts and sells them in other countries where the metric system is used. Convert all the measurements you have made for this project to metric measures.

2. Science. Determine which types of forces act upon the joint you have created. If you were making an actual joint, how would this information influence your choice of materials?

Credit: Robert A. Daiber

Glossary

A

abutments The braces at the ends of a bridge that support the bridge and the earth at its ends.

acceptance sampling Randomly selecting a few products from a manufacturing production run and inspecting them to see whether they meet specified standards.

adaptation In bio-related technology, an organism's ability to change.

adhesives Materials that are used to bond together, or adhere, two objects.

advertising The methods a company uses to persuade, inform, or influence consumers to buy a product.

aerodynamic drag The strong force created when a moving object meets air resistance, which slows it down.

aerodynamics The science that deals with the interaction of air and moving objects.

agriculture The science of producing crops and raising livestock.

air-supported structure Structure in which air pressure supplied by a fan supports the walls and roof.

alloy The new material created when two or more metals or a metal and a nonmetal are combined.

American National Standards Institute (ANSI) The organization that coordinates and organizes qualified groups who then establish manufacturing standards.

amplitude The measurement of the intensity, or strength, of an electromagnetic wave.

AMTRAK The AMerican TRavel trAcK system, which is owned by the federal government and operated by the railroad lines that own the track the trains are using.

analog In computer science, a term used to describe electronic signals with an infinite number of levels, variations, or measurements; continuous signals.

antibodies Proteins that attack foreign substances in the human body.

aquaculture The process of rearing fish, shellfish, or plants in water under controlled conditions.

arch bridge Bridge in which the curved portion or arch carries the weight of the load to the supports at the end.

architects People who design structures and develop the plans for building them.

artificial intelligence A type of software computers use to solve problems and make decisions that are commonly solved or made by humans.

assemblies In manufacturing, components or parts of a product that have been put together in a planned way.

assembly line In factories, an arrangement in which the product being made moves from one workstation to the next while parts are added.

assets Anything a company owns that has value.

automated storage and retrieval system (AS/RS) Type of materials-handling system in which a computer-controlled crane travels between sets of tall storage racks to store and retrieve materials.

automatic guided vehicle system (AGVS) Type of materials-handling system in which specially built computer-controlled driverless cars that carry materials follow a wire "path" installed in the floor.

B

bar code A striped code printed on most products, that gives information about the product a computer scanner can read.

barges Large, flat-bottomed vessels used to carry bulk cargo, such as petroleum products and grain, on rivers, canals, and lakes.

batter boards Boards held horizontally by stakes driven into the ground to mark the boundaries of a proposed building.

beam bridge Bridge in which a steel-reinforced concrete roadway is supported by steel or concrete girders (beams).

bid In construction, a price quote for how much a contractor will charge for a proposed building project.

bill of materials In manufacturing a complete list of the materials or parts, as well as the quantities of each, needed to make one product.

binary code Coded information consisting of a series of two digits, 1 and 0, that a computer can "understand."

bioethics Rules or standards that apply to people and companies who engage in bio-related science and technology.

biomaterials Human-made materials designed to be placed within the human body.

bio-related technology All the technology connected with plant and animal life.

bit Each on or off pulse of the binary code read by a computer.

booster rockets Rockets used to push a payload into space.

break bulk cargo In transportation, single units or cartons of freight.

building codes Regulations that specify the methods and materials that can or must be used for each aspect of construction.

bulk cargo In transportation, loose cargo such as sand, oil, or gas.

byte Eight bits of information that are combined and processed together in a computer.

C

cable-stayed bridge Bridge supported by cables that are connected directly to the roadway.

cantilever bridge Bridge consisting of two beams, or cantilevers, that extend from each end and are joined in the middle by a connecting section called a suspended span.

capital All the money, land, and equipment needed to set up a technological system.

cargo Everything other than people that is transported in vehicles. Also called *freight*.

CD-ROM (compact disc-read only memory) A special plastic disc for computers that stores large amounts of information.

cell (1) In biology, the smallest structural unit of an organism that can function independently. (2) In telephone communication, a specified operating area that has its own transmission tower linked to an electronic switching office.

central processing unit (CPU) The heart of the computer that fetches the instructions, analyzes the input data, executes the necessary operations, and sends the information to storage. Also known as the *microprocessor*.

chain of distribution The path that goods take in moving from the manufacturer to the consumer.

channels In transportation, deep paths of water in which vessels travel.

chemotherapy The use of chemicals to treat or control disease.

chromosomal analysis Evaluation of human cells to detect abnormalities in chromosomes and predict the risk of inherited disorders.

city planners People hired to create plans for community development.

classification yards Places where trains are disconnected, cars are sorted according to their destinations, and new trains are made up of cars going in the same direction. Also called *switch yards*.

cloning In bio-related technology, a propagation method wherein a cell from a plant or animal is used to create a clone or duplicate.

coaxial cable Cable for carrying electrical signals; consists of an outer tube made of electrical-conducting material (usually copper) that surrounds an insulated central conductor (also copper).

cofferdams Temporary watertight walls built to shift the flow of water around the construction area as a permanent dam is being built.

combining In manufacturing, joining two or more materials, such as welding pieces of metal or mixing chemicals to make a product.

commercial construction Building structures used for business.

commission A percentage of the selling price earned by salespersons.

communication channel In communication, the path over which a message must travel to get from the sender to the receiver.

communication satellite A device placed into orbit above the earth to receive messages from one location and transmit them to another.

communication technology All the ways people have developed to send and receive messages.

commuter service Regular back-and-forth passenger transportation service.

component In manufacturing, each individual part of a product.

composite A new material made by combining two or more materials to produce more desirable qualities.

computer An electronic device that can store, retrieve, and process data.

computer-aided drafting (CAD) A computer program used to create drawings used by production systems.

computer-aided engineering (CAE) A computer program used by engineers to perform needed math calculations, to generate working drawings, and to analyze parts.

computer-aided manufacturing (CAM) A computer program used to tell the machines in a factory what to do. Also called *computer-aided machining*.

computer-aided production planning (CAPP) In manufacturing, a computer program used to quickly determine the best processes for obtaining materials and scheduling production.

computer-integrated manufacturing (CIM) Manufacturing done with the aid of computer programs that help tie all the phases of manufacturing (planning, production, and control) together to make a unified whole.

computer numerical control (CNC) Manufacturing process in which numerical directions contained in a computer program control or monitor machine operations.

concurrent engineering In manufacturing, a problem-solving approach in which groups of specialists from all areas of design and production work as teams.

conditioning In manufacturing, changing the properties of a material.

construction technology All the technology used in designing and building structures.

consumers People who buy products for their own personal use.

container on flatcar (COFC) Method of transporting containers by rail in which containers are fastened directly to a flatcar.

containerization An efficient method of handling cargo in which it is loaded into large boxes or other containers before it is transported. Also called *containerized shipping*.

containerships Oceangoing ships specially designed and built to carry containers.

continuous production In manufacturing, a production system in which a large quantity of the same product is made using assembly lines. Also called *line production* and *mass production*.

contractor Person who owns and operates a construction company.

controlled environment agriculture (CEA) A type of agriculture in which a plant's or animal's surroundings are carefully monitored and adjusted for the right amount of humidity, temperature, light, and nutrients.

conversion In bio-related technology, the process of altering the form of an organism in some way to prepare it for use.

conveyor A continuous chain or belt that moves materials, parts, and products from one place to another over a fixed path.

coordinate measuring machine (CMM) In manufacturing, a very accurate computer-controlled device used to measure "hard-to-measure" parts.

custom production Type of manufacturing in which products are made one at a time according to a customer's specifications.

cyberspace A computer's artificial environment.

D

data bank A central computer that stores the information from many smaller computers.

debugging Finding and correcting problems in the operation of a technological system.

demolition The destruction of a structure by tearing it down or blasting it with explosives.

design for manufacturability The process of designing product parts so that they are easy to manufacture and assemble.

detail drawings In construction, special drawings of any features that cannot be shown clearly on floor plans or elevations or that require more information. In manufacturing, drawings that show the details of a particular part.

die In manufacturing, a piece of metal with a cut-out or raised area used to shape a part.

diesel-electric locomotives Trains having diesel engines that turn electrical generators which produce the electricity that moves the train.

diesel engine A special type of two- or four-stroke-cycle engine that does not need a spark plug and can be used to pull heavy loads.

digital Term used to refer to electronic signals having only two variations: on and off.

digital image processor In communication, a device that converts pictures into thousands of tiny points of light, or pixels.

digitize To change information into a number code that can then be transmitted as electronic data.

direct sales Type of sales in which a manufacturer sells its product directly to the customer rather than to a wholesaler.

distribution The process of getting goods to the purchaser.

DNA (deoxyribonucleic acid) In bio-related technology, a long chain of four simple molecules in a gene that carry all the genetic information.

downlink The process in which a communication satellite transmits signals it has received from one earth station to another.

drafting The process of accurately representing three-dimensional objects and structures on a two-dimensional surface, usually paper.

E

earth station A large pie-shaped antenna that receives signals from and transmits them to communication satellites. Also called a *ground station*.

efficiency The ability to bring about a desired result with the least waste of time, energy, or materials.

electromagnet A magnet that is created temporarily when a soft iron core is surrounded by a coil of wire and electric current is sent through it.

electromagnetic compatibility (EMC) The ability of a device or component to resist electromagnetic interference.

electromagnetic interference (EMI) Interference caused by electromagnetic radiation that disrupts communication devices.

electromagnetic radiation An invisible source of energy given off by the movement of electrons when electrical devices are in operation.

electromagnetic waves Waves that travel through the atmosphere and make communication without a connecting wire possible.

electronic mail (e-mail) Messages, letters, or documents that are sent by means of computers and telephone lines.

electrostatic printing A printing process that relies upon a charge of static electricity to transfer the message from the plate to the paper.

elevations Drawings that show the finished appearance of the outside of a structure, as viewed from the ground level.

emerging technologies New technologies that are just coming into use.

energy The ability to do work.

engine A machine that produces its own energy from fuel.

engineered wood materials Building materials developed from woods and wood waste that would otherwise be unusable.

engineers People who make sure the design of a structure or product is sound and determine how the structure will be built or the product will be made and what materials should be used.

entrepreneur A person who starts his or her own business.

ergonomics See *human factors engineering*.

escalator A moving stairway, used for transportation.

ethics A set of rules or standards that guide the conduct of a person or group of people.

excavating Digging for the foundation of a structure that is usually done with heavy equipment.

external-combustion engine Engine powered by fuel that is burned outside the engine.

F

failure analysis In manufacturing, evaluation of a failed prototype to see why it didn't work as it should.

fax (facsimile system) System for sending pictures and words by means of electronic signals over telephone lines. The fax machine at the receiving end converts the signals back into pictures and words.

feedback Information about the outputs of a technological system that is sent back to the system to help determine whether the system is doing what it is supposed to do.

finite element analysis (FEA) In manufacturing, an evaluation that predicts how a specific component or assembly will react to environmental factors such as force, heat, or vibration.

fixture In manufacturing, a special device that holds parts in place during processing.

flexible machining center (FMC) In manufacturing, a computer-controlled combination machine tool capable of drilling, turning, milling, and other processing.

floor plan Drawing that shows the locations of rooms, walls, windows, doors, stairs, and other features of a structure.

footing The part of a structure below the foundation wall that distributes the structure's weight to the ground.

forming In manufacturing, the process of changing the shape of a material without adding or taking anything away.

foundation The part of a structure that rests upon the earth and supports the superstructure.

freight Everything, other than people, that is transported in vehicles. Also called *cargo*.

frequency In communication, the number of electromagnetic waves that pass a given point in one second.

fuel cell A battery that generates power by means of the interaction between two chemicals such as hydrogen and oxygen.

functional analysis In manufacturing, the evaluation of a prototype to see if it works as predicted.

G

gauges In manufacturing, tools used to compare or measure sizes of parts and depths of holes.

genes In bio-related technology, factors or elements in cells that are carriers of heredity.

gene splicing In bio-related technology, the procedure in which a gene is removed from one strand of DNA and "glued" onto another strand.

gene therapy In bio-related technology, a medical treatment that involves the transfer of a normal gene into an individual who was born with a defective or absent gene.

genetic engineering In bio-related technology, the process of giving organisms traits they never had naturally by changing their normal genes.

genome In bio-related technology, all the DNA in an organism.

geotextiles In construction, large pieces of material used as a bottom layer for roadbeds and on slopes to help prevent erosion. Also called *engineered fabrics*.

global market In manufacturing, a worldwide market.

Global Positioning System (GPS) A system of communication satellites that orbit the Earth and are used to track an object's location.

graphic communication The method of transmitting messages by means of illustrations and printed words.

gravure printing A printing process in which images are transferred from plates that have recessed areas. Also called *intaglio printing*.

gridlock Traffic jam in which no vehicles can move in any direction.

groundwater Water located below the earth's surface that supplies wells and springs.

group technology In manufacturing, a computerized method of keeping track of information about product parts so that the design for an old part can be revised to develop plans for a new, similar part.

H

harvesting In bio-related technology, gathering materials or organisms and preparing them for use.

hazardous waste Any waste that can harm human health or the environment.

holography The use of lasers to project realistic three-dimensional images of objects.

human factors engineering In bio-related technology, the science of designing equipment and environments to promote human health, safety, and well-being. Also called *ergonomics*.

hybrid In bio-related technology, the offspring of two plants or animals of different varieties, breeds, or species.

hybrid vehicles Vehicles designed to run on more than one kind of fuel, such as both gasoline and compressed natural gas.

hydraulics The science of exerting a force on fluid under pressure to produce motion.

hydroponics Cultivating plants in nutrient solutions without soil.

hypertext markup language (HTML) In communication technology, the special language used for the World Wide Web.

hypothesis An explanation that can be tested.

I

immunization In bio-related technology, the process of making the body able to resist disease by causing the production of antibodies.

industrial construction The building and/or remodeling of factories and other industrial structures.

Industrial Revolution The great changes in society and the economy caused by switching from making products by hand at home to making them by machine in factories. The Industrial Revolution took place in the 1800s.

ink-jet printing A computer-controlled printing process in which tiny nozzles spray ink droplets onto the surface to be printed.

innovation A new device, process, or idea.

input Anything that comes from a technological system's resources and is put into the system.

insulation Material that is applied to walls and ceilings of a building to help keep heat from penetrating the building in the summer and cold from penetrating in the winter.

integrated circuit (IC) In communication technology, a tiny piece of silicon that contains thousands of interconnected electrical circuits that work together to process information. Also called *microchip* or *chip*.

integrated pest management (IPM) In bio-related technology, an approach to pest control that combines several different techniques.

interchangeable parts In manufacturing, identical parts, any of which that will fit the product.

intermittent production In manufacturing, the production of limited quantities of different products by retooling and making adjustments in the same assembly line.

intermodal transportation The process of combining two or more modes of transportation to efficiently transport passengers and/or cargo.

internal-combustion engine Engine in which the fuel is burned inside the engine itself.

International Organization for Standardization (ISO) Organization that promotes and coordinates worldwide standards for many products.

Internet A huge, interconnected, worldwide network of smaller computer networks.

inventory In manufacturing, the quantity of items a company has on hand.

inventory control In manufacturing, keeping track of raw materials, purchased parts, supplies, and finished products on hand.

irrigation canals In agriculture, canals that carry water from a place where water is plentiful to fields where water is needed.

J

jig In manufacturing, a special device that holds a part being processed and guides the tool doing the work.

just-in-time (JIT) In manufacturing, a system of delivering parts and materials just as they are needed for use in production.

L

landscaping In construction, the process of changing the natural features of a site to make it more attractive.

laptop computers Small portable computers that can do most of the same tasks as a larger personal computer.

laser A narrow, high-energy beam of light used for many purposes.

laser printing Printing process in which a laser beam is used to help transfer an image to the paper in a manner similar to that of an electrostatic copier.

leachate A toxic liquid produced when water trickles through decaying waste and capable of polluting streams and underground water supplies.

linear motion Motion in a straight line.

load-bearing-wall structures Structure in which the heavy outer walls support the weight of the structure; there is no frame.

lock In transportation, an enclosed part of a canal or other waterway in which the level of the water can be changed to raise or lower ships.

M

machine language Language in the form of electric pulses that is used by computers.

maglev systems Railroad systems that operate on the principle that like poles of a magnet repel each other. Magnets are used to both levitate and propel the train.

Glossary

manufacturing resource planning (MRP II) A computer program that helps manufactuerers plan for materials, people, time, and money requirements.

manufacturing technology All the technologies people use to make the things they want and need.

market A specific group of people who might buy a product.

marketing All the activities involved in selling a product.

market research All the activities used to determine what people want to buy and how much they will pay for it. The results indicate how well a company can expect its product to sell.

mass communication Communication with many people at the same time.

material requirements planning (MRP) In manufacturing, a computer program that analyzes information from the bill of materials, the master production schedule, and inventory cards and makes calculations and recommendations about resources and costs.

materials handling In manufacturing, moving and storing parts and materials.

materials processing In manufacturing, changing the size and shape of a material in order to increase its usability or value.

microgravity The very low pull of gravity and the resulting near weightlessness that occurs in outer space.

microwaves In communication, very short electromagnetic waves that can be used to carry telephone messages through the atmosphere.

mock-up In manufacturing, a three-dimensional model of a proposed product that looks real but has no working parts.

modem A device that sends computer data over telephone lines to other computers.

modular construction Type of construction in which standardized sections (modules) of a structure are built at factories and shipped to the construction site, where they are assembled to produce a finished building.

modular design In manufacturing, the process of designing products made of standardized sections, or pre-assembled, standardized modules.

module A part or section.

mold A hollow form.

monitor To watch over and inspect a structure or process to ensure safety and quality.

motor A machine that uses energy supplied by another source.

movable bridges Bridges designed so that a portion of the roadway can be moved to allow large water vessels to pass underneath.

multiview drawing In manufacturing, a drawing that shows two or more different views of an object drawn at right angles, or perpendicular, to one another.

N

network In communication, a system of computers and printers hooked together by wires or fiber optic cables.

numerical control (NC) In manufacturing, the use of a number code on a punched paper tape to give directions to a machine.

O

Occupational Safety and Health Administration (OSHA) Agency of the federal government that sets standards for safety and health in the workplace and sees that those standards are met.

on-site transportation Transportation within a building or group of buildings.

optical fibers Thin, flexible fibers of pure glass used to carry signals in the form of pulses of light.

original equipment manufacturer (OEM) A manufacturer who sells products directly to another manufacturer.

output The result of a technological system.

P

pager A tiny radio receiver that enables its user to receive messages anytime, anywhere.

panelized construction Type of construction in which prefabricated floors, walls, and roofs are made in factories and then assembled at the site.

part print analysis In manufacturing, the use of working drawings by process planners to get ideas about how a part can be made.

payload Anything transported.

personal communications services (PCS) All the services linked to a cellular phone.

pesticides Poisons that destroy insects.

phosphor A substance that emits light when given energy. Used in television and computer screens.

photographic printing A printing process in which light is projected through a plate, or negative, onto a light-sensitive material.

photovoltaic cells Devices that can change the energy from the sun's light into electricity.

pica In printing, a unit of measure equal to 12 points, or one-sixth inch.

pictorial drawing In drafting, a type of drawing that shows objects as they appear to the human eye.

piers The vertical supports placed between abutments in longer bridges.

piggyback service Transporting truck trailers on railroad flatcars. Also called *trailer on flatcar* (TOFC).

piles In construction, wood, metal, or concrete foundation members that are driven down into the earth to a solid base.

pilot run In manufacturing, a practice production run done to find and correct problems before actual production begins.

pixels Glowing phosphor dots that make up the image on a TV or computer screen.

planned obsolescence In manufacturing, the process of designing products to last only a certain period of time or under certain conditions.

planographic printing Any printing process that involves the transfer of a message from a flat surface.

plant layout The arrangement of machinery, equipment, materials, and traffic flow in a manufacturing plant.

porous printing processes Printing processes in which ink or dye is passed through an image plate or stencil and transferred onto the material being printed.

power of eminent domain A law that states the government has the right to buy private property for public use.

prefabrication Making parts of structures in factories so that the structure can be assembled quickly at the construction site.

primary processes The processes used to convert raw materials into industrial materials.

principles of design Factors that help to determine the effectiveness of a design. These factors include balance, proportion, contrast, variety, harmony, and unity.

private sector Ordinary people; the private part of our economy.

process All the activities that must take place for a technological system to give the expected result.

process chart A chart that shows the order of manufacturing steps.

product engineering Planning and designing to make sure a product will work properly, will withstand extensive use, and can be manufactured with a minimum of problems.

production In manufacturing, the multi-step process of making parts and assembling them into products.

production control In manufacturing, methods used to control what is made and when it is made.

productivity The comparison of the amount of goods produced to the amount of resources used to produce them.

profit The amount of money a business makes after all expenses have been paid.

program Set of instructions that control the operation of a computer.

programmable controller A small, self-contained computer used to run a machine. Workers can reprogram the controller to change the way a machine functions.

programming languages Languages used to give instructions to a computer.

propagation In bio-related technology, the technique of reproducing a living organism.

prosthetics The artificial replacement of missing, diseased, or injured body parts.

prototype In manufacturing, a full-size working model of a product. It is the first of its kind and may be built by hand, part by part.

public works construction Building structures intended for public use or benefit.

public sector City, county, state, and federal governments.

pumping station Point at which a pump helps to move cargo through a pipeline.

purchasing agent Buyer or person responsible for obtaining the right materials for a project at the right price.

Q

quality assurance In manufacturing, methods used to make sure that a product is produced according to plan and meets all specifications. Also called *quality control*.

R

rafters Sloping roof-framing members cut from individual pieces of lumber.

rapid prototyping In manufacturing, using CAD and CAE software to make a three-dimensional model of a product that is used for evaluation.

reciprocating motion The up-and-down or back-and-forth motion inside an engine.

recombinant DNA In bio-related technology, new DNA formed by combining pieces of DNA from two different species.

relief printing processes Printing processes in which an image is transferred from a raised surface.

resource Anything that provides support or supplies for a technological system.

retailers Companies that buy products from manufacturers or wholesalers and then sell them directly to consumers.

ridge The horizontal beam along a roof's peak.

robots Special computerized machines programmed to automatically do tasks that people usually do.

rolling stock Railroad transportation vehicles.

roof trusses Preassembled triangular frames used to frame the roof of a structure.

rotary motion A circular motion.

rough-in In construction, the first stage in the installation of utilities when parts, such as wires, pipes, and ductwork, are placed within the walls, floors, and ceilings before the surfaces are enclosed.

routes In transportation, particular courses over which vehicles travel.

S

sales forecast A prediction of how many products a company will sell.

sanitary landfills Landfills in which a liner has been placed at the bottom to keep waste from polluting surface water or groundwater.

scale drawings Drawings in which a small measurement is used to represent a large measurement.

schedule A plan of action that lists what must be done, in what order, and when.

scientific method Approach to a problem in which researchers make an observation, collect information, form a hypothesis, peform experiments, and analyze the results.

sea-lanes Shipping routes across oceans.

secondary processes In manufacturing, processes that convert industrial materials into finished products.

secure landfill Landfill that uses double liners and a system for collecting and treating leachate.

sensors Mechanical devices that sense such things as light, temperature, sound, or pressure.

separating In manufacturing, a way of changing the shape of a material by taking some away.

sewage The liquid and solid wastes that are carried by sewers.

sheathing A layer of material, such as plywood or insulating board, that is placed between the framing and the finished exterior of a building.

simulation A computer program or other testing environment that imitates as closely as possible the real-life circumstances for which a solution or product is designed to be used.

single-unit trucks One-piece vehicles. Also called *straight trucks*.

site The land on which a structure is built.

site plan Plan that shows where a structure will be located on its lot.

sludge Solids left behind when sewage settles.

slurry In transportation, a rough solution made by mixing liquids with solids that have been ground into small particles so that the material can be sent through a pipeline.

smart buildings Buildings in which sensors and computer systems control mechanical, electrical, and security systems to improve efficiency and reduce costs.

space shuttles Space ships that take off, maneuver in space, and fly back through the atmosphere for landing.

span In construction, the distance between each pier in a bridge or between a pier and an abutment.

specifications In manufacturing, the detailed descriptions of the design standards for a part or product, including the type and amount of materials. In architecture, the written details about what materials are to be used for a project as well as the standards and government regulations that must be followed.

standardization In manufacturing, agreement on a uniform or common size for certain parts.

standard stock Materials in a widely used size, shape, or amount.

statistical process control (SPC) In manufacturing, a technique based on mathematics that is used to improve a production process.

statistical quality control (SQC) Quality control method in which a computer uses sampling to determine how well the parts are being made and then predicts the percentage of parts not meeting specifications.

statistics The science of collecting and arranging facts in the form of numbers to show certain information.

structural materials In construction, materials used to support heavy loads or to hold a structure rigid.

subassembly In manufacturing, an assembly of components that is used as part of another product.

subcontractors People or companies who specialize in certain types of construction work.

superconductor A material that will carry electrical current with virtually no loss of energy.

superstructure The part of a building or other structure that rests on the foundation.

survey Measurement of the exact size and shape of a piece of property, its position in relation to other properties and to roads and streets, its elevation, and any special land features.

suspension bridge Bridge that has two tall towers supporting main cables that run the entire length of the bridge and are secured by heavy concrete anchorages at each end.

system A group of parts that work together to achieve a goal.

T

technology The way people use resources to meet their wants and needs.

technology assessment A study of the effects of a technology.

telecommunication Communication over a long distance.

telemarketing The use of the telephone to sell goods and services.

test marketing In manufacturing, trying out a new product in a limited area to get consumers' reactions and opinions.

thrust The forward force resulting from escaping gases in linear-motion vehicles.

thumbnail sketches Small drawings done quickly.

time and place utility The utility, or value, of a product that has been added because transportation has brought it to the right place at the right time.

tolerance In manufacturing, the amount that a part can vary from the specified design size and still be used.

ton mile In transportation, the measure involved in moving one ton a distance of one mile.

tooling-up In manufacturing, the process of getting the tools and equipment ready for production.

topography A construction site's surface features.

torque Any force that produces or tends to produce rotation.

total quality management A company's approach to its employees in which the employees are expected to meet a performance standard.

tractor-semi-trailer rig In transportation, a combination of a tractor and a semi-trailer. The tractor is the base unit that pulls the semi-trailer, which contains the cargo.

trailer on flatcar (TOFC) An intermodal method of transporting cargo in which semi-trailers full of cargo are carried on railroad flatcars. Also called *piggyback service*.

transceiver In communications, a transmitter and receiver combined into a single unit.

transportation The movement of people, animals, or things from one place to another using vehicles.

transportation technology All the means used to help move people or cargo through the air, in water, and over land.

trend A general movement or inclination toward something.

truss bridge Bridge supported by a steel or wooden triangular framework above or below the roadway.

U

uniform resource locator (URL) In communication, a specific address for a site on the Internet.

unit train Train that carries the same type of freight in the same type of car to the same place time after time.

uplink In communication, a process in which an earth station transmits signals to and receives signals from a satellite.

utilities In construction, services such as electricity, natural gas, and telephone. Used to refer to these systems in a building.

V

vaccine In medicine, the use of killed or weakened microbes or purified toxins in order to stimulate the body's immune response and prevent disease.

value analysis In manufacturing, the study of each part of a product to determine whether the most functional, yet lowest cost, material has been used.

variables Conditions that may change during testing.

vehicle Any means or device used to transport people, animals, or things.

vessels Water transportation vehicles.

virtual reality (VR) A realistic computer simulation presented in three-dimensional graphics that requires the use of special equipment such as a viewer and a glove or a hand-held wand.

W

warehouse In manufacturing, a building where products are stored temporarily.

waste management All the operations involved in the collection, storage, and treatment of waste.

water culture In agriculture, the process of growing a plant in water.

water table The level of groundwater.

ways The spaces set aside for use by a given transportation system.

wholesalers People or companies who buy large quantities of products from manufacturers and then sell the products to commercial, professional, retail, or other types of institutions.

working drawings Drawings that provide the information needed to make a product or construct a project.

Z

zoning laws In construction, regulations that tell what kinds of structures can be built in specific parts of a community.

Credits

Cover Image: ©William Whitehurst/The Stock Market

Interior Design: Pun Nio, Nio-Graphics

AP/Wide World Photos, 325, 419
   Kevin Bartram, 541
   Michael Caulfield, 419
   Mike Fiala, 305
   Kathy Willens, 539
AT&T Bell Laboratories, 128
AMP Incorporated, 238
Apple Computer, Inc., 188
Archer Daniels Midland Company, 236
Arco Solar, Inc., 199
Arkenstone, Inc., 461
Arnold & Brown, 30, 38, 41, 43, 44, 58, 65, 74, 76, 86, 89, 93, 96, 101, 102, 104, 110, 118, 123, 140, 146, 159, 160, 168, 184, 190, 198, 206, 212, 213, 214, 215, 220, 227, 228, 239, 241, 246-247, 249, 251, 254, 262, 263, 264, 268, 276, 277, 282, 292, 294, 309, 312, 314, 321, 324, 326, 331, 332, 335, 338, 339, 350, 351, 355, 369, 372, 374, 382, 400, 429, 431, 435, 438, 462, 478, 482, 530, 542, 550, 563, 568
Art MacDillo's/Gary Skillestad, 21, 40, 54, 69, 79, 105, 106, 112, 113, 122, 124, 125, 126, 127, 129, 130, 131, 135, 136, 138, 139, 150, 151, 152, 153, 154, 156, 157, 161, 163, 164, 165, 174, 183, 194, 197, 200, 234, 235, 237, 241, 242, 250, 255, 256, 269, 279, 302, 303, 315, 325, 336, 342, 343, 344, 352, 365, 366, 368, 370, 371, 377, 383, 384, 386, 388, 390, 405, 417, 427, 430, 443, 453, 457, 483, 485, 487, 488, 491, 492, 493, 494, 502, 512, 515, 516, 517, 518, 519, 520, 521, 522, 523, 565, 571
Ashland Oil Inc., 189
Association of American Railroads, 474
Roger B. Bean, 36, 414
Dan Breitenbach, 120-121, 328-329, 334
Bridgeport Machines, Inc., 199
CSX Corporation, 467
Calgene, Inc., 547
California Energy Commission, 433
Canadian National Tower, 305
Jim Carlson, 62, 148
Carnegie Institution of Washington/Christiane Nawrath, Yves Poirier, Chris Somerville, 540,
Caterpillar Inc., 144, 191, 285, 322
Chicago Convention and Tourism Bureau, 24
Chrysler Corporation, 144
Cincinnati Milacron, 241, 281, 287
Colonial Williamsburg Foundation, 186
Conversion Products, Inc., 319
Corbis-Bettmann, 19, 187, 444
The Cousteau Society, 499
Crossfield Electronics, Ltd., 101
Custom Medical Stock Photos, 556, 559
   B.S.I.P., 555
   Kevin Beebe, 555
   CDC, 529
   J. Croyle, 557
   K. Glaser & Associates, 561
   Martin, 556
   NIH, 556, 562

Brad Nelson, 525
Pete Saloutos, 555
Strange/Porter, 557
Dallas/Fort Worth International Airport, 477
Deere & Company, Moline, Illinois, 220
Denver International Airport, 440-441
Department of Entomology, Colorado State University/Whitney Cranshaw, 552
Department of Plant Pathology, Ohio State University, 552
Department of the Treasury, 241
Digital Art, 14-15
DigitalVision, 70-71, 182-183, 304-305
Dover Elevators, 549
Duchossois Industries, Inc., 464-465
Dykon, Inc., 362
Emerson Electric Company/Obata Design Inc., 248
©FPG International
  Paul Ambrose, 272
  Robert Bennett, 375
  Telegraph Colour Library, 83
Donna Faull, 160
Federal Express®/Arnold & Brown, 437
Scott Fishel, 36
Flow Systems Inc., 278
Ford Motor Company, 484
Frazer-Nash Research Limited, 500
David R. Frazier Photolibrary, Inc., 25, 26, 34, 55, 91, 196, 197, 201, 205, 308, 353, 395, 422, 444, 447, 448, 450, 454, 467, 469, 470, 472, 496, 526, 528, 536, 537, 539, 566
  Mark Burnett, 46-47, 109
  Aaron Haupt, 552
  Ed Young, 26, 94
GM Media Archives/General Motors Corporation, 39, 504
GT Bicycles, 219
Jim Gaffney, 87

Bob Gangloff, 145
Garrett Auxiliary Power Division/Allied Signal Aerospace Co., 505
Ann Garvin, 71, 214, 312
General Motors Corporation, 79, 225, 248, 280
Genzyme Corporation, 560
Giddings & Lewis, AMCA International, 286
©Jeff Isaac Greenberg/PhotoEdit, 468
Grumman Corporation, 217
Jim Gund/Allsport USA, 142, 143
Harnischfeger Industries Inc., 285
©Harris Corporation, RF Communications Division, Rochester, New York, 132
HelpMate Robotics Inc., 280
Hewlett-Packard Company, 107, 154
High Speed Ground Transportation Association, 496, 503
Ingersoll-Rand Company, 220
Intel Corporation, 107
International Business Machines Corporation, 71, 222, 273
©Bruce Iverson, 102, 530
Journal Star (Peoria, IL) Photo/Photographer: David Zalaznick, 363
Kellogg Company, 144
KETIV Technologies, Inc., 340, 342
Kohler Company, 482
Komatsu, 480-481
LA-Z-Boy Chair Company, 226
Laser Alignment Inc., 335, 364
Lexmark International Incorporated, Lexington, KY, 222
Lindal Cedar Homes, Inc., 340
Mall of America, Minneapolis, MN, 306-307
Maryland Port Administration/Dan Breitenbach, 471
Master Builders, Inc., 318
Kathy L. McClosky, 28
McDonnell Douglas Corporation, 275, 279
  NASA, 288

Microscribe by Immersion Corp., 167
Minolta Corporation, 108
Ted Mishima, 16-17, 38, 39, 71, 82, 89, 98, 165, 253, 265, 271, 330, 476, 561
Mason Morfit/Lone Star Industries, 321
Michael Morrow, 12, 13
Motorola Advanced Messaging Systems Division, 72-73
NASA, 15, 275, 289, 398, 399, 415, 419, 434, 459, 460
National Center for Supercomputing Applications/University of Illinois at Urbana-Champaign, 117
National Medal of Technology, 230-231
National Semiconductor, 111
New United Motors Manufacturing, Inc., 192
NIKE, Inc., 144
North Wind Picture Archives, 14, 70, 182, 183, 304, 305, 418, 419, 524, 531
©J. Nourok/PhotoEdit, 528
Odyssey of the Mind/Eric Schenk, 32, 33
Olympus America Inc., 158
Otis Elevator, 423
P&W Builders/Roger B. Bean, 312, 324
Tom Pantages, 217, 375
Parker-Hannifin, 129
Brent Phelps, 85, 266, 357
Philips International B.V., 137
Philips Mobile Computing Group, 75
Photodisc Inc., 1, 418-419, 524-525
Photri Inc., 87, 100, 397, 443
Picker International, Inc., 115
Potlatch Corporation/Tom Tracy, 149
QA Photos Ltd., 391
Questar Corporation, 393, 456
©Radisson Seven Seas Cruises, 449
Raychem, 213
Reuters/HO/Archive Photos, 15, 20
Rockwell International Company/Allen-Bradley, 283
Rohr Industries, Inc., 486

Scitex America Corp./©Brian Smith, 148
Jill Seiler, 147, 162, 216
Smith Floor Covering Co./Ann Garvin, 373
Smithsonian Institute, 133
Spectra Physics, 322
Wolfgang Spunbarg/Photo Edit, 84
Sonoco Products Co./Terry Parke, 204
Starnet Structures Inc., 317
©The Stock Market
   Roger Ball, 316
   James Barnett, 205
   Peter Beck, 312, 341
   James Blank, 396
   Nigel Francis, 14
   Ned Gillette, 304
   Craig Hammell, 85
   Brownie Harris, 233, 256
   Ted Horowitz, 420
   Michael Keller, 98, 348-349
   Mark M. Lawrence, 18
   John Madere, 208-209
   Kunio Owaki, 184-185
   Jose Pelaez Photography, 83
   Alon Reininger, 304
   Pete Saloutos, 544
   Masahiro Sano, 15
   Chris Sorensen, 424
   David Stoecklein, 260-261
   William Taufic, 425
Jeff Stoecker, 22, 27, 48, 49, 52, 53, 56, 57, 78, 80, 104, 156, 195, 202, 210, 211, 221, 222, 223, 224, 267, 284, 310, 367, 390, 392, 420, 424, 426, 428, 442, 452, 459, 483, 488, 489, 490, 498, 500, 506, 507, 525, 532, 533, 534, 546, 547, 549, 554
TJI® Joist, 316
The Tensar Corporation, 319, 383
Texas Cooperative Fish and Wildlife Research Unit, Texas Tech University, 537, 551

Texas Instruments Inc., 214
3D Systems Corporation, 270-271
3M Company, 35, 49
Toyota Motor Corporation, 183, 193, 503
Ken Trevarthan, 309
Tribune Media Services, Inc., ©1996, 73
UPI/Corbis-Bettmann, 214
U. S. Bureau of Reclamation/E. Herzog, 380-381
U.S.G.S. Photo Library, Denver, Colorado, 77
Unicorn Stock Photos/Jeff Isaac Greenberg, 243
Union Pacific Railroad, 468
Uniphoto Picture Agency/©Dany Krist, 434
Unity Systems, 360-361
Vaughn Agricultural Research Services Ltd., 552
Virtual i-O, All Rights Reserved, ©1996, 116
Wagner Spray Tech Corporation, 324
Jervis B. Webb Co., 239
Westlight
  Craig Aurness, 447
  Lois Ellen Frank, 81
  Kenneth Garrett, 425
  Larry Lee, 26, 317
  Bill Lisenby, 450
  R. Ian Lloyd, 240
  Chuck O'Rear, 445, 468
  Jim Pickerell, 446
  Bill Ross, 476
  Lucille Sardegna, 368
  Gene Stein, 436
  Chase Swift, 320
  William James Warren, 26, 130
  Doug Wilson, 387, 389, 432
  Mike Yamshita, 466
Westmoreland Coal Company, 473
Whirlpool Corporation, 564
Dana White Productions/Dana C. White, 23, 50, 88, 300, 429
The Williams Companies, 457
T. D. Williamson, Inc., 458
Frank Lloyd Wright, Falling Water, 333
©Elizabeth Zuckerman/PhotoEdit, 528

Special thanks to the following individuals, businesses, and organizations for their assistance with photographs in this book.

In Ontario, Canada—Otto Bock Orthopedic Industries, Oakville.
In Chicago, Illinois—Jack Uellendahl/Rehabilitation Institute of Chicago, Baxter Health Care Corporation.
In Central Illinois—A & E Blueprint Company, Peoria; Advanced Medical Transport of Central Illinois, Peoria; Austin Engineering Co. Inc., Peoria; Precision Laser Manufacturing, East Peoria; Yates City Hydrofarm, Yates City.

Models have been used to portray examples in this text.

Index

A

Abutments, 384, 385, 386 (illus.), 576
Acceptance sampling, 257, 576
Accident, on-the-job, 394
Activities
    advertisement, multimedia, 299-300
    aerodynamics of gliders, 519-523
    alternative transportation system, 510-513
    cellular phone network, designing, 173-175
    chair, cardboard, 68-69
    community construction, learning about, 402-403
    composite, product of, designing, 292-294
    composite, testing prototypes of, 295-296
    composted soil, creating and marketing, 570-572
    digital editor, 176-178
    environmental problem, solving, 62-63
    estimating and bidding on construction, 404-409
    gliders, aerodynamics of, 519-523
    highway improvement, 412-414
    home page, making own, 179-181
    Internet, 170-172, 178-181
    inventory management system, 297-298
    joint, replacement, making, 573-575
    modular hotel rooms, 410-411
    occupational survey, 60-61
    paint job, estimating and bidding on, 404-409
    recycling, 64-67
    solar-powered vehicle, 514-518
    space station, model of, 415-417
    standard stock, buying, 301-303
    survey, conducting a, 60-61
    testing composites and prototypes, 295-296
    using URLs, 170-172
    videotape, editing, 176-178
ADA (computer language), 103
Adaptation, of organisms, 538, 576
Address, web page, 113
Adhesives, 35, 320, 576
Advertisement, multimedia, creating, 299-300
Advertising, 78, 264, 576
Aerodynamic drag, 504, 576
Aerodynamics, 504, 519-523, 576
Aeroponics, 549
Agriculture
    aquaculture, 550-551
    and cloning, 21, 540, 548, 549 (illus.)
    controlled environment, 548-551
    early, 19, 182, 524
    and genetic engineering, 547-548
    and harvesting crops, 537, 576
    and hybrids, 546
    hydroponics, 549-550
    and integrated pest management, 527, 551-552
AGVS, 284, 285 (illus.)
Air
    compressed, as power source, 219, 220 (illus.)
    pollution of, 121, 204, 205, 436
Air bags, 37, 38
Aircraft. *Also see* Airplanes
    and aerodynamics, 504, 519-523
    glider, making and testing, 519-523
    and lift, 443, 519
    operating statistics for, 442 (chart)
    plastics in, 505
    power used in, 486-487, 498
    rockets, 458, 459 (illus.)
    types of, 443-444
Aircraft mechanic, career as, 508
Air lock, 396
Airplane
    combined technology in, 25-27
    and containerized shipping, 469
    engines in, 486, 487 (illus.)
    first, 419, 436
    glider, making and testing, 519-523
    instrument panels in, 91 (illus.)
    and lift, 443
    manufacturing of, 26
    types of, operating statistics for, 442
Airport
    automated baggage handling at, 441
    Denver, 320 (illus.), 441
    general discussion, 445
    passenger links to, 447
Airspace, defined, 444
Air-supported structures, 395-396, 576
Air transportation, 442-445, 469
Alaska Highway, 25
Alloys, 196-197, 289, 576
Alpha interferon, 559
ALS, 545
Alternative energy sources, 199, 499-503
Aluminum, as structural material, 317
Aluminum cans, recycling, 64-67
Ambulances, 421
American National Standards Institute, 254, 576
Amperage, calculating, 509

Amplitude, of wave, 130, 131 (illus.), 576
AM radio, 130, 133
AMTRAK, 446, 506, 576
Amyotrophic lateral sclerosis, 545
Analog signal
    for cellular phone, 127, 576
    in computers, 104-105
    for television broadcasts, 136-137
Analysis
    chromosomal, 557
    failure, 226
    finite element, 274
    functional, 225-226
    part print, 232, 233 (illus.)
    prototypes for, 225-226
    simulation, 227
    value, 226-227
Analyst, computer systems, 96
Animals
    antibiotics to promote growth in, 536
    cloning, 20, 21, 536, 540, 548
    genetically engineered, 547, 548
    insects, controlling, 527, 538, 551-552
    transgenic, 547
Animation, computer, 90 (illus.)
ANSI, 254
Antennas
    for cellular phone, 127, 130
    earth station, 139
    for microwave transmissions, 129, 130 (illus.)
    for radio signals, 130, 132
    radio telescope as, 139
    for satellite systems, 87 (illus.), 138 (illus.), 139
    for telephone transmissions, 129-130
    for TV transmissions, 134, 138, 139
Antibodies, 553, 559, 576
Antibiotics, to promote growth, 536
Aperture, lens, 157
Apple Computer, 100, 231
Aquaculture, 537 (illus.), 550-551, 576
Art, camera-ready, 147, 148
Arch bridge, 385, 386 (illus.), 576
Architect
    career as, 326
    information needed by, 311, 337-338, 339
    plans developed by, 338, 340-345
    role of, 311, 576
    Wright, Frank Lloyd, 333 (illus.)
Architectural plans, 338, 340 (illus.), 341-345
Army Corps of Engineers, 451
Arm, digitizing, 166, 167 (illus.)
Artificial body parts
    designing and making, 573-575
    types of, 561-562
Artificial intelligence, 114-115, 576

Artist, graphic, 143, 147-148
Asphalt paving, 382, 383 (illus.)
AS/RS, 284-285
Assemblies (components), 248-249, 576
Assembly
    automated, 279
    design for, 222-223, 274
    process of, 189, 248-249
Assembly drawing, 223-224
Assembly line, 183, 189-190, 243, 576
Assessment, of technology, 29
Assets, defined, 200, 252, 576. *Also see* Capital
Asthma, occupational, 394
Automated assembly, 279
Automated factory, 286-287
Automated storage and retrieval system, 284-285, 577
Automatic baggage handling system, 441
Automatic factory, 287-288
Automatic guided vehicle system (AGVS), 284, 577
Automatic identification, 282. *Also see* Bar code
Automation, islands of, 286
Automobile. *See* Car
Award, National Medal of Technology, 231

B

Bacteria
    aerobic, 565
    for cheese-making, 535
    in compost, 565
    discovery of, 530, 531
    E. coli, 559
    and food spoilage, 267
    and infected wounds, 530-531
    interferon produced by, 559
    for oil spill cleanup, 541 (illus.)
    and pasteurization of milk, 530
    sterilization to destroy, 531
    vaccines against, 553, 554 (illus.)
    and waste management, 541, 565, 567
Baggage handling, automatic, 441
Balance
    in layout, 145
    in photographs, 159
Balance sheet, 252
Balloon, hot-air, 444
Balloon-popping car, 33
Ball parks, 329
Bar code, 111, 250, 282, 577
Barge, 450, 577
Bar graph, reading a, 40
Bascule bridge, 385, 386 (illus.)
Batter boards, 364, 365 (illus.), 577
Battery
    for electric cars, 496, 499-500

fuel cells, 502-503
and solar power, 499-500
Beam bridge, 385, 386 (illus.), 577
Bearing walls, 369
Bell, Alexander Graham, 20, 125
Bible, the, 14, 90
Bicycle
assembly drawing of, 223 (illus.)
first model of, 418
Bid, on construction project, 350, 404-409, 577
Billing clerk, career as, 462
Bill of materials, 232, 577
Binary code, 102, 103-104, 123, 577
Bioethics, 539-540, 577
Biological control of pests, 527
Biomaterials, 562, 577
Biopsy, 555-556
Bio-related technology
and agriculture. *See* Agriculture
artificial body parts, 561-562, 573-575
careers in, 542, 568
and cloning, 20, 21, 540, 548, 549 (illus.)
defined, 24-25, 528, 577
development of, 524-525, 530-535
and DNA. *See* DNA
and environment, 527, 541
and ethics, 21, 539-540
in everyday life, 534-535
genetic engineering, 547-548, 551
and health care. *See* Health care
human factors engineering, 563-564
hydroponics, 549-550
impacts of, 534-535, 538
and pest control, 527, 538, 551-552
processes in, 561-562, 573-575
as system, 535-541
waste management, 564-567
Bioremediation, 567
Biosphere, defined, 543
Bit (in binary code), 104, 577
Blimps, 444
Blood, lab analysis of, 553
Board feet, calculating, 303
Board lumber, 314, 315
Boards
batter, 364, 365 (illus.)
sizes and types of, 304 (illus.), 314-315
Boats, 450-451. *Also see* Ships
Bomb, nuclear, 15
Books, digitizing, 116
Booster rockets, 458, 459 (illus.), 577
Bose Corporation, and JIT, 185
Brainstorming, 35, 36, 212-213
Break bulk cargo, 425, 577

Breast cancer, 540
Breeding, selective, 531, 546-547
Bridge
Brooklyn, 304
collapse of, 324-325
Golden Gate, 331, 385
longest, 384
parts of, 384-385
span of, 325, 385
suspension, 325, 385, 386 (illus.)
Tacoma Narrows, 324-325
types of, 385-386
Bronze Age, 18
Browser, web, 113
Building codes, 314, 332, 356, 577
Buildings
constructing, 363-375, 378
demolition of, 362 (illus.), 363
foundation of, 365-366
framework of, 367-368
prefabricated, 374-375
skyscrapers, 37, 305, 317
smart, 361, 564
tallest, 305, 357
Bulk cargo, 425, 577
Burn in, 257
Burns, skin patches for, 560
Business
growth in, 435
starting own, 42
stock in, 42
subsidiaries of, 191
Byte, 104, 577

C

Cable, coaxial, 128, 134
Cable-stayed bridge, 385, 386 (illus.), 577
Cable television, 134, 135
CAD
and architectural planning, 340 (illus.), 341
careers involving, 228
and digitizing arm, 166, 167 (illus.)
general discussion, 166-167
modeling with, 166, 167 (illus.), 209, 216
and product development, 209, 216, 271, 272, 274
CAD/CAE programs, 272, 274, 276
CAD drafter/technician, 228
CAE, 218, 272, 274, 276
CAM, 209, 277-280
Camera
digital, 158
film for, 157-158
integrated circuit in, 108 (illus.)

parts of, 156-157
Camera-ready art, 147, 148
Canal
   irrigation, 389
   navigation, 389, 390 (illus.)
   Panama, 24, 49
Cancer
   breast, and genetic risk, 540
   chemotherapy for, 115
   and monoclonal antibodies, 559
Cans, aluminum, recycling, 64-67
Cantilever bridge, 385, 386 (illus.), 577
Capital
   in communication systems, 87-88
   in construction systems, 323
   defined, 55, 577
   fixed, 200, 431
   for manufacturing, 200
   for transportation systems, 431
   working, 200, 431
CAPP, 276
Car
   accidents in, simulating, 38
   aerodynamic design of, 504
   air bags in, 37, 38
   balloon-popping, 33
   Chrysler Concorde, 209
   computerized navigation system in, 461
   crash testing, 38
   design of, and aerodynamics, 504
   for disabled, 34
   electric, 496, 499-500
   and environment, 204, 205
   and ergonomic design, 563
   first, 418
   hand-operated, for disabled, 34
   and human factors engineering, 563
   hybrid, 503
   impacts of, 204
   lightweight materials used in, 505, 506, 507
   modular design of, 221-222
   passenger safety in, 37, 38
   plastics in, 505
   and pollution, 204, 205
   quality standards (QS 9000) for, 254
   recycling, 204
   seat belts in, 37, 38
   smart, 461
   solar-powered, 499-500
   steam-powered, 418
   and traffic jams, 436
   Ultralite by GM, 504 (illus.)
Carburetor, 488 (illus.), 489
Careers
   aircraft mechanic, 508
   analyst, computer systems, 96
   architect, 326
   billing clerk, 462
   in bio-related technology, 542, 568
   CAD drafter/technician, 228
   carpenter, 378
   chemical technician, 30
   civil engineering technician, 400
   clinical lab technician, 542
   in communications, 81-82, 96, 118, 140, 168
   computer programmer, 118
   computer repairer, 140
   computer systems analyst, 96
   in construction, 326, 346, 349, 358, 378, 400
   construction supervisor, 358
   consumer safety inspector, 44
   cost estimator, 346
   and creativity, 82
   dispatcher, 478
   and drafting, knowledge of, 162, 206, 228, 326, 378, 400
   electronics technician, 58
   employee training specialist, 244
   engineering, 206, 400
   forester, 568
   industrial engineer, 206
   inspector, safety, 44
   machine operator, 258
   in management, 82
   manufacturer's sales representative, 268
   in manufacturing, 30, 206, 228, 244, 258, 268, 290
   mechanic, aircraft, 508
   painting and coating machine operator, 258
   repairer, computer, 140
   repairer, industrial machinery, 290
   sales representative, 268
   survey of, making a, 60
   technical writer, 168
   technicians, 30, 58, 400, 542
   in training, employee, 244
   in transportation, 428, 438, 462, 478, 508
   truck driver, 438
Cargo
   break bulk, defined, 425
   bulk, defined, 425
   containerized shipping of, 468-471
   defined, 577
   exporting, 449
   importing, 449
   intermodal transportation of, 465-477
   payload as, 458, 460
   in pipelines, 455-457
   on ships, 449, 450, 469

in spacecraft, 458, 460
    on trains, 446-447, 467-469, 473
    on trucks, 451-455, 465, 468-469
Carlson, Chester, 144
Carpal tunnel syndrome, 212 (illus.)
Carpenter, 378
Cartoons, making, 90
Castle, Windsor, 376
Caterpillar®, 191 (illus.)
CATIA, 209
CAT scan, 556
CAVE, 117
CD
    advantages of, 75
    materials used for, 51
    musical, making of, 17
    pits burned into, 76, 102
CD player, 51, 56, 57
CD-ROM, 75, 92, 93 (illus.), 102, 577
CEA, 548-551
Ceiling joists, 367, 368
Cells (living)
    and cloning, 21, 548, 549 (illus.)
    defined, 530, 577
    hereditary information in, 531, 533, 557
    T, 554 (illus.)
Cells, manufacturing, 286, 287
Cells (power)
    fuel, 502-503
    photovoltaic, 199 (illus.), 499, 500 (illus.)
Cellular telephone
    analog, 127
    digital, 127
    and interference with medical devices, 94-95
    network for, designing, 173-175
    and radio waves, 127, 130
    workings of, 73, 126-127
Central processing unit, 101-102, 107, 108-109, 578
Ceramics
    advantages of, 197
    earliest, 182
    ingredients to make, 52
    in vehicles, 505
CFCs, 29
Chain of distribution, 266, 426 (illus.), 578
Chair, cardboard, making a, 68-69
Channel, communication, defined, 79, 578
Charts, types of
    control, 254-255
    mileage estimating, 430
    pie, 40

    process, 233-234
    time (production schedule), 250
Chat rooms, 113-114
Cheese, inputs needed for, 535
Chemicals
    for cancer treatment, 115
    for insect control, 527, 538, 551, 552
    in plastics, 313
    and pollution of groundwater, 371
Chemical technician, 30
Chemotherapy, 115, 578
Chess, with computers, 273
China, Great Wall of, 309
Chlorofluorocarbons, 29
Chromosomal analysis, 557, 559 (illus.), 578
Chromosomes, 533 (illus.), 548, 557, 559 (illus.)
Chrysler Concorde, designing and engineering, 209
Chunnel, the, 391, 392 (illus.)
CIM, 270-289
Circle graph, reading a, 40
Circuit
    defined, 107
    integrated, 107-108, 126
City planners, 330-331, 578
Civil engineering technician, 400
Clark, Barney, 525
Classification yards, 448, 578
Clear-span structures, 317
Clerk, billing, 462
Clinical laboratory technician, 542
Cloning, 20, 21, 536, 540, 548, 549 (illus.), 578
Clothing, insect damage to, 267
CMM, 281
CNC, 277-278
Coal
    as nonrenewable energy source, 199
    transporting, 472-473
Coating machine operator, 258
Coaxial cable, 128, 134, 578
Cochlear implant, 562
Code, building, 314, 332
Codes (encryptions)
    on airport baggage, 441
    for automatic identification, 282
    bar, 111, 250, 282
    binary, 102, 103-104, 123
    for cellular phone signals, 127
    genetic, 533-534
    Morse, 124
    in pits on CDs, 102
    for satellite messages, 139

in World War II, 74
COFC, 469
Cofferdams, 387-388, 578
Colossus (computer), and codes, 74
Combining
   mechanical fasteners for, 302 (illus.)
   as process, 203, 578
Comic strip, and technology, 73
Commercial
   multimedia, making a, 299-300
   purpose of, 78, 264
Commercial construction, 307, 310-311, 578
Commission, 264, 578
Communication. *Also see* Communication technology
   careers involving, 81-82, 96, 118, 140, 162, 168
   channels for, defined, 79
   definition of, 22
   with drawings. *See* Drawings
   and electromagnetic waves. *See* Electromagnetic waves
   with E-mail, 114
   graphic, 142-168
   over Internet, 99, 112-114, 170-172
   over long distances, 92, 120-141
   mass, defined, 132
   with photography, 155-160
   printed, 142-168
   reasons for, 22, 77-79
   by satellite, 137-139
   and spread of technology, 19-20
   systems of, 79-95
   tele-, 120-141
Communication channel, defined, 79, 578
Communication satellites, 137-139, 460-461, 578
Communication systems, 72-95
Communication technology
   and advertising, 264, 299-300
   artificial intelligence, 114-115
   capital for, 87-88
   CDs. *See* CD; CD-ROM
   cellular phones, 73, 126-127, 173
   in comic strip (Dick Tracy), 73
   and communication systems, 73-95
   computers. *See* Computers
   to control machines, 23, 154, 272-288
   defined, 22, 578
   digital image processing, 143
   digital libraries, 116
   efficiency and speed of, 76
   electronic mail, 114
   energy for, 87
   and entertainment, 78
   and environment, 85, 95
   and ethics, 110, 143
   expert systems, 114-115
   fax systems, 110 (illus.), 111
   general discussion, 22-23, 74-79
   Global Positioning System, 460-461
   holography, 160
   impacts of, 92-95, 121, 138
   integrated circuit, 107-109, 126
   materials for, 84
   need for, 77-79
   pagers, 73
   photography, 155-160
   and portability, 73, 76
   to predict natural disasters, 77
   printing processes, 149-155
   and privacy, 93
   quality of, 75
   radio, 130-133
   satellites, 137-139, 460-461
   in schools, 88, 92, 114
   speech recognition, 115, 282
   speed and efficiency of, 76
   telegraph, 123-125
   telephones, 73, 125-131
   television, 133-137
   tools and machines for, 73, 86
   trends in, 75-76, 91
   video conference, 283
   virtual reality, 116 (illus.), 117, 275, 341, 342 (illus.)
Commuter service, 446, 578
Compact disc. *See* CD
Company, 190-191. *Also see* Corporation
Competition, in business, 193, 435
Competitions (contests)
   Odyssey of the Mind, 33
   Technology Student Association, 43
Component, defined, 248, 578
Composites
   advantages of, 197, 292
   creating, for designed product, 292-294
   defined, 578
   examples of, 52, 292
   plastic, for construction, 319
   testing, 295-296
   in vehicles, 52, 506
Composition, photographic, 158-160
Compost, 565, 570-572

Computer
- altering photos with, 109-110, 143, 147-148
- animation with, 90 (illus.)
- Apple, 100, 231
- artificial intelligence in, 114-115
- breaking codes with, 74
- burn in time for, 257
- careers using, 82, 96, 118, 140, 228
- and carpal tunnel syndrome, 212 (illus.)
- for cartoon-making, 90 (illus.)
- CD-ROM for, 75, 92, 93 (illus.), 102
- central processing unit for, 101-102, 107, 108-109
- chess with, 273
- *Colossus*, 74
- controlling machines with, 154, 277-288
- data bank, 111
- Deep Blue, 273
- defined, 100, 578
- designing with, 143, 147-148. *Also see* CAD
- and desktop publishing, 147-148
- as diagnostic tool, 115, 556
- drafting with, 166-167
- for engineering products, 218, 227, 272, 274, 276
- and fax systems, 111
- graphic design with, 143, 147-148
- and health concerns, 93-94, 212 (illus.), 563
- and home offices, 121
- and human factors engineering, 563
- impacts of, 92, 114, 121
- and ink-jet printing, 154
- input units for, 101
- and integrated circuit, 107-109, 126
- and Internet. *See* Internet
- invention of, 183
- keyboards for, 101, 212 (illus.)
- language of, 103-106
- laptop, 51, 73
- in law enforcement, 109 (illus.)
- layout with, 143, 147-148
- machine language for, 103-106
- in manufacturing, 209, 270-289
- Mark I, 51
- memory for, 102
- microprocessor for, 101-102, 107, 108-109
- modeling with, 166-167, 209, 216
- modem for, 73, 92, 103, 113
- mouse for, 101
- for navigation, 460-461
- networks of, 99, 112-114, 283-284. *Also see* Internet
- and on-line services, 99, 112-114
- output units for, 103
- personal, development of, 100, 231
- in printing, 154-155
- and privacy concerns, 93
- in product design, 274-275
- in production, 277-288
- for production planning, 276-277
- programmable controller, 282
- and programming language, 103
- programs for. *See* Program
- prototypes with, making, 271, 276
- in quality control, 281
- in schools, 88, 92, 114
- sending images with, 110-111
- simulation with, 117, 227, 274, 275, 341
- software for. *See* Program
- in Space Shuttle, 100 (illus.)
- and speech recognition, 115, 282
- and telecommuting, 121
- and telephone, 110, 126
- in telephone exchange systems, 131
- and traffic control, 460
- in trucks, 454-455
- in vehicles, 454-455, 461, 506
- weight of, 51

Computer-aided drafting, 166-167, 578. *Also see* CAD
Computer-aided engineering, 218, 272, 274, 276, 578
Computer-aided manufacturing, 209, 277-280, 578
Computer-aided production planning, 276, 578
Computer-controlled machines, 154, 272-288
Computer-integrated manufacturing, 270-289, 578
Computerized tomography, 556
Computer models, 166-167 (illus.), 209, 216
Computer numerical control, 277-278, 579
Computer programmer, 118
Computer repairer, 140
Computer systems analyst, 96
Concrete, 318, 382, 384
Concurrent engineering, 209, 218, 579
Condemnation, 337
Conditioning, of materials, 203, 579
Construction. *Also see* Buildings; Structures
- bidding on, 350, 404-409
- bridge, 384-386. *Also see* Bridge
- canal, 49, 389, 390 (illus.)
- capital for, 323
- careers in, 326, 346, 349, 358, 378, 400
- commercial, 307, 310-311
- dam, 381, 387-388
- design process in, 337-345
- energy for, 322
- and environment, 319, 323-324
- estimating costs of, 346, 404-409
- highway, 382-384
- and human needs, 308

impacts of, 323-324, 331
industrial, 310
initiating, 330-331
inspecting, 332, 353-357
laws regulating, 313-314, 332
of malls, shopping, 307
managing, 350-358
materials for, 314-320, 382-384
modular, 374-375
monitoring, 353-356
panelized, 374, 375 (illus.)
pipeline, 392-393
planning, 330-346
prefabricated, 374-375
public works, 311, 330
quality in, 355-356
residential, 310. *Also see* Buildings
scheduling, 323, 351-353
site for, selecting, 332-335
in space, 397-399
of Stonehenge, 47
surveying site for, 334-335, 382
as system, 306-326
time in, scheduling, 323, 351-353
tools and machines in, 321-322
tunnel, 389-391
women employed in, 349
Construction supervisor, 358
Construction systems, 306-326
Construction technology, 23, 579.
  Also see Construction
Consumer, defined, 263, 579
Consumer safety inspector, 44
Contact lenses, biomaterials for, 562
Containerization, 468-471, 579
Container on flatcar, 469, 579
Containerships, 469, 579
Continuous production, 189-190, 579
Contractor, 350-353, 579
Contrast, in design, 146
Control
  defined, 579
  biological, for insects, 527, 551
  computer numerical, 277-278
  inventory, 251
  production, 250-251
  quality, 253-257
  shop floor, 251
  statistical process, 254-255
Control chart, 254-255
Controlled environment agriculture, 548-551, 579
Controller, programmable, 282, 283 (illus.)
Conversion, as bio-related process, 538, 579
Conveyor
  defined, 579

in mining, 472-473
for people, moving, 475, 476 (illus.)
types of, 238
Cooling system, of engine, 489, 490 (illus.)
Coordinate measuring machine, 281, 579
Copier machines, 144, 153, 154 (illus.)
Corn smut, 552
Corporation
  assets of, 200, 252
  defined, 190
  liabilities of, 252
  stock shares in, 42, 200
  subsidiaries of, 191
Corps of Engineers, 451
Cost estimator, 346
Costs
  of bulk vs. break bulk cargo, 425
  estimating, 346, 404-409
  and "make-or-buy" decisions, 233
  operating, of airplanes, 442
  of painting rooms, estimating, 404-409
  of product, and competition, 193
  of water transportation, 449
  of work-related illness and injury, 394
Cottage industry, 186 (illus.), 187
Cotton gin, 182, 213
Court, DNA evidence in, 42
CPU, 101-102, 107, 108-109
Crane, 238, 239 (illus.), 284-285
Crankshaft, 483, 487, 488 (illus.), 489
Crash testing, 38
Creative design process. *See* Design process
Creative thinking skills, 36, 42
Crick, Francis, 533
Crime, and DNA evidence, 42
Critical thinking skills, 36, 42
Cruise ships, 451
Cryosurgery, 561
CT scan, 556
Curtain, laser, 281
Custom production, 189, 579
Cutting
  laser, 202, 278
  waterjet, 278
Cyberspace, and virtual reality, 117, 341, 579
Cystic fibrosis, 533, 560

D

Dams, 381, 387-388
Data bank, 111, 579
DDT, 538, 539 (illus.)
Debugging, 243, 579
Decision-making skills, 36
Deductions, on paycheck, 194
Deep Blue, 273

Delivery, of materials, scheduling, 185, 282, 353, 355
Demodulation, of radio waves, 133
Demolition, 362 (illus.), 363, 579
Denver International Airport, 320 (illus.), 441
Design. *Also see* Design process
 and aerodynamics, 504, 519-523
 analyzing, 224-227
 for assembly, 222-223, 274, 275 (illus.)
 balance in, 145, 159
 with computers, 143, 147-148, 209, 216, 274
 contrast in, 146
 effectiveness of, 144-146, 149, 158-160
 and ergonomics, 563-564
 functional, 218-219
 graphic, 144-148
 harmony in, 146
 and human factors engineering, 563-564
 management approval of, 217
 for manufacturability, 219-223, 579
 modular, 221-222
 in photography, 158-160
 principles of, 145-146
 of printed matter, 143, 146-148
 product, 209, 215-223, 274-275
 proportion in, 146
 and recyclability, 204
 unity in, 146
 variety in, 146
 of vehicles, and efficiency, 504-507
 and virtual reality simulation, 275, 341, 342 (illus.)
 visual, 144-146, 158-160
Design process
 in construction, 337-344
 for printed product, 146-148
 in product development, 209, 212-227
Designer, graphic, 82, 143, 146-148
Desktop publishing, 147, 148 (illus.)
Detail drawing, 222 (illus.), 223, 343, 344 (illus.), 579
DFA/DFM, 274
Diagnostic testing, 553-557, 559
Diaphragm, in telephone, 125, 126 (illus.)
Die punching, 241
Diesel-electric engine, 481, 497, 498 (illus.)
Diesel-electric locomotives, 447, 497, 579
Diesel engines, 492-493, 580
Diesel-hydraulic locomotives, 497
Die stamping, 242
Diet drugs, dangers of, 538
Die (tooling), 241-242, 579
Digital audio broadcasting, 133
Digital editor, using a, 176-178

Digital imaging, 143, 580
Digital library, 116
Digital photography, 158
Digital radio, 133
Digital signal
 for cellular phones, 127
 in computers, 104-106
 for radio, 133
 for television broadcasts, 136-137
Digital television, 136-137
Dimensional lumber, 314-315
Direct sales, 265, 580
Dirigibles, 444
Disabled persons
 artificial body parts for, 561-562, 573
 cars for, 34
 hearing impaired, implants for, 562
 homes designed for, 564
 implants for, 562
 prostheses for, 561-562, 573
 quadriplegics, implants for, 562
 and telecommuting, 121
 vision impaired, and GPS, 461
Disc, compact
 CD-ROM, 102
 musical recording, 17
Disease
 ALS, 545
 and antibodies, 553
 and bacteria, 530-531
 cancer, 115, 540, 559
 cystic fibrosis, 557, 560
 diagnostic testing for, 553-557
 genetic testing for, 540, 556-557, 559
 Huntington's, 545
 immunity to, and exercise, 50
 immunizations against, 553
 Lou Gehrig's, 545
 orphan, 545
 in plants, 552 (illus.)
 preventing, 553, 554 (illus.)
 rare, 545
 risk of, and genetics, 540, 556-557, 559
 skin, and chemical irritants, 394
 vaccines against, 553, 554 (illus.)
 and viruses, 529, 553, 554 (illus.)
Disk, floppy
 for computers, 102
 for digital camera, 158
 and magnetism, 123
Dispatcher, 448
Distribution, chain of, 17, 266, 426 (illus.), 580
Diversification, of companies, 191
DMAPS, 209

DNA
    as court evidence, 42
    defined, 580
    discovery of, 533
    and gene splicing, 547
    recombinant, 547, 559
    in vaccines, 553, 554 (illus.)
Doctors Without Borders, 421
Dolly, the cloned sheep, 20 (illus.), 21
Downlink, 139, 580
Drafter, 162
Drafting, 161-167, 580
Drag, aerodynamic, 504
Drawings
    architectural, 341 (illus.), 342-344
    assembly, 223-224
    in CAD, 166, 167 (illus.), 340 (illus.), 341
    detail, 222 (illus.), 223, 343, 344 (illus.)
    dimensions on, 164
    in drafting, 162-167
    elevations, 343, 344 (illus.)
    exploded view, 166, 223 (illus.)
    floor plans, 161 (illus.), 343
    fractions on, working with, 242
    isometric, 165
    lines on, 164
    multiview, 162-164
    oblique, 165
    orthographic, 162-164
    perspective, 165
    pictorial, 164-166
    plotter to print, 86, 276
    police, of criminals, 109 (illus.)
    in product design, 215-216, 271
    prototype from, 271, 276
    renderings, 216, 340 (illus.), 341
    scale, defined, 341
    schematic, 224
    site plan, 341, 342 (illus.)
    sketches. *See* Sketch
    survey, 334
    technical illustrations, 165-166
    thumbnail sketch, 147, 215
    tolerances indicated on, 222 (illus.), 223
    tools for, 86 (illus.)
    working, 164, 223-224, 232, 341-344
Drugs
    diet, dangers of, 538
    interferon, 559
Drywall, 373
Duplex, 310

E

Ears
    cochlear implants for, 562

    transfer of sound through, 78
Earthmoving, 364, 382, 383 (illus.)
Earthquake, tools to predict, 77
Earth station, 139, 580
Eclipse, predicting, and Stonehenge, 47
Economy
    global, 192
    manufacturing's effects on, 186
Editor, digital, using, 176-178
Edison, Thomas, 35, 212
EDM, 334, 335 (illus.)
Education, computers used in, 88, 92, 113-114
Efficiency
    in manufacturing, 201, 219-222, 580. *Also see* Production planning
    and portable products, 89
    and vehicle design, 504-507
Eggs, for developing photos, 155
Egyptian pyramids, 352
Elasticity, 33
EKG, 554-555
Electret microphone, 126
Electricity
    cars powered by, 496, 499-500
    and hydroelectric energy, 381, 387
    and semiconductors, 106
    and solar energy, 55, 199 (illus.), 499-500
    and superconductors, 200
Electric motors, 496-497, 498
Electrocardiograph, 554-555
Electromagnet
    in telegraph, 124, 580
    in telephone, 125, 126 (illus.)
Electromagnetic compatibility, 95, 580
Electromagnetic interference, 94-95, 580
Electromagnetic radiation, 94, 556, 580
Electromagnetic waves
    and cellular phones, 129-130
    defined, 580
    frequency of, 130, 131 (illus.), 132, 135
    modulation of, 130-131 (illus.)
    radiation from, 94
    and radio transmissions, 132-133
    and satellites, 138 (illus.), 139
    and TV transmissions, 135
Electromagnetism
    and computer's hard disk drive, 102
    and floppy disks, 102, 123
    and maglev propulsion, 501-502
    and telegraphs, 124
    and telephones, 125, 126 (illus.), 131
Electronic distance meter, 334, 335 (illus.)
Electronic mail, 114, 580
Electronic pagination system, 148
Electronics technician, 58

Index **603**

Electron microscope, 28
Electrostatic printing, 153, 154 (illus.), 580
Elevation (drawing), 343, 344 (illus.), 580
Elevations, surveying to determine, 334
Elevator, invention of, 37
E-mail, 114
EMC, 95
Emerging technologies, defined, 20, 580
Emerging viruses, 529
EMI, 94-95
Eminent domain, power of, 337
Empowerment, 201
EMT, 421
End effector, of robot, 279
Energy
   alternative sources of, 199, 499-500
   for communication systems, 87
   in construction systems, 322
   defined, 54, 580
   electrical, 55. *Also see* Electricity
   forms of, 55
   from fossil fuels, 55, 199
   hydroelectric, 381, 387
   for manufacturing, 199-200
   nonrenewable sources of, 55, 199
   radiant, 87. *Also see* Electromagnetic waves
   renewable sources of, 55
   for satellite, 87, 138 (illus.)
   solar, 55, 138 (illus.), 199 (illus.), 499-500
   for transportation, 431, 499-503
   from water, 19, 381, 387
   wind, 499
Engine (software program)
   inference, 114
   search, 99, 113, 170
Engine (vehicle)
   in airplanes, 486-487, 498
   in cars, 483-484, 498
   ceramics in, 505
   computers in, 454-455, 506
   defined, 482, 580
   diesel, 492-493, 498
   diesel-electric, 447, 481, 497, 498
   external-combustion, 493-495
   four-stroke-cycle, 491
   and friction, 483, 487
   horsepower of, 453, 484
   internal-combustion, 482-493
   jet, 486-487
   motion in, types of, 483-487
   parts of, 487-489
   rocket, 486-487
   RPMs of, 484, 486
   in ships, 485, 493, 495, 498
   steam-driven, 493-495, 498
   systems in, 487-489
   train, 447, 485, 495, 497-498
   in trucks, 454-455, 481, 493, 498
   two-stroke-cycle, 491-492
Engineer. *Also see* Engineering
   civil, 400
   in construction, 311, 339, 580
   manufacturing (or industrial), 206, 232-233, 235, 243
   product, 218, 224, 225-227
Engineered fabric, 319
Engineered wood materials, 315-316, 580
Engineering
   careers in, 206, 400
   computer-aided, 209, 218, 272, 274-276
   concurrent, 209, 218
   and ergonomics, 563
   genetic, 547-548, 551
   human factors, 563-564
   product, 209, 218-227, 274-276
   re-, 201
   simultaneous, 209, 218
   tissue, 560
Engineers, Army Corps of, 451
Enigma, and codes, 74
Enterprise resource planning, 277
Entrepreneurs, 42, 580
Envelope, work, 279
Environment
   and air pollution, 121, 204, 205, 436
   artificial, in virtual reality, 117, 275, 341
   and bio-related technology, 527, 541, 548, 551-552
   built, and human factors engineering, 564
   and cars, 204, 205
   and CFCs, 29
   and communication systems, 95
   in community, solving problems with, 62-63
   and composting, 565, 570
   and construction, 323-324
   controlled, for agriculture, 548-551
   controlled, for experiments, 211
   and electric-powered vehicles, 491
   and erosion, preventing, 319
   and genetic engineering, 548
   and hazardous waste, 567
   and landfills, 64, 565-566, 567
   and manufacturing, 204, 205
   microbes to remove pollutants from, 541, 567
   natural resources in, preserving, 84-85, 315, 316
   ozone layer, and CFCs, 29
   and pesticides, 527, 538, 539 (illus.)
   and recycling, 64, 85, 203, 204, 566
   and telecommuting, 121
   and transportation systems, 204, 205, 436

and waste management, 564-567
and water pollution, 371, 565-566
Enzymes, and food storage, 267
Equity, owner's, 252
Ergonomics, 563-564, 580
Erosion, preventing, 319
ERP, 277
Escalator, 476, 581
Estimating paint needs, 404-409
Estimator, cost, 346
Ethics
   and altering photos, 110, 143
   and bio-related technology, 539-540, 581
European Common Market, 192
Evaluation, in problem solving, 37, 39
Eye, human, 156
Excavating, 365-366, 581
Exercise, and job performance, 50
Experiment
   conducting an, 211
   to test hypothesis, 41
Expert system, 114-115
Exploded view drawing, 166, 223 (illus.)
External-combustion engines, 493-495, 581

F

FAA, 444
Fabric, engineered, 319
Facilities
   for air transportation, 441, 445
   for highway transportation, 455
   manufacturing plant, planning, 235-238
   for pipeline transportation, 456-457
   for rail transportation, 448
   for water transportation, 451
Facsimile system, 110 (illus.), 111
Factory
   automated, 286-287
   automatic, 287-288
   distribution of product from 266, 426 (illus.)
   early, 187
   hydroponic, 549-550
   layout of, 236 (illus.), 237
   location of, planning, 235, 236 (illus.)
   and prefabricated construction, 374-375
   in space, 288-289
Failure analysis, 226, 581
Falling Water, 333 (illus.)
Farm
   fish, 550-551
   solar, 199 (illus.)
Farming. *See* Agriculture
Fasteners, standard types of, 302 (illus.)
Fax, 110 (illus.), 111, 581
FCC, 136-137

FEA, 274
Federal Aviation Administration, 444
Feedback
   in bio-related technology, 538-541
   in communication systems, 95
   in construction systems, 324-325
   defined, 57, 581
   as input, 57
   in manufacturing systems, 205
   in transportation systems, 437
Fibers, optical, 128-129, 507
FICA, 194
Film, photographic, 157-158
Financial schedule, 353
Finite element analysis, 274, 581
Fish farm, 536, 537 (illus.), 550-551
Fixed capital, 200, 431
Fixture (for processing), 240, 241 (illus.), 581
Flatcar, container on, 469
Flatcar, trailer on, 467
Fleming, Alexander, 524
Flexible-fuel vehicles, 503
Flexible machining center, 278, 279 (illus.), 287, 581
Flexible manufacturing systems, 287
Flexible roadbeds, 382, 383 (illus.)
Flexography, 151
Floor, installing a, 369, 370 (illus.)
Floor coverings, 307, 373
Floor joist, 367, 370 (illus.)
Floor plan, 161 (illus.), 343, 581
Floppy disk, 102, 123, 158
Fluid-powered motors, 497
Flu virus, and DNA vaccine, 554 (illus.)
"Flying doctor" service, 421
Flywheel, 489
FMC, 278, 279 (illus.), 287
FM radio, 130, 133
FMS, 287
Footing, 366, 581
Forces
   and aerodynamics, 504
   friction, 483, 485, 487, 490
   thrust, 486, 487 (illus.)
   torque, 497
Ford, Henry, 243
Forecast, sales, 263
Forester, career as, 568
Forklift, 238, 239 (illus.)
Formal balance, 145
Forming processes, 202, 203, 240-242, 581
Fossil fuel, 55, 482
Foundation, of structures, 365-366, 581
Four-stroke-cycle engines, 491
Fractions, working with, 242

Framed structures, 367-368
Framing, in photography, 159
Framing members, 367-368
Freight, 581. *Also see* Cargo
Freight trains, 446-447
Frequency, of electromagnetic waves
    modulating, 130, 131 (illus.), 581
    for radio stations, 132-133
    for TV, 135
Fry, Arthur (3M chemist), 35
F-stop, 157
Fuel
    alternative, 500-501
    diesel, 493
    fossil, 55, 482
    recycled, 501
    for rocket engines, 486
Fuel cells, 502-503, 581
Fuel systems, for engines, 488 (illus.), 489
Functional analysis, 225-226
Functional design, 218-219, 581
Fungi, 267, 552

G

Gages (gauges), 256, 581
Galileo, 28
Garbage
    composting, 565, 570-572
    and landfills, 64, 565-566
    recycling, 64-65, 204
Gastroscope, 554, 555 (illus.)
Gate, in IC, 107
Gates, Bill, 231
Gauge, of railroad tracks, 447
Genes
    altering, 533-534, 547
    and breast cancer risks, 540
    and chromosomes, 533, 548
    and cystic fibrosis, 560
    defective, replacing, 560
    defined, 531, 581
    and disease risks, 540, 556-557, 559
    DNA in, 533, 547
    in DNA vaccines, 553, 554 (illus.)
    and genetic engineering, 547
    and Human Genome Project, 533
    and heredity, 531, 533, 557, 559
    Mendel's discoveries about, 531-532
    splicing, 547-548, 581
    testing of, 556-557, 559
    in vaccines, 553, 554 (illus.)
Gene therapy, 560, 581
Genetic engineering, 547-548, 551, 581
Genome, 533, 540, 582
Geotextiles, 319, 320 (illus.), 582

Glass, safety, 320
Gliders, making and testing, 519-523
Global economy, 192, 193
Global market, 191-193
Global Positioning System, 460-461, 582
Glue, in construction, 320
Glulam, 315
GM Sunraycer, 500 (illus.)
Golden Gate Bridge, 331, 385
Go/no go gage, 256
Gorges Dam, 387
Gould, Chester, 73
GPS, 460-461
Graphic artist, 143, 147-148
Graphic communication, 142-169, 582
Graphic designer, 82, 143, 146-148
Graphs, reading, 40
Gravity
    discovery of, 27
    micro-, manufacturing in, 288-289
Gravure printing, 152-153, 582
Great Wall of China, 309
Gridlock, 436, 582
Groundwater, 371, 582
Group technology, 235, 582
Guitars, manufacturing, 247
Gutenberg, Johann, 90, 151

H

Hand
    implants to help use, 562
    myoelectric, 561-562
    on robot, 279
Handcrafted products, 189
Handicapped. *See* Disabled
Hand tools, 198, 321
Hard disk drive, 102
Harmony, in design, 146
Harvesting, defined, 537, 582
Hawkins, Gerald, 47
Hazardous waste, managing, 567, 582
HDTV, 75, 136, 137 (illus.)
Health, worker
    and exercise, 94
    and work-related illness, 394
Health care. *Also see* Health concerns
    artificial body parts, 561-562, 573-575
    and biomaterials, 562
    biopsy, 555-556
    brain disorders, diagnosing, 556
    burns, skin patches for, 560
    cancer treatments, 115, 559
    career in, 542
    CAT scan, 556
    chemotherapy, 115, 559

diagnostic testing, 553-557
disease, treatment of, 115, 559-560
disease prevention, 530-531, 553
EKG, 554-555
expert systems for, 115
gastroscope, 554, 555 (illus.)
gene therapy, 560
genetic testing, 556-557, 559
heart, EKG of activity in, 554-555
heart valves, 562
immunizations, 553
implants, 562
insurance for, and genetic testing, 540
interferon, 559
laboratory testing, 553
laser surgery, 560, 561 (illus.)
medical imaging, 555, 556, 557
and microorganisms, 529, 530-531, 553
and monoclonal antibodies, 559
MRI, 25 (illus.), 557
and NORD, 545
orphan diseases, 545
PET scan, 556
prosthetics, 561-562, 573-575
recombinant DNA, 559
robots for, 280
sonography, 557
stomach ailments, diagnostic testing for, 554, 555 (illus.)
surgery, 540, 555-556, 560-561
technician in, 542
tissue engineering, 560
ultrasound, 557, 561
vaccines, 530, 553, 554 (illus.)
X-rays, 556
Health concerns
AIDS, 421
amyotrophic lateral sclerosis (ALS), 545
asthma, occupational, 394
bacteria, 530-531
cancer, 115, 540, 559
and computer use, 93-94, 212 (illus.), 563
cystic fibrosis, 533, 560
diet drugs, 538
disease, spread of, 421, 529, 530-531, 553
electromagnetic radiation, 94
Lou Gehrig's disease, 545
microorganisms, 421, 529, 530-531, 553
rare (orphan) diseases, 545
seasickness, 437
skin diseases, 394
viruses, 421, 529, 553
work-related illness and injury, 394

Hearing
and human ear, 78
implant to improve, 562
Heart
artificial, first, 525
artificial valves for, 562
electrical activity in, measuring, 554-555
Heart valves, artificial, 562
Helium, in lighter-than-air vehicles, 444
Hepatitis, interferon to treat, 559
Heredity
and disease risk, 557, 559
and genes, 531, 533, 557, 559
Mendel's findings about, 531-532
Hertz, of radio waves, 130 (illus.)
High definition TV, 75, 136, 137 (illus.)
High-speed rail systems, 446, 477, 501, 506
Highways
African, and spread of AIDS, 421
Alaska, 25
congestion on, 436, 460, 461
construction of, 382-384
early, 418
improvement of, designing, 412-414
interstate, 424 (illus.), 455
map of, calculating distances on, 430
numbering of, 477
smart, 461
taxes for, 455
toll, 454 (illus.), 455
traffic congestion on, 436, 460, 461
U.S., map of, 424
Highway transportation, 451-455. Also see Car; Intermodal transportation
Hindenburg, the, 444 (illus.)
HIV, 421
Hoist, 238
Holography, 160, 582
Home, working at, 121
Home page, Internet, developing a, 179-181
Hoover Dam, 381
Hopper cars, 473
Horsepower, 453, 484
Hospital, robots in, 280
Hotel rooms, modular, designing and building, 410-411
House
constructing, 363-375, 378
custom-built, designing, 337-345
framing members of, 367-368
insulating, 372-373
smart, 361
town, 310

HTML, 112, 179-181
Human factors engineering, 563-564, 582
Human resources department, 240
HVAC systems, 370
Hybrids, 536, 546, 582
Hybrid vehicles, 503, 582
Hydraulic power, 219, 497, 582
Hydroelectric energy, 381, 387
Hydrophone, 77
Hydroponics, 549-550, 582
Hypertext markup language, 112, 179-181, 582
Hypothesis, 41, 582

I

IBOC, 133
IC (integrated circuit), 107-108, 126. *Also see* Microchip
Identification, automatic, 282. *Also see* Bar code
Ignition system, of engine, 489
Illness. *See* Disease; Health concerns
Imaging
    digital, 143
    medical, 555, 556, 557
Immune system, and exercise, 50
Immunization, 553, 582
Impacts
    of bio-related technology, 534-535, 538
    of cars, 204
    of CDs and CD players, 56
    of communication technology, 92-95, 114, 121, 138
    of computers, 92, 114, 121
    of construction, 323-324, 331
    environmental. *See* Environment
    of genetic engineering, 548
    of global economy, 193
    of Internet, 114
    of manufacturing, 186, 204, 205
    of satellites, 138
    social, of technology, 19, 20
    of television, 92-93
    of transportation systems, 421, 434-436
Implants, prosthetic, 562
Income, taxes on, 194
Industry
    cottage, 186 (illus.), 187
    modern. *See* Manufacturing
Industrial construction, 310, 582
Industrial engineer, 206
Industrial machinery repairer, 290
Industrial materials, 51-53, 196-197
Industrial Revolution, 14, 187, 582
Industrial truck, 238, 239 (illus.), 422

Inference engine, 114
Informal balance, 145
Information
    on CD-ROM, 92, 93 (illus.), 102
    collecting, and problem solving, 34-35
    in communication systems, 82-83
    in construction systems, 311
    downloading, from Internet, 99
    genetic, and DNA, 533
    genetic, and insurance, 540
    on Internet, 99, 112-114, 170
    in library, electronic access to, 116
    on magnetic media, 102, 123
    in manufacturing systems, 200
    on optical media, 102
    as resource for system (general), 51
    sources of, 34-35
    in transportation systems, 429
    up-to-date, need for, 83
Information superhighway, 99. *Also see* Internet
Injury, work-related, 394
Ink
    in printing processes, 151-152, 154
    transporting, 472
Ink-jet printing, 154, 582
Innovation, award for, 231, 582
Inputs
    in bio-related technology, 535
    in cheese-making, 535
    in communication systems, 80-89
    in construction systems, 311-323
    feedback as, 57
    general discussion, 48-49, 51-56, 582
    in manufacturing systems, 195-201
    and productivity, 193
    in transportation systems, 427-429, 431-432
Input units, for computer, 101
Insect pests
    controlling, 527, 538, 551-552
    damage caused by, 267
Inspection
    careers in, 44
    computers for, 281
    in construction, 332, 356-357, 376
    gages for, 256-257
    in manufacturing, 255-257, 281
    by OSHA, 357
    tools and machines for, 256-257, 281
Insulation, 372-373, 582
Insurance, and genetic information, 540
Integrated circuit, 107-108, 126, 583. *Also see* Microchip
Integrated Pest Management, 527, 551-552, 583

Intelligence, artificial, 114-115, 461
Intelligent Vehicle Highway Systems, 461
Interchangeable parts, 220, 583
Interference, electromagnetic, 94-95
Interferon, 559
Intermittent production, 189, 583
Intermodal transportation, 464-477, 583
Internal-combustion engines, 482-493, 583
International Organization for Standardization (ISO), 253-254, 583
International Space Station, 399
Internet
    business sites on, 88, 114
    chat rooms on, 113-114
    for distance learning, 113-114
    and education, 88, 113-114
    and E-mail, 114
    finding information on, 113, 170-172
    home page for, making a, 179-181
    and hypertext markup language, 112, 179-181
    as information superhighway, 99, 583
    as marketplace, 114, 183
    nodes on, 112
    on-line services for, 99, 112-114
    in schools, 88, 113-114
    search engines for, 99, 113, 170
    web browsers for, 113, 170
    web sites on, 112-113
Interstate highways, 424 (illus.), 455
Interview, and market research, 263
Inventory control, 251, 276, 297-298 (activity), 583
Inventory records, using, 354
IPM, 527, 551-552
Iron, processing into steel, 53
Iron Age, 18, 19
Irrigation canals, 389, 583
Islands of automation, 286
Isometric drawing, 165
ISO standards, 253-254
IVHS, 461

J

Jacobs Field, 329
Jenkins, Charles Francis, 133 (illus.)
Jet engines, 486-487
Jetway, 476
Jig, 240, 583
JIT and JIT II, 185, 282
Job lot production, 189
Job performance, and exercise, 50
Jobs, Steven, 100, 231
Joint, artificial, designing and making, 573-575
Joists, 367, 368 370 (illus.)
Just-in-time scheduling, 185, 282, 583

K

Kasparov, Garry, 273
Kidney stones, ultrasound to destroy, 561
Kitchen, designed for wheelchairs, 564
Kreplin, Terry, 271

L

Laboratory tests, to diagnose disease, 553
Labor unions, 182, 187
Lab technician, career as, 542
Laminated veneer lumber, 315-316
Landfarming, 567
Landfills, 64, 565-566, 567, 587
Landscaping, 376, 377 (illus.), 583
Language
    ADA, 103
    of computers, 103-106, 109
    hypertext markup (HTML), 112, 179-181
    machine, 103-106, 109
    programming, 103
Language arts, and technology, 28
Laptop computer, 73, 583
Laser
    and bar codes, 111
    and CD-ROM, 102
    cutting materials with, 278
    to encode data, 102, 583
    in holography, 160
    for measurement, 281
    and optical fibers, 128
    for printing, 155
    in quality control, 281
    ruby, 336
    in stereo lithography, 271
    surgery with, 560, 561 (illus.)
    and surveying equipment, 334, 335 (illus.)
    workings of, 336
Laser curtain, 281
Laser cutting, 278
Laser printing, 155, 583
Law enforcement, and DNA, 42
Laws
    building codes, 314, 332
    power of eminent domain, 337
    zoning, 332, 334
Laying out the site, 334, 364, 365 (illus.)
Layout
    plant (factory), 236 (illus.), 237
    of printed matter, 143, 147-148
Leachate, 565-566, 583
Leading lines, 160
Learning, distance, 113-114
Lens, camera, 156 (illus.), 157
Letterpress printing, 151

Level stick, for surveying, 334, 335 (illus.)
Liabilities, 252
Library, digital, 116
Life cycle, product, 210
Lighting, in malls, 307
Lift, and aircraft, 443
Light, laser, 336. *Also see* Lasers
Linear motion engines, 486-487, 583
Line graph, reading a, 40
Line production, 189-190
Lines
    in drawings, 164
    leading, 160
Lippershey, Hans, 28
Liquids, transporting, 455-457, 471-472
Lister, Joseph, 530-531
Lithography
    offset, 152
    stereo, 271
Load-bearing wall structure, 369, 583
Lock, of canal, 389, 390 (illus.), 583
Locomotives, train, 447, 495, 497-498
Logos, 144
Lou Gehrig's disease, 545
Lovelace, Augusta Ada, 103
Lubrication system, of engine, 487
Lumber
    engineered, 315-316
    from recycled plastic, 319
    sizes and types of, 302 (illus.), 314-316

M

Machine language, 103-106, 108, 583
Machine operator, career as, 258
Machines
    as capital, 200
    computer-controlled, 154, 272-288
    and human factors engineering, 563
    as resource, 54, 86, 198, 321
    tooling-up, 240-242
Maglev systems, 501-502, 583
MagLifter, 501-502
Magnetic resonance imaging, 25 (illus.), 557
Magnetic storage media, 102, 123
Magnetism. *Also see* Electromagnetism
    and electrostatic printing, 153
    and floppy disks, 102, 123
    and telegraph, 124
    and telephones, 125, 126 (illus.)
Mail
    electronic, 114
    RoadRailer delivery of, 465
Mall, shopping, 307
Management
    in communication systems, 82
    in construction, 350-356, 358
    in manufacturing, 201, 217, 232
    total quality, 201
    in transportation, 432-433
Management teams, 217
Manufacturability, design for, 219-222, 274-275
Manufactured housing, 375
Manufacturer's sales representative, 268
Manufacturing. *Also see* Production
    of airplanes, 26
    automation in, 247, 277-288
    capital for, 200
    careers in, 30, 206, 228, 244, 258, 268, 290
    of CDs, 17
    and competition, 193
    of composite product (activities), 292-296
    computer-aided, 209, 277-280
    computer-integrated, 270-288
    on demand, 188
    development of, 186-188
    and economy, 186
    and efficiency, 201, 222-223
    energy for, 199-200
    and environment, 204, 205
    and feedback, 205
    and global market, 191-193
    of guitars, 247
    impacts of, 204
    inventory control in, 251, 276
    JIT and JIT II, 185, 282
    machines for, 198-199, 277-281, 286-287
    managing, 201, 217, 232
    and marketing, 260-266, 299-300
    materials for, 23, 51-53, 196-197
    in microgravity, 288-289
    of plastic, from plants, 540
    processes in, 201-203, 232-234, 277-281
    and product development, 208-228
    productivity in, 193
    quality assurance in, 253-257, 281
    and recycling, 203, 204
    robots in, 279-280
    in space, 288-289
    as system, 195-205
Manufacturing cells, 286, 287
Manufacturing executive systems, 282
Manufacturing resource planning, 276-277, 584
Manufacturing systems, 184-205
Manufacturing technology, 23, 584. *Also see* Manufacturing
Map
    calculating distances on, 430
    first, 418

of interstate highway system, 424 (illus.)
Marconi, Guglielmo, 132
Market
   creating a, 261
   defined, 262, 584
   European Common, 192
   global, 191-193
   target, 264
   test, 263
Marketing, 93, 250, 260-268, 584
Market research, 263, 264, 584
Market share, 263
Mars, *Sojourner* vehicle on, 460, 507 (illus.)
Masonry, 318
Mass communication, defined, 132, 584
Mass production, 189-190. *Also see* Manufacturing; Production
Material requirements planning, 276, 584
Materials
   adhesives, 35, 320
   alloys, 196-197, 289
   aluminum, 317
   for artificial body parts, 562, 573
   bill of, 232
   bio-, 562
   in cars, 505
   for CDs, 51
   ceramics, 52, 197, 505
   combining, 203
   for communication systems, 84-85
   composite. *See* Composites
   concrete, 318, 382-384
   conditioning, 203
   in construction, 314-320, 382-384
   for contact lenses, 562
   cutting (separating), 202, 203, 278
   delivery of, scheduling, 185, 282, 353, 355
   engineered wood, 315-316
   estimating costs of, 346, 404-409
   fabric, engineered, 319
   for flooring, 307, 373
   forming, 202, 203
   and functional design, 218
   geotextiles, 319, 320 (illus.)
   glass, safety, 320
   handling, in manufacturing, 238-239, 284, 584
   for highway construction, 382-384
   industrial, 51-53, 196-197, 204
   insulation, 372-373
   for joint replacement, 573
   lightweight, 317, 505
   lumber, 314-316
   for manufacturing, 23, 51-53, 196-197
   masonry, 318
   metal. *See* Metals
   natural, 51, 84
   nonrenewable, 51
   optical fibers, 128-129, 507
   paint, estimating need for, 404-407
   planning, needs for, 232, 276
   plastic. *See* Plastic
   processing, 52-53, 202-203
   and product engineering, 218
   for prostheses, 562, 573
   raw, 23, 51-53, 196
   recycling, 64-67, 85, 566
   renewable, 51
   separating (cutting), 202, 203, 278
   silicon, 106, 108 (illus.)
   standard stock, 196, 197 (illus.), 301-302
   steel, 53, 316, 317
   structural, 314-320, 382-384
   synthetic, 51, 52
   in transportation systems, 431
   transporting, as cargo, 425-426, 465-477
   and value analysis, 226-227
   for vehicles, 505-506
   wood, 314-316
Materials delivery schedule, 353, 355
Materials-handling systems, 238-239, 284-285
Materials processing, 202-203, 584
Mathematics
   amperage, calculating, 509
   balance sheet, using a, 252
   board feet, calculating, 303
   costs of workplace illness, 394
   distance, on map, calculating, 430
   electric current, calculating amperage of, 509
   fractions, working with, 242
   graphs, reading, 40
   horsepower, computing, 453
   importance of, in technology, 28
   inventory records, using, 354
   map, calculating distances on, 430
   mileage, calculating, 430
   paycheck deductions, understanding, 194
   speed, calculating, 475
McDonald's Happy Meal® promotion, 262
Measles, 552
Measurement
   and accuracy, need for, 28
   coordinate measuring machine for, 281
   laser curtain for, 281
   tolerances for, 220, 254-255, 256
   tools for, 54, 150, 256
   of type, 150
Mechanic, aircraft, 508
Mechanical fastening, 203, 302 (illus.)

Mechanical system, of engine, 487, 488 (illus.)
Médecins Sans Frontières, 421
Media
   defined, 144
   graphic, 143-168
   magnetic storage, 102, 123
   multimedia ad, creating a, 299-300
   optical storage, 102
Medical equipment
   diagnostic, 25 (illus.), 554-555, 556, 557
   and electromagnetic interference, 94-95
   pacemakers, 95
Medical imaging, 555, 556, 557
Medical technician, 542
Medicine, field of. *See* Health care
Memory, computer, 102
Mendel, Gregor, 531-532
MES, 282
Metals
   alloys of, 196-197, 289
   aluminum, 317
   in construction, 316-317
   iron, processing into steel, 53
   lightweight and ultrastrong, 317
   and oxidation, 267
   standard shapes of, 197 (illus.)
   steel, 53, 316, 317
Methane gas, 501, 566
Methanol, 501
Metrodome, 396
Micklus, Sam, 33
Microbes. *See* Microorganisms
Microchip
   development of, 214
   general discussion, 107-108
   and nanotechnology, 277
   semiconductors in, 106
Microgravity, 288-289, 584
Microorganisms
   and antibodies, 553
   bacteria. *See* Bacteria
   corn smut, 552
   discovery of, by Pasteur, 530
   and food spoilage, 267
   fungi, 267, 552
   and infected wounds, 530-531
   and pasteurization, 530
   and plant diseases, 552
   sterilization against, 531
   vaccines against, 530, 553, 554 (illus.)
   viruses, 421, 524, 529, 553
   and waste management, 541, 565, 567
Microphone
   as amplifier, 105
   in telephone, 125, 126
Microprocessor
   first, 15
   integrated circuit in, 107-108
   role of, in computer, 101-102
   workings of, 108-109
Microscope, electron, 28
Microwaves, 129, 584
Mileage, determining, from map, 430
Military aircraft, 442, 443 (illus.)
Milk
   pasteurization of, 530
   transporting, 472
Mills, grain, 19
Mind, Odyssey of the, 33
Miniload, 285
Mining, vehicles for, 472-473, 481
Mir space station, 398
Mock-up, 216, 584
Model
   computer, 166, 167 (illus.), 209
   mock-up, 216
   plastic, and stereo lithography, 271, 276
   prototype, 225-226
   rapid prototype, 271, 276
   scale, 216, 217 (illus.)
   wireframe, 166, 167 (illus.)
Modeling, with computer, 166-167, 209, 216
Modem
   defined, 73, 584
   and home offices, 121
   and Internet, 99, 113
   for laptops, 73
Modular construction
   general discussion, 374-375, 584
   of hotel rooms, designing and building, 410-411
Modular design, in manufacturing, 221-222, 584
Modulation, of waves, 130, 131 (illus.)
Module, 374, 584
Mold (for shaping), 240-241, 584
Mold (organism), and food spoilage, 267, 584
Monitor, video, for computer, 103, 584
Morse, Samuel, 123-125
Morse Code, 124
Motion (in engines), types of, 483-487
Motors, 482, 495-497, 584
Mouse, for computer, 101
Movable bridge, 385, 386 (illus.), 584
Movies
   advertising in, 264
   frame movement in, 90
MRI, 25 (illus.), 557
MRP, 276, 282
MRP II, 276-277, 282

Index

Multimedia ad, developing a, 299-300
Multiview drawings, 162-164, 584
Music, recording, 17
Myoelectric hand, 561

N

NAFTA, 192
Nails, types of, 302 (illus.)
Nanotechnology, 277
NASA's space program
    booster rockets, 458, 459 (illus.)
    founding of, 458
    International Space Station, 399
    MagLifter, 501-502
    Skylab, 397, 398 (illus.)
    Sojourner, on Mars, 434 (illus.), 460, 507 (illus.)
    space shuttles, 458-459
National Medal of Technology, 231
National Organization for Rare Disorders, 545
Natural materials, 51
Natural resources, 84, 85, 95, 315
Navigation, computerized, 460-461
Navigation canals, 389, 390 (illus.)
NAWIC, 349
Needs, and building design, 337-338, 339
Network, cell phone, designing a, 173-175
Network, computer, 584.
    Also see Internet
    in manufacturing, 283-284
Newspapers
    and expert systems, 115
    and satellite transmissions, 138
Newton, Isaac, 27
Nitrogen, liquid, in cryosurgery, 561
NMT award, 231
Nominal size of lumber, 314
Nonrenewable energy sources, 55, 199
NORD, 545
Nuclear waste, disposing of, 567
Numerical control, 277-278, 584

O

Oblique drawing, 165
Obsolescence, planned, 224
Occupational Safety and Health Administration, 357, 584
Odyssey of the Mind, 33
OEM, 265
Office, home, 121
Offset lithography, 152
Ohm's Law, 509
Oil
    as engine lubricant, 487
    as nonrenewable energy source, 199
    transporting, 455-457, 472

Oil spills, microbes to clean up, 541
On-line services, 99, 112-114
On-site transportation, 422, 472-473, 476, 584.
    Also see AGVs
Optical fibers, 128, 507, 585
Optical storage media, 102. *Also see* CDs, CD-ROM
Oriented-strand board, 315
Original equipment manufacturer, 265, 585
Oriole Park, 329
Orphan diseases, 545
Orthographic projection, 162-164
OSHA, 357
Otis, Elisha, 37
Outlet stores, 265
Outlet works, of dam, 387, 388 (illus.)
Outputs
    of bio-related technology, 538
    of communication systems, 91-95
    of construction systems, 323-324
    general discussion, 56
    of manufacturing systems, 203-204
    and productivity, 193
    of transportation systems, 434-436
Output units, computer, 103
Oxidation, effects of, 267
Ozone layer, 29

P

Pacemakers, 95
Packaging, 249-250
Paddlewheel boats, 451
Pager, 73, 91, 585
Pagination system, electronic, 148
Paint, estimating needs of, 404-409
Painting machine operator, 258
Panama Canal, 24, 49
Panelized construction, 374, 375 (illus.), 585
Part print analysis, 232, 233 (illus.), 585
Parts, manufactured
    assemblies of, 248-249
    detail and assembly drawings of, 222 (illus.), 223-224
    interchangeable, 220
    make-or-buy decisions about, 233
    and modular design, 221-222
    quality standards for, 253, 254-257
    similar, and group technology, 235
    specifications for, 255-256
    standardization of, 221
    tolerances for, 220, 254, 256
Parthenon, 344
Passenger service
    airline, 442, 477

intermodal, 475-477
　rail, 446, 506
　ships and boats, 451
Passenger trains, 446, 506
Paste-up, 147
Pasteur, Louis, 530
Pasteurization, 530
Patents, 125, 212, 213
Pathfinder, 460
Pavement, highway, 382-384
Paving, 376
Paycheck, deductions on, 194
Payload, defined, 458, 585
PCS, 127
Penicillin
　discovery of, 524
　to promote growth, 536
People
　in bio-related technology, 535, 542, 568
　in communication systems, 81-82, 96, 118, 140, 168
　in construction systems, 311, 326, 346, 349, 358, 378, 400
　general discussion, 48-49
　intermodal transportation of, 475-477
　in manufacturing systems, 30, 195-196, 206, 228, 244, 258, 268, 290
　in transportation systems, 427-428, 438, 462, 478, 508
Personal communications services, 127, 585
Perspective drawing, 165
Pesticides
　alternatives to, 527, 551-552
　dangers of, 527, 538, 551
　DDT, 538, 539 (illus.)
　defined, 585
Pests, insect, controlling, 267, 527, 538, 551-552
Petroleum. *Also see* Oil
　in plastics, 53, 84, 313
　as waste, landfarming, 567
Petronas Towers, 305, 357
PET scan, 556
Phone. *See* Telephone
Phosphors, 135, 136 (illus.), 585
Photocopier, 144, 153, 154 (illus.)
Photograph
　altering, with computer, 109-110, 143, 148
　composing a, 158-160
　digital, 158
　first, 70
Photographic composition, 158-160

Photographic film, 157-158
Photographic printing, 153, 585
Photography, 155-160
Photopolymerization, 271
Photovoltaic cells, 199 (illus.), 499-500, 585
Pica, 150, 585
Pickup tube, 134, 135 (illus.)
Pictorial drawings, 164-166, 585
Picture, altering with computer, 109-110, 143, 148
Picture tube, 136
Pie chart, reading a, 40
Piers, in bridge construction, 384 (illus.), 385, 585
Pig, pipeline, 457, 458 (illus.)
Piggyback service, 467, 585
Piles (bridge supports), 384 (illus.), 385
Pilot run, 243, 585
Pipelines
　cleaning, with "pigs," 457, 458 (illus.)
　construction of, 392-393
　facilities for, 456-457
　and intermodal transportation, 472
　power for, 497, 498
　pumping stations for, 456-457, 497
　"rolling," 471
　types of, 456
Piston, 483, 491-492
Pixels, 134, 135, 143, 585
Plan
　floor, 161 (illus.), 343
　landscape, 376, 377 (illus.)
　sampling, 257
　site, 341, 342 (illus.)
Planned obsolescence, 224, 585
Planner
　city, 330-331
　layout, 236 (illus.), 237
　process, 232, 276
　production, 240
Planning
　community, 330-331
　computerized, 272, 274-277
　construction, 328-346
　enterprise resource (ERP), 277
　layout, plant, 236 (illus.), 237
　manufacturing resource (MRP II), 276-277
　material requirements (MRP), 276
　process, 232-235, 276
　and product engineering, 218-223
　production, 230-243, 276-277
　and schedules. *See* Schedule; Scheduling
　in transportation, 432

Planning commissions, 331
Planographic printing, 152, 585
Plant layout, 237, 585
Plants
   cloning, 548, 549 (illus.)
   diseases in, types of, 552
   first, 538
   genetically engineered, 547-548
   harvesting, 537
   hybrid, 536, 546
   hydroponically grown, 549-550
   Mendel's experiments with, 531-532
   plastics grown from, 540
   reproduction of, 533, 536, 546, 548, 549 (illus.)
   and selective breeding, 531, 546-547
   strawberries, growing in wastewater, 550
   tissue culture of, 548, 549 (illus.)
   transgenic, 547
   wastewater-grown, 550
   water-grown, 549-550
Plasmids, 554 (illus.)
Plastics
   for artificial body parts, 561, 562, 573
   chemicals in, 52, 53, 313
   for communication devices, 84
   as construction materials, 319
   dangers of, 313
   for geotextiles, 319
   grown from plants, 540
   invention of, 215
   in joint replacements, 573
   manufacture of, 53, 313
   and petroleum, 53, 84, 313
   processes used to make, 53
   for prostheses, 561, 562, 573
   for prototyping, rapid, 271
   recycled, lumber from, 319
   recycling, 204, 269
Plates, in building frame, 367
Plotter, 86, 276
Plywood, 315
Point size of type, 150
Pollution
   air, 121, 204, 205, 436
   and cars, 204, 205
   and electric-powered vehicles, 496
   and landfills, 565-566, 567
   and manufacturing, 205
   microorganisms to clean up, 541, 567
   oil spills, 541 (illus.)
   from vehicles, 204, 205, 436
   water, 371, 565-566
Porous printing processes, 151, 152 (illus.), 585
Portable power tools, 198
Post-it® notes, 35
Postprocessing, 203
Power. *Also see* Engine
   alternative sources of, 499-503
   defined, 482
   effects of design on, 504-507
   hydraulic, 219, 497
   solar, 499-500
   source of, and product engineering, 219
   in transportation, 431, 480-507
   wind, 418, 499
Power of eminent domain, 337, 585
Power tools, 198, 321
Prefabricated structures, 374-375, 397, 585
Preprocessing, 202
Pricing, competitive, 435
Primary processes 52, 53, 585
Prime mover, defined, 482
Principles of design, 145-146, 585
Printer
   ink-jet, 154
   laser, 155
Printing on demand, 155
Printing press, and spread of technology, 19-20
Printing processes
   advances in, 90-91
   types of, 149-155
Privacy, right to, and computers, 93
Private sector, defined, 330, 585
Problem-solving process, 34-39, 41
Process chart, 233-234, 237, 586
Processes
   in bio-related technology, 536-538
   in communication systems, 90-91
   in construction. *See* Construction
   design. *See* Design process
   forming, 202, 203, 240-242
   improvement of, in production, 254-255
   for industrial materials, creating, 52, 53
   laying out the site, 335, 364
   in manufacturing systems, 201-203
   primary, 52, 53
   printing, 90-91, 149-155
   problem-solving, 33-39, 41
   production, 201-203, 233-234, 277-281
   scientific method, 41

Index

secondary, 52, 53, 587
  in technological systems, 56, 586
  in transportation systems, 432-434
Processor, digital image, 143
Product
  advertising a, 264, 299-300
  analyzing design of, 224-227, 274-275
  cars. *See* Cars
  CDs, 17
  of composite material, developing and testing, 292-296
  custom-made, 189
  designing a, 215-223, 272, 274-275
  developing a, 208-228, 271-272, 274-277
  distribution of, 17, 266, 426 (illus.)
  engineering of, 209, 218-227, 274-277
  functional design of, 218-219
  handcrafted, 189
  innovative, award for, 231
  life cycle of, 210
  marketing a, 260-268, 299-300
  mechanical elements of, 219
  models of. *See* Model
  packaging for, 249-250
  patents on, 125, 212, 213
  planned obsolescence of, 224
  prototypes of, 225-226, 271, 274
  quality of, assuring, 253-257
  reliability of, testing, 224-227
  seasonal, 189
  selling, 264-265
  structural elements of, 219
  testing and analyzing, 224-227
  test market for, 263
Product development, 208-228
Production control, 250-251
Product engineering, 218-227, 274-277, 586
Production. *Also see* Manufacturing
  computerized, 277-288
  continuous, 189-190
  controlling, 250-251, 586
  custom, 189
  defined, 248, 586
  flow of, and plant layout, 236 (illus.), 237
  intermittent, 189
  job-lot, 189
  line, 189-190
  mass, 189-190
  on-demand, 188
  pilot run for, 243
  planning, 230-243, 276-277
  processes in, 201-203, 233-234, 277-281
  quality assurance in, 253-257, 281

  schedule for, 250
  of seasonal products, 189
  tooling up for, 240-242
Production processes, 202-203, 232-234, 277-281
Productivity
  and competition, 193
  and cost control, 193
  and industrial engineering, 206
  and time, 56, 586
  of workers, and health habits, 50
Product research 211, 212-215
Profit, 193, 586
Program, computer
  architectural, 338, 341, 342 (illus.)
  and artificial intelligence, 114-115
  CAD. *See* CAD
  CAD/CAE, 272, 274, 276
  CAE, 218, 272, 274, 276
  CAPP, 276
  CATIA, 209
  on CD-ROM, 93 (illus.), 102
  defined, 102, 586
  for desktop publishing, 147-148
  DFA/DFM, 274
  DMAPS, 209
  ERP, 277
  expert systems, 114-115
  on floppy disk, 102, 123
  language for, 103
  MRP, 276
  MRP II, 276-277
  for production planning, 276-277
  for rapid prototyping, 271, 276
  search engines, 99, 113, 170
  virtual reality, 117, 275, 341, 342 (illus.)
  web browser, 113
  writing, career in, 118
Programmable controller, 282, 283 (illus.), 586
Programmer, computer, career as, 118
Programming language, 103, 586
Propagation, 536. *Also see* Cloning, 586
Proportion, in design, 146
Propulsion system
  elasticity as, 33
  in jet engines, 486, 487 (illus.)
  for maglev vehicles, 501-502
  for rocket engines, 486
  for solar-powered vehicle, 515-516
Prosthetics, 561-562, 573-574, 586
Prototype, 225-226, 586
Prototyping, rapid, 271, 276
Public sector, defined, 330
Public transportation, 418, 435, 446,

Public works construction, 311, 330, 586. *Also see specific types* (Bridges; Dams; *etc.*)
Publishing, desktop, 147, 148 (illus.)
Pumping station, for pipelines, 456-457, 586
Punch and die, 241
Purchasing agent, 251, 353, 355, 586
Pyramids, Egyptian, 14, 352

Q

Quadriplegics, implants for, 562
Quality
    and competition, 193
    in construction, 355-356
    ISO 9000 standards for, 253
    loss of, during storage, 267
    in manufacturing, 253-257, 281
    QS 9000 standards for, 254
Quality assurance, 253-257, 281, 585
Quality control, 253-257, 281
Questionnaire, sample, 61
QS 9000, 254

R

R&D groups, 212-215
Radiation, electromagnetic, 94
Radio
    AM and FM, 130
    digital, 133
    miniaturizing of components in, 214
    workings of, 130, 132-133
Radiography, 556
Radio telescope, 139
Radio waves. *Also see* Electromagnetic waves
    amplitude of, 130
    frequency of, 130, 132-133
    modulating, 130, 131 (illus.)
    telescopes that pick up, 139
Rafters, 367 (illus.), 368, 586
Railroad, Transcontinental, 418
Railroad cars, 447
Rail transportation
    advantages of, 445
    AMTRAK, 446, 506
    automated guided vehicles, 284, 285 (illus.)
    in Chunnel, 391, 392 (illus.)
    commuter service, 446
    container on flatcar, 469
    electric-powered, 496
    facilities for, 448
    gauge in, 447
    high-speed, 446, 477, 501, 506
    in intermodal transportation, 465, 467-469, 473, 475, 477
    locomotives for, 447, 495, 497-498
    maglev, 501-502
    passenger, 446, 496, 501, 506
    piggyback, 467
    power for, 495, 496, 497, 498, 501
    RoadRailer, 465
    rolling stock, 447
    trailer on flatcar, 467
    unit trains, 447, 473
RAM, 102
Random Access Memory, 102
Rapid prototyping, 271, 276, 586
Raw materials, 23, 51-53, 196
Read-Only Memory, 102
Reality, virtual. *See* Virtual reality
Reciprocating motion, 483-484, 586
Recombinant DNA, 547, 559, 586
Recording music, 17
Recycled fuel, 501
Recycling
    aluminum cans, 64-67
    cars and car parts, 204, 205 (illus.)
    and design, 204
    fuel from, 501
    and landfills, reducing waste at, 566
    and natural resources, preserving, 85
    paper, 566
    plastics, 204, 269, 319
    as postprocessing, 203
    school programs for, 64-67
    soda bottles, 204
Reengineering, 201
Relief printing processes, 151, 586
Renderings
    of product, 216
    of structure, 340 (illus.), 341
Repairer, computer, 140
Repairer, industrial machinery, 290
Research
    market, 263, 264
    product, 211, 212-215
Reservoirs, 381, 387, 388 (illus.)
Residential construction
    custom-built, designing, 337-345
    types of, 310
Resin, and plastics, 313
Resources
    natural, 84, 85, 95, 315
    for systems, 48, 586. *Also see* Inputs

Retailers, defined, 265
Revolution, Industrial, 14, 187
Ridge, 368, 586
Roadbeds, 382-384
Road
    highways. *See* Highways
    Roman, 466
Road map, calculating distances on, 430
Robots, 183, 279-280, 586
Rocket
    booster, 458, 459-460
    first, 14
    power for, 486-487
Rocket engine, 486-487
Rolling stock, 447, 587
ROM, 102
Roman roads, 466
Roof construction, 367 (illus.), 368, 369
Roof truss, 368, 587
Rotary motion, 484-486, 587
Rotor, 485
Rough-in, 370, 587
Routes, transportation, 423, 587. *Also see* Highways
Row house, 310
RPMs, 483, 486
Ruby laser, 336. *Also see* Lasers
Rule, typographer's, 150
Rule of thirds, 159
Russian *Mir* space station, 398
R-value, 372

S

Safety
    and air bags, 37, 38
    careers related to, 44
    of car passengers, 37, 38
    in construction, inspecting for, 357
    of elevators, 37
    inspecting for, 44, 357
    workplace, and OSHA, 357
Safety glass, 320
Safety inspector, 44
Salary, tax deductions from, 194
Sales
    careers in, 268
    forecasting, 263
    types of, 264-265
Sales forecast, 263, 587
Sales representative, career as, 268
Sampling, acceptance, 257
Sanitary landfills, 565-566, 587
Satellites
    for communication, 137-139
    for navigation, 460-461
Scale, on map, 430
Scale drawings, defined, 341, 587
Scale model, 216, 217 (illus.)
Scan
    CAT (or CT), 556
    PET, 556
Scanner
    to digitize books, 116
    to digitize photos, 148
    laser, and bar codes, 111
Schedule. *Also see* Scheduling
    defined, 351, 587
    financial, 353
    materials delivery, 353
    production, 250
    work, 352
Scheduling. *Also see* Schedule
    in construction, 323, 351-353
    just-in-time, 185, 282
    in manufacturing, 185, 250, 276, 282
Schematic drawing, 224
Schools, computers in, 88, 92, 113-114
Science connections. *Also see* Bio-related technology
    aerodynamics, 504, 519-523
    cloning, 20, 21, 540, 548, 549 (illus.)
    DNA. *See* DNA
    emerging viruses, 529
    exercise and job performance, 50
    experiment, designing a, 211
    food spoilage, causes of, 267
    friction, 490
    general discussion, 27-28
    genetics. *See* Genes
    groundwater, 371
    integrated pest management, 527, 551-552
    lasers, workings of, 336
    magnetism, and data storage, 102, 123
    microorganisms. *See* Microorganisms
    oxidation, effects of, 267
    plastics, chemistry of, 313
    scientific method, 41
    scientific notation, 558
    semiconductors, 106
    viruses, emerging, 529
Scientific method, 41, 587
Scientific notation, 558
Screen printing, 17, 151, 152 (illus.)
Screws, types of, 302 (illus.)
Seagirt Marine Terminal, 471
Sea-lanes, defined, 451, 587

Search engine, 99, 113
Sears Tower, 357, 367
Sea sickness, 437
Seat belts, 37, 38
Seismometers, 77
Selective breeding, 546-547
Semiconductors, 106
Sensors
   in myoelectric hand, 561
   and traffic control, 78-79, 461, 587
Separating processes, 202, 203, 587
Services
   commuter, 446
   on-line, 99, 112, 113
   personal communications, 127
Seven wonders of the world, 49
Sewage, treatment of, 566-567, 587
Shape, and aerodynamics, 504
Share
   market, 263
   of stock, 42, 200
Sheathing, 367 (illus.), 369, 587
Sheep, cloning, 20 (illus.) 21
Shipping, containerized, 468-471
Ships
   and containerized shipping, 469
   power for, 493, 495, 498, 499
   types of, 450-451, 469, 499
   wind-powered, 499
Shutter, on camera, 157
Shuttles, space, 458-459
Signals
   amplified, 105 (illus.)
   analog. *See* Analog signal
   digital. *See* Digital signal
   radio, 132-133
   and satellite transmissions, 137-139
   television, 134-135
Silicon, 106, 108 (illus.)
Silk-screen printing, 151
Simulation
   of car accidents, 38
   for product analysis, 38, 227, 274, 275, 587
   3D, with virtual reality, 117, 275, 341, 342 (illus.)
Simultaneous engineering, 209, 218
Single unit trucks, 453, 587
Site, construction
   acquiring the, 337
   clearing the, 363-364, 382
   information needed about, 313, 333-334, 587
   laying out the, 334, 364, 365 (illus.)
   selecting a, 332-335
   survey (drawing) of, 334
   surveying the, 334-335, 382
Site plan, 341, 342 (illus.), 364, 587
Sketch
   in architectural design, 338
   of criminal, with computer, 109 (illus.)
   in drafting, 162
   of graphic design, 147
   of product design, 215-216
   rough, 162, 216
   thumbnail, 147, 215
Skilled trades workers, 311
Skills, thinking, 36-37
Skin disease, work-related, 394
Skin patches, 560
Skylab, 397, 398 (illus.)
Skyscraper
   and elevator, invention of, 37
   first, 305
   materials used for, 317
Sludge, 567, 587
Slurry, 456, 587
Smart buildings, 564, 587
Smart cars, 461
Smart highways, 461
Snowboards, marketing, 261
Social Security tax, 194
Social studies, and technology, 29
Society, impacts of technology, on, 19-20, 29. *Also see* Impacts
Software programs. *See* Program
Soil
   compacting, 366, 382
   composted, creating and selling, 570-572
   excavating, 365-366
   growing plants without, 549-550
   leveling, 364
   tunnels in, 391
Sojourner, 434 (illus.), 460, 507 (illus.)
Solar energy
   for heating, 55
   for satellites, (138 (illus.)
   for *Sojourner* on Mars, 507 (illus.)
   for solar farm, 199 (illus.)
   for vehicles, 499-500, 507 (illus.), 514-518
Solar-powered vehicle
   designing and building a, 514-518
   types of, 499-500
Sonography, 557
Sound waves
   amplitude of, 130, 131 (illus.)
   diagnostic testing with, 557

frequency of, 130, 131 (illus.)
and human ear, 78
kidney stones destroyed with, 561
modulating, 130, 131 (illus.)
in sonography, 557
and telephone, 125, 126, 128, 130
Space
  construction in, 397-399
  manufacturing in, 288-289
Spacecraft, 397-399, 458-460, 507 (illus.)
Space shuttles, 458-459, 587
Space station
  general discussion, 397-399
    model of, designing and building, 415-417
Span, of bridge, 325, 385, 587
Spark plug, 489, 491, 492
SPC, 254-255
Specifications
  in architecture, 345, 587
  in manufacturing, 255-257, 587
Speech recognition system, 115, 282
Speed
  average, calculating, 475
  engine, and part sizes, 483-484
  of train, early fears about, 501
Spillway, of dam, 387, 388 (illus.)
Splicing, gene, 547-548
Sports stadiums, 329
Sputnik I, 419
SQC, 281
Stamping, die, 242
Standardization, International Organization for, 253-254
Standardization, of parts, 221, 588
Standard stock
  defined, 196, 301, 588
  fasteners, mechanical, 302 (illus.)
  industrial materials, 196, 197 (illus.)
  lumber, 302 (illus.), 314-315
  metal, 197 (illus.)
  selecting and buying, 301-303
Standards, quality, 253-254
Stanley, J. K., 418
Starting system, of engine, 489
Statistical process control, 254-255, 588
Statistical quality control, 281, 588
Statistics, 254, 588
Steam-driven engines, 493-495
Steel
  framework of, for structures, 368
  processing from iron, 53
  as structural material, 316, 317
Stencil, for printing, 151, 152 (illus.)

Stereo lithography, 271
Stethoscope, invention of, 553
Sterilization, to destroy germs, 531
Stock, company, shares of, 42, 200
Stock (materials), standard. *See* Standard stock
Stone Age, 18, 19 (illus.)
Stonehenge, 47
Storage, and product quality, 267
Strauss, Joseph, 331
Strawberries, growing in wastewater, 550
Structural elements, 219
Structural materials, 314-320, 588
Structures. *Also see* Buildings; Construction
  air-supported, 395-396
  bridges. *See* Bridge
  Chunnel, the, 391, 392 (illus.)
  clear-span, 317
  demolishing, 362 (illus.), 363
  foundation of, 365-366
  framed, 367-368
  Hoover Dam, 381
  load-bearing wall, 369
  Panama Canal, 24, 49
  Parthenon, 344
  prefabricated, 374-375, 397
  pyramids, 14, 352
  Sears Tower, 357, 367
  Stonehenge, 47
  super-, 366-369, 385-386
  tallest, 357
  tepees, 356
  Windsor Castle, 376
Studs, 367
Subassembly, 248, 588
Subcontractors, 350, 588
Subfloor, 369, 370 (illus.)
Subsidiaries, 191
Substructure, of bridges, 384-385
Sunraycer, 500 (illus.)
Superconductors, 200, 588
Superhighway, information, 99
Superstructure
  bridge, types of, 385-386
  of buildings, 365, 366-370, 372-375, 588
Supervisor, construction, 358
Surgery
  breast removal, 540
  cryo-, 561
  diagnostic, 555-556
  implants, 562
  laser, 560, 561 (illus.)
Survey, occupational, conducting, 60-61, 588
Survey drawing of site, 334

Surveying, 334-335, 382
Suspension bridge, 325, 385, 386 (illus.), 588
Switching system, for telephones, 131
Switch yards, 448
Symbols, on drawings, 164
Synthetic materials, 51, 52
Systems
    bio-related, 526-541
    communication, 79-95
    construction, 306-326
    defined, 48, 588
    general discussion, 46-57
    manufacturing, 184-205
    transportation, 420-437

T

Tacoma Narrows Bridge, 324-325
Tank trucks, 472
Target market, 264
Taxes
    deductions for, from paycheck, 194
    for highways, 455
    for public works construction, 311, 330, 455
T cells, 554 (illus.)
Technical illustrations, 165-166
Technical writer, 168
Technician
    chemical, 30
    civil engineering, 400
    clinical laboratory, 542
    electronics, 58
Technology. *Also see individual areas, e.g.,*
        Communication
    areas of, 22-25
    assessment of, 29, 588
    combined, 25-27
    definition of, 17, 18, 588
    emerging, defined, 20
    and environment. *See* Environment
    group, 235
    history of, 18-19
    impacts of. *See* Impacts
    and language arts, 28
    and mathematics, 28. *Also see* Mathematics
        Connections
    nano-, 277
    physical, 22
    and science, 27-28. *Also see* Science
        Connections
    and social studies, 29
    spread of, 19-20
Technology for All Americans, 18
Technology Student Association, 43

Teenie Beanie Babies®, 262
Telecommunication, 120-141, 588
Telecommuting, 121
Telegraph
    electronic, 123-125
    transatlantic, 70
    visual, 122
    wireless, 132
Telemarketing, 93, 588
Telephone
    cellular. *See* Cellular telephone
    cordless, first, 71
    and fax systems, 111
    impacts of, 20, 29
    integrated circuits in, 107, 126
    switching systems for, 131
    as system, 92
    and telemarketing, 93
    transmission channels for, 128-130
    wrist, in comic strip, 73
    workings of, 125-126, 128
Telepresence, 117
Telescope
    development of, 28
    radio, 139
Television
    cable, 134, 135
    commercials on, 264
    development of, 133-134
    digital, 136-137
    high definition, 136, 137 (illus.)
    impacts of, 92-93, 134
    miniature, 74 (illus.)
    satellite transmissions for, 138
    violence on, 92
Television camera, 134, 135 (illus.)
Tepees, 356
Terminals
    air, 444 (illus.), 445
    rail, 448
    Seagirt Marine, 471
Testing
    and acceptance sampling, 257
    of blood, 553
    burn-in, 257
    of composite product, 295-296
    crash, 38
    destructive, 38, 225, 227
    diagnostic, 553-557, 559
    with experiment, 41, 211
    genetic, 556-557, 559
    laboratory, 553
    in product engineering, 224-227, 274

and simulation, 38, 274
  of solution, in problem solving, 37-38
Test marketing, 263, 588
Therapy, gene, 560
Thinking skills, types of, 36-37
Thirds, rule of, 159
3D simulations, 117, 275, 341
Thrust, 486, 487 (illus.), 588
Thumbnail sketch, 147, 215, 588
Time
  and productivity, 56
  as resource 56, 88-89, 201, 323, 431-432
  and scheduling. *See* Schedule; Scheduling
Time and place utility, 427, 588
Tissue culture, plant, 548, 549 (illus.)
Tissue engineering, 560
TOFC, 467
Tolerance, 220, 222 (illus.), 223, 254, 256, 588
Toll roads, 454 (illus.), 455
Tomato, genetically engineered, 547
Tomography, 556
Ton mile, defined 449, 588
Tooling up, 240-241, 588
Tools
  agricultural, early, 182, 524
  in construction, 321-322
  dies, 241
  for drawing, 86
  early development of, 18-19
  as fixed capital, 200
  fixtures, 240, 241 (illus.)
  gages, for inspecting parts, 256
  hand, 198, 321
  jig, 240
  for measuring type, 150
  portable power, 198, 321
  as resource, 54, 86, 198, 321
  surveyor's, 334-335
  typographer's rule, 150
Topography, 333, 382, 588
Torque, defined, 497, 588
Total quality management, 201, 588
Towboat, 450
Tower, Sears, 357, 367
Town house, 310
Track, train, 447, 473. *Also see* Rail transportation
Tractor-semi-trailer rig
  and containerized shipping, 468-469
  general discussion, 452 (illus.), 454, 588
    piggyback service, 467
    and RoadRailer, 467
Tracy, Dick, 73
Trade agreements, 192
Trade barriers, reducing, 192
Traffic congestion, 436, 460, 461

Traffic signals, sensors for, 78-79
Trailer on flatcar, 467, 589
Trailer units, 452 (illus.), 454
Train. *See* Rail transportation
Training, employee, career in, 244
Transceiver, 127, 589
Transcontinental Railroad, 418
Transferring ownership, 376
Transistors, 107, 214
Transit (surveyor's tool), 335 (illus.)
Transmission channels. *Also see* Electromagnetic
    waves; IC
  radio waves, 130, 131 (illus.)
  for telephone, 128-130
  for television, 134-135
Transmitter, in telephone, 125, 126 (illus.)
Transportation. *Also see* Vehicles
  air, 442-445, 469
  and business growth, 435
  careers in, 428, 438, 462, 478
  of cargo. *See* Cargo
  and chain of distribution, 266, 426 (illus.)
  containerized, 468-471
  defined, 589
  and environment, 204, 205, 436
  facilities for. *See* Facilities
  highway, 451-455. *Also see* Cars; Trucks
  impacts of, 421, 434-436
  intermodal, 464-477
  interstate and intrastate, defined, 451-452
  of liquids, 455-457, 471-472
  and lifesaving, 421
  mass, 496
  modes of, 440-460
  on-site, 422, 472-473, 476
  piggyback, 467
  pipeline, 455-457, 458 (illus.)
  power for, 431, 480-503
  of product to consumer, 266, 426 (illus.)
  rail. *See* Rail transportation
  space, 397-399, 458-460, 507 (illus.)
  system of, 420-437
  water, 449-451, 469
Transportation system
  alternative, for future, developing, 510-513
  general discussion, 420-437
Transportation technology, 24, 589. *Also see*
    Transportation; Vehicles
Travel sickness, 437
Trends
  in communication, 75-76, 91
  defined, 589
Trial and error, 35
Trim, installing, 373

Truck
    ceramic engine parts for, 505
    computers in, 454-455
    design of, and efficiency, 505
    engines in, 481, 493, 498
    haul, 481
    industrial, 238, 239 (illus.), 422
    in mining, 481
    and piggyback service, 467
    RoadRailer, 465
    tank, 472
    tractor-semi-trailer rig, 452 (illus.), 454, 467, 468-469
Truck driver, career as, 438
Trucking, commercial, 451-455
Truss, roof, 368
Truss bridge, 385, 386 (illus.), 589
TSA, 43
Tube
    pickup, 134, 135 (illus.)
    picture, 136
    vacuum, 132
Tugboat, 450
Tunnels, 305, 389-391
Turbine
    in diesel-hydraulic locomotive, 497
    in rotary engine, 485
    in steam locomotive, 495
Turbosails, 499
TV. *See* Television
Two-stroke-cycle engines, 491-492
Type
    measuring, 150
    movable, invention of, 70, 151
Typographer's rule, 150

U

UHF television, 135
Ultralite, GM car, 504 (illus.)
Ultrasound, 557, 561
Union, labor, 182, 187
Unit train, 447, 473, 589
Unity, in design, 146
Universal resource locator, 113, 170-172
Uplink, 139, 589
URL, 113, 170-172, 589
U.S. Army Corps of Engineers, 451
Utilities
    installing, 370, 589
    and site selection, 333-334
Utility, time and place, 427

V

Vaccines, 553, 554 (illus.), 589

Value added, 186
Value analysis, 226-227, 589
Valve, heart, artificial, 562
Variables, in experiments, 211, 589
Variety, in design, 146
Vehicles
    aircraft, 443-444
    automatic guided (AGVs), 284, 285 (illus.)
    automobiles. *See* Car
    bicycle, first, 418
    blimps, 444
    Chrysler Concorde, 209
    commuter service, 446
    defined, 422, 589
    design of, and efficiency, 504-507
    dirigibles, 444
    electric-powered, 495-497
    electronic controls in, 506
    fiber optics in, 507
    GM Sunraycer, 500 (illus.)
    highway, 453-455. *Also see* Car; Truck
    horsepower in, 453, 484
    hybrid, 503
    lighter-than-air, 443-444
    maglev, 501-502
    Mars *Sojourner*, 434 (illus.), 460, 507 (illus.)
    for materials handling, 284-285
    for mining, 472-473
    operation of, processes used in, 433
    and pollution, 204, 205, 436
    power for, 480-503
    rail. *See* Rail transportation
    RoadRailer, 465
    Sojourner, 434 (illus.), 460, 507 (illus.)
    solar-powered, 499-500, 514-518
    spacecraft, 458-460
    speed of, calculating, 475
    trains. *See* Rail transportation
    truck. *See* Truck
    water, 450-451, 469
    wind-powered, 499
Very Large Array, 139
Vessels, water, 450-451, 589. *Also see* Ships
VHF television, 135
Video conference, 283
Videotape, editing, 176-178
Viewfinder, on camera, 156 (illus.), 157
Views, on drawings, 163-164
Violence, and TV viewing, 92
Virtual reality
    in architectural planning, 341, 342 (illus.)
    general discussion, 116 (illus.), 117, 589
    in product design, 275

Viruses
    AIDS, 421
    discovery of, 524
    and DNA vaccines, 553, 554 (illus.)
    emerging, 529
    spread of, and transportation, 421
Vision, impaired
    and Global Positioning System, 461
    laser surgery for, 560, 561 (illus.)
Visual design, 144-146
Vitamins, and disease prevention, 553
VLA, 139
Volcanic activity, monitoring, 77
V/STOL aircraft, 443

W

Waferboard, 315
Walkway, moving, 475, 476 (illus.)
Walls
    finishing, 373
    foundation, 366
    framed, 367
    load-bearing, 369
Warehouse, 266, 426 (illus.), 589
Waste-eating microbes, 541, 565, 567
Waste management, 564-567, 589. *Also see* Recycling
Wastewater, growing strawberries in, 550
Water culture, 549, 589
Water
    ground-, 371
    growing plants in, 549-550
    and hydroelectric energy, 381, 387
    for irrigation, 381, 389
    in paper manufacturing, 84, 95
    pollution of, 371, 541 (illus.), 565-566
    waste-, growing strawberries in, 550
Water culture, 549, 589
Waterjet cutting, 278
Water table, 371, 589
Water transportation
    general discussion, 449-451
    power for, 493, 495, 498, 499
    vehicles for, 450-451, 469, 499
Waterways, 451
Waterwheel, 14, 19
Watson, James, 15, 533
Watt, James, 453
Waves
    electromagnetic. *See* Electromagnetic waves
    light, and lasers, 336
    micro-, 129
    radio, 130
    sound. *See* Sound waves

Ways, transportation, 423, 444, 451, 589
Web, World Wide, 99, 112-113
Web browser, 113
Web page, 112-113
Web site, 112-113
Whitney, Eli, 182
Wholesalers, 265, 426 (illus.), 589
Wilkins, Maurice, 533
Wilmut, Ian, 21
Wind power, 418, 499
Windshield wipers, invention of, 448
Windsor Castle, 376
Wing shape, and lift, 443
WIP, 251
Wire, copper, for telephone, 128
Wireframe model, 166, 167 (illus.)
Withholding tax, 194
Women, in construction, 349
Wood, as structural material, 314-316
Wood-frame structure, 367-368
Work envelope, 279
Working capital, 200, 431
Working drawings
    analyzing, for process planning, 232
    in construction, 341-344
    defined, 589
    drafting of, 164
    for manufacturing, 223-224
Work in process (WIP), 251
Work-related illness and injury, 394
Work schedule, 352
World Wide Web, 99, 112-113
Wozniak, Steve, 100, 231
Wright, Frank Lloyd, 333 (illus.)
Wright, Orville, 436
Wrigley Field, 329
Wristwatch phone, 73
Writer, technical, 168

X

Xerox Corporation, 144
X-ray
    for diagnostic testing, 556
    for quality control, 281

Y

Yeager, Charles, 419

Z

Zoning laws, 332, 334, 589